Barker 9/5/94

Graduate Texts in Mathematics 106

Graduate Texts in Mathematics

continued after index

Joseph H. Silverman

The Arithmetic
of Elliptic Curves

With 13 Illustrations

Springer-Verlag
New York Berlin Heidelberg London Paris
Tokyo Hong Kong Barcelona Budapest

Joseph H. Silverman
Department of Mathematics
Brown University
Providence, RI 02912
U.S.A.

AMS Subject Classifications (1991): 14-01, 11GXX, 14GXX, 14H52

Library of Congress Cataloging-in-Publication Data
Silverman, Joseph H.
  The arithmetic of elliptic curves.
  (Graduate texts in mathematics ; 106)
  Bibliography: p.
  Includes index.
  1. Curves, Elliptic.   2. Curves, Algebraic.
3. Arithmetic—1961–   . I. Title. II. Series.
QA567.S44   1985   516.3'5   85-17182

Production managed by Dimitry L. Loseff; manufacturing supervised by Vincent Scelta.
Typeset by Asco Trade Typesetting Ltd., Hong Kong.
Printed and bound by R. R. Donnelley & Sons, Harrisonburg, Virginia.
Printed in the United States of America.

9 8 7 6 5 4 3 2 (corrected 2nd printing)

ISBN 0-387-96203-4 Springer-Verlag New York Berlin Heidelberg
ISBN 3-540-96203-4 Springer-Verlag Berlin Heidelberg New York

Preface

The preface to a textbook frequently contains the author's justification for offering the public "another book" on the given subject. For our chosen topic, the arithmetic of elliptic curves, there is little need for such an apologia. Considering the vast amount of research currently being done in this area, the paucity of introductory texts is somewhat surprising. Parts of the theory are contained in various books of Lang (especially [La 3] and [La 5]); and there are books of Koblitz ([Ko]) and Robert ([Rob], now out of print) which concentrate mostly on the analytic and modular theory. In addition, survey articles have been written by Cassels ([Ca 7], really a short book) and Tate ([Ta 5], which is beautifully written, but includes no proofs). Thus the author hopes that this volume will fill a real need, both for the serious student who wishes to learn the basic facts about the arithmetic of elliptic curves; and for the research mathematician who needs a reference source for those same basic facts.

Our approach is more algebraic than that taken in, say, [La 3] or [La 5], where many of the basic theorems are derived using complex analytic methods and the Lefschetz principle. For this reason, we have had to rely somewhat more on techniques from algebraic geometry. However, the geometry of (smooth) curves, which is essentially all that we use, does not require a great deal of machinery. And the small price paid in learning a little bit of algebraic geometry is amply repaid in a unity of exposition which (to the author) seems to be lacking when one makes extensive use of either the Lefschetz principle or lengthy (but elementary) calculations with explicit polynomial equations.

This last point is worth amplifying. It has been the author's experience that "elementary" proofs requiring page after page of algebra tend to be quite uninstructive. A student may be able to verify such a proof, line by line, and at the end will agree that the proof is complete. But little true understanding results from such a procedure. In this book, our policy is always to state when a result can be proven by such an elementary calculation, indicate briefly how that calculation might be done, and then give a more enlightening proof which is based on general principles.

The basic (global) theorems in the arithmetic of elliptic curves are the Mordell–Weil theorem, which is proven in chapter VIII and analyzed more closely in chapter X; and Siegel's theorem, which is proven in chapter IX. The reader desiring to reach these results fairly rapidly might take the following path:

I and II (briefly review), III (§1–8), IV (§1–6), V (§1),
VII (§1–5), VIII (§1–6), IX (§1–7), X (§1–6).

This material also makes a good one-semester course, possibly with some time left at the end for special topics. The present volume is built around the notes for such a course, taught by the author at M.I.T. during the spring term of 1983. [Of course, there are many other possibilities. For example, one might include all of chapters V and VI, skipping IX and (if pressed for time) X.] Other important topics in the arithmetic of elliptic curves, which do not appear in this volume due to time and space limitations, are briefly discussed in appendix C.

It is certainly true that some of the deepest results in this subject, such as Mazur's theorem bounding torsion over $\mathbb{Q}$ and Faltings' proof of the isogeny conjecture, require many of the resources of modern "SGA-style" algebraic geometry. On the other hand, one needs no machinery at all to write down the equation of an elliptic curve and to do explicit computations with it; and so there are many important theorems whose proof requires nothing more than cleverness and hard work. Whether your inclination leans toward heavy machinery or imaginative calculations, you will find much that remains to be discovered in the arithmetic theory of elliptic curves. Happy hunting!

Acknowledgments

In writing this book, I have consulted a great many sources. Citations have been included for major theorems, but many results which are now considered "standard" have been presented as such. In any case, I can claim no originality for any of the unlabeled theorems in this book, and apologize in advance to anyone who may feel slighted. The excellent survey articles of Cassels [Ca 7] and Tate [Ta 5] served as guidelines for organizing the material. (The reader is especially urged to peruse the latter.) In addition to [Ca 7] and [Ta 5], other sources which were extensively consulted include [La 5], [La 7], [Mum], [Rob], and [Se 10].

It would not be possible to catalogue all of the mathematicians from whom

I learned this beautiful subject; but to all of them, my deepest thanks. I would especially like to thank John Tate, Barry Mazur, Serge Lang, and the "Elliptic Curves Seminar" group at Harvard (1977–1982), whose help and inspiration set me on the road which led to this book. I would also like to thank David Rohrlich and Bill McCallum for their careful reading of the original draft, Gary Cornell and the editorial staff of Springer-Verlag for encouraging me to undertake this project in the first place, and Ann Clee for her meticulous preparation of the manuscript. Finally, I would like to thank my wife, Susan, for her patience and understanding through the turbulent times during which this book was written; and also Deborah and Daniel, for providing most of the turbulence.

Cambridge, Massachusetts   JOSEPH H. SILVERMAN
September, 1985

Acknowledgements For The Second Printing

I would like to thank the following people who kindly provided corrections which have been incorporated in this second revised printing: Andrew Baker, Arthur Baragar, Wah Keung Chan, Yen-Mei (Julia) Chen, Bob Coleman, Fred Diamond, David Fried, Dick Gross, Ron Jacobwitz, Kevin Keating, Masato Kuwata, Peter Landweber, H. W. Lenstra Jr., San Ling, Bill McCallum, David Masser, Hwasin Park, Elisabeth Pyle, Ken Ribet, John Rhodes, David Rohrlich, Mike Rosen, Rene Schoof, Udi de Shalit, Alice Silverberg, Glenn Stevens, John Tate, Jaap Top, Paul van Mulbregt, Larry Washington, Don Zagier.

It has unfortunately not been possible to include in this second printing many important results proven during the past six years, such as the work of Kolyvagin and Rubin on the Birch and Swinnerton-Dyer conjectures (C.16.5) and the finiteness of the Tate-Shafarevich group (X.4.13), Ribet's proof that the conjecture of Shimura-Taniyama-Weil (C.16.4) implies Fermat's Last Theorem, and recent work of Mestre on elliptic curves of high rank (C §20). The inclusion of such material (and more) will have to await an eventual second edition, so the reader should be aware that some of our general discussion, especially in Appendix C, is out of date. In spite of this obsolescence, it is our hope that this book will continue to provide a useful introduction to the study of the arithmetic of elliptic curves.

Providence, Rhode Island   JOSEPH H. SILVERMAN
August, 1992

Contents

APPENDIX C

Further Topics: An Overview   338

Introduction

The study of Diophantine equations, that is the solution of polynomial equations in integers or rational numbers, has a history stretching back to ancient Greece and beyond. The term *Diophantine geometry* is of more recent origin, and refers to the study of Diophantine equations through a combination of techniques from algebraic number theory and algebraic geometry. On the one hand, the problem of finding integer and rational solutions to polynomial equations calls into play the tools of algebraic number theory, which describes the rings and fields wherein those solutions lie. On the other hand, such a system of polynomial equations describes an algebraic variety, which is a geometric object. It is the interplay between these two points of view which is the subject of Diophantine geometry.

The simplest sort of equation is linear:

$$aX + bY = c \qquad a, b, c \in \mathbb{Z}, \qquad a \text{ or } b \neq 0.$$

Such an equation always has rational solutions. It will have integer solutions if and only if the greatest common divisor of a and b divides c; and if this occurs, then one can find all solutions by using the Euclidean algorithm.

Next in order of difficulty come quadratic equations:

$$aX^2 + bXY + cY^2 + dX + eY + f = 0 \qquad a, \ldots, f \in \mathbb{Z}, \qquad a, b, \text{ or } c \neq 0.$$

They describe conic sections, and by a suitable linear change of coordinates *with rational coefficients*, one can transform a given equation into one of the following forms:

$$AX^2 + BY^2 = C \qquad \text{ellipse}$$

$$AX^2 - BY^2 = C \qquad \text{hyperbola}$$

$$AX + BY^2 = 0 \qquad \text{parabola.}$$

For quadratic equations, one has the following powerful theorem which aids in their solution.

Hasse–Minkowski Theorem ([Se 7, IV Thm. 8]). *Let* $f(X, Y) \in \mathbb{Q}[X, Y]$ *be a quadratic polynomial. Then the equation* $f(X, Y) = 0$ *has a solution* $(x, y) \in \mathbb{Q}^2$ *if and only if it has a solution* $(x, y) \in \mathbb{R}^2$ *and a solution* $(x, y) \in \mathbb{Q}_p^2$ *for every prime p. (Here* $\mathbb{Q}_p$ *is the field of p-adic numbers.)*

In other words, a quadratic polynomial has a solution in $\mathbb{Q}$ if and only if it has a solution in every completion of $\mathbb{Q}$. Now checking for solutions in $\mathbb{Q}_p$ will, by Hensel's lemma, be more or less the same as checking for solutions in the finite field $\mathbb{Z}/p\mathbb{Z}$; and this, in turn, is easily accomplished by using quadratic reciprocity. Let us summarize the steps which go into the Diophantine analysis of quadratic equations.

(1) Analyze the equations over finite fields. [Quadratic reciprocity]
(2) Use this information to study the equations over complete local fields $\mathbb{Q}_p$. [Hensel's lemma] (We must also analyze them over $\mathbb{R}$.)
(3) Piece together all the local information to obtain results for the global field $\mathbb{Q}$. [Hasse principle]

Where does the geometry appear? Linear and quadratic equations in two variables define curves of genus 0. The above discussion says that we have a fairly good understanding of the arithmetic of curves of genus 0. The next simplest case, namely the arithmetic properties of curves of genus 1 (which are given by cubic equations in two variables), is our object of study in this book. The arithmetic of these so-called *elliptic curves* already presents complexities on which much current research is centered. Further, they provide a standard testing ground for conjectures and techniques which can then be fruitfully applied to the study of curves of higher genus and (abelian) varieties of higher dimension.

Briefly, the organization of this book is as follows. After two introductory chapters giving basic material on algebraic geometry, we start by studying the geometry of elliptic curves over algebraically closed fields (chapter III). We then follow the program outlined above and investigate the properties of elliptic curves over finite fields (chapter V), local fields (chapters VI, VII), and global (number) fields (chapters VIII, IX, X). Our understanding of elliptic curves over finite and local fields will be fairly satisfactory. However, it turns out that the analogue of the Hasse–Minkowski theorem is false for polynomials of degree greater than 2; this means that the transition from local to global is far more tenuous than in the degree 2 case. We study this problem in some detail in chapter X.

The theory of elliptic curves is rich, varied, and amazingly vast. The original aim of this book was to provide an essentially self-contained introduction to the basic arithmetic properties of elliptic curves. Even such a limited goal proved to be too ambitious. The material described above is approximately half of what the author had hoped to include. The reader will find a brief discussion and list of references for the omitted topics in appendix C.

Our other goal, that of being self-contained, has been more successful. We have, of course, felt free to state results that every reader should be aware of, even when the proofs are far beyond the scope of this book. However, we have endeavored not to use such results for making further deductions. There are three major exceptions to this general policy. First, we have not proven that every elliptic curve over $\mathbb{C}$ is uniformized by elliptic functions (VI.5.1). This result fits most naturally into a discussion of modular functions, which is one of the topics which had to be omitted. Second, we have not proven that over a complete local field, the "non-singular" points sit with finite index inside the set of all points (VII.6.2). This can actually be proven by quite explicit polynomial computations (cf. [Ta 6]), but they are rather lengthy, and again have not been included due to lack of space. Finally, in the study of integral points on elliptic curves, we have made use of Roth's theorem (IX.1.4) without giving a proof. However, a brief discussion of the proof has been given in (IX §8), and the reader who then wishes to see the myriad details can proceed to one of the references listed there.

The prerequisites for reading this book are fairly modest. We assume that the reader has had a first course in algebraic number theory, and so is acquainted with number fields, rings of integers, prime ideals, ramification, absolute values, completions, etc. The contents of any basic text in algebraic number theory, such as [La 2, Part I] or [Bo–Sh], should more than suffice. Chapter VI, which deals with elliptic curves over $\mathbb{C}$, assumes a familiarity with the basic principles of complex analysis. In chapter X we will need a little bit of group cohomology, but just H^0 and H^1. The reader will find the cohomological facts needed to read chapter X given in appendix B. Finally, since our approach is mainly algebraic, there is the question of background material in algebraic geometry. On the one hand, since much of the theory of elliptic curves can be obtained through the use of explicit equations and calculations, we do not want to require the reader to already know a great deal of algebraic geometry. On the other hand, this being a book on number theory and not algebraic geometry, it would not be reasonable to spend half of the book developing from first principles the algebro-geometric facts that we will use. As a compromise, the first two chapters give an introduction to the algebraic geometry of varieties and curves, stating all of the facts which we will need, giving complete references, and providing enough proofs so that the reader can gain a flavor for some of the basic techniques used in algebraic geometry.

Numerous exercises have been included at the end of each chapter. The reader desiring to gain a real understanding of the subject is urged to attempt as many as possible. Some of these exercises are (special cases of) results which have appeared in the literature. A list of comments and citations for the exercises will be found at the end of the book. Exercises with a single asterisk are somewhat more difficult, and two asterisks signal an unsolved problem.

References

Bibliographical references are enclosed in square brackets, e.g. [Ta 5, thm. 6]. Cross references to theorems, propositions, lemmas within the same chapter are given by number in parentheses, e.g. (4.3). Reference to an exercise is given by (exer. 3.6). References from within one chapter to another chapter or an appendix are preceded by the appropriate Roman numeral or letter, e.g. (IV.3.1), (B.2.1).

Standard Notation

Throughout this book, we use the symbols

$$\mathbb{Z}, \mathbb{Q}, \mathbb{R}, \mathbb{C}, \mathbb{F}_q, \text{ and } \mathbb{Z}_\ell$$

to represent the integers, rational numbers, real numbers, complex numbers, field with q elements, and ℓ-adic integers respectively. Further, if R is any ring, then R^* denotes the group of invertible elements of R; and if A is an abelian group, then $A[m]$ denotes the subgroup of A consisting of elements of order dividing m. A more complete list of notation is included on p. 379.

CHAPTER I

Algebraic Varieties

In this chapter we describe the basic objects which arise in the study of algebraic geometry. We set the following notation, which will be used throughout this book.

K a perfect field (i.e. every algebraic extension of K is separable).
$\bar{K}$ a fixed algebraic closure of K
$G_{\bar{K}/K}$ the Galois group of $\bar{K}/K$

For this chapter, we also let m and n denote positive integers.

The assumption that K is a perfect field is made solely to simplify our exposition. However, since our eventual goal is to do arithmetic, the field K will eventually be taken as an algebraic extension of $\mathbb{Q}$, $\mathbb{Q}_p$, or $\mathbb{F}_p$. Thus this restriction on K need not concern us unduly.

For a more extensive exposition of the basic concepts which appear in this chapter, we refer the reader to any introductory book on algebraic geometry, such as [Har], [Sha 2], [Ful].

§1. Affine Varieties

We begin our study of algebraic geometry with Cartesian (or affine) n-space and its subsets defined by zeros of polynomials.

Definition. *Affine n-space (over K)* is the set of n-tuples

$$\mathbb{A}^n = \mathbb{A}^n(\bar{K}) = \{P = (x_1, \ldots, x_n) : x_i \in \bar{K}\}.$$

Similarly, the *set of K-rational points in* $\mathbb{A}^n$ is the set

$$\mathbb{A}^n(K) = \{P = (x_1, \ldots, x_n) \in \mathbb{A}^n : x_i \in K\}.$$

Notice that the Galois group $G_{\bar{K}/K}$ acts on $\mathbb{A}^n$; for $\sigma \in G_{\bar{K}/K}$ and $P \in \mathbb{A}^n$,

$$P^\sigma = (x_1^\sigma, \ldots, x_n^\sigma).$$

Then $\mathbb{A}^n(K)$ may be characterized by

$$\mathbb{A}^n(K) = \{P \in \mathbb{A}^n : P^\sigma = P \text{ for all } \sigma \in G_{\bar{K}/K}\}.$$

Let $\bar{K}[X] = \bar{K}[X_1, \ldots, X_n]$ be a polynomial ring in n variables, and let $I \subset \bar{K}[X]$ be an ideal. To each such I we associate a subset of $\mathbb{A}^n$,

$$V_I = \{P \in \mathbb{A}^n : f(P) = 0 \text{ for all } f \in I\}.$$

Definition. An *(affine) algebraic set* is any set of the form V_I. If V is an algebraic set, the *ideal of V* is given by

$$I(V) = \{f \in \bar{K}[X] : f(P) = 0 \text{ for all } P \in V\}.$$

An algebraic set V is *defined over K* if its ideal $I(V)$ can be generated by polynomials in $K[X]$. We denote this by V/K. If V is defined over K, the *set of K-rational points of V* is the set

$$V(K) = V \cap \mathbb{A}^n(K).$$

Remark 1.1. Note that by the Hilbert basis theorem ([A–M, 7.6]), all ideals in $\bar{K}[X]$ and $K[X]$ are finitely generated.

Remark 1.2. Let V be an algebraic set, and consider the ideal

$$I(V/K) = \{f \in K[X] : f(P) = 0 \text{ for all } P \in V\} = I(V) \cap K[X].$$

Then we see that V is defined over K if and only if

$$I(V) = I(V/K)\bar{K}[X].$$

Now suppose V is defined over K, and let $f_1, \ldots, f_m \in K[X]$ be generators for $I(V/K)$. Then $V(K)$ is precisely the set of solutions $(x_1, \ldots, x_n)$ to the polynomial equations

$$f_1(X) = \cdots = f_m(X) = 0$$

with $x_1, \ldots, x_n \in K$. Thus one of the fundamental problems in the subject of *Diophantine geometry*, namely the solution of polynomial equations in rational numbers, may be said to be the problem of describing sets of the form $V(K)$ when K is a number field.

Notice that if $f(X) \in K[X]$ and $P \in \mathbb{A}^n$, then for any $\sigma \in G_{\bar{K}/K}$,

$$f(P^\sigma) = f(P)^\sigma.$$

Hence if V is defined over K, then the action of $G_{\bar{K}/K}$ on $\mathbb{A}^n$ induces an action on V, and clearly

$$V(K) = \{P \in V : P^\sigma = P \text{ for all } \sigma \in G_{\bar{K}/K}\}.$$

Example 1.3.1. Let V be the algebraic set in $\mathbb{A}^2$ given by the single equation

$$X^2 - Y^2 = 1.$$

Clearly V is defined over K for any field K. Let us assume that char$(K) \neq 2$. Then the set $V(K)$ is in one-to-one correspondence with $\mathbb{A}^1(K) - \{0\}$, one possible map being

$$\mathbb{A}^1(K) - \{0\} \to V(K)$$

$$t \to ((t^2 + 1)/2t, (t^2 - 1)/2t).$$

Example 1.3.2. The algebraic set

$$V : X^n + Y^n = 1$$

is defined over $\mathbb{Q}$. Fermat's last "theorem" states that for all $n \geqslant 3$,

$$V(\mathbb{Q}) = \begin{cases} \{(1, 0), (0, 1)\} & n \text{ odd} \\ \{(\pm 1, 0), (0, \pm 1)\} & n \text{ even}. \end{cases}$$

Example 1.3.3. The algebraic set

$$V : Y^2 = X^3 + 17$$

has many $\mathbb{Q}$-rational points, for example

$$(-2, 3) \qquad (5234, 378661) \qquad (137/64, 2651/512).$$

In fact, $V(\mathbb{Q})$ is infinite. See (2.8) and (III.2.4) for further discussion of this example.

Definition. An affine algebraic set V is called an *(affine) variety* if $I(V)$ is a prime ideal in $\bar{K}[X]$. (Note that if V is defined over K, it is not enough to check that $I(V/K)$ is prime. For example, consider the ideal $(X_1^2 - 2X_2^2)$ in $\mathbb{Q}[X_1, X_2]$.) Let V/K be a variety (i.e. V is a variety defined over K). Then the *affine coordinate ring of V/K* is defined by

$$K[V] = \frac{K[X]}{I(V/K)}.$$

It is an integral domain; and its quotient field, denoted $K(V)$, is called the *function field of V/K*. Similarly $\bar{K}[V]$ and $\bar{K}(V)$ are defined by replacing K with $\bar{K}$.

Note that since an element $f \in \bar{K}[V]$ is well-defined up to a polynomial vanishing on V, it induces a well-defined function $f : V \to \bar{K}$. Now if $f(X) \in \bar{K}[X]$, then $G_{\bar{K}/K}$ acts on f by acting on its coefficients. Hence if V is defined over K, so $G_{\bar{K}/K}$ takes $I(V)$ into itself, then we obtain an action of $G_{\bar{K}/K}$ on $\bar{K}[V]$ and $\bar{K}(V)$. One can check (exer. 1.12) that $K[V]$ and $K(V)$ are respectively the subsets of $\bar{K}[V]$ and $\bar{K}(V)$ fixed by $G_{\bar{K}/K}$. We denote the action of σ on f by $f \to f^\sigma$. Then for all points $P \in V$,

$$(f(P))^\sigma = f^\sigma(P^\sigma).$$

Definition. Let V be a variety. The *dimension of* V, denoted by $\dim(V)$, is the transcendence degree of $\bar{K}(V)$ over $\bar{K}$.

Example 1.4. The dimension of $\mathbb{A}^n$ is n, since $\bar{K}(\mathbb{A}^n) = \bar{K}(X_1, \ldots, X_n)$. Similarly, if $V \subset \mathbb{A}^n$ is given by a single non-constant polynomial equation

$$f(X_1, \ldots, X_n) = 0,$$

then $\dim(V) = n - 1$. (The converse is also true, cf. [Har, I.1.3].) In particular, the examples (1.3.1), (1.3.2), and (1.3.3) all have dimension 1.

In studying any geometric object, one is naturally interested in knowing whether it looks reasonably "smooth". The next definition formalizes this notion in terms of the usual Jacobian criterion for the existence of a tangent plane.

Definition. Let V be a variety, $P \in V$, and $f_1, \ldots, f_m \in \bar{K}[X]$ a set of generators for $I(V)$. Then V is *non-singular* (or *smooth*) at P if the $m \times n$ matrix

$$(\partial f_i / \partial X_j(P))_{1 \leqslant i \leqslant m, 1 \leqslant j \leqslant n}$$

has rank $n - \dim(V)$. If V is non-singular at every point, then we say that V is *non-singular* (or *smooth*).

Example 1.5. Let V be given by a single non-constant polynomial equation

$$f(X_1, \ldots, X_n) = 0.$$

Then $\dim V = n - 1$ (1.4), so $P \in V$ is a singular point if and only if

$$\partial f / \partial X_1(P) = \cdots = \partial f / \partial X_n(P) = 0.$$

Since P also satisfies $f(P) = 0$, this gives $n + 1$ equations for the n coordinates of any singular point. Thus for a "randomly chosen" f, one would expect V to be non-singular. We will not pursue this idea further, but see (exer. 1.1).

Example 1.6. Consider the two varieties

$$V_1 : Y^2 = X^3 + X \quad \text{and} \quad V_2 : Y^2 = X^3 + X^2.$$

Using (1.5) we see that any singular points on V_1 and V_2 satisfy respectively:

$$V_1 : 3X^2 + 1 = 2Y = 0;$$

$$V_2 : 3X^2 + 2X = 2Y = 0.$$

Thus V_1 is non-singular, while V_2 has one singular point, namely $(0, 0)$. The graphs of $V_1(\mathbb{R})$ and $V_2(\mathbb{R})$ (Figure 1.1) illustrate the difference.

There is another characterization of smoothness, in terms of the functions

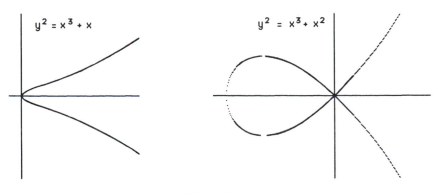

Figure 1.1

on the variety V, which is often quite useful. Let $P \in V$, and define an ideal M_P of $\bar{K}[V]$ by

$$M_P = \{f \in \bar{K}[V] : f(P) = 0\}.$$

Notice that M_P is a maximal ideal, since there is an isomorphism

$$\bar{K}[V]/M_P \to \bar{K} \quad \text{given by } f \to f(P).$$

The quotient M_P/M_P^2 is a finite dimensional $\bar{K}$-vector space.

Proposition 1.7. *Let V be a variety. A point $P \in V$ is non-singular if and only if*

$$\dim_{\bar{K}} M_P/M_P^2 = \dim V.$$

PROOF. [Har, I.5.1]. (See exer. 1.3 for a special case.) □

Example 1.8. Consider the point $P = (0, 0)$ on the varieties V_1 and V_2 of (1.6). In both cases, M_P is the ideal of $\bar{K}[V]$ generated by X, Y; and M_P^2 is the ideal generated by X^2, XY, Y^2. Now for V_1, we have

$$X = Y^2 - X^3 \equiv 0 \,(\text{mod } M_P^2),$$

so M_P/M_P^2 is generated by Y alone. On the other hand, for V_2 there is no non-trivial relationship between X and Y modulo M_P^2, so M_P/M_P^2 requires both X and Y as generators. Since each V_i has dimension 1, (1.7) implies that V_1 is smooth at P and V_2 is not.

Definition. The *local ring of V at P*, denoted $\bar{K}[V]_P$, is the localization of $\bar{K}[V]$ at M_P. In other words,

$$\bar{K}[V]_P = \{F \in \bar{K}(V) : F = f/g \text{ for some } f, g \in \bar{K}[V] \text{ with } g(P) \neq 0\}.$$

Notice that if $F = f/g \in \bar{K}[V]_P$, then $F(P) = f(P)/g(P)$ is well-defined. The functions in $\bar{K}[V]_P$ are said to be *regular* (or *defined*) at P.

§2. Projective Varieties

Historically, projective space arose through the process of adding "points at infinity" to affine space. We define projective space as the collection of lines in affine space of one higher dimension.

Definition. *Projective n-space (over K)*, denoted $\mathbb{P}^n$ or $\mathbb{P}^n(\bar{K})$, is the set of all $(n + 1)$-tuples

$$(x_0, \ldots, x_n) \in \mathbb{A}^{n+1}$$

such that at least one x_i is non-zero, modulo the equivalence relation given by

$$(x_0, \ldots, x_n) \sim (y_0, \ldots, y_n)$$

if there exists a $\lambda \in \bar{K}^*$ with $x_i = \lambda y_i$ for all i. An equivalence class $\{(\lambda x_0, \ldots, \lambda x_n)\}$ is denoted $[x_0, \ldots, x_n]$, and $x_0, \ldots, x_n$ are called *homogeneous coordinates* for the corresponding point in $\mathbb{P}^n$. The *set of K-rational points in $\mathbb{P}^n$* is the set

$$\mathbb{P}^n(K) = \{[x_0, \ldots, x_n] \in \mathbb{P}^n : \text{all } x_i \in K\}.$$

Remark 2.1. Note that if $P = [x_0, \ldots, x_n] \in \mathbb{P}^n(K)$, it does not follow that each $x_i \in K$. However, choosing some i with $x_i \neq 0$, it does follow that each $x_j/x_i \in K$.

Definition. Let $P = [x_0, \ldots, x_n] \in \mathbb{P}^n(\bar{K})$. The *minimal field of definition for P (over K)*, denoted $K(P)$, is the field

$$K(P) = K(x_0/x_i, \ldots, x_n/x_i) \text{ for any } i \text{ with } x_i \neq 0.$$

The Galois group $G_{\bar{K}/K}$ acts on $\mathbb{P}^n$ by acting on homogeneous coordinates,

$$[x_0, \ldots, x_n]^\sigma = [x_0^\sigma, \ldots, x_n^\sigma].$$

(This clearly respects the equivalence relation defining $\mathbb{P}^n$.) Then one checks (exer. 1.12) that

$$\mathbb{P}^n(K) = \{P \in \mathbb{P}^n : P^\sigma = P \text{ for all } \sigma \in G_{\bar{K}/K}\},$$

and

$$K(P) = \text{fixed field of } \{\sigma \in G_{\bar{K}/K} : P^\sigma = P\}.$$

Definition. A polynomial $f \in \bar{K}[X] = \bar{K}[X_0, \ldots, X_n]$ is *homogeneous of degree d* if

$$f(\lambda X_0, \ldots, \lambda X_n) = \lambda^d f(X_0, \ldots, X_n)$$

for all $\lambda \in \bar{K}$. An ideal $I \subset \bar{K}[X]$ is *homogeneous* if it is generated by homogeneous polynomials.

Note that for a homogeneous polynomial f, it makes sense to ask whether $f(P) = 0$ for a point $P \in \mathbb{P}^n$. To each homogeneous ideal I we associate a subset of $\mathbb{P}^n$,

$$V_I = \{P \in \mathbb{P}^n : f(P) = 0 \text{ for all homogeneous } f \in I\}.$$

Definition. A (*projective*) *algebraic set* is any set of the form V_I. If V is a projective algebraic set, the (*homogeneous*) *ideal of V*, denoted $I(V)$, is the ideal in $\bar{K}[X]$ generated by

$$\{f \in \bar{K}[X] : f \text{ is homogeneous and } f(P) = 0 \text{ for all } P \in V\}.$$

Such a V is *defined over K*, denoted by V/K, if its ideal $I(V)$ can be generated by homogeneous polynomials in $K[X]$. If V is defined over K, the *set of K-rational points of V* is the set

$$V(K) = V \cap \mathbb{P}^n(K).$$

As usual, $V(K)$ may also be described by

$$V(K) = \{P \in V : P^\sigma = P \text{ for all } \sigma \in G_{\bar{K}/K}\}.$$

Example 2.2. A *line* in $\mathbb{P}^2$ is an algebraic set given by a linear equation

$$aX + bY + cZ = 0$$

with $a, b, c \in \bar{K}$ not all zero. If, say, $c \neq 0$, then such a line is defined over any field containing a/c and b/c. More generally, a *hyperplane* in $\mathbb{P}^n$ is given by an equation

$$a_0 X_0 + a_1 X_1 + \cdots + a_n X_n = 0$$

with $a_i \in \bar{K}$ not all zero.

Example 2.3. Let V be the algebraic set in $\mathbb{P}^2$ given by the single equation

$$X^2 + Y^2 = Z^2.$$

Then for any field K with $\mathrm{char}(K) \neq 2$, the set $V(K)$ is isomorphic (i.e. structurally identical, see (3.5)) to $\mathbb{P}^1(K)$, for example by the map

$$\mathbb{P}^1(K) \to V(K)$$

$$[s, t] \to [s^2 - t^2, 2st, s^2 + t^2].$$

Remark 2.4. A point of $\mathbb{P}^n(\mathbb{Q})$ has the form $[x_0, \ldots, x_n]$ with $x_i \in \mathbb{Q}$. Multiplying by an appropriate $\lambda \in \mathbb{Q}$, one can clear denominators and common factors from the x_i's. In other words, every $P \in \mathbb{P}^n(\mathbb{Q})$ may be written with homogeneous coordinates $[x_0, \ldots, x_n]$ satisfying

$$x_0, \ldots, x_n \in \mathbb{Z} \quad \text{and} \quad \gcd(x_0, \ldots, x_n) = 1.$$

(Notice the x_i's are actually determined by P up to multiplication by -1.)

Thus if the ideal of an algebraic set $V/\mathbb{Q}$ is generated by homogeneous polynomials $f_1, \ldots, f_m \in \mathbb{Q}[X]$, then to describe $V(\mathbb{Q})$ means to find the solutions to the homogeneous equations

$$f_1(X_0, \ldots, X_n) = \cdots = f_m(X_0, \ldots, X_n) = 0$$

in relatively prime integers $x_0, \ldots, x_n$.

Example 2.5. The algebraic set

$$V : X^2 + Y^2 = 3Z^2$$

is defined over $\mathbb{Q}$. However, $V(\mathbb{Q}) = \emptyset$. To see this, suppose $[x, y, z] \in V(\mathbb{Q})$ with $x, y, z \in \mathbb{Z}$ and $\gcd(x, y, z) = 1$. Then

$$x^2 + y^2 \equiv 0 \,(\text{mod } 3),$$

so

$$x \equiv y \equiv 0 \,(\text{mod } 3).$$

(Note -1 is not a square modulo 3.) Hence x^2 and y^2 are divisible by 3^2, so from the equation for V it follows that 3 also divides z, which contradicts the assumption that $\gcd(x, y, z) = 1$. This example illustrates one of the fundamental tools used in the study of Diophantine equations.

> *In order to show that an algebraic set $V/\mathbb{Q}$ has no $\mathbb{Q}$-rational points, it suffices to show that the corresponding homogeneous polynomial equations have no non-zero solutions modulo p for any one prime p (or even one prime power p^r).*

A more succinct way to phrase this is to say that if $V(\mathbb{Q})$ is non-empty, then $V(\mathbb{Q}_p)$ is non-empty for every p-adic field $\mathbb{Q}_p$. Similarly, $V(\mathbb{R})$ would also be non-empty. One of the reasons that the study of Diophantine equations is so difficult is because the converse to this statement, the so-called "Hasse principle", does not in general hold. An example, due to Selmer [Sel 1], is the equation

$$V : 3X^3 + 4Y^3 + 5Z^3 = 0.$$

Onc can check that $V(\mathbb{Q}_p)$ is non-empty for every prime p, yet $V(\mathbb{Q})$ is empty. (See, e.g., [Ca 7, §4]. For other examples, see (X.6.5).)

Definition. A projective algebraic set is called a (*projective*) *variety* if its homogeneous ideal $I(V)$ is a prime ideal in $\bar{K}[X]$.

It is clear that $\mathbb{P}^n$ contains many copies of $\mathbb{A}^n$. For example, for each $0 \leqslant i \leqslant n$, there is an inclusion

$$\phi_i : \mathbb{A}^n \to \mathbb{P}^n$$

$$(y_1, \ldots, y_n) \to [y_1, y_2, \ldots, y_{i-1}, 1, y_i, \ldots, y_n].$$

If we let H_i denote the hyperplane in $\mathbb{P}^n$ given by $X_i = 0$,

$$H_i = \{P = [x_0, \ldots, x_n] \in \mathbb{P}^n : x_i = 0\};$$

and let U_i be the complement of H_i,

$$U_i = \{P = [x_0, \ldots, x_n] \in \mathbb{P}^n : x_i \neq 0\};$$

then there is a natural bijection

$$\phi_i^{-1} : U_i \to \mathbb{A}^n$$

$$[x_0, \ldots, x_n] \to \left(\frac{x_0}{x_i}, \frac{x_1}{x_i}, \ldots, \frac{x_{i-1}}{x_i}, \frac{x_{i+1}}{x_i}, \ldots, \frac{x_n}{x_i}\right).$$

(Note that for any point of $\mathbb{P}^n$ with $x_i \neq 0$, the quantities x_j/x_i are well-defined.) Having fixed an i, we will normally identify $\mathbb{A}^n$ with the set U_i in $\mathbb{P}^n$ via the map ϕ_i.

Now let V be a projective algebraic set with homogeneous ideal $I(V) \subset \bar{K}[X]$. Then $V \cap \mathbb{A}^n$ (by which we mean $\phi_i^{-1}(V \cap U_i)$) is an affine algebraic set with ideal $I(V \cap \mathbb{A}^n) \subset \bar{K}[Y]$ given by

$$I(V \cap \mathbb{A}^n) = \{f(Y_1, \ldots, Y_{i-1}, 1, Y_i, \ldots, Y_n) : f(X_0, \ldots, X_n) \in I(V)\}.$$

Notice that the sets $U_0, \ldots, U_n$ cover all of $\mathbb{P}^n$, so any projective variety V is covered by subsets $V \cap U_0, \ldots, V \cap U_n$, each of which is an affine variety (via the appropriate ϕ_i^{-1}). The process of replacing $f(X_0, \ldots, X_n)$ by $f(Y_1, \ldots, Y_{i-1}, 1, Y_i, \ldots, Y_n)$ is called *dehomogenization with respect to* X_i.

This process can be reversed. For any $f(Y) \in \bar{K}[Y]$, let

$$f^*(X_0, \ldots, X_n) = X_i^d f\left(\frac{X_0}{X_i}, \frac{X_1}{X_i}, \ldots, \frac{X_{i-1}}{X_i}, \frac{X_{i+1}}{X_i}, \ldots, \frac{X_n}{X_i}\right),$$

where $d = \deg(f)$ is the smallest integer for which f^* is a polynomial. We say that f^* is the *homogenization of* f *with respect to* X_i.

Definition. Let V be an affine algebraic set with ideal $I(V)$, and consider V as a subset of $\mathbb{P}^n$ via the map

$$V \subset \mathbb{A}^n \xrightarrow{\phi_i} \mathbb{P}^n.$$

The *projective closure of* V, denoted $\bar{V}$, is the projective algebraic set whose homogeneous ideal $I(\bar{V})$ is generated by

$$\{f^*(X) : f \in I(V)\}.$$

Proposition 2.6. (a) *Let V be an affine variety. Then $\bar{V}$ is a projective variety, and*

$$V = \bar{V} \cap \mathbb{A}^n.$$

(b) *Let V be a projective variety. Then $V \cap \mathbb{A}^n$ is an affine variety, and either*

$$V \cap \mathbb{A}^n = \emptyset \quad \text{or} \quad V = \overline{V \cap \mathbb{A}^n}.$$

(c) *If an affine (respectively projective) variety V is defined over K, then $\overline{V}$ (respectively $V \cap \mathbb{A}^n$) is also defined over K.*

PROOF. [Har, I.2.3] for (a) and (b); and (c) is clear from the definitions.   □

Remark 2.7. In view of (2.6), each affine variety can be identified with a unique projective variety. Notationally, it is easier to deal with affine coordinates, so we will often say "let V be a projective variety" and write down some non-homogeneous equations, with the understanding that V is the projective closure of the indicated affine variety W. The points of $V-W$ are called the *points at infinity* on V.

Example 2.8. Let V be the *projective* variety given by the equation

$$V : Y^2 = X^3 + 17.$$

Thus we really mean the variety in $\mathbb{P}^2$ given by the homogeneous equation

$$\overline{Y}^2 \overline{Z} = \overline{X}^3 + 17\overline{Z}^3,$$

the identification being

$$X = \overline{X}/\overline{Z} \qquad Y = \overline{Y}/\overline{Z}.$$

This variety has one point at infinity, namely $[0, 1, 0]$, obtained by setting $\overline{Z} = 0$. Thus, for example,

$$V(\mathbb{Q}) = \{(x, y) \in \mathbb{A}^2(\mathbb{Q}) : y^2 = x^3 + 17\} \cup \{[0, 1, 0]\}.$$

In (1.3.3) we listed several points of $V(\mathbb{Q})$. The reader may verify that the line connecting any two points of $V(\mathbb{Q})$ will intersect V in a third point of $V(\mathbb{Q})$ (provided the line is not tangent to V). (See exer. 1.5.) Using this secant-line procedure, one can actually produce infinitely many points in $V(\mathbb{Q})$, although this is by no means obvious. The variety V is called an *elliptic curve*, and as such it provides the first example of the varieties which will be our principal object of study in this book. See (III.2.4) for further discussion of this example.

Most of the important properties of a projective variety V may now be defined in terms of the affine subvariety $V \cap \mathbb{A}^n$.

Definition. Let V/K be a projective variety, and choose $\mathbb{A}^n \subset \mathbb{P}^n$ so that $V \cap \mathbb{A}^n \neq \varnothing$. The *dimension of V* is the dimension of $V \cap \mathbb{A}^n$. The *function field of V*, denoted $K(V)$, is the function field of $V \cap \mathbb{A}^n$; and similarly for $\overline{K}(V)$. (Note that for different choices of $\mathbb{A}^n$, the different $K(V)$'s are canonically isomorphic, so we will always identify them. See (2.9) for another description of $K(V)$.)

Definition. Let V be a projective variety, $P \in V$, and choose $\mathbb{A}^n \subset \mathbb{P}^n$ with $P \in \mathbb{A}^n$. Then V is *non-singular* (or *smooth*) at P if $V \cap \mathbb{A}^n$ is non-singular at P.

The *local ring of V at P*, denoted $\bar{K}[V]_P$, is the local ring of $V \cap \mathbb{A}^n$ at P. A function $F \in \bar{K}(V)$ is *regular* (or *defined*) at P if it is in $\bar{K}[V]_P$; in this case, it makes sense to evaluate F at P.

Remark 2.9. The function field of $\mathbb{P}^n$ may also be described as the subfield of $\bar{K}(X_0, \ldots, X_n)$ consisting of rational functions $F(X) = f(X)/g(X)$ for which f and g are *homogeneous* polynomials of the *same degree*. Such an expression gives a well-defined function on $\mathbb{P}^n$ at all points P where $g(P) \neq 0$. Similarly, the function field of a projective variety V is the field of rational functions $F(X) = f(X)/g(X)$ such that:

(i) f and g are homogeneous of the same degree:
(ii) $g \notin I(V)$;
(iii) two functions f/g and f'/g' are identified if $fg' - f'g \in I(V)$.

§3. Maps between Varieties

In this section we look at algebraic maps between projective varieties, which are those maps defined by rational functions.

Definition. Let V_1 and $V_2 \subset \mathbb{P}^n$ be projective varieties. A *rational map from V_1 to V_2* is a map of the form

$$\phi : V_1 \to V_2$$

$$\phi = [f_0, \ldots, f_n],$$

where $f_0, \ldots, f_n \in \bar{K}(V_1)$ have the property that for every point $P \in V_1$ at which $f_0, \ldots, f_n$ are all defined,

$$\phi(P) = [f_0(P), \ldots, f_n(P)] \in V_2.$$

If V_1 and V_2 are defined over K, then $G_{\bar{K}/K}$ acts on ϕ in the obvious way:

$$\phi^\sigma(P) = [f_0^\sigma(P), \ldots, f_n^\sigma(P)].$$

Notice that we have the formula

$$\phi(P)^\sigma = \phi^\sigma(P^\sigma) \quad \text{for all } \sigma \in G_{\bar{K}/K} \text{ and } P \in V_1.$$

Now if there is some $\lambda \in \bar{K}^*$ so that $\lambda f_0, \ldots, \lambda f_n \in K(V_1)$, then ϕ is said to be *defined over K*. (Notice that $[f_0, \ldots, f_n]$ and $[\lambda f_0, \ldots, \lambda f_n]$ give the same map on points.) As usual, it is true that ϕ is defined over K if and only if $\phi = \phi^\sigma$ for all $\sigma \in G_{\bar{K}/K}$ (cf. exer. 1.12c).

Remark 3.1. Note that a rational map $\phi : V_1 \to V_2$ is not necessarily a function on all of V_1. However, it is sometimes possible to evaluate $\phi(P)$ at points P of V_1 where some f_i is not regular by replacing each f_i with gf_i for an appropriate $g \in \bar{K}(V_1)$.

Definition. A rational map

$$\phi = [f_0, \ldots, f_n] : V_1 \to V_2$$

is *regular* (or *defined*) *at* $P \in V_1$ if there is a function $g \in \bar{K}(V_1)$ such that

(i) each gf_i is regular at P; and
(ii) for some i, $(gf_i)(P) \neq 0$.

If such a g exists, we set

$$\phi(P) = [(gf_0)(P), \ldots, (gf_n)(P)].$$

(N.B. It may be necessary to take different g's for different points.) A rational map which is regular at every point is called a *morphism*.

Remark 3.2. Let $V_1 \subset \mathbb{P}^m$ and $V_2 \subset \mathbb{P}^n$ be projective varieties. Recall (2.9) that the functions in $\bar{K}(V_1)$ may be described as quotients of homogeneous polynomials in $K[X_0, \ldots, X_m]$ having the same degree. Thus by multiplying a rational map $\phi = [f_0, \ldots, f_n]$ by a homogeneous polynomial which clears the "denominators" of the f_i's, we obtained the following alternative definition:
    A *rational map* $\phi : V_1 \to V_2$ is a map of the form

$$\phi = [\phi_0(X), \ldots, \phi_n(X)],$$

where

(i) $\phi_i(X) \in \bar{K}[X] = \bar{K}[X_0, \ldots, X_m]$ are homogeneous polynomials, not all in $I(V_1)$, having the same degree; and
(ii) for every $f \in I(V_2)$,

$$f(\phi_0(X), \ldots, \phi_n(X)) \in I(V_1).$$

Clearly, $\phi(P)$ is well-defined provided some $\phi_i(P) \neq 0$. However, even if all $\phi_i(P) = 0$, it may be possible to "alter" ϕ so as to make sense of $\phi(P)$. We make this precise as follows:
    A rational map $\phi = [\phi_0, \ldots, \phi_n] : V_1 \to V_2$ as above is *regular* (or *defined*) at $P \in V_1$ if there exist homogeneous polynomials $\psi_0, \ldots, \psi_n \in \bar{K}[X]$ such that

(i) $\psi_0, \ldots, \psi_n$ have the same degree,
(ii) $\phi_i \psi_j \equiv \phi_j \psi_i \pmod{I(V_1)}$ for $0 \leq i, j \leq n$, and
(iii) $\psi_i(P) \neq 0$ for some i.

If this occurs, we set

$$\phi(P) = [\psi_0(P), \ldots, \psi_n(P)].$$

As above, a rational map which is everywhere regular is called a *morphism*.

Remark 3.3. Let $\phi = [\phi_0, \ldots, \phi_n] : \mathbb{P}^m \to \mathbb{P}^n$ be a rational map as in (3.2), where $\phi_i \in \bar{K}[X]$ are homogeneous polynomials of the same degree. Since $\bar{K}[X]$ is a UFD, we may assume that the ϕ_i's have no common factor. Then ϕ is regular at a point $P \in \mathbb{P}^m$ if and only if some $\phi_i(P) \neq 0$. (Note that $I(\mathbb{P}^m) = (0)$, so there is no way to alter the ϕ_i's.) Hence ϕ is a morphism if and only if the ϕ_i's have no common zero in $\mathbb{P}^m$.

Definition. Let V_1 and V_2 be varieties. We say that V_1 and V_2 are *isomorphic*, and write $V_1 \simeq V_2$, if there are morphisms $\phi : V_1 \to V_2$ and $\psi : V_2 \to V_1$ such that $\psi \circ \phi$ and $\phi \circ \psi$ are the identity maps on V_1 and V_2 respectively. V_1/K and V_2/K are *isomorphic over K* if such ϕ and ψ can be defined over K. [N.B. ϕ and ψ must be morphisms, not merely rational maps.]

Remark 3.4. If $\phi : V_1 \to V_2$ is an isomorphism defined over K, then ϕ identifies $V_1(K)$ with $V_2(K)$. Hence for Diophantine problems, it suffices to study any one variety in a given K-isomorphism class of varieties.

Example 3.5. Assume $\mathrm{char}(K) \neq 2$, and let V be the variety from (2.3),

$$V : X^2 + Y^2 = Z^2.$$

Consider the rational map

$$\phi : V \to \mathbb{P}^1$$

$$\phi = [X + Z, Y].$$

Clearly ϕ is regular at every point of V except possibly $[1, 0, -1]$ (i.e. where $X + Z = Y = 0$). But using

$$(X + Z)(X - Z) \equiv -Y^2 \pmod{I(V)},$$

we have

$$\phi = [X + Z, Y] = [X^2 - Z^2, Y(X - Z)]$$

$$= [-Y^2, Y(X - Z)] = [-Y, X - Z].$$

Thus

$$\phi([1, 0, -1]) = [0, 2] = [0, 1],$$

so ϕ is regular at every point of V. (I.e. ϕ is a morphism.) One easily checks that the map

$$\psi : \mathbb{P}^1 \to V$$

$$\psi = [S^2 - T^2, 2ST, S^2 + T^2]$$

is a morphism and provides an inverse for ϕ, so V and $\mathbb{P}^1$ are isomorphic.

Example 3.6. The rational map

$$\phi : \mathbb{P}^2 \to \mathbb{P}^2$$

$$\phi = [X^2, XY, Z^2]$$

is regular everywhere except at the point [0, 1, 0], where it is not regular (cf. 3.3).

Example 3.7. Let V be the variety

$$V : Y^2 Z = X^3 + X^2 Z,$$

and consider the rational maps

$$\psi : \mathbb{P}^1 \to V \qquad\qquad\qquad \phi : V \to \mathbb{P}^1$$
$$\psi = [(S^2 - T^2)T, (S^2 - T^2)S, T^3] \qquad \phi = [Y, X].$$

Here ψ is a morphism, while ϕ is not regular at [0, 0, 1]. Not coincidently (see II.2.1), [0, 0, 1] is a singular point of V. Notice that the compositions $\phi \circ \psi$ and $\psi \circ \phi$ are the identity map whenever they are defined, but nonetheless ϕ and ψ are not isomorphisms.

Example 3.8. Consider the varieties

$$V_1 : X^2 + Y^2 = Z^2 \qquad V_2 : X^2 + Y^2 = 3Z^2.$$

They are not isomorphic over $\mathbb{Q}$, since $V_2(\mathbb{Q}) = \emptyset$ (2.5), while $V_1(\mathbb{Q})$ contains lots of points. (More precisely, $V_1(\mathbb{Q}) \simeq \mathbb{P}^1(\mathbb{Q})$ from (3.5).) However, V_1 and V_2 are isomorphic over $\mathbb{Q}(\sqrt{3})$, an isomorphism being given by

$$\phi : V_2 \to V_1$$
$$\phi = [X, Y, \sqrt{3}Z].$$

EXERCISES

1.1. Let $A, B \in \bar{K}$. Characterize the values of A and B for which each of the following varieties is singular. In particular, as (A, B) ranges over $\mathbb{A}^2$, the "singular values" lie on a one-dimensional subset of $\mathbb{A}^2$, so "most" values of (A, B) give a non-singular variety.
 (a) $V : Y^2 Z + AXYZ + BYZ^2 = X^3$.
 (b) $V : Y^2 Z = X^3 + AXZ^2 + BZ^3$ (char $K \neq 2$).

1.2. Find the singular point(s) on each of the following varieties, and sketch $V(\mathbb{R})$.
 (a) $V : Y^2 = X^3$ in $\mathbb{A}^2$.
 (b) $V : 4X^2 Y^2 = (X^2 + Y^2)^3$ in $\mathbb{A}^2$.
 (c) $V : Y^2 = X^4 + Y^4$ in $\mathbb{A}^2$.
 (d) $V : X^2 + Y^2 = (Z - 1)^2$ in $\mathbb{A}^3$.

1.3. Let $V \subset \mathbb{A}^n$ be a variety given by a single equation (cf. 1.4). Prove that a point $P \in V$ is non-singular if and only if

$$\dim_{\bar{K}} M_P / M_P^2 = \dim V.$$

[*Hint:* Let $f = 0$ be the equation of V, and define the *tangent plane to V at P* by

$$T = \{(y_1, \ldots, y_n) \in \mathbb{A}^n : \sum (\partial f / \partial X_i(P)) y_i = 0\}.$$

Show that the map

$$M_P/M_P^2 \times T \to \bar{K}, \qquad (g, y) \to \sum (\partial g/\partial X_i(P)) y_i$$

is a well-defined perfect pairing of $\bar{K}$-vector spaces. Now use (1.5).]

1.4. Let $V/\mathbb{Q}$ be the variety

$$V : 5X^2 + 6XY + 2Y^2 = 2YZ + Z^2.$$

Prove that $V(\mathbb{Q}) = \varnothing$.

1.5. Let $V/\mathbb{Q}$ be the projective variety

$$V : Y^2 = X^3 + 17,$$

and let $P_1 = (x_1, y_1)$ and $P_2 = (x_2, y_2)$ be distinct points of V. Let L be the line through P_1 and P_2.
(a) Show that $V \cap L = \{P_1, P_2, P_3\}$, and express $P_3 = (x_3, y_3)$ in terms of P_1 and P_2. (If L is tangent to V, then P_3 may equal P_1 or P_2.)
(b) Calculate P_3 for $P_1 = (-1, 4)$ and $P_2 = (2, 5)$.
(c) Show that if $P_1, P_2 \in V(\mathbb{Q})$, then $P_3 \in V(\mathbb{Q})$.

1.6. Let V be the variety

$$V : Y^2 Z = X^3 + Z^3.$$

Show that the map

$$\phi : V \to \mathbb{P}^2, \qquad \phi = [X^2, XY, Z^2]$$

is a morphism. (Notice ϕ does not give a morphism $\mathbb{P}^2 \to \mathbb{P}^2$.)

1.7. Let V be the variety

$$V : Y^2 Z = X^3,$$

and let ϕ be the map

$$\phi : \mathbb{P}^1 \to V, \qquad \phi = [S^2 T, S^3, T^3].$$

(a) Show that ϕ is a morphism.
(b) Find a rational map $\psi : V \to \mathbb{P}^1$ so that $\phi \circ \psi$ and $\psi \circ \phi$ are the identity map wherever they are defined.
(c) Is ϕ an isomorphism?

1.8. Let $K = \mathbb{F}_q$, and let $V \subset \mathbb{P}^n$ be a variety which is defined over K.
(a) Show that the qth-power map

$$\phi = [X_0^q, \dots, X_n^q]$$

is a morphism $\phi : V \to V$. It is called the *Frobenius morphism*.
(b) Show that ϕ is one-to-one and onto.
(c) Show that ϕ is not an isomorphism.
(d) Show that

$$\{P \in V : \phi(P) = P\} = V(K).$$

1.9. If $m > n$, prove that there are no non-constant morphisms $\mathbb{P}^m \to \mathbb{P}^n$. [*Hint:* Use the dimension theorem [Har, I.7.2].]

1.10. For each prime $p \geqslant 3$, let V_p be the variety in $\mathbb{P}^2$ given by the equation

$$V_p : X^2 + Y^2 = pZ^2.$$

(a) Prove that V_p is isomorphic to $\mathbb{P}^1$ over $\mathbb{Q}$ if and only if $p \equiv 1 \pmod 4$.
(b) Prove that for $p \equiv 3 \pmod 4$, no two of the V_p's are isomorphic over $\mathbb{Q}$.

1.11. (a) Let $f \in K[X_0, \ldots, X_n]$ be a homogeneous polynomial, and let

$$V = \{P \in \mathbb{P}^n : f(P) = 0\}$$

be the hypersurface defined by f. Prove that if a point $P \in V$ is singular, then

$$\partial f / \partial X_0(P) = \cdots = \partial f / \partial X_n(P) = 0.$$

(Thus in projective space, one can check for smoothness using homogeneous coordinates.)

(b) Let $W \subset \mathbb{P}^n$ be a smooth algebraic set of dimension $n - 1$. Prove that W is a variety. [*Hint*: First use Krull's Hauptidealsatz ([A–M] p. 122) to show that W is the zero set of a single homogeneous polynomial.]

1.12. (a) Let V/K be an affine variety. Prove that

$$K[V] = \{ f \in \bar{K}[V] : f^\sigma = f \text{ for all } \sigma \in G_{\bar{K}/K} \}.$$

[*Hint*: One inclusion is clear. For the other, choose some $F \in \bar{K}[X]$ with $F \equiv f \pmod{I(V)}$. Show that the map $G_{\bar{K}/K} \to I(V)$ defined by $\sigma \to F^\sigma - F$ is a 1-cocycle (cf. B §2). Now use (B.2.5a) to conclude that there exists a $G \in I(V)$ such that $F + G \in K[X]$.]

(b) Prove that

$$\mathbb{P}^n(K) = \{ P \in \mathbb{P}^n(\bar{K}) : P^\sigma = P \text{ for all } \sigma \in G_{\bar{K}/K} \}.$$

[*Hint*: Write $P = [x_0, \ldots, x_n]$. If $P = P^\sigma$, then there is a $\lambda_\sigma \in \bar{K}^*$ such that $x_i^\sigma = \lambda_\sigma x_i$ for $0 \leqslant i \leqslant n$. Show that the map $\sigma \to \lambda_\sigma$ gives a 1-cocycle from $G_{\bar{K}/K}$ to $\bar{K}^*$. Now use Hilbert's theorem 90 (B.2.5b) to find an $\alpha \in \bar{K}^*$ so that $[\alpha x_0, \ldots, \alpha x_n] \in \mathbb{P}^n(K)$.]

(c) Let $\phi : V_1 \to V_2$ be a rational map of projective varieties. Prove that ϕ is defined over K if and only if $\phi^\sigma = \phi$ for every $\sigma \in G_{\bar{K}/K}$. [*Hint*: Use (a) and (b).]

CHAPTER II

Algebraic Curves

In this chapter we present the basic facts about algebraic curves (i.e. projective varieties of dimension 1) which will be needed for our study of elliptic curves. (Actually, since elliptic curves are curves of genus 1, one of our tasks will be to define the genus of a curve.) As in Chapter I, we give references for those proofs which are not included. There are many books where the reader can find more material on the subject of algebraic curves, for example [Har, Ch. IV], [Sha 2], [G–H, Ch. 2], [Wa].

We recall the following notation from Chapter I, which will be used in this chapter. (C is a curve and $P \in C$.)

C/K	C is defined over K
$K(C), \bar{K}(C)$	the function field of C
$\bar{K}[C]_P$	the local ring of C at P
M_P	the maximal ideal of $\bar{K}[C]_P$

§1. Curves

By a *curve* we will always mean a projective variety of dimension 1. We will generally deal with curves which are smooth. Examples of smooth curves are provided by $\mathbb{P}^1$, (I.2.3), and (I.2.8). We start by describing the local rings of a smooth curve.

Proposition 1.1. *Let C be a curve and $P \in C$ a smooth point. Then $\bar{K}[C]_P$ is a discrete valuation ring.*

PROOF. From (I.1.7), M_P/M_P^2 has dimension 1 over $\bar{K} = \bar{K}[C]_P/M_P$. Now use [A–M, Prop. 9.2] (or exer. 2.1).   □

Definition. Let C be a curve and $P \in C$ a smooth point. The (*normalized*) *valuation* on $\bar{K}[C]_P$ is given by

$$\text{ord}_P : \bar{K}[C]_P \to \{0, 1, 2, \dots\} \cup \{\infty\}$$

$$\text{ord}_P(f) = \max\{d \in \mathbb{Z} : f \in M_P^d\}.$$

Using $\text{ord}_P(f/g) = \text{ord}_P(f) - \text{ord}_P(g)$, we extend ord_P to $\bar{K}(C)$,

$$\text{ord}_P : \bar{K}(C) \to \mathbb{Z} \cup \{\infty\}.$$

A *uniformizer for C at P* is a function $t \in \bar{K}(C)$ with $\text{ord}_P(t) = 1$ (i.e. a generator for M_P).

Definition. Let C, P be as above and $f \in \bar{K}(C)$. The *order of f at P* is $\text{ord}_P(f)$. If $\text{ord}_P(f) > 0$, then f has a *zero* at P; if $\text{ord}_P(f) < 0$, then f has a *pole* at P. If $\text{ord}_P(f) \geqslant 0$, then f is *regular* (or *defined*) at P, and we can calculate $f(P)$. Otherwise f has a pole at P, and we write $f(P) = \infty$.

Proposition 1.2. *Let C be a smooth curve and $f \in \bar{K}(C)$. Then there are only finitely many points of C at which f has a pole or a zero. Further, if f has no poles, then $f \in \bar{K}$.*

PROOF. [Har, I.6.5], [Har, II.6.1], or [Sha 2, III §1] for the finiteness of the number of poles. To deal with the zeros, look instead at $1/f$. The last statement is [Har, I.3.4a] or [Sha 2, I §5, cor. 1].   □

Example 1.3. Consider the two curves

$$C_1 : Y^2 = X^3 + X \quad \text{and} \quad C_2 : Y^2 = X^3 + X^2.$$

(Remember our convention (I.2.7) concerning affine equations for projective varieties. Each of C_1 and C_2 has a single point at infinity.) Let $P = (0,0)$. Then C_1 is smooth at P and C_2 is not (I.1.6). The maximal ideal M_P of $\bar{K}[C_1]_P$ has the property that M_P/M_P^2 is generated by Y (I.1.8), so for example

$$\text{ord}_P(Y) = 1 \qquad \text{ord}_P(X) = 2 \qquad \text{ord}_P(2Y^2 - X) = 2.$$

(For the last, note that $2Y^2 - X = 2X^3 + X$.) On the other hand, $K[C_2]_P$ is not a discrete valuation ring.

The next proposition is useful when dealing with curves over fields of characteristic $p > 0$. (See also exer. 2.15.)

Proposition 1.4. *Let C/K be a curve, and let $t \in K(C)$ be a uniformizer at some non-singular point $P \in C$. Then $K(C)$ is a finite separable extension of $K(t)$.*

PROOF. $K(C)$ is clearly a finite (algebraic) extension of $K(t)$, since it is finitely generated over K, has transcendence degree 1 over K, and $t \notin K$. Now let $x \in K(C)$. We will show that x is separable over $K(t)$.

In any case, x is algebraic over $K(t)$, so it satisfies some polynomial relation

$$\sum a_{ij} t^i x^j = 0, \qquad \text{where } \Phi(T, X) = \sum a_{ij} T^i X^j \in K[X, T].$$

We may further assume that Φ is chosen so as to have minimal degree in X. (I.e. $\Phi(t, X)$ is a minimal polynomial for x over $K(t)$.) Let $p = \text{char}(K)$. If Φ contains a non-zero term $a_{ij} T^i X^j$ with $j \not\equiv 0 \pmod{p}$, then $\partial \Phi(t, X)/\partial X$ is not identically 0, so x is separable over $K(t)$. Suppose now that $\Phi(T, X) = \Psi(T, X^p)$. We proceed to derive a contradiction.

The main point to note is that if $F(T, X) \in K[T, X]$ is any polynomial, then $F(T^p, X^p)$ is a pth-power. This is true because we have assumed that K is perfect, which implies that every element of K is a pth-power. Thus if $F(T, X) = \Sigma \alpha_{ij} T^i X^j$, then writing $\alpha_{ij} = \beta_{ij}^p$ gives $F(T^p, X^p) = (\Sigma \beta_{ij} T^i X^j)^p$. We now regroup the terms in $\Phi(T, X) = \Psi(T, X^p)$ according to powers of T (modulo p):

$$\Phi(T, X) = \Psi(T, X^p) = \sum_{k=0}^{p-1} \left(\sum_{i,j} b_{ijk} T^{ip} X^{jp} \right) T^k = \sum_{k=0}^{p-1} \phi_k(T, X)^p T^k.$$

Now by assumption, $\Phi(t, x) = 0$. On the other hand, since t is a uniformizer at P, we have

$$\text{ord}_P(\phi_k(t, x)^p t^k) = p \, \text{ord}_P(\phi_k(t, x)) + k \, \text{ord}_P(t) \equiv k \pmod{p}.$$

Thus each of the terms in the sum $\Sigma \phi_k(t, x)^p t^k$ has a distinct order at P, so every term must vanish:

$$\phi_0(t, x) = \phi_1(t, x) = \cdots = \phi_{p-1}(t, x) = 0.$$

But one of the $\phi_k(T, X)$'s must involve X; and for that k, the relation $\phi_k(t, x) = 0$ contradicts the fact that we chose $\Phi(t, X)$ to be a minimal polynomial for x over $K(t)$. (Note that $\deg_X(\phi_k(T, X)) \leqslant \deg_X(\Phi(T, X))/p$.) This contradiction completes the proof that x is separable over $K(t)$. $\qquad\square$

§2. Maps between Curves

We start with the fundamental result that for smooth curves, a rational map is always defined at every point.

Proposition 2.1. Let C be a curve, $V \subset \mathbb{P}^N$ a variety, $P \in C$ a smooth point, and $\phi : C \to V$ a rational map. Then ϕ is regular at P. In particular, if C is smooth, then ϕ is a morphism.

PROOF. Write $\phi = [f_0, \ldots, f_N]$ with $f_i \in \bar{K}(C)$, and choose a uniformizer

$t \in \bar{K}(C)$ for C at P. Let

$$n = \min_{0 \leqslant i \leqslant N} \{\mathrm{ord}_P\, f_i\}.$$

Then

$$\mathrm{ord}_P(t^{-n}f_i) \geqslant 0 \text{ for all } i \quad \text{and} \quad \mathrm{ord}_P(t^{-n}f_j) = 0 \text{ for some } j,$$

so each $t^{-n}f_i$ is regular at P and $(t^{-n}f_j)(P) \neq 0$. Therefore ϕ is regular at P. $\square$

For examples where (2.1) is false if P is not smooth or C has dimension greater than 1, see (I.3.6) and (I.3.7).

Example 2.2. Let C/K be a smooth curve and $f \in K(C)$ a function. Then f defines a rational map, which we also denote by f,

$$f : C \to \mathbb{P}^1$$

$$P \to [f(P), 1].$$

From (2.1), this map is actually a morphism. It is given explicitly by

$$f(P) = \begin{cases} [f(P), 1] & \text{if } f \text{ is regular at } P \\ [1, 0] & \text{if } f \text{ has a pole at } P. \end{cases}$$

Conversely, let

$$\phi : C \to \mathbb{P}^1$$

$$\phi = [f, g]$$

be a rational map defined over K. Then either $g = 0$, in which case ϕ is the constant map $\phi = [1, 0]$; or else ϕ is the map corresponding to the function $f/g \in K(C)$. Denoting the former map by ∞, we thus have a one-to-one correspondence

$$K(C) \cup \{\infty\} \leftrightarrow \{\text{maps } C \to \mathbb{P}^1 \text{ defined over } K\}.$$

We will often implicitly identify these two sets.

Theorem 2.3. *Let* $\phi : C_1 \to C_2$ *be a morphism of curves. Then* ϕ *is either constant or surjective.*

PROOF. [Har, II.6.8] or [Sha 2, I §5, thm. 4]. $\square$

Now let C_1/K and C_2/K be curves and $\phi : C_1 \to C_2$ a non-constant rational map defined over K. Then composition with ϕ induces an injection of function fields fixing K,

$$\phi^* : K(C_2) \to K(C_1)$$

$$\phi^* f = f \circ \phi.$$

Theorem 2.4. *Let C_1/K and C_2/K be curves.*
(a) *Let $\phi: C_1 \to C_2$ be a non-constant map defined over K. Then $K(C_1)$ is a finite extension of $\phi^* K(C_2)$.*
(b) *Let $\iota: K(C_2) \to K(C_1)$ be an injection of function fields fixing K. Then there exists a unique non-constant map $\phi: C_1 \to C_2$ (defined over K) such that $\phi^* = \iota$.*
(c) *Let $\mathbb{K} \subset K(C_1)$ be a subfield of finite index containing K. Then there exists a smooth curve C'/K, unique up to K-isomorphism, and a non-constant map $\phi: C_1 \to C'$ defined over K, so that $\phi^* K(C') = \mathbb{K}$.*

PROOF. (a) [Har, II.6.8].
(b) Let $C_2 \subset \mathbb{P}^N$; and for each i, let $g_i \in K(C_2)$ be the function on C_2 corresponding to X_i/X_0. (Relabeling if necessary, we will assume that C_2 is not contained in the hyperplane $X_0 = 0$.) Then

$$\phi = [1, \iota g_1, \ldots, \iota g_N]$$

gives a map $\phi: C_1 \to C_2$ with $\phi^* = \iota$. (Note ϕ is not constant, since the g_i's cannot all be constant and ι is injective.) Finally, if $\psi = [f_0, \ldots, f_N]$ is another map with $\psi^* = \iota$, then for each i,

$$f_i/f_0 = \psi^* g_i = \phi^* g_i = \iota g_i,$$

which shows that $\psi = \phi$.
(c) [Har, I.6.12] for the case that K is algebraically closed. The general case may be done similarly, or it may be deduced from the algebraically closed case by examining $G_{\bar{K}/K}$-invariants.   □

Definition. Let $\phi: C_1 \to C_2$ be a map of curves defined over K. If ϕ is constant, we define the *degree of ϕ* to be 0; otherwise we say that ϕ is *finite*, and define its *degree* by

$$\deg \phi = [K(C_1) : \phi^* K(C_2)].$$

We say that ϕ is *separable (inseparable, purely inseparable)* if the extension $K(C_1)/\phi^* K(C_2)$ has the corresponding property, and we denote the separable and inseparable degrees of the extension by $\deg_s \phi$ and $\deg_i \phi$ respectively.

Definition. Let $\phi: C_1 \to C_2$ be a non-constant map of curves defined over K. From (2.4a), $K(C_1)$ is a finite extension of $\phi^* K(C_2)$. We define

$$\phi_*: K(C_1) \to K(C_2)$$

by using the norm map relative to ϕ^*,

$$\phi_* = (\phi^*)^{-1} \circ N_{K(C_1)/\phi^* K(C_2)}.$$

Corollary 2.4.1. *Let C_1 and C_2 be smooth curves, and let $\phi: C_1 \to C_2$ be a map of degree 1. Then ϕ is an isomorphism.*

PROOF. By definition, $\deg \phi = 1$ means that $\phi^* \bar{K}(C_2) = \bar{K}(C_1)$, so ϕ^* is an isomorphism of function fields. Hence from (2.4b), corresponding to the inverse map $(\phi^*)^{-1} : \bar{K}(C_1) \xrightarrow{\sim} \bar{K}(C_2)$, there is a rational map $\psi : C_2 \to C_1$ such that $\psi^* = (\phi^*)^{-1}$; and since C_2 is smooth, ψ is actually a morphism (2.1). Finally, since $(\phi \circ \psi)^* = \psi^* \circ \phi^*$ and $(\psi \circ \phi)^* = \phi^* \circ \psi^*$ are the identity maps on $\bar{K}(C_2)$ and $\bar{K}(C_1)$ respectively, the uniqueness assertion of (2.4b) implies that $\phi \circ \psi$ and $\psi \circ \phi$ are the identity maps on C_2 and C_1, so ϕ and ψ are isomorphisms.   □

Remark 2.5. The above result (2.4) shows the close connection between curves and their function fields. This can be made precise by stating that the following map is an equivalence of categories. (See [Har, I §6] for details.)

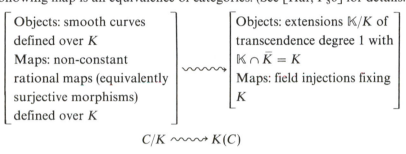

$$C/K \rightsquigarrow K(C)$$

$$\phi : C_1 \to C_2 \rightsquigarrow \phi^* : K(C_2) \to K(C_1).$$

Example 2.5.1. *Hyperelliptic Curves.* We assume $\text{char}(K) \neq 2$. Let $f(x) \in K[x]$ be a polynomial of degree d, and consider the *affine* curve C_0/K given by the equation

$$C_0 : y^2 = f(x) = a_0 x^d + a_1 x^{d-1} + \cdots + a_d.$$

Suppose that the point $P = (x_0, y_0) \in C_0$ is singular. Then

$$2y_0 = f'(x_0) = 0,$$

which means that $y_0 = 0$ and x_0 is a double root of $f(x)$. Hence if we assume that $\text{disc}(f) \neq 0$, then the affine curve $y^2 = f(x)$ will be non-singular.

Now, if we treat C_0 as giving a curve in $\mathbb{P}^2$ by homogenizing its affine equation, then one easily checks that the point at infinity will be singular whenever $d \geq 4$. On the other hand, (2.4c) assures us that there exists some smooth projective curve C/K whose function field equals $K(C_0) = K(x, y)$. The problem is that this smooth curve is not a subset of $\mathbb{P}^2$.

For example, let us consider the case $d = 4$. (See also exer. 2.14.) Then C_0 has an affine equation

$$C_0 : y^2 = a_0 x^4 + a_1 x^3 + a_2 x^2 + a_3 x + a_4.$$

Consider the map

$$[1, x, y, x^2] : C_0 \to \mathbb{P}^3.$$

Letting $[X_0, X_1, X_2, X_3] = [1, x, y, x^2]$, the ideal of the image clearly contains the homogeneous polynomials

$$F = X_3 X_0 - X_1^2$$

and

$$G = X_2^2 X_0^2 - a_0 X_1^4 - a_1 X_1^3 X_0 - a_2 X_1^2 X_0^2 - a_3 X_1 X_0^3 - a_4 X_0^4.$$

However, the zero set of these two polynomials cannot be the desired curve C, since it includes the line $X_0 = X_1 = 0$. But if we substitute $X_1^2 = X_0 X_3$ into G and cancel an X_0^2, we obtain the quadratic polynomial

$$H = X_2^2 - a_0 X_3^2 - a_1 X_1 X_3 - a_2 X_0 X_3 - a_3 X_0 X_1 - a_4 X_0^2.$$

Now we claim that the ideal generated by F and H will give a smooth curve C.

To see this, note first that if $X_0 \neq 0$, then dehomogenization with respect to X_0 gives the affine curve (setting $x = X_1/X_0$, $y = X_2/X_0$, $z = X_3/X_0$)

$$z = x^2 \qquad y^2 = a_0 z^2 + a_1 x z + a_2 z + a_3 x + a_4.$$

Substituting the first equation into the second gives us back the original curve C_0. Thus $C_0 \cong C \cap \{X_0 \neq 0\}$.

Next, if $X_0 = 0$, then necessarily $X_1 = 0$, and then $X_2 = \pm\sqrt{a_0} X_3$. Thus C has two points $[0, 0, \pm\sqrt{a_0}, 1]$ on the hyperplane $X_0 = 0$. (Note that $a_0 \neq 0$, since we have assumed that $f(x)$ has degree exactly 4.) To check that C is non-singular at these points, we dehomogenize with respect to X_3, setting $u = X_0/X_3$, $v = X_1/X_3$, $w = X_2/X_3$. This gives the equations

$$u = v^2 \qquad w^2 = a_0 + a_1 v + a_2 u + a_3 uv + a_4 u^2,$$

from which we obtain the single affine equation

$$w^2 = a_0 + a_1 v + a_2 v^2 + a_3 v^3 + a_4 v^4.$$

Since $a_0 \neq 0$, the points $(v, w) = (0, \pm\sqrt{a_0})$ are non-singular. We summarize the above discussion in the following proposition, which will be used in chapter X.

Proposition 2.5.2. *Let* $f(x) \in K[x]$ *be a polynomial of degree* 4 *with* $\text{disc}(f) \neq 0$. *There exists a smooth projective curve* $C \subset \mathbb{P}^3$ *with the following properties.*

(i) *The intersection of* C *with* $\mathbb{A}^3 = \{X_0 \neq 0\}$ *is isomorphic to the affine curve* $y^2 = f(x)$.

(ii) *Let* $f(x) = a_0 x^4 + \cdots + a_4$. *Then the intersection of* C *with the hyperplane* $\{X_0 = 0\}$ *consists of the two points* $\{[0, 0, \pm\sqrt{a_0}, 1]\}$.

We next look at the behavior of a map in the neighborhood of a point.

Definition. Let $\phi : C_1 \to C_2$ be a non-constant map of smooth curves, and let $P \in C_1$. The *ramification index of ϕ at P*, denoted $e_\phi(P)$, is given by

$$e_\phi(P) = \mathrm{ord}_P(\phi^* t_{\phi(P)}),$$

where $t_{\phi(P)} \in K(C_2)$ is a uniformizer at $\phi(P)$. Note that $e_\phi(P) \geqslant 1$. We say that ϕ is *unramified at P* if $e_\phi(P) = 1$; and ϕ is *unramified* if it is unramified at every point of C_1.

Proposition 2.6. *Let $\phi : C_1 \to C_2$ be a non-constant map of smooth curves.*
(a) *For every $Q \in C_2$,*

$$\sum_{P \in \phi^{-1}(Q)} e_\phi(P) = \deg \phi.$$

(b) *For all but finitely many $Q \in C_2$,*

$$\#\phi^{-1}(Q) = \deg_s(\phi).$$

(c) *Let $\phi : C_2 \to C_3$ be another non-constant map. Then for all $P \in C_1$,*

$$e_{\psi \circ \phi}(P) = e_\phi(P) e_\psi(\phi P).$$

PROOF. (a) [Har, II.6.9] (take $Y = \mathbb{P}^1$ and $D = (0)$), [La 2, I, prop. 21], [Se 9, I, Prop. 10], or [Sha 2, III §2, thm. 1].
(b) [Har, II.6.8].
(c) Let $t_{\phi P}$ and $t_{\psi \phi P}$ be uniformizers at the indicated points. By definition,

$$t_{\phi P}^{e_\psi(\phi P)} \quad \text{and} \quad \psi^* t_{\psi \phi P}$$

have the same order at $\phi(P)$. Applying ϕ^* and taking orders at P yields

$$\mathrm{ord}_P(\phi^* t_{\phi P}^{e_\psi(\phi P)}) = \mathrm{ord}_P((\psi\phi)^* t_{\psi \phi P}),$$

which is the desired result. $\square$

Corollary 2.7. *A map $\phi : C_1 \to C_2$ is unramified if and only if $\#\phi^{-1}(Q) = \deg(\phi)$ for all $Q \in C_2$.*

PROOF. From (2.6a), $\#\phi^{-1}(Q) = \deg \phi$ is equivalent to

$$\sum_{P \in \phi^{-1}(Q)} e_\phi(P) = \#\phi^{-1}(Q).$$

Since $e_\phi(P) \geqslant 1$, this occurs if and only if each $e_\phi(P) = 1$. $\square$

Remark 2.8. Proposition 2.6 is exactly analogous to the theorems describing the ramification of primes in number fields. Thus let L/K be number fields. Then (2.6a) is the analogue of the $\Sigma e_i f_i = [K : \mathbb{Q}]$ theorem ([La 2, I, prop. 21], [Se 9, I, prop. 10]), (2.6b) is similar to the fact that only finitely many primes of K ramify in L, and (2.6c) gives the multiplicativity of ramification degrees in towers of fields. Of course, (2.6) and the analogous results for number fields are both merely special cases of the basic theorems describing finite extensions of Dedekind domains.

Example 2.9. Consider the map

$$\phi : \mathbb{P}^1 \to \mathbb{P}^1$$

$$\phi([X, Y]) = [X^3(X - Y)^2, Y^5].$$

Then ϕ is ramified at the points $[0, 1]$ and $[1, 1]$. Further,

$$e_\phi([0, 1]) = 3 \quad \text{and} \quad e_\phi([1, 1]) = 2;$$

so

$$\sum_{P \in \phi^{-1}([0, 1])} e_\phi(P) = e_\phi([0, 1]) + e_\phi([1, 1]) = 5 = \deg \phi,$$

which is in accordance with (2.6a).

The Frobenius Map

Assume that $\text{char}(K) = p > 0$, and let $q = p^r$. For any polynomial $f \in K[X]$, let $f^{(q)}$ be the polynomial obtained from f by raising each coefficient of f to the qth power. Then for any curve C/K we can define a new curve $C^{(q)}/K$ by describing its homogeneous ideal as follows:

$$I(C^{(q)}) = \text{ideal generated by } \{ f^{(q)} : f \in I(C)\}.$$

Further, there is a natural map from C to $C^{(q)}$, called the qth-*power Frobenius morphism*, given by

$$\phi : C \to C^{(q)}$$

$$\phi([x_0, \ldots, x_n]) = [x_0^q, \ldots, x_n^q].$$

To see that ϕ actually maps C to $C^{(q)}$, it suffices to show that for every $P = [x_0, \ldots, x_n] \in C$, $\phi(P)$ is a zero of each generator $f^{(q)}$ of $I(C^{(q)})$. But

$$f^{(q)}(\phi(P)) = f^{(q)}(x_0^q, \ldots, x_n^q)$$

$$= (f(x_0, \ldots, x_n))^q \qquad \text{since } \text{char}(K) = p$$

$$= 0 \qquad \text{since } f(P) = 0,$$

which gives the desired result.

Example 2.10. Let C be the curve in $\mathbb{P}^2$ given by the single equation

$$C : Y^2Z = X^3 + aX^2Z + bZ^3.$$

Then $C^{(q)}$ has the equation

$$C^{(q)} : Y^2Z = X^3 + a^qX^2Z + b^qZ^3.$$

The next proposition describes the basic properties of the Frobenius map.

Proposition 2.11. *Let K be a field of characteristic $p > 0$, $q = p^r$, C/K a curve, and $\phi : C \to C^{(q)}$ the qth-power Frobenius morphism described above.*

(a) $\phi^* K(C^{(q)}) = K(C)^q \, (= \{ f^q : f \in K(C) \})$.

(b) ϕ is purely inseparable.

(c) $\deg \phi = q$.

[*N.B. We are assuming that K is perfect. If K is not perfect, (b) and (c) remain true, but (a) must be modified.*]

PROOF. (a) Using the description (I.2.9) of $K(C)$ as consisting of quotients f/g of homogeneous polynomials of the same degree, we see that $\phi^* K(C^{(q)})$ is the subfield given by quotients

$$\phi^*(f/g) = f(X_0^q, \ldots, X_n^q)/g(X_0^q, \ldots, X_n^q).$$

Similarly, $K(C)^q$ is the subfield given by quotients

$$f(X_0, \ldots, X_n)^q / g(X_0, \ldots, X_n)^q.$$

But since K is perfect, we know that every element of K is a qth-power, so

$$(K[X_0, \ldots, X_n])^q = K[X_0^q, \ldots, X_n^q].$$

Thus the set of quotients $f(X_i^q)/g(X_i^q)$ and the set of quotients $f(X_i)^q/g(X_i)^q$ give the exact same subfield of $K(C)$.

(b) Immediate from (a).

(c) Choose $t \in K(C)$ to be a uniformizer at some smooth point $P \in C$, so $K(C)$ is separable over $K(t)$ (1.4). Consider the tower of fields

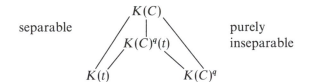

It follows that $K(C) = K(C)^q(t)$, so from (a),

$$\deg \phi = [K(C)^q(t) : K(C)^q].$$

Now $t^q \in K(C)^q$, so in order to prove that $\deg \phi = q$, we need merely show that $t^{q/p} \notin K(C)^q$. But if $t^{q/p} = f^q$ for some $f \in K(C)$, then

$$q/p = \text{ord}_P(t^{q/p}) = q \, \text{ord}_P(f),$$

which is clearly impossible. $\square$

Corollary 2.12. *Every map $\psi : C_1 \to C_2$ of (smooth) curves over a field of characteristic $p > 0$ factors as*

$$C_1 \xrightarrow{\phi} C_1^{(q)} \xrightarrow{\lambda} C_2,$$

where $q = \deg_i(\psi)$, ϕ is the qth-power Frobenius map, and λ is separable.

PROOF. Let $\mathbb{K}$ be the separable closure of $\psi^*K(C_2)$ in $K(C_1)$. Then $K(C_1)/\mathbb{K}$ is purely inseparable of degree q, so $K(C_1)^q \subset \mathbb{K}$. But from (2.11a, c),

$$K(C_1)^q = \phi^*(K(C_1^{(q)})) \quad \text{and} \quad [K(C_1):\phi^*(K(C_1^{(q)}))] = q.$$

Hence by comparing degrees, $\mathbb{K} = \phi^*K(C_1^{(q)})$. We now have the tower of function fields

$$K(C_1)/\phi^*K(C_1^{(q)})/\psi^*K(C_2),$$

and from (2.4b) this corresponds to maps

$$C_1 \xrightarrow{\phi} C_1^{(q)} \xrightarrow{\lambda} C_2$$

$$\underbrace{\qquad\qquad\qquad}_{\psi} \qquad .\qquad\qquad\qquad \square$$

§3. Divisors

The *divisor group of a curve* C, denoted $\mathrm{Div}(C)$, is the free abelian group generated by the points of C. Thus a divisor $D \in \mathrm{Div}(C)$ is a formal sum

$$D = \sum_{P \in C} n_P(P)$$

with $n_P \in \mathbb{Z}$ and $n_P = 0$ for all but finitely many $P \in C$. The *degree of* D is defined by

$$\deg D = \sum_{P \in C} n_P.$$

The *divisors of degree* 0 form a subgroup of $\mathrm{Div}(C)$, which we denote by

$$\mathrm{Div}^0(C) = \{D \in \mathrm{Div}(C) : \deg D = 0\}.$$

If C is defined over K, we let $G_{\bar{K}/K}$ act on $\mathrm{Div}(C)$ (and $\mathrm{Div}^0(C)$) in the obvious way,

$$D^\sigma = \sum_{P \in C} n_P(P^\sigma).$$

Then D is *defined over* K if $D^\sigma = D$ for all $\sigma \in G_{\bar{K}/K}$. (N.B. If $D = n_1(P_1) + \cdots + n_r(P_r)$ with $n_1, \ldots, n_r \neq 0$, then to say that D is defined over K does *not* mean that $P_1, \ldots, P_r \in C(K)$. It suffices for $G_{\bar{K}/K}$ to permute the P_i's in an appropriate fashion.) We denote the *group of divisors defined over K* by $\mathrm{Div}_K(C)$, and similarly for $\mathrm{Div}_K^0(C)$.

Assume now that the curve C is smooth, and let $f \in \bar{K}(C)^*$. Then we can associate to f the divisor $\mathrm{div}(f)$ given by

$$\mathrm{div}(f) = \sum_{P \in C} \mathrm{ord}_P(f)(P).$$

(This is a divisor by (1.2).) Now if $\sigma \in G_{\bar{K}/K}$, then one easily sees that

$$\mathrm{div}(f^\sigma) = \mathrm{div}(f)^\sigma.$$

In particular, if $f \in K(C)$, then $\mathrm{div}(f) \in \mathrm{Div}_K(C)$.

Since each ord_P is a valuation, we see that the map

$$\mathrm{div} : \bar{K}(C)^* \to \mathrm{Div}(C)$$

is a homomorphism of abelian groups. It is analogous to the map which sends an element of a number field to the corresponding fractional ideal. This prompts the following definitions.

Definition. A divisor $D \in \mathrm{Div}(C)$ is *principal* if it has the form $D = \mathrm{div}(f)$ for some $f \in \bar{K}(C)^*$. Two divisors D_1, D_2 are *linearly equivalent*, denoted $D_1 \sim D_2$, if $D_1 - D_2$ is principal. The *divisor class group* (or *Picard group*) of C, denoted $\mathrm{Pic}(C)$, is the quotient of $\mathrm{Div}(C)$ by the subgroup of principal divisors. We let $\mathrm{Pic}_K(C)$ be the subgroup of $\mathrm{Pic}(C)$ fixed by $G_{\bar{K}/K}$. [N.B. In general, $\mathrm{Pic}_K(C)$ is not the quotient of $\mathrm{Div}_K(C)$ by its subgroup of principal divisors. But see (exer. 2.13).]

Proposition 3.1. *Let C be a smooth curve and $f \in \bar{K}(C)^*$.*
(a) $\mathrm{div}(f) = 0$ *if and only if $f \in \bar{K}^*$.*
(b) $\deg(\mathrm{div}(f)) = 0.$

PROOF. (a) If $\mathrm{div}(f) = 0$, then f has no poles, so the corresponding map $f : C \to \mathbb{P}^1$ (cf. 2.2) is not surjective. Therefore it is constant (2.3), so $f \in \bar{K}^*$. The converse is clear.
(b) [Har, II.6.10], [Sha 2, III 2, cor. to thm. 1], or see (3.7) below.   □

Example 3.2. On $\mathbb{P}^1$, every divisor of degree 0 is principal. To see this, suppose that $D = \Sigma n_P(P)$ has degree 0. Writing $P = [\alpha_P, \beta_P] \in \mathbb{P}^1$, we see that D is the divisor of the function

$$\prod_{P \in \mathbb{P}^1} (\beta_P X - \alpha_P Y)^{n_P}.$$

(Note this function is in $K(\mathbb{P}^1)$ because $\Sigma n_P = 0$.) We have thus proven that for $\mathbb{P}^1$, the degree map

$$\deg : \mathrm{Pic}(\mathbb{P}^1) \to \mathbb{Z}$$

is an isomorphism. It turns out that the converse is also true: if C is a smooth curve and $\mathrm{Pic}(C) \simeq \mathbb{Z}$, then C is isomorphic to $\mathbb{P}^1$.

Example 3.3. Assume that $\mathrm{char}(K) \neq 2$. Let $e_1, e_2, e_3 \in \bar{K}$ be distinct, and consider the curve

$$C : y^2 = (x - e_1)(x - e_2)(x - e_3).$$

One can check that C is smooth; and it has a single point at infinity, which we will denote by P_∞. For $i = 1, 2, 3$, let $P_i = (e_i, 0) \in C$. Then

$$\mathrm{div}(x - e_i) = 2(P_i) - 2(P_\infty)$$

and

$$\mathrm{div}(y) = (P_1) + (P_2) + (P_3) - 3(P_\infty).$$

From (3.1b) we see that the principal divisors form a subgroup of $\mathrm{Div}^0(C)$.

Definition. The *degree 0 part of the divisor class group of* C, which we denote by $\mathrm{Pic}^0(C)$, is the quotient of $\mathrm{Div}^0(C)$ by the subgroup of principal divisors. Further, $\mathrm{Pic}^0_K(C)$ is the subgroup of $\mathrm{Pic}^0(C)$ fixed by $G_{\bar{K}/K}$.

Remark 3.4. Proposition 3.1 and the above definitions may be summarized by saying that there is an exact sequence

$$1 \to \bar{K}^* \to \bar{K}(C)^* \overset{\mathrm{div}}{\to} \mathrm{Div}^0(C) \to \mathrm{Pic}^0(C) \to 0.$$

This sequence is the function field analogue of the fundamental exact sequence in algebraic number theory, which for a number field K reads

$$1 \to \begin{pmatrix} \text{units} \\ \text{of } K \end{pmatrix} \to K^* \to \begin{pmatrix} \text{fractional} \\ \text{ideals of } K \end{pmatrix} \to \begin{pmatrix} \text{ideal class} \\ \text{group of } K \end{pmatrix} \to 1.$$

Now let $\phi : C_1 \to C_2$ be a non-constant map of smooth curves. As we have seen, ϕ induces maps on the function fields of C_1 and C_2,

$$\phi^* : \bar{K}(C_2) \to \bar{K}(C_1) \quad \text{and} \quad \phi_* : \bar{K}(C_1) \to \bar{K}(C_2).$$

We similarly define maps on the divisor groups as follows.

$$\phi^* : \mathrm{Div}(C_2) \to \mathrm{Div}(C_1) \qquad\qquad \phi_* : \mathrm{Div}(C_1) \to \mathrm{Div}(C_2)$$

$$(Q) \to \sum_{P \in \phi^{-1}(Q)} e_\phi(P)(P) \qquad\qquad (P) \to (\phi P),$$

and extend $\mathbb{Z}$-linearly to arbitrary divisors.

Example 3.5. Let C be a smooth curve, $f \in \bar{K}(C)$ a non-constant function, and $f : C \to \mathbb{P}^1$ the corresponding map (2.2). Then directly from the definitions,

$$\mathrm{div}(f) = f^*((0) - (\infty)).$$

Proposition 3.6. *Let* $\phi : C_1 \to C_2$ *be a non-constant map of smooth curves.*

(a) $\deg(\phi^* D) = (\deg \phi)(\deg D)$ *for all* $D \in \mathrm{Div}(C_2)$.

(b) $\phi^*(\mathrm{div}\, f) = \mathrm{div}(\phi^* f)$ *for all* $f \in \bar{K}(C_2)^*$.

(c) $\deg(\phi_* D) = \deg D$ for all $D \in \mathrm{Div}(C_1)$.

(d) $\phi_*(\mathrm{div}\ f) = \mathrm{div}(\phi_* f)$ for all $f \in \bar{K}(C_1)^*$.

(e) $\phi_* \circ \phi^*$ acts as multiplication by $\deg \phi$ on $\mathrm{Div}(C_2)$.

(f) If $\psi : C_2 \to C_3$ is another such map, then

$$(\psi \circ \phi)^* = \phi^* \circ \psi^* \quad and \quad (\psi \circ \phi)_* = \psi_* \circ \phi_*.$$

PROOF. (a) Follows directly from (2.6a).
(b) Follows from the definitions and the easy fact (exer. 2.2) that for all $P \in C_1$,

$$\mathrm{ord}_P(\phi^* f) = e_\phi(P)\, \mathrm{ord}_{\phi P}(f).$$

(c) Clear from the definitions.
(d) [La 2, ch. 1, prop. 22] or [Se 9, I, prop. 14].
(e) Follows directly from (2.6a).
(f) The first equality follows from (2.6c). The second is obvious. □

Remark 3.7. From (3.6) we see that ϕ^* and ϕ_* take divisors of degree 0 to divisors of degree 0, and principal divisors to principal divisors. They thus induce maps

$$\phi^* : \mathrm{Pic}^0(C_2) \to \mathrm{Pic}^0(C_1) \quad and \quad \phi_* : \mathrm{Pic}^0(C_1) \to \mathrm{Pic}^0(C_2).$$

In particular, if $f \in \bar{K}(C)$ gives the map $f : C \to \mathbb{P}^1$, then

$$\deg \mathrm{div}(f) = \deg f^*((0) - (\infty)) = \deg f - \deg f = 0.$$

This provides a proof of (3.1b).

§4. Differentials

In this section we will discuss the vector space of differential forms on a curve. This vector space will be useful for two different purposes. First, it will perform the traditional calculus role of linearization. (See (III §5), especially (III.5.2).) Second, it will give a useful criterion for determining when an algebraic map is separable. (See (4.2c) below and its utilization in the proof of (III.5.5).) Of course, this latter is also a familiar use of calculus, since a field extension is separable if and only if the minimal polynomial of each element has non-zero derivative.

Definition. Let C be a curve. The *space of (meromorphic) differential forms* on C, denoted Ω_C, is the $\bar{K}(C)$-vector space generated by symbols of the form dx for $x \in \bar{K}(C)$, subject to the usual relations:
(i) $d(x + y) = dx + dy$ for all $x, y \in \bar{K}(C)$;

(ii) $d(xy) = xdy + ydx$ for all $x, y \in \bar{K}(C)$;
(iii) $da = 0$ for all $a \in \bar{K}$.

Remark 4.1. There is, of course, a more functorial definition of Ω_C. See, for example, [Mat, ch. 10], [Har, II.8], or [Rob, II §3].

Let $\phi : C_1 \to C_2$ be a non-constant map of curves. Then the natural map $\phi^* : \bar{K}(C_2) \to \bar{K}(C_1)$ induces a map on differentials

$$\phi^* : \Omega_{C_2} \to \Omega_{C_1}$$

$$\phi^* \left(\sum f_i dx_i \right) = \sum (\phi^* f_i) d(\phi^* x_i).$$

This map will provide a useful criterion for determining when ϕ is separable.

Proposition 4.2. *Let C be a curve.*
(a) Ω_C *is a 1-dimensional $\bar{K}(C)$-vector space.*
(b) *Let $x \in \bar{K}(C)$. Then dx is a $\bar{K}(C)$ basis for Ω_C if and only if $\bar{K}(C)/\bar{K}(x)$ is a finite separable extension.*
(c) *Let $\phi : C_1 \to C_2$ be a non-constant map of curves. Then ϕ is separable if and only if the map*

$$\phi^* : \Omega_{C_2} \to \Omega_{C_1}$$

is injective (equivalently, non-zero.)

PROOF. (a) [Mat, 27.A, B], [Rob, II.3.4], or [Sha 2, III §4, thm. 3].
(b) [Mat, 27.A, B] or [Sha 2, III §4, thm. 4].
(c) Using (a) and (b), choose $y \in \bar{K}(C_2)$ so that $\Omega_{C_2} = \bar{K}(C_2) dy$ and $\bar{K}(C_2)/\bar{K}(y)$ is a separable extension. Note $\phi^* \bar{K}(C_2)$ is then separable over $\phi^* \bar{K}(y) = \bar{K}(\phi^* y)$. Now

$$\phi^* \text{ is injective} \Leftrightarrow d(\phi^* y) \neq 0$$

$$\Leftrightarrow d(\phi^* y) \text{ is a basis for } \Omega_{C_1} \text{ (from (a))}$$

$$\Leftrightarrow \bar{K}(C_1)/\bar{K}(\phi^* y) \text{ is separable (from (b))}$$

$$\Leftrightarrow \bar{K}(C_1)/\phi^* \bar{K}(C_2) \text{ is separable,}$$

where the last equivalence follows because we already know that $\phi^* \bar{K}(C_2)/\bar{K}(\phi^* y)$ is separable.   □

Proposition 4.3. *Let $P \in C$, and let $t \in \bar{K}(C)$ be a uniformizer at P.*
(a) *For every $\omega \in \Omega_C$ there exists a unique function $g \in \bar{K}(C)$, depending on ω and t, such that*

$$\omega = g \, dt.$$

We denote g by ω/dt.
(b) *Let $f \in \bar{K}(C)$ be regular at P. Then df/dt is also regular at P.*

(c) *The quantity*

$$\mathrm{ord}_P(\omega/dt)$$

depends only on ω and P, independent of the choice of uniformizer t. We call this value the order of ω at P, *and denote it by* $\mathrm{ord}_P(\omega)$.
(d) *Let $x, f \in \bar{K}(C)$ with $x(P) = 0$, and let $p = \mathrm{char}\, K$. Then*

$$\mathrm{ord}_P(f\, dx) = \mathrm{ord}_P(f) + \mathrm{ord}_P(x) - 1, \qquad \textit{if } p = 0 \textit{ or } p \nmid \mathrm{ord}_P(x),$$
$$\mathrm{ord}_P(f\, dx) \geqslant \mathrm{ord}_P(f) + \mathrm{ord}_P(x), \qquad \textit{if } p > 0 \textit{ and } p \mid \mathrm{ord}_P(x).$$

(e) *For all but finitely many $P \in C$,*

$$\mathrm{ord}_P(\omega) = 0.$$

PROOF. (a) This follows from (1.4) and (4.2a, b).
(b) [Har, comment following IV.2.1] or [Rob, II.3.10].
(c) Let t' be another uniformizer at P. Then from (b), dt/dt' and dt'/dt are both regular at P, so $\mathrm{ord}_P(dt'/dt) = 0$. Since

$$\omega = g\, dt' = g(dt'/dt)dt,$$

the desired result follows.
(d) Write $x = ut^n$ with $n = \mathrm{ord}_P(x) \geqslant 1$, so $\mathrm{ord}_P(u) = 0$. Then

$$dx = [nut^{n-1} + (du/dt)t^n]dt.$$

Now from (b), du/dt is regular at P, so provided $n \neq 0$, the first term dominates. Hence we obtain the desired equality

$$\mathrm{ord}_P(f\, dx) = \mathrm{ord}_P(fnut^{n-1}\, dt) = \mathrm{ord}_P(f) + n - 1.$$

Finally, if $p > 0$ and $p \mid n$, then the first term vanishes, which yields

$$\mathrm{ord}_P(f\, dx) = \mathrm{ord}_P(f(du/dt)t^n\, dt) \geq \mathrm{ord}_P(f) + n.$$

(e) Let $x \in \bar{K}(C)$ so that $\bar{K}(C)/\bar{K}(x)$ is separable, and write $\omega = f\, dx$. From [Har, IV.2.2a], the map $x: C \to \mathbb{P}^1$ ramifies at only finitely many points of C. Hence discarding finitely many points, we may restrict our attention to points $P \in C$ such that $f(P) \neq 0, \infty$, $x(P) \neq \infty$, and $x: C \to \mathbb{P}^1$ is unramified at P. But the latter two conditions imply that $x - x(P)$ is a uniformizer at P, so

$$\mathrm{ord}_P(\omega) = \mathrm{ord}_P(fd(x - x(P))) = 0. \qquad \square$$

Definition. Let $\omega \in \Omega_C$. The *divisor associated to ω* is

$$\mathrm{div}(\omega) = \sum_{P \in C} \mathrm{ord}_P(\omega)(P) \in \mathrm{Div}(C).$$

Definition. A differential $\omega \in \Omega_C$ is *regular* (or *holomorphic*) if

$$\mathrm{ord}_P(\omega) \geqslant 0 \qquad \text{for all } P \in C.$$

It is *non-vanishing* if

$$\mathrm{ord}_P(\omega) \leqslant 0 \qquad \text{for all } P \in C.$$

Remark 4.4. If ω_1, $\omega_2 \in \Omega_C$ are non-zero differentials, then (4.2a) implies that there is a function $f \in \bar{K}(C)^*$ so that $\omega_1 = f\omega_2$. Thus

$$\operatorname{div}(\omega_1) = \operatorname{div}(f) + \operatorname{div}(\omega_2),$$

which shows that the following definition makes sense.

Definition. The *canonical divisor class on C* is the image in $\operatorname{Pic}(C)$ of $\operatorname{div}(\omega)$ for any non-zero differential $\omega \in \Omega_C$. Any divisor in this divisor class is called a *canonical divisor*.

Example 4.5. Let us show that there are no holomorphic differentials on $\mathbb{P}^1$. First, if t is a coordinate function on $\mathbb{P}^1$, then

$$\operatorname{div}(dt) = -2(\infty).$$

(To see this, note that for all $\alpha \in K$, $t - \alpha$ is a uniformizer at α, so

$$\operatorname{ord}_\alpha(dt) = \operatorname{ord}_\alpha(d(t - \alpha)) = 0.$$

However, at $\infty \in \mathbb{P}^1$, $1/t$ is a uniformizer, so

$$\operatorname{ord}_\infty(dt) = \operatorname{ord}_\infty(-t^2 d(1/t)) = -2.)$$

Thus dt is not holomorphic. But now for any non-zero $\omega \in \Omega_{\mathbb{P}^1}$, (4.3a) implies that

$$\deg \operatorname{div}(\omega) = \deg \operatorname{div}(dt) = -2,$$

so ω cannot be holomorphic either.

Example 4.6. Let C be the curve

$$C: y^2 = (x - e_1)(x - e_2)(x - e_3),$$

where we continue with the notation of (3.3). Then

$$\operatorname{div}(dx) = (P_1) + (P_2) + (P_3) - 3(P_\infty).$$

(Note $dx = d(x - e_i) = -x^2 d(1/x)$.) We thus see that

$$\operatorname{div}(dx/y) = 0.$$

Hence dx/y is both holomorphic and non-vanishing.

§5. The Riemann–Roch Theorem

Let C be a curve. We put a partial order on $\operatorname{Div}(C)$ as follows.

Definition. A divisor $D = \Sigma n_P(P) \in \operatorname{Div}(C)$ is *positive* (or *effective*), denoted by

$$D \geqslant 0,$$

if $n_P \geq 0$ for every $P \in C$. Similarly, if $D_1, D_2 \in \mathrm{Div}(C)$, then we write

$$D_1 \geq D_2$$

to indicate that $D_1 - D_2$ is positive.

Example 5.1. Let $f \in \bar{K}(C)^*$ be a function which is regular everywhere except at one point $P \in C$, and such that it has a pole of order at most n at P. These requirements on f may be succinctly summarized by the inequality

$$\mathrm{div}(f) \geq -n(P).$$

Similarly,

$$\mathrm{div}(f) \geq (Q) - n(P)$$

says that in addition, f has a zero at Q. Thus divisorial inequalities are a useful tool for describing poles and/or zeros of functions.

Definition. Let $D \in \mathrm{Div}(C)$. We associate to D the set of functions

$$\mathscr{L}(D) = \{f \in \bar{K}(C)^* : \mathrm{div}(f) \geq -D\} \cup \{0\}.$$

$\mathscr{L}(D)$ is a finite-dimensional $\bar{K}$-vector space (see (5.2b) below), and we denote its dimension by

$$\ell(D) = \dim_{\bar{K}} \mathscr{L}(D).$$

Proposition 5.2. *Let* $D \in \mathrm{Div}(C)$.
(a) *If* $\deg D < 0$, *then*

$$\mathscr{L}(D) = \{0\} \quad and \quad \ell(D) = 0.$$

(b) $\mathscr{L}(D)$ *is a finite-dimensional* $\bar{K}$-*vector space.*
(c) *If* $D' \in \mathrm{Div}(C)$ *is linearly equivalent to* D, *then*

$$\mathscr{L}(D) \simeq \mathscr{L}(D'); \quad and \ so \ \ell(D) = \ell(D').$$

PROOF. (a) Let $f \in \mathscr{L}(D)$ with $f \neq 0$. Then using (3.1b),

$$0 = \deg \mathrm{div}(f) \geq \deg(-D) = -\deg D,$$

so $\deg D \geq 0$.
(b) [Har, II.5.19] or (exer. 2.4).
(c) If $D = D' + \mathrm{div}(g)$, then the map

$$\mathscr{L}(D) \to \mathscr{L}(D')$$

$$f \to fg$$

is an isomorphism.   □

Example 5.3. Let $K_C \in \mathrm{Div}(C)$ be a canonical divisor on C, say

$$K_C = \mathrm{div}(\omega).$$

Then each function $f \in \mathcal{L}(K_C)$ has the property that

$$\operatorname{div}(f) \geq -\operatorname{div}(\omega), \qquad \text{so } \operatorname{div}(f\omega) \geq 0.$$

In other words, $f\omega$ is holomorphic. Conversely, if $f\omega$ is holomorphic, then $f \in \mathcal{L}(K_C)$. Since every differential on C has the form $f\omega$ for some f, we have thus established an isomorphism of $\bar{K}$-vector spaces

$$\mathcal{L}(K_C) \simeq \{\omega \in \Omega_C : \omega \text{ is holomorphic}\}.$$

The dimension $\ell(K_C)$ of these spaces is an important invariant of the curve C.

We are now ready to state one of the most fundamental results in the algebraic geometry of curves. Its importance, as we will see amply demonstrated (cf. III §3), lies in its potential for allowing us to describe the functions on C having prescribed zeros and poles.

Theorem 5.4 (Riemann–Roch). *Let C be a smooth curve and K_C a canonical divisor on C. There is an integer $g \geq 0$, called the* genus *of C, such that for every divisor $D \in \operatorname{Div}(C)$,*

$$\ell(D) - \ell(K_C - D) = \deg D - g + 1.$$

PROOF. For a fancy proof using Serre duality, see [Har, IV §1]. A more elementary proof, due to Weil, is given in [La 6, Ch. I].   □

Corollary 5.5. (a) $\ell(K_C) = g$.
(b) $\deg K_C = 2g - 2$.
(c) *If* $\deg D > 2g - 2$, *then*

$$\ell(D) = \deg D - g + 1.$$

PROOF. (a) Use (5.4) with $D = 0$. Note that $\mathcal{L}(0) = \bar{K}$ from (1.2), so $\ell(0) = 1$.
(b) Use (a) and (5.4) with $D = K_C$.
(c) From (b), $\deg(K_C - D) < 0$. Now use (5.4) and (5.2a).   □

Example 5.6. Let $C = \mathbb{P}^1$. Then there are no holomorphic differentials on C (4.5), so using the identification from (5.3), we see that $\ell(K_C) = 0$. Thus by (5.5a), $\mathbb{P}^1$ has genus 0, and the Riemann–Roch theorem reads

$$\ell(D) - \ell(-2(\infty) - D) = \deg D + 1.$$

In particular, if $\deg D \geq -1$, then

$$\ell(D) = \deg D + 1.$$

(See exer. 2.3b.)

Example 5.7. Let C be the curve

$$C : y^2 = (x - e_1)(x - e_2)(x - e_3),$$

where we continue with the notation of (3.3) and (4.6). We have seen (4.6) that

$$\text{div}(dx/y) = 0,$$

so the canonical class on C is trivial (i.e. we may take $K_C = 0$). Hence from (5.5a),

$$g = \ell(K_C) = \ell(0) = 1,$$

so C has genus 1. The Riemann–Roch theorem (actually (5.5c)) then reads

$$\ell(D) = \deg D \qquad \text{provided } \deg D \geqslant 1.$$

Let's look at several special cases.

(i) Let $P \in C$. Then $\ell((P)) = 1$. But $\mathscr{L}((P))$ certainly contains the constant functions. This shows that there are no functions on C having a single simple pole.

(ii) Recall P_∞ is the point at infinity on C. Then $\ell(2(P_\infty)) = 2$, and $\{1, x\}$ provides a basis for $\mathscr{L}(2(P_\infty))$.

(iii) Similarly $\{1, x, y\}$ is a basis for $\mathscr{L}(3(P_\infty))$, and $\{1, x, y, x^2\}$ is a basis for $\mathscr{L}(4(P_\infty))$.

(iv) Now the functions $1, x, y, x^2, xy, x^3, y^2$ are all in $\mathscr{L}(6(P_\infty))$. But $\ell(6(P_\infty)) = 6$, so it follows that these functions are $\bar{K}$-linearly dependent. Of course, the original equation used above to define C gives an equation of linear dependence among them.

The next result says that if C and D are defined over K, then so is $\mathscr{L}(D)$.

Proposition 5.8. *Let C/K be a smooth curve, and let $D \in \text{Div}_K(C)$. Then $\mathscr{L}(D)$ has a basis consisting of functions in $K(C)$.*

PROOF. Since D is defined over K, we have

$$f^\sigma \in \mathscr{L}(D^\sigma) = \mathscr{L}(D) \qquad \text{for all } f \in \mathscr{L}(D) \text{ and } \sigma \in G_{\bar{K}/K}.$$

Thus $G_{\bar{K}/K}$ acts on $\mathscr{L}(D)$, and the desired conclusion follows from the following general lemma. $\square$

Lemma 5.8.1. *Let V be a $\bar{K}$-vector space, and assume that $G_{\bar{K}/K}$ acts continuously on V in a manner compatible with its action on $\bar{K}$. Let*

$$V_K = V^{G_{\bar{K}/K}} = \{v \in V : v^\sigma = v \text{ for all } \sigma \in G_{\bar{K}/K}\}.$$

Then

$$V \cong \bar{K} \otimes_K V_K.$$

[I.e. V has a basis of $G_{\bar{K}/K}$-invariant vectors.]

PROOF. It suffices to show that every $v \in V$ is a $\bar{K}$-linear combination of vectors in V_K. Choose a $v \in V$, and let L/K be a finite Galois extension such that v is fixed by $G_{\bar{K}/L}$. (The fact that $G_{\bar{K}/K}$ acts continuously on V means precisely that the subgroup $\{\sigma \in G_{\bar{K}/K} : v^\sigma = v\}$ has finite index in $G_{\bar{K}/K}$. We take L to be the Galois closure of its fixed field.) Let $\{\alpha_1, \ldots, \alpha_n\}$ be a basis for L/K, and let $\{\sigma_1, \ldots, \sigma_n\} = G_{L/K}$. For each $1 \leqslant i \leqslant n$, consider the vector

$$w_i = \sum_{j=1}^{n} (\alpha_i v)^{\sigma_j} = \text{Trace}_{L/K}(\alpha_i v).$$

It is clearly $G_{\bar{K}/K}$ invariant, so $w_i \in V_K$. Now a basic result in field theory [La 2, III, prop. 9] says that the matrix $(\alpha_i^{\sigma_j})_{1 \leqslant i,j \leqslant n}$ is non-singular, so each v^{σ_j} (and in particular v) is an L-linear combination of the w_i's. (For a fancier proof, see exer. 2.12.) □

We conclude by giving the classic relationship connecting the genus of curves linked by a non-constant map.

Theorem 5.9 (Hurwitz). *Let $\phi : C_1 \to C_2$ be a non-constant separable map of smooth curves. Then*

$$2g_1 - 2 \geqslant (\deg \phi)(2g_2 - 2) + \sum_{P \in C_1} (e_\phi(P) - 1),$$

where g_i is the genus of C_i. Further, equality holds if and only if either:

(i) $\text{char}(K) = 0$; *or*
(ii) $\text{char}(K) = p > 0$ *and p does not divide $e_\phi(P)$ for all $P \in C_1$.*

PROOF. Let $\omega \in \Omega_{C_2}$, $\omega \neq 0$, let $P \in C_1$, and let $Q = \phi(P)$. Since ϕ is separable, $\phi^*\omega \neq 0$ (4.2c); we wish to relate $\text{ord}_P(\phi^*\omega)$ and $\text{ord}_Q(\omega)$. Write $\omega = f \, dt$ with $t \in \bar{K}(C_2)$ a uniformizer at Q. Then letting $e = e_\phi(P)$, we have $\phi^*t = us^e$, where s is a uniformizer at P and $u(P) \neq 0, \infty$. Hence

$$\phi^*\omega = (\phi^*f)d(\phi^*t) = (\phi^*f)d(us^e) = (\phi^*f)[eus^{e-1} + (du/ds)s^e]ds.$$

Now $\text{ord}_P(du/ds) \geqslant 0$ (4.3b), so we see that

$$\text{ord}_P(\phi^*\omega) \geqslant \text{ord}_P(\phi^*f) + e - 1,$$

with equality if and only if $e \neq 0$ in K. Further,

$$\text{ord}_P(\phi^*f) = e_\phi(P) \, \text{ord}_Q(f) = e_\phi(P) \, \text{ord}_Q(\omega).$$

Hence adding over $P \in C_1$ yields

$$\deg \text{div}(\phi^*\omega) \geqslant \sum_{P \in C_1} [e_\phi(P) \, \text{ord}_{\phi(P)}(\omega) + e_\phi(P) - 1]$$

$$= \sum_{Q \in C_2} \sum_{P \in \phi^{-1}(Q)} e_\phi(P) \, \text{ord}_Q(\omega) + \sum_{P \in C_1} (e_\phi(P) - 1)$$

$$= (\deg \phi)(\deg \text{div}(\omega)) + \sum_{P \in C_1} (e_\phi(P) - 1),$$

where the last equality follows from (2.6a). Now Hurwitz' theorem is a consequence of (5.5b), which says that on a curve of genus g, the divisor of any non-zero differential has degree $2g - 2$. □

EXERCISES

2.1. Let R be a Noetherian local domain, M its maximal ideal, and $k = R/M$. Prove that the following are equivalent:
(i) R is a discrete valuation ring.
(ii) M is principal.
(iii) $\dim_k M/M^2 = 1$.
(Note this lemma was used in (1.1) to show that on a smooth curve, the local rings $\bar{K}[C]_P$ are discrete valuation rings.)

2.2. Let $\phi : C_1 \to C_2$ be a non-constant map of smooth curves, $f \in \bar{K}(C_2)^*$, $P \in C_1$. Then
$$\operatorname{ord}_P(\phi^* f) = e_\phi(P) \operatorname{ord}_{\phi(P)}(f).$$

2.3. Verify directly that each of the following theorems is true for the particular case of the curve $C = \mathbb{P}^1$ and a non-constant map $\phi : \mathbb{P}^1 \to \mathbb{P}^1$.
(a) (proposition 2.6) Prove that
$$\sum_{P \in \phi^{-1}(Q)} e_\phi(P) = \deg \phi \qquad \text{for all } Q \in \mathbb{P}^1; \text{ and}$$
$$\#\phi^{-1}(Q) = \deg_s(\phi) \qquad \text{for all but finitely many } Q \in \mathbb{P}^1.$$

(b) Prove the Riemann–Roch theorem (5.4) for $\mathbb{P}^1$.
(c) Prove Hurwitz' theorem (5.9) for $\phi : \mathbb{P}^1 \to \mathbb{P}^1$.

2.4. Let C be a smooth curve and $D \in \operatorname{Div}(C)$. Without using the Riemann–Roch theorem, prove:
(a) $\mathcal{L}(D)$ is a $\bar{K}$-vector space.
(b) If $\deg D \geqslant 0$, then
$$\ell(D) \leqslant \deg D + 1.$$

2.5. Let C be a smooth curve. Prove that the following are equivalent (over $\bar{K}$):
(i) C is isomorphic to $\mathbb{P}^1$.
(ii) C has genus 0.
(iii) There exist distinct points $P, Q \in C$ with $(P) \sim (Q)$.

2.6. Let C be a smooth curve of genus 1. Fix a basepoint $P_0 \in C$. Prove the following.
(a) For all $P, Q \in C$ there exists a unique $R \in C$ such that
$$(P) + (Q) \sim (R) + (P_0).$$

Denote this R by $\sigma(P, Q)$.
(b) The map $\sigma : C \times C \to C$ from (a) makes C into an abelian group with identity P_0.
(c) Define a map
$$\kappa : C \to \operatorname{Pic}^0(C)$$
$$P \to \text{divisor class of } (P) - (P_0).$$

Prove that κ is a bijection of sets. Hence κ can be used to make C into a group,
$$P + Q = \kappa^{-1}(\kappa(P) + \kappa(Q)).$$

(d) Prove that the group operations on C defined in (b) and (c) are the same.

2.7. Let $F(X, Y, Z) \in K[X, Y, Z]$ be homogeneous of degree $d \geqslant 1$, and suppose that the curve C in $\mathbb{P}^2$ given by the equation $F = 0$ is non-singular. Prove that

$$\text{genus } (C) = (d - 1)(d - 2)/2.$$

[*Hint:* Define a map $C \to \mathbb{P}^1$ and use (5.9).]

2.8. Let $\phi : C_1 \to C_2$ be a non-constant separable map of smooth curves.
   (a) Prove that genus $(C_1) \geqslant$ genus (C_2).
   (b) Prove that if C_1 and C_2 both have genus g, then one of the following is true.
      (i) $g = 0$.
      (ii) $g = 1$ and ϕ is unramified.
      (iii) $g \geqslant 2$ and ϕ is an isomorphism.

2.9. Let a, b, c, d be square free integers with $a > b > c > 0$, and let C be the curve in $\mathbb{P}^2$ given by the equation

$$C : aX^3 + bY^3 + cZ^3 + dXYZ = 0.$$

Let $P = [x, y, z] \in C$ and let L be the tangent line to C at P.
   (a)   Show that $C \cap L = \{P, P'\}$, and calculate $P' = [x', y', z']$ in terms of a, b, c, d, x, y, z.
   (b)   Show that if $P \in C(\mathbb{Q})$, then $P' \in C(\mathbb{Q})$.
   (c)   Let $P \in C(\mathbb{Q})$. Choose homogeneous coordinates for P and P' which are integers satisfying $\gcd(x, y, z) = 1$ and $\gcd(x', y', z') = 1$. Prove that

$$|x'y'z'| > |xyz|.$$

   (Note the strict inequality.)
   (d)   Conclude that either $C(\mathbb{Q}) = \varnothing$, or else $C(\mathbb{Q})$ is infinite.
   (e)** Characterize, in terms of a, b, c, d, whether $C(\mathbb{Q})$ contains any points.

2.10. Let C be a smooth curve. The *support* of a divisor $D = \Sigma n_P(P) \in \text{Div}(C)$ is the set of points $P \in C$ for which $n_P \neq 0$. Now let $f \in \bar{K}(C)^*$ be a function such that $\text{div}(f)$ and D have disjoint supports. Then it makes sense to define

$$f(D) = \prod_{P \in C} f(P)^{n_P}.$$

Next let $\phi : C_1 \to C_2$ be a non-constant map of smooth curves. Prove that the following two equalities are valid in the sense that if one side is well-defined, then so is the other, and they are equal.
   (a) $f(\phi^*D) = (\phi_* f)(D)$ for $f \in K(C_1)^*, D \in \text{Div}(C_2)$.
   (b) $f(\phi_* D) = (\phi^* f)(D)$ for $f \in K(C_2)^*, D \in \text{Div}(C_1)$.

2.11. Let C be a smooth curve and $f, g \in \bar{K}(C)^*$ functions such that $\text{div}(f)$ and $\text{div}(g)$ have disjoint support. (See exer. 2.10.) Prove *Weil's reciprocity law*

$$f(\text{div}(g)) = g(\text{div}(f))$$

in two steps.
   (a) Verify it directly for $C = \mathbb{P}^1$.
   (b) Now prove it for arbitrary C by using the map $g : C \to \mathbb{P}^1$ to reduce to the case already done.

2.12. Use the extension of Hilbert's theorem 90 (B.3.2) which says that

$$H^1(G_{\bar{K}/K}, GL_n(\bar{K})) = 0$$

to give another proof of (5.8.1).

2.13. Let C/K be a curve.
   (a) Prove that the following sequence is exact.

$$1 \to K^* \to K(C)^* \to \mathrm{Div}^0_K(C) \to \mathrm{Pic}^0_K(C).$$

   (b) Suppose that C has genus 1 and that $C(K) \neq \varnothing$. Prove that the map $\mathrm{Div}^0_K(C) \to \mathrm{Pic}^0_K(C)$ is surjective.

2.14. Let $f(x) \in K[x]$ be a polynomial of degree $d \geq 1$ with $\mathrm{disc}(f) \neq 0$, let C_0/K be the affine curve given by the equation

$$C_0 : y^2 = f(x) = a_0 x^d + a_1 x^{d-1} + \cdots + a_{d-1} x + a_d,$$

and let g be the unique integer satisfying $d - 3 < 2g \leq d - 1$.
   (a) Let C be the closure of the image of C_0 under the map

$$[1, x, x^2, \ldots, x^{g+1}, y] : C \to \mathbb{P}^{g+2}.$$

   Prove that C is smooth and that $C \cap \{X_0 \neq 0\}$ is isomorphic to C_0. C is called a *hyperelliptic curve*.
   (b) Let $f^*(v) = a_0 + a_1 v + \cdots + a_{d-1} v^{d-1} + a_d v^d = v^d f(1/v)$. Show that C consists of two affine pieces

$$C_0 : y^2 = f(x) \quad \text{and} \quad C_1 : w^2 = f^*(v),$$

   "glued" together via the maps

$$C_0 \to C_1 \qquad\qquad\qquad C_1 \to C_0$$
$$(x, y) \to (1/x, y/x^{g+1}) \qquad (v, w) \to (1/v, w/v^{g+1}).$$

   (c) Calculate the divisor of the differential dx/y on C, and use the result to show that C has genus g. Check your answer by applying Hurwitz' formula (5.9) to the map $[1, x] : C \to \mathbb{P}^1$. (Note exercise 2.7 does not apply, since $C \not\subset \mathbb{P}^2$.)
   (d) Find a basis for the holomorphic differentials on C. [*Hint:* Consider the set $\{x^i \, dx/y : i = 0, 1, 2, \ldots\}$. How many elements are holomorphic?]

2.15. Let C/K be a smooth curve defined over a field of characteristic $p > 0$, and let $t \in K(C)$. Prove that the following are equivalent:
   (i) $K(C)$ is a finite separable extension of $K(t)$.
   (ii) There exists a point $P \in C$ such that $\mathrm{ord}_P(t)$ is relatively prime to p.
   (iii) For all but finitely many points $P \in C, t - t(P)$ is a uniformizer at P.
   (iv) $t \notin K(C)^p$.

CHAPTER III

The Geometry of Elliptic Curves

Elliptic curves, our principal object of study in this book, are curves of genus 1 having a specified basepoint. Our ultimate goal, as the title of the book indicates, is to study the arithmetic properties of these curves. In other words, we will be interested in analyzing their points defined over arithmetically interesting fields, such as finite fields, local (p-adic) fields, and global (number) fields. Before doing so, however, we are well-advised to study the properties of these curves in the simpler situation of an algebraically closed field (i.e. their geometry). This reflects the general principle in Diophantine geometry that in attempting to study any significant problem, it is essential to have a thorough understanding of the geometry before one can hope to make progress on the number theory. It is the purpose of this chapter to make an intensive study of the geometry of elliptic curves over arbitrary algebraically closed fields. (The particular case of the complex numbers is studied in more detail in chapter VI.)

We start in the first two sections by looking at elliptic curves given by explicit polynomial equations, called Weierstrass equations. Using these explicit equations, we show (among other things) that the set of points of an elliptic curve forms an abelian group, and the group law is given by rational functions. Then in section 3 we use the Riemann–Roch theorem to study arbitrary elliptic curves and to show, in particular, that every elliptic curve has a Weierstrass equation, so the results of the first two sections in fact apply generally. The remainder of the chapter studies (in various guises) the algebraic maps between elliptic curves. In particular, since the points of an elliptic curve form a group, for each integer m there is always a multiplication-by-m map from the curve to itself. As will become apparent throughout this book, it would be difficult to overestimate the importance of these multiplication maps in any attempt to study the arithmetic of elliptic curves (which will perhaps explain why we devote so much space to them in this chapter).

§1. Weierstrass Equations

Our main object of study will be *elliptic curves*, which are curves of genus 1 having a specified basepoint. As we will see in section 3, every such curve can be written as the locus in $\mathbb{P}^2$ of a cubic equation with only one point (the basepoint) on the line at ∞; i.e., after scaling X and Y, as an equation of the form

$$Y^2 Z + a_1 XYZ + a_3 YZ^2 = X^3 + a_2 X^2 Z + a_4 XZ^2 + a_6 Z^3.$$

Here $O = [0, 1, 0]$ is the basepoint and $a_1, \ldots, a_6 \in \bar{K}$. (It will become clear later why the coefficients are labeled in this way.) In this section and the next, we will study the curves given by such *Weierstrass equations*, using explicit formulas as much as possible to replace the need for general theory.

To ease notation, we will usually write the Weierstrass equation for our elliptic curve using non-homogeneous coordinates $x = X/Z$ and $y = Y/Z$,

$$E : y^2 + a_1 xy + a_3 y = x^3 + a_2 x^2 + a_4 x + a_6,$$

always remembering that there is the extra point $O = [0, 1, 0]$ out at infinity. As usual, if $a_1, \ldots, a_6 \in K$, then E is said to be *defined over* K.

If $\text{char}(\bar{K}) \neq 2$, then we can simplify the equation by completing the square. Thus replacing y by $\frac{1}{2}(y - a_1 x - a_3)$ gives an equation of the form

$$E : y^2 = 4x^3 + b_2 x^2 + 2b_4 x + b_6,$$

where

$$b_2 = a_1^2 + 4a_2,$$

$$b_4 = 2a_4 + a_1 a_3,$$

$$b_6 = a_3^2 + 4a_6.$$

We also define quantities

$$b_8 = a_1^2 a_6 + 4a_2 a_6 - a_1 a_3 a_4 + a_2 a_3^2 - a_4^2,$$

$$c_4 = b_2^2 - 24b_4,$$

$$c_6 = -b_2^3 + 36b_2 b_4 - 216b_6,$$

$$\Delta = -b_2^2 b_8 - 8b_4^3 - 27b_6^2 + 9b_2 b_4 b_6,$$

$$j = c_4^3/\Delta,$$

$$\omega = dx/(2y + a_1 x + a_3) = dy/(3x^2 + 2a_2 x + a_4 - a_1 y).$$

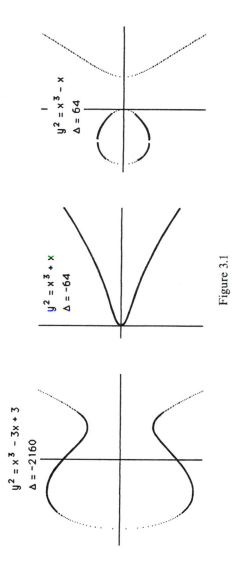

$$y^2 = x^3 - x$$
$$\Delta = 64$$

$$y^2 = x^3 + x$$
$$\Delta = -64$$

$$y^2 = x^3 - 3x + 3$$
$$\Delta = -2160$$

Figure 3.1

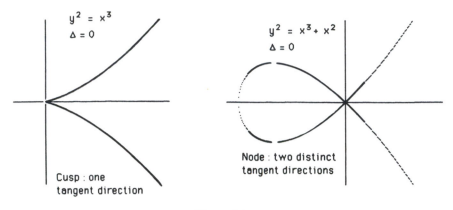

Figure 3.2

One easily verifies that they satisfy the relations

$$4b_8 = b_2 b_6 - b_4^2 \quad \text{and} \quad 1728\Delta = c_4^3 - c_6^2.$$

If further $\text{char}(\bar{K}) \neq 2, 3$, then replacing (x, y) by $((x - 3b_2)/36, y/108)$ eliminates the x^2 term, yielding the simpler equation

$$E : y^2 = x^3 - 27c_4 x - 54c_6.$$

Definition. The quantity Δ given above is called the *discriminant* of the Weierstrass equation, j is called the *j-invariant* of the elliptic curve E, and ω is the *invariant differential* associated with the Weierstrass equation.

Example 1.1. It is easy to graph the real locus of a Weierstrass equation. Some representative examples are shown in Figure 3.1. If $\Delta = 0$, then we will see that the curve is singular (1.4). Two sorts of behavior can occur, as illustrated in Figure 3.2.

With this example in mind, we consider the following situation. Let $P = (x_0, y_0)$ be a point satisfying a Weierstrass equation

$$f(x, y) = y^2 + a_1 xy + a_3 y - x^3 - a_2 x^2 - a_4 x - a_6 = 0.$$

Assume that P is a singular point on the curve $f(x, y) = 0$, so (I.1.5)

$$\partial f/\partial x(P) = \partial f/\partial y(P) = 0.$$

Then the Taylor expansion of $f(x, y)$ at P has the form

$$f(x, y) - f(x_0, y_0)$$
$$= [(y - y_0) - \alpha(x - x_0)][(y - y_0) - \beta(x - x_0)] - (x - x_0)^3$$

for some $\alpha, \beta \in \bar{K}$.

Definition. With notation as above, the singular point P is a *node* if $\alpha \neq \beta$. In this case, the lines

$$y - y_0 = \alpha(x - x_0) \quad \text{and} \quad y - y_0 = \beta(x - x_0)$$

are the *tangent lines* at P. Similarly, P is a *cusp* if $\alpha = \beta$, in which case the *tangent line* at P is given by

$$y - y_0 = \alpha(x - x_0).$$

To what extent is the Weierstrass equation for an elliptic curve unique? As we will see (3.1(b)), assuming the line at infinity (i.e. the line $Z = 0$ in $\mathbb{P}^2$) is to intersect E only at $[0, 1, 0]$, then the only change of variables fixing $[0, 1, 0]$ and preserving the Weierstrass form of the equation is

$$x = u^2 x' + r,$$

$$y = u^3 y' + u^2 s x' + t,$$

with $u, r, s, t \in \bar{K}$, $u \neq 0$. It is now a simple (but tedious) matter to make this substitution and compute the a_i' coefficients (and associated quantities) for the new equation. The results are compiled in Table 1.2.

It is now clear why the j-invariant has been so named; it is an invariant of the isomorphism class of the curve, and does not depend on the particular equation chosen. For algebraically closed fields, the converse is true, a fact which we will establish below (1.4b).

Remark 1.3. As we have seen, if the characteristic of K is different from 2 and 3, then any elliptic curve over K has a Weierstrass equation of a particularly simple kind. Thus any proof which involves extensive algebraic manipulation with Weierstrass equations (such as (1.4) below) tends to be much shorter if K is so restricted. On the other hand, even if one is primarily interested in characteristic 0 (e.g. $K = \mathbb{Q}$), an important tool is the process of reducing the

Table 1.2

$$ua_1' = a_1 + 2s$$
$$u^2 a_2' = a_2 - sa_1 + 3r - s^2$$
$$u^3 a_3' = a_3 + ra_1 + 2t$$
$$u^4 a_4' = a_4 - sa_3 + 2ra_2 - (t + rs)a_1 + 3r^2 - 2st$$
$$u^6 a_6' = a_6 + ra_4 + r^2 a_2 + r^3 - ta_3 - t^2 - rta_1$$

$$u^2 b_2' = b_2 + 12r$$
$$u^4 b_4' = b_4 + rb_2 + 6r^2$$
$$u^6 b_6' = b_6 + 2rb_4 + r^2 b_2 + 4r^3$$
$$u^8 b_8' = b_8 + 3rb_6 + 3r^2 b_4 + r^3 b_2 + 3r^4$$

$$u^4 c_4' = c_4$$
$$u^6 c_6' = c_6$$
$$u^{12} \Delta' = \Delta$$
$$j' = j$$
$$u^{-1} \omega' = \omega$$

coefficients of an equation modulo p for various primes p (including $p = 2$ and $p = 3$). So even for $K = \mathbb{Q}$ it is important to understand elliptic curves in all characteristics. Consequently, we will adopt the following policy. All theorems will be stated for a general Weierstrass equation. However, if it makes the proof substantially shorter, we will make the assumption that the characteristic of K is not 2 or 3, and give the proof in this case. Then, in the interest of completeness, we will return to these theorems in appendix A and give the proofs for general Weierstrass equations and arbitrary characteristic.

Now if the characteristic of K is not 2 or 3, we may assume that our elliptic curve(s) have Weierstrass equations(s) of the form

$$E : y^2 = x^3 + Ax + B.$$

This equation has associated quantities

$$\Delta = -16(4A^3 + 27B^2), \qquad j = -1728(4A)^3/\Delta.$$

The only change of variables preserving this form of the equation is

$$x = u^2 x', \qquad y = u^3 y' \qquad \text{for some } u \in \bar{K}^*;$$

and then

$$u^4 A' = A, \qquad u^6 B' = B, \qquad u^{12} \Delta' = \Delta.$$

Proposition 1.4. (a) *The curve given by a Weierstrass equation can be classified as follows.*

(i) *It is non-singular if and only if $\Delta \neq 0$.*
(ii) *It has a node if and only if $\Delta = 0$ and $c_4 \neq 0$.*
(iii) *It has a cusp if and only if $\Delta = c_4 = 0$.*

(*In case* (ii) *and* (iii), *there is only the one singular point.*)
(b) *Two elliptic curves are isomorphic (over $\bar{K}$) if and only if they have the same j-invariant.*
(c) *Let $j_0 \in \bar{K}$. Then there exists an elliptic curve (defined over $K(j_0)$) with j-invariant equal to j_0.*

PROOF. (a) Let E be given by the Weierstrass equation

$$E : f(x, y) = y^2 + a_1 xy + a_3 y - x^3 - a_2 x^2 - a_4 x - a_6 = 0.$$

We start by showing that the point at infinity is never singular. Thus we look at the curve in $\mathbb{P}^2$ with homogeneous equation

$$F(X, Y, Z) = Y^2 Z + a_1 XYZ + a_3 YZ^2 - X^3 - a_2 X^2 Z - a_4 XZ^2 - a_6 Z^3$$
$$= 0$$

and at the point $O = [0, 1, 0]$. Since

$$\partial F/\partial Z(O) = 1 \neq 0,$$

we see that O is a non-singular point on E.

Next suppose that E is singular, say at $P_0 = (x_0, y_0)$. The substitution

$$x = x' + x_0 \qquad y = y' + y_0$$

leaves Δ and c_4 invariant (1.2), so without loss of generality we may assume that E is singular at $(0, 0)$. Then

$$a_6 = f(0, 0) = 0 \qquad a_4 = \partial f/\partial x(0, 0) = 0 \qquad a_3 = \partial f/\partial y(0, 0) = 0,$$

so the equation for E takes the form

$$E : f(x, y) = y^2 + a_1 xy - a_2 x^2 - x^3 = 0.$$

This equation has associated quantities

$$c_4 = (a_1^2 + 4a_2)^2 \quad \text{and} \quad \Delta = 0.$$

Now by definition, E has a node (respectively cusp) at $(0, 0)$ if the quadratic form $y^2 + a_1 xy - a_2 x^2$ has distinct (respectively equal) factors, which occurs if and only if its discriminant

$$a_1^2 + 4a_2 \neq 0 \qquad \text{(respectively} = 0).$$

This proves the "only if" part of (ii) and (iii).

To complete the proof of (i)–(iii), it remains to show that if E is non-singular, then $\Delta \neq 0$. To simplify the computation, we will assume that $\text{char}(K) \neq 2$, and consider a Weierstrass equation of the form

$$E : y^2 = 4x^3 + b_2 x^2 + 2b_4 x + b_6.$$

(Cf. (1.3) and (A.1.2a).) Now E is singular if and only if there is a point $(x_0, y_0) \in E$ satisfying

$$2y_0 = 12x_0^2 + 2b_2 x_0 + 2b_4 = 0.$$

In other words, the singular points are exactly points of the form $(x_0, 0)$ with x_0 a double root of $4x^3 + b_2 x^2 + 2b_4 x + b_6 = 0$. This cubic polynomial has a double root if and only if its discriminant (which equals 16Δ) vanishes, which completes the proof of (i)–(iii). Further, since a cubic polynomial cannot have two double roots, E can have at most one singular point.

(b) If two elliptic curves are isomorphic, then the transformation formulas (1.2) show that they have the same j-invariant. For the converse, we will assume that $\text{char}(K) \neq 2, 3$ (cf. (1.3) and (A.1.2b)). Let E and E' be elliptic curves with the same j-invariant, say with Weierstrass equations

$$E : y^2 = x^3 + Ax + B,$$

$$E' : (y')^2 = (x')^3 + A'x' + B'.$$

Then

$$(4A)^3/(4A^3 + 27B^2) = (4A')^3/(4A'^3 + 27B'^2),$$

which yields

$$A^3 B'^2 = A'^3 B^2.$$

We look for an isomorphism of the form $(x, y) = (u^2 x', u^3 y')$, and consider three cases.

Case 1. $A = 0$ ($j = 0$). Then $B \neq 0$ (since $\Delta \neq 0$), so $A' = 0$, and we obtain an isomorphism using $u = (B/B')^{1/6}$.

Case 2. $B = 0$ ($j = 1728$). Then $A \neq 0$, so $B' = 0$, and we take $u = (A/A')^{1/4}$.

Case 3. $AB \neq 0$ ($j \neq 0$, 1728). Then $A'B' \neq 0$ (since if one of them is zero, then they both are, contradicting $\Delta' \neq 0$.) Hence taking $u = (A/A')^{1/4} = (B/B')^{1/6}$ gives the desired isomorphism.

(c) Assume that $j_0 \neq 0$, 1728, and look at the curve

$$E : y^2 + xy = x^3 - \frac{36}{j_0 - 1728} x - \frac{1}{j_0 - 1728}.$$

One computes

$$\Delta = j_0^2/(j_0 - 1728)^3 \quad \text{and} \quad j = j_0.$$

Thus E gives the desired elliptic curve (in any characteristic) provided $j_0 \neq 0$, 1728. To complete the list we use the two curves

$$E : y^2 + y = x^3 \qquad \Delta = -27, \qquad j = 0;$$

$$E : y^2 = x^3 + x \qquad \Delta = -64, \qquad j = 1728.$$

(Notice that in characteristic 2 or 3, 1728 equals 0, so even in these cases one of the two curves will be non-singular, and so fill in the one missing value of j.) $\square$

Proposition 1.5. *Let E be an elliptic curve. Then the invariant differential ω associated to a Weierstrass equation for E is holomorphic and non-vanishing (i.e. $\mathrm{div}(\omega) = 0$).*

PROOF. Let $P = (x_0, y_0) \in E$ and

$$F(x, y) = y^2 + a_1 xy + a_3 y - x^3 - a_2 x^2 - a_4 x - a_6,$$

so

$$\omega = d(x - x_0)/F_y(x, y) = -d(y - y_0)/F_x(x, y).$$

Thus P cannot be a pole of ω, since otherwise $F_y(P) = F_x(P) = 0$, which would say that P is a singular point. Now the map

$$E \to \mathbb{P}^1$$

$$[x, y, 1] \to [x, 1]$$

is of degree 2, so $\mathrm{ord}_P(x - x_0) \leqslant 2$; and $\mathrm{ord}_P(x - x_0) = 2$ if and only if the quadratic polynomial $F(x_0, y)$ has a double root. In other words, either $\mathrm{ord}_P(x - x_0) = 1$, or else $\mathrm{ord}_P(x - x_0) = 2$ and $F_y(x_0, y_0) = 0$. Thus in both cases (II.4.3)

$$\operatorname{ord}_P(\omega) = \operatorname{ord}_P(x - x_0) - \operatorname{ord}_P(F_y) - 1 = 0.$$

Finally, we must check the point $P = O$. Let t be a uniformizer at O. Since $\operatorname{ord}_P(x) = 2$ and $\operatorname{ord}_P(y) = 3$, $x = t^{-2}f$ and $y = t^{-3}g$ for functions f and g satisfying $f(O) \neq 0, \infty$ and $g(O) \neq 0, \infty$. Now

$$\omega = dx/F_y(x, y) = ((-2t^{-3}f + t^{-2}f')/(2t^{-3}g + a_1 t^{-2}f + a_3)) dt$$

$$= ((-2f + tf')/(2g + a_1 tf + a_3 t^3)) dt.$$

(Here $f' = df/dt$ (cf. II.4.3). In particular, (II.4.3b) tells us that f' is regular at O.) Assuming that $\operatorname{char}(K) \neq 2$, the function $(-2f + tf')/(2g + a_1 tf + a_3 t^3)$ is regular and non-vanishing at O; and so

$$\operatorname{ord}_O(\omega) = 0.$$

If $\operatorname{char}(K) = 2$, then the same result follows from a similar calculation (using $\omega = dy/F_x(x, y)$) which we will leave for the reader.   □

Next we look at what happens when a Weierstrass equation is singular.

Proposition 1.6. *If a curve E given by a Weierstrass equation is singular, then there exists a rational map $\phi : E \to \mathbb{P}^1$ of degree 1. (I.e. E is birational to $\mathbb{P}^1$. Note that since E is singular, we cannot use (II.2.4.1) to conclude that $E \simeq \mathbb{P}^1$.)*

PROOF. Making a linear change of variables, we may assume that the singular point is $(x, y) = (0, 0)$. Then checking partial derivatives, we see that the Weierstrass equation will have the form

$$E : y^2 + a_1 xy = x^3 + a_2 x^2.$$

Hence the rational map

$$E \to \mathbb{P}^1 \qquad (x, y) \to [x, y]$$

has degree 1, with inverse

$$\mathbb{P}^1 \to E \qquad [1, t] \to (t^2 + a_1 t - a_2, t^3 + a_1 t^2 - a_2 t).$$

[I.e. Use $t = y/x$ to map to $\mathbb{P}^1$, and note that dividing the equation for E by x^2 yields $t^2 + a_1 t = x + a_2$, so x and $y = xt$ are both in $\bar{K}(t)$.]   □

Legendre Form

There is another form of Weierstrass equation which is sometimes convenient.

Definition. A Weierstrass equation is in *Legendre form* if it can be written as

$$y^2 = x(x - 1)(x - \lambda).$$

Proposition 1.7. *Assume* $\text{char}(K) \neq 2$.
(a) *Every elliptic curve E/K is isomorphic (over $\bar{K}$) to an elliptic curve in Legendre form*

$$E_\lambda : y^2 = x(x-1)(x-\lambda)$$

for some $\lambda \in \bar{K}$, $\lambda \neq 0, 1$.
(b) *The j-invariant of E_λ is*

$$j(E_\lambda) = 2^8 \frac{(\lambda^2 - \lambda + 1)^3}{\lambda^2(\lambda - 1)^2}.$$

(c) *The association*

$$\bar{K} - \{0, 1\} \to \bar{K}$$

$$\lambda \to j(E_\lambda)$$

is surjective and exactly six-to-one except above $j = 0$ and $j = 1728$, where it is two-to-one and three-to-one respectively.

PROOF. (a) Since $\text{char}(K) \neq 2$, we know that E has a Weierstrass equation of the form

$$y^2 = 4x^3 + b_2 x^2 + 2b_4 x + b_6.$$

Replacing (x, y) by $(x, 2y)$ and factoring the cubic yields an equation

$$y^2 = (x - e_1)(x - e_2)(x - e_3),$$

where $e_1, e_2, e_3 \in \bar{K}$. Further, since

$$\Delta = 16(e_1 - e_2)^2(e_1 - e_3)^2(e_2 - e_3)^2 \neq 0,$$

the e_i's are seen to be distinct. Now the substitution

$$x = (e_2 - e_1)x' + e_1 \qquad y = (e_2 - e_1)^{3/2} y'$$

gives an equation in Legendre form with

$$\lambda = \frac{e_3 - e_1}{e_2 - e_1} \in \bar{K}, \qquad \lambda \neq 0, 1.$$

(b) Calculation.
(c) One can work directly from the formula for $j(E_\lambda)$ in (b), an approach that we leave for the reader. Instead we use the fact that the j-invariant classifies an elliptic curve up to isomorphism (1.4b). Thus suppose $j(E_\lambda) = j(E_\mu)$. Then $E_\lambda \cong E_\mu$, so their Weierstrass equations (in Legendre form) are related by a change of variables

$$x = u^2 x' + r \qquad y = u^3 y'.$$

Equating

$$x(x-1)(x-\mu) = \left(x + \frac{r}{u^2}\right)\left(x + \frac{r-1}{u^2}\right)\left(x + \frac{r-\lambda}{u^2}\right),$$

there are six ways of assigning the linear terms to one another, and one easily checks that these lead to six possibilities

$$\mu \in \left\{ \lambda, \frac{1}{\lambda}, 1 - \lambda, \frac{1}{1 - \lambda}, \frac{\lambda}{\lambda - 1}, \frac{\lambda - 1}{\lambda} \right\}.$$

Hence $\lambda \rightarrow j(E_\lambda)$ is exactly six-to-one unless two or more of these values for μ coincide. Equating them by pairs shows that this only occurs for $\lambda = -1, 2, 1/2$ and $\lambda^2 - \lambda + 1 = 0$, for which the set has respectively three and two elements; these values of λ correspond respectively to $j = 1728$ and $j = 0$.   □

§2. The Group Law

Let E be an elliptic curve given by a Weierstrass equation. Remember that $E \subset \mathbb{P}^2$ consists of the points $P = (x, y)$ satisfying the equation together with the point $O = [0, 1, 0]$ at infinity. Let $L \subset \mathbb{P}^2$ be a line. Then since the equation has degree three, L intersects E at exactly 3 points, say P, Q, R. (Note if L is tangent to E, then P, Q, R may not be distinct. The fact that $L \cap E$, taken with multiplicities, consists of three points, is a special case of Bezout's theorem [Har, I.7.8]. But since we will give explicit formulas below, there is no need to use a general theorem.)

Define a composition law $\oplus$ on E by the following rule.

Composition Law 2.1. *Let $P, Q \in E$, L the line connecting P and Q (tangent line to E if $P = Q$), and R the third point of intersection of L with E. Let L' be the line connecting R and O. Then $P \oplus Q$ is the point such that L' intersects E at R, O, and $P \oplus Q$.*

The following diagrams illustrates this rule (Figure 3.3).
   We now justify the use of the symbol $\oplus$.

Proposition 2.2. *The composition law (2.1) has the following properties:*
(a) *If a line L intersects E at the (not necessarily distinct) points P, Q, R, then*

$$(P \oplus Q) \oplus R = O.$$

(b) *$P \oplus O = P$ for all $P \in E$.*
(c) *$P \oplus Q = Q \oplus P$ for all $P, Q \in E$.*
(d) *Let $P \in E$. There is a point of E, denoted $\ominus P$, so that*

$$P \oplus (\ominus P) = O.$$

(e) *Let $P, Q, R \in E$. Then*

$$(P \oplus Q) \oplus R = P \oplus (Q \oplus R).$$

In other words, the composition law (2.1) makes E into an abelian group with

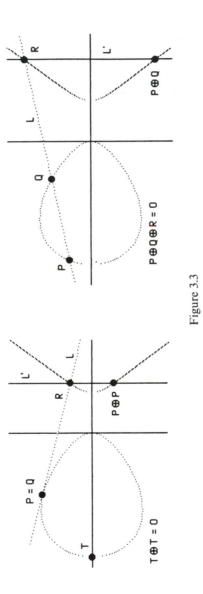

Figure 3.3

identity element O. We further have:
(f) *Suppose E is defined over K. Then*

$$E(K) = \{(x, y) \in K^2 : y^2 + a_1 xy + a_3 y = x^3 + a_2 x^2 + a_4 x + a_6\} \cup \{O\}$$

is a subgroup of E.

PROOF. All of this is easy except for the associativity (e).
(a) Obvious from (2.1). (Or look at Figure 3.3. Note that the tangent line to E at O intersects E with multiplicity 3 at O.)
(b) Taking $Q = O$ in (2.1), we see that the lines L and L' coincide. The former intersects E at P, O, R, and the latter at $R, O, P \oplus O$, so $P \oplus O = P$.
(c) Clear, since the construction in (2.1) is symmetric in P and Q.
(d) Let the line through P and O also intersect E at R. Then using (a) and (b),

$$O = (P \oplus O) \oplus R = P \oplus R.$$

(e) Using the explicit formulas given below (2.3), one can laboriously verify the associative law case by case. We leave this task for the reader. A more enlightening proof, using the Riemann–Roch theorem, will be given in the next section (3.4e). For a more geometric proof, see [Ful].
(f) If P and Q have coordinates in K, then the equation of the line connecting them has coefficients in K. If further E is defined over K, then the third point of intersection will have coordinates given by a rational combination of the coefficients of the line and of E, so will be in K. (See (2.3) below for explicit formulas.)   □

Notation. From here on, we will drop the special symbols $\oplus$ and $\ominus$ and simply use $+$ and $-$ for the group operations on an elliptic curve E. For $m \in \mathbb{Z}$ and $P \in E$, we let

$$[m]P = P + \cdots + P \ (m \text{ terms}) \text{ for } m > 0,$$

$$[0]P = O, \quad \text{and} \quad [m]P = [-m](-P) \text{ for } m < 0.$$

As promised above, we now derive explicit formulas for the group operations. Let E be an elliptic curve with the usual Weierstrass equation

$$F(x, y) = y^2 + a_1 xy + a_3 y - x^3 - a_2 x^2 - a_4 x - a_6 = 0,$$

and let $P_0 = (x_0, y_0) \in E$. Following the proof of (2.2d), to calculate $-P_0$ we take the line L through P_0 and O and find its third point of intersection with E. The line L is given by:

$$L: x - x_0 = 0.$$

Substituting this into the equation for E, we see that the quadratic polynomial $F(x_0, y)$ has roots y_0 and y_0', where $-P_0 = (x_0, y_0')$. Writing out

$$F(x_0, y) = c(y - y_0)(y - y_0')$$

and comparing the coefficients of y^2 gives $c = 1$, and then the coefficients of y give $y'_0 = -y_0 - a_1 x_0 - a_3$. This yields

$$-P_0 = (x_0, -y_0 - a_1 x_0 - a_3).$$

Next we derive a formula for the addition law. Let $P_1 = (x_1, y_1)$ and $P_2 = (x_2, y_2)$ be points of E. If $x_1 = x_2$ and $y_1 + y_2 + a_1 x_2 + a_3 = 0$, then from the above formula $P_1 + P_2 = O$. Otherwise the line L through P_1 and P_2 (tangent line to E if $P_1 = P_2$) has an equation of the form

$$L : y = \lambda x + v.$$

(Formulas for λ and v are given below.) Substituting into the equation for E, we see that $F(x, \lambda x + v)$ has roots x_1, x_2, x_3, where $P_3 = (x_3, y_3)$ is the third point of $L \cap E$. From (2.2a),

$$P_1 + P_2 + P_3 = O;$$

while writing out

$$F(x, \lambda x + v) = c(x - x_1)(x - x_2)(x - x_3)$$

and equating coefficients of x^3 and x^2 yields $c = -1$ and

$$x_1 + x_2 + x_3 = \lambda^2 + a_1 \lambda - a_2.$$

This gives the formula for x_3, and substituting into the equation for L gives $y_3 = \lambda x_3 + v$. Finally, to find $P_1 + P_2 = -P_3$, we apply the negation formula found above to P_3. All of this is summarized in the following.

Group Law Algorithm 2.3. *Let E be an elliptic curve given by a Weierstrass equation*

$$E : y^2 + a_1 xy + a_3 y = x^3 + a_2 x^2 + a_4 x + a_6.$$

(a) *Let $P_0 = (x_0, y_0) \in E$. Then*

$$-P_0 = (x_0, -y_0 - a_1 x_0 - a_3).$$

Now let

$$P_1 + P_2 = P_3 \quad with \quad P_i = (x_i, y_i) \in E.$$

(b) *If $x_1 = x_2$ and $y_1 + y_2 + a_1 x_2 + a_3 = 0$, then*

$$P_1 + P_2 = O.$$

Otherwise, let

$$\lambda = \frac{y_2 - y_1}{x_2 - x_1}, \qquad v = \frac{y_1 x_2 - y_2 x_1}{x_2 - x_1} \qquad if\ x_1 \neq x_2;$$

$$\lambda = \frac{3x_1^2 + 2a_2 x_1 + a_4 - a_1 y_1}{2y_1 + a_1 x_1 + a_3},$$

$$v = \frac{-x_1^3 + a_4 x_1 + 2a_6 - a_3 y_1}{2y_1 + a_1 x_1 + a_3} \qquad if\ x_1 = x_2.$$

(*Then $y = \lambda x + v$ is the line through P_1 and P_2, or tangent to E if $P_1 = P_2$.*)

(c) $P_3 = P_1 + P_2$ is given by

$$x_3 = \lambda^2 + a_1\lambda - a_2 - x_1 - x_2,$$

$$y_3 = -(\lambda + a_1)x_3 - v - a_3.$$

(d) *As special cases of (c), we have for* $P_1 \neq \pm P_2,$

$$x(P_1 + P_2) = \left(\frac{y_2 - y_1}{x_2 - x_1}\right)^2 + a_1\left(\frac{y_2 - y_1}{x_2 - x_1}\right) - a_2 - x_1 - x_2;$$

and the duplication formula *for* $P = (x, y) \in E,$

$$x([2]P) = \frac{x^4 - b_4x^2 - 2b_6x - b_8}{4x^3 + b_2x^2 + 2b_4x + b_6},$$

where b_2, b_4, b_6, b_8 *are the polynomials in the* a_i's *given in section 1.*

Corollary 2.3.1. *With notation as in (2.3), we say that a function* $f \in \bar{K}(E) = \bar{K}(x, y)$ *is even if* $f(P) = f(-P)$ *for all* $P \in E$. *Then*

$$f \text{ is even} \qquad \text{if and only if} \qquad f \in \bar{K}(x).$$

PROOF. From (2.3), if $P = (x_0, y_0)$, then $-P = (x_0, -y_0 - a_1x_0 - a_3)$. It is thus clear that every element of $\bar{K}(x)$ is even. Suppose now that $f \in \bar{K}(x, y)$ is even. Using the Weierstrass equation for E, we can write f as

$$f(x, y) = g(x) + h(x)y \qquad \text{for some } g, h \in \bar{K}(x).$$

Then

$$g(x) + h(x)y = f(x, y) = f(x, -y - a_1x - a_3)$$

$$= g(x) - (y + a_1x + a_3)h(x).$$

Thus

$$(2y + a_1x + a_3)h(x) = 0.$$

Since this holds for all $(x, y) \in E$, it follows that either $h = 0$, or else $2 = a_1 = a_3 = 0$. But the latter implies that the discriminant $\Delta = 0$, contradicting our assumption that the Weierstrass equation is non-singular (1.4a). Therefore $h = 0$, so $f(x, y) = g(x) \in \bar{K}(x)$. $\square$

Example 2.4. Let $E/\mathbb{Q}$ be the elliptic curve

$$E : y^2 = x^3 + 17.$$

A brief inspection reveals some points with integer coordinates

$$P_1 = (-2, 3) \quad P_2 = (-1, 4) \quad P_3 = (2, 5) \quad P_4 = (4, 9) \quad P_5 = (8, 23),$$

and a short computer search gives some others

$$P_6 = (43, 282) \qquad P_7 = (52, 375) \qquad P_8 = (5234, 378661).$$

Using the above formulas, one easily verifies relations such as

$$P_5 = [-2]P_1, \qquad P_4 = P_1 - P_3, \qquad [3]P_1 - P_3 = P_7.$$

Of course, there are lots of rational points, too, for example

$$[2]P_2 = \left(\frac{137}{64}, -\frac{2651}{512}\right) \qquad P_2 + P_3 = \left(-\frac{8}{9}, -\frac{109}{27}\right).$$

Now it is true (but not easy to prove) that every rational point $P \in E(\mathbb{Q})$ can be written in the form

$$P = [m]P_1 + [n]P_3 \quad \text{with} \quad m, n \in \mathbb{Z};$$

and with this identification the group $E(\mathbb{Q})$ is isomorphic to $\mathbb{Z} \times \mathbb{Z}$. Further, there are only 16 integral points $P = (x, y) \in E$ (i.e. with $x, y \in \mathbb{Z}$), namely $\{\pm P_1, \ldots, \pm P_8\}$. (See [Nag].) These facts illustrate two of the most fundamental theorems in the arithmetic of elliptic curves, namely that the group of rational points on an elliptic curve is finitely generated (the Mordell–Weil theorem, proven in chapter VIII) and that the set of integral points on an elliptic curve is finite (Siegel's theorem, proven in chapter IX).

Now suppose that a Weierstrass equation has discriminant $\Delta = 0$, so from (1.4a) it has a singular point. To what extent does the analysis of the composition law fail in this case? As we will see below, everything is fine provided we discard the singular point; and in fact the resulting group then has a particularly simple structure.

The reason we will be interested in this situation is best illustrated by an example. Consider again the elliptic curve of (2.4),

$$E : y^2 = x^3 + 17.$$

This is an elliptic curve, defined over $\mathbb{Q}$, with discriminant $\Delta = 2^4 3^3 17$. But we will also be interested in reducing the coefficients of this equation modulo p for various primes p, and considering it as a curve defined over the finite field $\mathbb{F}_p$. For almost all primes, namely those with $\Delta \not\equiv 0 \pmod{p}$, the "reduced" curve will still be non-singular, and so we will have an elliptic curve. But for $p \in \{2, 3, 17\}$, the "reduced" curve will have a singular point. Thus even when dealing with non-singular curves (for example, defined over $\mathbb{Q}$), one finds singular curves naturally appearing. We will return to this reduction process in more detail in chapter VII.

Definition. Let E be a (possibly singular) curve given by a Weierstrass equation. The *non-singular part of E*, denoted E_{ns}, is the set of non-singular points of E. Similarly, if E is defined over K, then $E_{ns}(K)$ is the set of non-singular points of $E(K)$.

Recall that if E is singular, then there are two possibilities for the singularity, namely a node or a cusp (determined by whether $c_4 = 0$ or $c_4 \neq 0$, see (1.4a)).

Proposition 2.5. *Let E be a curve given by a Weierstrass equation with discriminant $\Delta = 0$, so E has a singular point S. Then the composition law (2.1) makes E_{ns} into an abelian group.*

(a) *Suppose E has a node (so $c_4 \neq 0$), and let*

$$y = \alpha_1 x + \beta_1 \quad and \quad y = \alpha_2 x + \beta_2$$

be the two distinct tangent lines to E at S. Then the map

$$E_{ns} \to \bar{K}^*$$

$$(x, y) \to \frac{y - \alpha_1 x - \beta_1}{y - \alpha_2 x - \beta_2}$$

is an isomorphism (of abelian groups).

(b) *Suppose E has a cusp (so $c_4 = 0$), and let*

$$y = \alpha x + \beta$$

be the tangent line to E at S. Then the map

$$E_{ns} \to \bar{K}^+$$

$$(x, y) \to \frac{x - x(S)}{y - \alpha x - \beta}$$

is an isomorphism.

Remark 2.6. For a description of $E_{ns}(K)$ in case K is not algebraically closed, see (exer. 3.5).

PROOF. We will check that the maps in (a) and (b) are set bijections with the property that if a line L not hitting S intersects E_{ns} in three (not necessarily distinct) points, then the images of these three points in $\bar{K}^*$ (respectively $\bar{K}^+$) will multiply to 1 (respectively sum to 0). Using this, one easily verifies that the composition law (2.1) makes E_{ns} into abelian group and that the maps in (a) and (b) are group isomorphisms.

Since the composition law (2.1) and the maps in (a) and (b) are defined in terms of lines in $\mathbb{P}^2$, it suffices to prove the theorem after making a linear change of variables. We start by moving the singular point to $(0, 0)$, yielding a Weierstrass equation

$$y^2 + a_1 xy = x^3 + a_2 x^2.$$

Let $s \in \bar{K}$ be a root of $s^2 + a_1 s - a_2 = 0$. Then replacing y by $y + sx$ eliminates the x^2 term, giving the equation (which we now write using homogeneous coordinates)

$$E : Y^2 Z + AXYZ - X^3 = 0.$$

Note that E has a node (respectively cusp) if $A \neq 0$ (respectively $A = 0$).

(a) The tangent lines to E at $S = [0, 0, 1]$ are $Y = 0$ and $Y + AX = 0$, so we are looking at the map

$$E_{ns} \to \bar{K}^* \qquad [X, Y, Z] \to 1 + \frac{AX}{Y}.$$

It is convenient to make one more variable change, so let

$$X = A^2 X' - A^2 Y' \qquad Y = A^3 Y' \qquad Z = Z'.$$

Dropping the primes, this gives the equation

$$E : XYZ - (X - Y)^3 = 0;$$

and if we now dehomogenize by setting $Y = 1$ (i.e. $x = X/Y$ and $z = Z/Y$), this gives

$$E : xz - (x - 1)^3 = 0$$

with the map

$$E_{ns} \to \bar{K}^* \qquad (x, z) \to x.$$

(Notice the singular point is now out at infinity.) The inverse map is clearly given by

$$\bar{K}^* \to E_{ns} \qquad t \to (t, (t - 1)^3/t),$$

so we have a bijection of sets. It remains to show that if a line (not hitting $[0, 0, 1]$) intersects E at $(x_1, z_1), (x_2, z_2), (x_3, z_3)$, then $x_1 x_2 x_3 = 1$. (See Figure 3.4.) But such a line has the form $z = ax + b$, and so the three x-coordinates x_1, x_2, x_3 are the roots of the cubic polynomial

$$x(ax + b) - (x - 1)^3 = 0.$$

Looking at the constant term, we see that $x_1 x_2 x_3 = 1$, as desired.

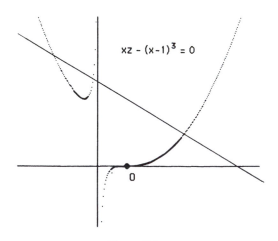

$$xz - (x-1)^3 = 0$$

Figure 3.4

(b) In this case $A = 0$, and the tangent line to E at $S = [0, 0, 1]$ is $Y = 0$, so we are looking at the map

$$E_{ns} \to \bar{K}^+ \qquad [X, Y, Z] \to X/Y.$$

Again dehomogenizing by setting $Y = 1$, we obtain

$$E : z - x^3 = 0$$

$$E_{ns} \to \bar{K}^+ \qquad (x, z) \to x.$$

The inverse map is clearly $t \to (t, t^3)$. Finally, if the line $z = ax + b$ intersects E in the three points (x_i, y_i), $1 \le i \le 3$, then from the lack of an x^2 term in

$$(ax + b) - x^3 = 0,$$

we see that $x_1 + x_2 + x_3 = 0$. $\square$

§3. Elliptic Curves

Let E be a smooth curve of genus 1. For example, the non-singular Weierstrass equations studied in sections 1 and 2 will define curves with this property. As we have seen, such Weierstrass curves have a group law associated to them. Now in order to make a set into a group, clearly an initial requirement is to choose a distinguished (identity) element. This leads us to make the following definition.

Definition. An *elliptic curve* is a pair (E, O), where E is a curve of genus 1 and $O \in E$. (We often just write E for the elliptic curve, the point O being understood.) The elliptic curve E is *defined over* K, written E/K, if E is defined over K as a curve and $O \in E(K)$.

In order to connect this definition with the material of sections 1 and 2, we begin by using the Riemann–Roch theorem to show that every elliptic curve can be written as a plane cubic; and conversely, every smooth Weierstrass plane cubic curve is an elliptic curve.

Proposition 3.1. *Let E be an elliptic curve defined over K.*
(a) *There exist functions $x, y \in K(E)$ such that the map*

$$\phi : E \to \mathbb{P}^2$$

$$\phi = [x, y, 1]$$

gives an isomorphism of E/K onto a curve given by a Weierstrass equation

$$C : Y^2 + a_1 XY + a_3 Y = X^3 + a_2 X^2 + a_4 X + a_6$$

with coefficients $a_1, \ldots, a_6 \in K$; and such that $\phi(O) = [0, 1, 0]$. (We call x, y Weierstrass coordinate functions on E.)

(b) *Any two Weierstrass equations for E as in* (a) *are related by a linear change of variables of the form*

$$X = u^2 X' + r$$

$$Y = u^3 Y' + su^2 X' + t$$

with $u, r, s, t \in K$, $u \neq 0$.
(c) *Conversely, every smooth cubic curve C given by a Weierstrass equation as in* (a) *is an elliptic curve defined over K with origin* $O = [0, 1, 0]$.

PROOF. (a) We look at the vector spaces $\mathscr{L}(n(O))$ for $n = 1, 2, \ldots$. By the Riemann–Roch theorem (specifically (II.5.5c) with $g = 1$),

$$\ell(n(O)) = \dim \mathscr{L}(n(O)) = n \qquad \text{for all } n \geqslant 1.$$

Thus we can choose functions $x, y \in K(E)$ (II.5.8) so that $\{1, x\}$ is a basis for $\mathscr{L}(2(O))$ and $\{1, x, y\}$ is a basis for $\mathscr{L}(3(O))$. Note that x must have a pole of exact order 2 at O, and similarly y must have a pole of exact order 3.

Now $\mathscr{L}(6(O))$ has dimension 6, but it contains the seven functions 1, x, y, x^2, xy, y^2, x^3. It follows that there is a linear relation

$$A_1 + A_2 x + A_3 y + A_4 x^2 + A_5 xy + A_6 y^2 + A_7 x^3 = 0,$$

where by (II.5.8) we may take $A_1, \ldots, A_7 \in K$. Note that $A_6 A_7 \neq 0$, since otherwise every term would have a different order pole at O, and so all the A_j's would vanish. Replacing x, y by $-A_6 A_7 x$, $A_6 A_7^2 y$ and dividing by $A_6^3 A_7^4$ gives a cubic equation in Weierstrass form. This gives the desired map

$$\phi : E \to \mathbb{P}^2 \qquad \phi = [x, y, 1]$$

whose image lies in the locus C described by a Weierstrass equation. Note that $\phi : E \to C$ is a morphism (II.2.1) and surjective (II.2.3). Note also that $\phi(O) = [0, 1, 0]$, since y has a higher order pole than x at O.

The next step is to show that the map $\phi : E \to C \subset \mathbb{P}^2$ has degree 1, or equivalently, that $K(E) = K(x, y)$. Consider the map $[x, 1] : E \to \mathbb{P}^1$. Since x has a double pole at O and no other poles, (II.2.6a) says that this map has degree 2. Thus $[K(E) : K(x)] = 2$. Similarly, $[y, 1] : E \to \mathbb{P}^1$ has degree 3, so $[K(E) : K(y)] = 3$. Therefore $[K(E) : K(x, y)] = 1$, since it divides both 2 and 3, so $K(E) = K(x, y)$.

Next we show that C is smooth. Suppose C is singular. Then from (1.6) there is a rational map $\psi : C \to \mathbb{P}^1$ of degree 1. Hence the composition $\psi \circ \phi : E \to \mathbb{P}^1$ is a map of degree 1 between smooth curves, so from (II.2.4.1) it is an isomorphism. This contradicts the fact that E has genus 1 and $\mathbb{P}^1$ has genus 0 (II.5.6). Therefore C is smooth, and now another application of (II.2.4.1) shows that the degree 1 map $\phi : E \to C$ is an isomorphism.
(b) Let $\{x, y\}$ and $\{x', y'\}$ be two sets of Weierstrass coordinate functions on E. Then x and x' have poles of order 2 at O, and y and y' have poles of order 3. Hence $\{1, x\}$ and $\{1, x'\}$ are both bases for $\mathscr{L}(2(O))$, and similarly $\{1, x, y\}$

and $\{1, x', y'\}$ are both bases for $\mathscr{L}(3(O))$. It follows that there are constants $u_1, u_2, r, s_2, t \in K$ with $u_1 u_2 \neq 0$ such that

$$x = u_1 x' + r \qquad y = u_2 y' + s_2 x' + t.$$

But (x, y) and (x', y') both satisfy Weierstrass equations in which the Y^2 and X^3 terms have coefficient 1, so $u_1^3 = u_2^2$. Letting $u = u_2/u_1$ and $s = s_2/u^2$ puts the change of variable formula in the desired form.

(c) Let E be given by a non-singular Weierstrass equation. We have seen (1.5) that the differential

$$\omega = dx/(2y + a_1 x + a_3) \in \Omega_E$$

has neither zeros nor poles, so $\operatorname{div}(\omega) = 0$. But from the Riemann–Roch theorem (specifically II.5.5b),

$$2 \operatorname{genus}(E) - 2 = \deg \operatorname{div}(\omega).$$

Hence E has genus 1, so together with the point $[0, 1, 0]$, it is an elliptic curve. (For another proof of (c) using the Hurwitz genus formula, see exer. 2.7.)   □

Corollary 3.1.1. *Let E/K be an elliptic curve with Weierstrass coordinate functions x, y. Then*

$$K(E) = K(x, y) \quad and \quad [K(E) : K(x)] = 2.$$

PROOF. This was proven during the course of proving (3.1a).   □

Remark 3.2. Note that (3.1b) does *not* imply that if two Weierstrass equations have coefficients in a given field K, then every change of variables mapping one to the other has coefficients in K. A simple example is the equation

$$y^2 = x^3 + x,$$

which has coefficients in $\mathbb{Q}$. Yet it is mapped to itself by the substitution

$$x = -x' \qquad y = iy',$$

where $i^2 = -1$.

Next we use the Riemann–Roch theorem to describe a group law on the points of E. Of course, this will turn out to be the same group law already described by (2.1) when E is given by a Weierstrass equation. We start with a simple lemma, which serves to distinguish $\mathbb{P}^1$ from curves of genus 1. (For a generalization, see exer. 2.5.)

Lemma 3.3. *Let C be a curve of genus 1, and let $P, Q \in C$. Then*

$$(P) \sim (Q) \qquad if \ and \ only \ if \quad P = Q.$$

PROOF. Suppose $(P) \sim (Q)$, and choose $f \in \bar{K}(C)$ so that

$$\mathrm{div}(f) = (P) - (Q).$$

Then $f \in \mathcal{L}((Q))$, and by the Riemann–Roch theorem (II.5.5c),

$$\dim \mathcal{L}((Q)) = 1.$$

But $\mathcal{L}((Q))$ already contains the constant functions, hence $f \in \bar{K}$ and $P = Q$. $\square$

Proposition 3.4. *Let (E, O) be an elliptic curve.*
(a) *For every divisor $D \in \mathrm{Div}^0(E)$ there exists a unique point $P \in E$ so that*

$$D \sim (P) - (O).$$

Let

$$\sigma : \mathrm{Div}^0(E) \to E$$

be the map given by this association.
(b) *The map σ is surjective.*
(c) *Let $D_1, D_2 \in \mathrm{Div}^0(E)$. Then*

$$\sigma(D_1) = \sigma(D_2) \qquad \text{if and only if} \quad D_1 \sim D_2.$$

Thus σ induces a bijection of sets (which we also denote by σ)

$$\sigma : \mathrm{Pic}^0(E) \overset{\sim}{\to} E.$$

(d) *The inverse to σ is the map*

$$\kappa : E \overset{\sim}{\to} \mathrm{Pic}^0(E)$$

$$P \to \text{class of } (P) - (O).$$

(e) *If E is given by a Weierstrass equation, then the "geometric group law" on E arising from (2.1) and the group law induced from $\mathrm{Pic}^0(E)$ by using σ are the same.*

PROOF. (a) Since E has genus 1, the Riemann–Roch theorem (II.5.5c) says that

$$\dim \mathcal{L}(D + (O)) = 1.$$

Let $f \in \bar{K}(E)$ be a generator for $\mathcal{L}(D + (O))$. Since

$$\mathrm{div}(f) \geqslant -D - (O) \quad \text{and} \quad \deg(\mathrm{div}(f)) = 0,$$

it follows that

$$\mathrm{div}(f) = -D - (O) + (P)$$

for some $P \in E$. Hence

$$D \sim (P) - (O),$$

which gives the existence of a point with the desired property.

Next suppose $P' \in E$ has the same property. Then

$$(P) \sim D + (O) \sim (P'),$$

so $P = P'$ from (3.3). Hence P is unique.

(b) For any $P \in E$,

$$\sigma((P) - (O)) = P.$$

(c) Let $D_1, D_2 \in \mathrm{Div}^0(E)$, and set $P_i = \sigma(D_i)$. Then from the definition of σ,

$$(P_1) - (P_2) \sim D_1 - D_2.$$

Hence $P_1 = P_2$ certainly implies $D_1 \sim D_2$. Conversely, if $D_1 \sim D_2$, then $(P_1) \sim (P_2)$, so $P_1 = P_2$ from (3.3).

(d) Clear.

(e) Let E be given by a Weierstrass equation, and let $P, Q \in E$. It clearly suffices to show that

$$\kappa(P + Q) = \kappa(P) + \kappa(Q).$$

[N.B. The first $+$ is addition on E using (2.1), while the second is addition of divisor classes in $\mathrm{Pic}^0(E)$.]

Let

$$f(X, Y, Z) = \alpha X + \beta Y + \gamma Z = 0$$

give the line L in $\mathbb{P}^2$ going throught P and Q, let R be the third point of intersection of L with E, and let

$$f'(X, Y, Z) = \alpha' X + \beta' Y + \gamma' Z = 0$$

be the line L' through R and O. Then from the definition of addition on E (2.1) and the fact that the line $Z = 0$ intersects E at O with multiplicity 3, we have

$$\mathrm{div}(f/Z) = (P) + (Q) + (R) - 3(O)$$

and

$$\mathrm{div}(f'/Z) = (R) + (P + Q) - 2(O).$$

Hence

$$(P + Q) - (P) - (Q) + (O) = \mathrm{div}(f'/f) \sim 0,$$

so

$$\kappa(P + Q) - \kappa(P) - \kappa(Q) = 0. \qquad \square$$

Corollary 3.5. *Let E be an elliptic curve and $D = \Sigma n_P(P) \in \mathrm{Div}(E)$. Then D is principal if and only if $\Sigma n_P = 0$ and $\Sigma [n_P]P = O$. (Note the first sum is of integers, the second is addition on E.)*

PROOF. From (II.3.1b), every principal divisor has degree 0. Assuming now $D \in \mathrm{Div}^0(E)$, (3.4a, e) implies

$$D \sim 0 \Leftrightarrow \sigma(D) = 0 \Leftrightarrow \sum [n_P]\sigma((P) - (O)) = 0,$$

which is the desired result since $\sigma((P) - (O)) = P$.   □

Remark 3.5.1. If we combine (3.4) with (II.3.4), we see that every elliptic curve E/K fits into an exact sequence

$$1 \to \bar{K}^* \to \bar{K}(E)^* \xrightarrow{\mathrm{div}} \mathrm{Div}^0(E) \xrightarrow{\sigma} E \to 0,$$

where σ is the operation "sum up the divisor using the group law on E". Further, (exer. 2.13b) implies that the sequence remains exact if we take $G_{\bar{K}/K}$-invariants:

$$1 \to K^* \to K(E)^* \xrightarrow{\mathrm{div}} \mathrm{Div}_K^0(E) \xrightarrow{\sigma} E(K) \to 0.$$

(See also (X.3.8).)

We now prove the fundamental fact that the addition law on an elliptic curve is a *morphism*. Since addition is a map $E \times E \to E$, and $E \times E$ has dimension 2, we cannot use (II.2.1) directly; but it will play a crucial role in our proof. One can also give a proof using explicit equations, but the algebra is somewhat lengthy (see (3.6.1) below).

Theorem 3.6. *Let E/K be an elliptic curve. Then the equations* (2.3) *giving the group law on E define morphisms*

$$+ : E \times E \to E \qquad\qquad \text{and} \qquad - : E \to E$$

$$(P_1, P_2) \to P_1 + P_2 \qquad\qquad\qquad P \to -P.$$

PROOF. First, the subtraction map

$$(x, y) \to (x, -y - a_1 x - a_3)$$

is clearly a rational map $E \to E$. Since E is smooth, it is a morphism (II.2.1).
    Next we fix a point $Q \neq O$ on E, and consider the "translation-by-Q" map

$$\tau : E \to E \qquad \tau(P) = P + Q.$$

From the addition formula given in (2.3c), this is clearly a rational map; and so, again by (II.2.1), it is a morphism. In fact, since it has an inverse, namely $P \to P - Q$, it is isomorphism.
    Finally, we consider the general addition map $+ : E \times E \to E$. From (2.3c) we see that it is a morphism except possibly at points of the form (P, P), $(P, -P)$, (P, O), and (O, P), since for points not of this form, the rational functions

$$\lambda = (y_2 - y_1)/(x_2 - x_1) \quad \text{and} \quad v = (y_1 x_2 - y_2 x_1)/(x_2 - x_1)$$

on $E \times E$ are well-defined.
    To deal with the four exceptional cases, one can work directly with the definition of morphism. (See (3.6.1) below.) However, we prefer to let the

group law assist us. Thus let τ_1 and τ_2 be translation maps, as defined above, for points Q_1 and Q_2 respectively. Consider the composition of maps

$$\phi : E \times E \xrightarrow{\ \tau_1 \times \tau_2\ } E \times E \xrightarrow{\ +\ } E \xrightarrow{\ \tau_1^{-1}\ } E \xrightarrow{\ \tau_2^{-1}\ } E.$$

Since the group law on E is associative and commutative (2.2), the net effect of these maps is as follows:

$$(P_1, P_2) \rightarrow (P_1 + Q_1, P_2 + Q_2) \rightarrow P_1 + Q_1 + P_2 + Q_2$$
$$\rightarrow P_1 + P_2 + Q_2 \rightarrow P_1 + P_2.$$

Thus the rational map ϕ agrees with the addition map wherever they are both defined.

Further, since the τ_i's are isomorphisms, it follows from the discussion above that ϕ is a morphism except possibly at points of the form

$$(P - Q_1, P - Q_2) \quad (P - Q_1, -P - Q_2) \quad (P - Q_1, -Q_2) \quad (-Q_1, P - Q_2).$$

But Q_1 and Q_2 may be chosen essentially arbitrarily. Hence by varying Q_1 and Q_2, we can find a finite set of rational maps

$$\phi_1, \phi_2, \ldots, \phi_n : E \times E \rightarrow E$$

such that

(i) ϕ_1 is the addition map given in (2.3c).
(ii) For each $(P_1, P_2) \in E \times E$, some ϕ_i is defined at (P_1, P_2).
(iii) If ϕ_i and ϕ_j are both defined at (P_1, P_2), then $\phi_i(P_1, P_2) = \phi_j(P_1, P_2)$.

It follows that addition is defined on all of $E \times E$, so it is a morphism. $\qquad \square$

Remark 3.6.1. During the course of proving (3.6), we noted that the formulas in (2.3c) make it clear that the addition map $+ : E \times E \rightarrow E$ is a morphism except possibly at points of the form $(P, \pm P)$, (P, O), (O, P). Rather than using translation maps to circumvent this difficulty, one can work directly from the definition of morphism using explicit equations. It turns out that this involves consideration of quite a number of cases; we will do one to illustrate the method.

Thus let $(x_1, y_1; x_2, y_2)$ be Weierstrass coordinates on $E \times E$. We will show explicitly that addition is defined at points of the form (P, P) with $P \neq O$ and $[2]P \neq O$. Note that addition is defined in general by the formulas given in (2.3c):

$$\lambda = (y_2 - y_1)/(x_2 - x_1) \qquad\qquad v = (y_1 x_2 - y_2 x_1)/(x_2 - x_1) = y_1 - \lambda x_1$$
$$x_3 = \lambda^2 + a_1 \lambda - a_2 - x_1 - x_2 \qquad y_3 = -(\lambda + a_1)x_3 - v - a_3.$$

Here λ, v, x_3, y_3 are functions on $E \times E$, and addition is given by the map $[x_3, y_3, 1] : E \times E \rightarrow E$. Thus to show that addition is defined at (P, P), it suffices to show that λ is defined there. But by assumption, both pairs of functions (x_1, y_1) and (x_2, y_2) satisfy the same Weierstrass equation. Subtracting one equation from the other and factoring yields

$$(y_1 - y_2)(y_1 + y_2 + a_1 x_1 + a_3)$$
$$= (x_1 - x_2)(x_1^2 + x_1 x_2 + x_2^2 + a_2 x_1 + a_2 x_2 + a_4 - a_1 y_2).$$

Hence λ may also be written as

$$\lambda = \frac{x_1^2 + x_1 x_2 + x_2^2 + a_2(x_1 + x_2) + a_4 - a_1 y_2}{y_1 + y_2 + a_1 x_1 + a_3}.$$

Therefore, if $P = (x, y)$, then

$$\lambda(P, P) = \frac{3x^2 + 2a_2 x + a_4 - a_1 y}{2y + a_1 x + a_3};$$

and so λ is defined at (P, P) (unless $2y + a_1 x + a_3 = 0$, which is excluded by our assumption that $[2]P \neq O$). The reader may deal similarly with the other cases.

§4. Isogenies

Having now examined in some detail the geometry of individual elliptic curves, we turn to the study of the maps between them. Since an elliptic curve has a distinguished zero point, it is natural to single out those maps which respect this property.

Definition. Let E_1 and E_2 be elliptic curves. An *isogeny* between E_1 and E_2 is a morphism

$$\phi : E_1 \rightarrow E_2$$

satisfying $\phi(O) = O$. E_1 and E_2 are *isogenous* if there is an isogeny ϕ between them with $\phi(E_1) \neq \{O\}$.

Notice that from (II.2.3), an isogeny ϕ satisfies either $\phi(E_1) = \{O\}$ or $\phi(E_1) = E_2$. Thus except for the zero isogeny, defined by $[0](P) = O$ for all $P \in E_1$, every other isogeny is a finite map of curves. Hence we obtain the usual injection of function fields (II §2)

$$\phi^* : \bar{K}(E_2) \rightarrow \bar{K}(E_1);$$

and the degree of ϕ (deg ϕ), separable and inseparable degrees of ϕ (deg$_s \phi$ and deg$_i \phi$), and whether ϕ is separable, inseparable, or purely inseparable are defined by the corresponding property for the finite extension $\bar{K}(E_1)/\phi^* \bar{K}(E_2)$. By convention, we set

$$\deg[0] = 0.$$

Since elliptic curves are groups, the maps between them form groups.

Thus let

$$\text{Hom}(E_1, E_2) = \{\text{isogenies } \phi : E_1 \to E_2\}.$$

Then (3.6) implies that $\text{Hom}(E_1, E_2)$ is a group under the addition law

$$(\phi + \psi)(P) = \phi(P) + \psi(P).$$

If $E_1 = E_2$, then we can also compose isogenies. Thus if E is an elliptic curve, we let

$$\text{End}(E) = \text{Hom}(E, E)$$

be the ring with addition as above and multiplication given by composition,

$$(\phi\psi)(P) = \phi(\psi(P)).$$

(The fact that the distributive law holds follows from (4.8) proven below.) $\text{End}(E)$ is called the *endomorphism ring of* E. The invertible elements of $\text{End}(E)$ form the *automorphism group of* E, which is denoted $\text{Aut}(E)$. The endomorphism ring of an elliptic curve is an important invariant which we will study in some detail throughout the rest of this chapter.

Of course, if E_1, E_2, E are defined over a field K, then we can restrict attention to those isogenies defined over K. The corresponding groups of isogenies are denoted with the usual subscripts, thus

$$\text{Hom}_K(E_1, E_2), \qquad \text{End}_K(E), \qquad \text{Aut}_K(E).$$

Example 4.1. For each $m \in \mathbb{Z}$ we can define an isogeny *multiplication by* m

$$[m] : E \to E$$

in the natural way. If $m > 0$ then

$$[m](P) = P + P + \cdots + P \ (m \text{ terms});$$

if $m < 0$ then $[m](P) = [-m](-P)$; and we have already defined $[0](P) = O$. That $[m]$ is an isogeny follows easily by induction using (3.6). Notice that if E is defined over K, then $[m]$ is defined over K. We start our analysis of the group of isogenies by showing that the multiplication by m map is non-constant (provided, of course, that $m \neq 0$).

Proposition 4.2. (a) *Let E/K be an elliptic curve and let $m \in \mathbb{Z}$, $m \neq 0$. Then the multiplication by m map*

$$[m] : E \to E$$

is non-constant.
(b) *Let E_1 and E_2 be elliptic curves. Then the group of isogenies*

$$\text{Hom}(E_1, E_2)$$

is a torsion-free $\mathbb{Z}$-module.

(c) *Let E be an elliptic curve. Then the endomorphism ring* End(E) *is a (not necessarily commutative) ring of characteristic* 0 *with no zero divisors.*

PROOF. (a) We start by showing that $[2] \neq [0]$. From the duplication formula (2.3d), if a point $P = (x, y) \in E$ has order 2, then it must satisfy

$$4x^3 + b_2 x^2 + 2b_4 x + b_6 = 0.$$

If char$(K) \neq 2$, this shows immediately that there are only finitely many such points; and even for char$(K) = 2$, the only way to have $[2] = [0]$ is for this polynomial to be identically 0, which means $b_2 = b_6 = 0$, which in turn implies $\Delta = 0$. Hence in all cases $[2] \neq [0]$. Now, using the fact that $[mn] = [m] \circ [n]$, we are reduced to considering the case of odd m.

Assume now that char$(K) \neq 2$. Then using long division, one easily verifies that the polynomial

$$4x^3 + b_2 x^2 + 2b_4 x + b_6$$

does not divide

$$x^4 - b_4 x^2 - 2b_6 x - b_8.$$

(I.e., if it does, then one finds that $\Delta = 0$. In fact, these two polynomials are relatively prime (exer. 3.1).) Hence we can find an $x_0 \in \bar{K}$ so that the former vanishes to a higher order at $x = x_0$ than the latter. Choosing $y_0 \in \bar{K}$ so that $P_0 = (x_0, y_0) \in E$, the doubling formula then implies that $[2]P_0 = O$. In other words, we have shown that E has a non-trivial point of order 2. But then for m odd,

$$[m]P_0 = P_0 \neq O,$$

so clearly $[m] \neq [0]$.

Finally, if char$(K) = 2$, one can proceed as above using the "triplication formula" (exer. 3.2) to produce a point of order 3. We will leave this for the reader, since later in this chapter (5.4) we will prove a result which includes the case of char$(K) = 2$ and m odd.

(b) This follows immediately from (a). Suppose $\phi \in \text{Hom}(E_1, E_2)$ and $m \in \mathbb{Z}$ satisfy

$$[m] \circ \phi = [0].$$

Taking degrees gives

$$(\deg[m])(\deg \phi) = 0;$$

so either $m = 0$; or else (a) implies that $\deg[m] \geqslant 1$, in which case we must have $\phi = [0]$.

(c) From (b), End(E) has characteristic 0. Further, if $\phi, \psi \in \text{End}(E)$ satisfy $\phi \circ \psi = [0]$, then

$$\deg \phi \deg \psi = \deg \phi \circ \psi = 0.$$

It follows that either $\phi = [0]$ or $\psi = [0]$. Therefore $\text{End}(E)$ is an integral domain. $\qquad\qquad\qquad\qquad\qquad\qquad\qquad\qquad\qquad\qquad\quad\square$

For an arbitrary elliptic curve, the only isogenies which are immediately evident are the multiplication-by-m maps. As a consequence, these maps will provide one of the most powerful tools at our disposal for studying elliptic curves.

Definition. Let E be an elliptic curve and $m \in \mathbb{Z}$, $m \neq 0$. The *m-torsion subgroup of E*, denoted $E[m]$, is the set of points of order m in E,

$$E[m] = \{P \in E : [m]P = O\}.$$

The *torsion subgroup of E*, denoted E_{tors}, is the set of points of finite order,

$$E_{\text{tors}} = \bigcup_{m=1}^{\infty} E[m].$$

If E is defined over K, then $E_{\text{tors}}(K)$ will denote the points of finite order in $E(K)$.

The most important fact about the multiplication-by-m map is that it has degree m^2, from which one can deduce the structure of the finite group $E[m]$. We will not prove this result here, since it will be an immediate corollary of our work with "dual isogenies" (cf. §6). However, the reader should be aware that there is a completely elementary (but rather messy) proof of this fact using explicit formulas and induction. (See exer. 3.7. For some other approaches, see exers. 3.8, 3.9.)

Remark 4.3. Suppose that $\text{char}(K) = 0$. Then the map

$$[\ \ ] : \mathbb{Z} \to \text{End}(E)$$

is usually the whole story (i.e. $\text{End}(E) \cong \mathbb{Z}$). If $\text{End}(E)$ is strictly larger than $\mathbb{Z}$, then we say that E has *complex multiplication*. The elliptic curves with complex multiplication have many special properties. (See appendix C §11 for a brief discussion.) On the other hand, if K is a finite field, then $\text{End}(E)$ is always larger than $\mathbb{Z}$ (see V §3).

Example 4.4. Assume $\text{char}(K) \neq 2$ and let E/K be the elliptic curve

$$E : y^2 = x^3 - x.$$

Then, in addition to $\mathbb{Z}$, $\text{End}(E)$ contains an element which we denote $[i]$, given by

$$[i] : (x, y) \to (-x, iy).$$

(Here $i \in \bar{K}$ is a primitive fourth root of unity.) Thus E has complex multiplication. Clearly $[i]$ is defined over K if and only if $i \in K$. Hence even if E is defined over K, $\mathrm{End}_K(E)$ may be strictly smaller than $\mathrm{End}(E)$.

One immediately checks that $[i] \circ [i] = [-1]$, so we have a ring homomorphism

$$\mathbb{Z}[i] \to \mathrm{End}(E)$$

$$m + ni \to [m] + [n] \circ [i].$$

If $\mathrm{char}(K) = 0$, this is an isomorphism; and so for example

$$\mathrm{Aut}(E) = \mathbb{Z}[i]^* = \{\pm 1, \pm i\}$$

is a cyclic group of order 4.

Example 4.5. Again assume $\mathrm{char}(K) \neq 2$, and let $a, b \in K$ with $b \neq 0$ and $r = a^2 - 4b \neq 0$. Consider the two elliptic curves

$$E_1 : y^2 = x^3 + ax^2 + bx$$

$$E_2 : Y^2 = X^3 - 2aX^2 + rX.$$

There are isogenies (of degree 2)

$$\phi : E_1 \to E_2 \qquad\qquad\qquad \hat{\phi} : E_2 \to E_1$$

$$(x, y) \to \left(\frac{y^2}{x^2}, \frac{y(b - x^2)}{x^2} \right) \qquad (X, Y) \to \left(\frac{Y^2}{4X^2}, \frac{Y(r - X^2)}{8X^2} \right)$$

By a direct computation one can check that $\hat{\phi} \circ \phi = [2]$ on E_1 and $\phi \circ \hat{\phi} = [2]$ on E_2. This gives an example of *dual isogenies*, which we will discuss in section 6.

Example 4.6. Suppose K is a field of characteristic p with $p > 0$, and let $q = p^r$. If E/K is an elliptic curve given by a Weierstrass equation, recall (II §2) that the curve $E^{(q)}/K$ is defined by raising the coefficients of the equation for E to the qth-power; and the Frobenius morphism is given by

$$\phi_q : E \to E^{(q)}$$

$$(x, y) \to (x^q, y^q).$$

Since $E^{(q)}$ is the zero locus of a Weierstrass equation, it too will be an elliptic curve provided that the equation is non-singular. But writing everything out in terms of the Weierstrass coefficients and using the fact that the qth-power map $K \to K$ is a homomorphism, one readily sees that

$$\Delta(E^{(q)}) = \Delta(E)^q \quad \text{and} \quad j(E^{(q)}) = j(E)^q.$$

In particular, the equation for $E^{(q)}$ is non-singular.

Now suppose that $K = \mathbb{F}_q$ is a finite field. Then the qth-power map on K is the identity, so $E^{(q)} = E$, and ϕ_q is an endomorphism of E, called the *Frobenius endomorphism.* The set of points fixed by ϕ_q is exactly the finite group $E(K)$, a fact which lies at the heart of Hasse's proof for estimating $\# E(K)$. (See V §1.)

Example 4.7. Let E/K be an elliptic curve and $Q \in E$. Then we can define a *translation by Q map*

$$\tau_Q : E \to E$$
$$P \to P + Q.$$

This is clearly an isomorphism, since τ_{-Q} provides an inverse. Of course, it is not an isogeny unless $Q = O$.

Now let

$$F : E_1 \to E_2$$

be any morphism of elliptic curves. Then the map

$$\phi = \tau_{-F(O)} \circ F$$

is an isogeny (since $\phi(O) = O$). We have thus shown that any map

$$F = \tau_{F(O)} \circ \phi$$

is the composition of an isogeny and a translation.

An isogeny is a map between elliptic curves which sends O to O. Since an elliptic curve is a group, it might seem more natural to focus on those isogenies which are group homomorphisms. In fact, it turns out that every isogeny has this property.

Theorem 4.8. *Let*

$$\phi : E_1 \to E_2$$

be an isogeny. Then

$$\phi(P + Q) = \phi(P) + \phi(Q) \qquad \text{for all } P, Q \in E_1.$$

PROOF. If $\phi(P) = O$ for all $P \in E_1$, there is nothing to prove. Otherwise, ϕ is a finite map, so by (II.3.7) it induces a homomorphism

$$\phi_* : \text{Pic}^0(E_1) \to \text{Pic}^0(E_2)$$

defined by

$$\phi_*(\text{class of } \sum n_i(P_i)) = \text{class of } \sum n_i(\phi P_i).$$

On the other hand, from (3.4) we have *group* isomorphisms

$$\kappa_i : E_i \to \text{Pic}^0(E_i)$$

$$P \to \text{class of } (P) - (O).$$

Then, since $\phi(O) = O$, we have the following commutative diagram:

$$E_1 \xrightarrow[\kappa_1]{\cong} \text{Pic}^0(E_1)$$

$$\phi \downarrow \qquad\qquad \downarrow \phi_*$$

$$E_2 \xrightarrow[\kappa_2]{\cong} \text{Pic}^0(E_2).$$

Since κ_1, κ_2, and ϕ_* are all group homomorphisms, and κ_2 is injective, it follows that ϕ is also a homomorphism.   □

Corollary 4.9. *Let* $\phi : E_1 \to E_2$ *be a non-zero isogeny. Then*

$$\ker \phi = \phi^{-1}(O)$$

is a finite subgroup.

PROOF. It is a subgroup from (4.8) and finite (of order at most deg ϕ) from (II.2.6a).   □

The next three results (4.10, 4.11, 4.12) encompass the basic Galois theory of elliptic function fields.

Theorem 4.10. *Let* $\phi : E_1 \to E_2$ *be a non-constant isogeny.*
(a) *For every* $Q \in E_2$,

$$\# \phi^{-1}(Q) = \deg_s \phi.$$

Further, for every $P \in E_1$,

$$e_\phi(P) = \deg_i(\phi).$$

(b) *The map*

$$\ker \phi \to \text{Aut}[\bar{K}(E_1)/\phi^*\bar{K}(E_2)]$$

$$T \to \tau_T^*$$

is an isomorphism. (Here τ_T *is the translation-by-T map (4.7), and* τ_T^* *is the automorphism it induces on* $\bar{K}(E_1)$.)
(c) *Assume that* ϕ *is separable. Then* ϕ *is unramified,*

$$\# \ker \phi = \deg \phi,$$

and $\bar{K}(E_1)$ *is a Galois extension of* $\phi^*\bar{K}(E_2)$.

PROOF. (a) From (II.2.6.b) we know that

$$\# \phi^{-1}(Q) = \deg_s \phi$$

for all but finitely many $Q \in E_2$. But for any Q, $Q' \in E_2$, if we choose some $R \in E_1$ with $\phi(R) = Q' - Q$, then the fact that ϕ is a homomorphism implies that there is a one-to-one correspondence

$$\phi^{-1}(Q) \to \phi^{-1}(Q')$$

$$P \to P + R.$$

Hence

$$\#\phi^{-1}(Q) = \deg_s \phi \qquad \text{for all } Q \in E_2,$$

which proves the first assertion.

Now let P, $P' \in E_1$ with $\phi(P) = \phi(P') = Q$, and let $R = P' - P$. Then $\phi(R) = O$, so $\phi \circ \tau_R = \phi$. Therefore, using (II.2.6c) and the fact that τ_R is an isomorphism,

$$e_\phi(P) = e_{\phi \circ \tau_R}(P) = e_\phi(\tau_R(P))e_{\tau_R}(P) = e_\phi(P').$$

Hence every point of $\phi^{-1}(Q)$ has the same ramification index. We compute

$$(\deg_s \phi)(\deg_i \phi) = \deg \phi = \sum_{P \in \phi^{-1}(Q)} e_\phi(P) \qquad \text{(II.2.6a)}$$

$$= (\#\phi^{-1}(Q))e_\phi(P) \qquad \text{for any } P \in \phi^{-1}(Q)$$

$$= (\deg_s \phi)e_\phi(P) \qquad \text{from above.}$$

Cancelling $\deg_s \phi$ gives the second assertion.

(b) First, if $T \in \ker \phi$ and $f \in \bar{K}(E_2)$, then

$$\tau_T^*(\phi^* f) = (\phi \circ \tau_T)^* f = \phi^* f,$$

since $\phi \circ \tau_T = \phi$. Hence as an automorphism of $\bar{K}(E_1)$, τ_T^* does fix $\phi^* \bar{K}(E_2)$, so the indicated map is well-defined. Next, since

$$\tau_S \circ \tau_T = \tau_{S+T} = \tau_T \circ \tau_S,$$

the map is clearly a homomorphism. Finally, from (a) we have

$$\#\ker \phi = \deg_s \phi;$$

while from basic Galois theory,

$$\#\operatorname{Aut}(\bar{K}(E_1)/\phi^* \bar{K}(E_2)) \leqslant \deg_s \phi.$$

Hence to prove that the map $T \to \tau_T^*$ is an isomorphism, it suffices to show that it is injective. But if τ_T^* fixes $\bar{K}(E_1)$, then in particular every function on E_1 takes the same value at T and O. This clearly implies that $T = O$.

(c) If ϕ is separable, then from (a) we see that

$$\#\phi^{-1}(Q) = \deg \phi \qquad \text{for all } Q \in E_2.$$

Hence ϕ is unramified (II.2.7), and putting $Q = O$ gives

$$\#\ker \phi = \deg \phi.$$

Then from (b) we find that

$$\#\operatorname{Aut}(\bar{K}(E_1)/\phi^* \bar{K}(E_2)) = [\bar{K}(E_1) : \phi^* \bar{K}(E_2)],$$

so $\bar{K}(E_1)/\phi^* \bar{K}(E_2)$ is a Galois extension. $\square$

Corollary 4.11. *Let*

$$\phi : E_1 \to E_2 \quad \text{and} \quad \psi : E_1 \to E_3$$

be non-constant isogenies, and assume that ϕ is separable. If

$$\ker \phi \subset \ker \psi,$$

then there is a unique isogeny

$$\lambda : E_2 \to E_3$$

such that $\psi = \lambda \circ \phi$.

PROOF. Since ϕ is separable, (4.10c) says that $\bar{K}(E_1)$ is a Galois extension of $\phi^*\bar{K}(E_2)$. Then the inclusion $\ker \phi \subset \ker \psi$ and the identification given in (4.10b) implies that every element of $\mathrm{Gal}(\bar{K}(E_1)/\phi^*\bar{K}(E_2))$ fixes $\psi^*\bar{K}(E_3)$. Hence by Galois theory, there are field inclusions

$$\psi^*\bar{K}(E_3) \subset \phi^*\bar{K}(E_2) \subset \bar{K}(E_1).$$

Now (II.2.4b) gives a map

$$\lambda : E_2 \to E_3$$

satisfying

$$\phi^*(\lambda^*K(E_3)) = \psi^*K(E_3);$$

and this in turn implies that

$$\lambda \circ \phi = \psi.$$

Finally, λ is an isogeny, since

$$\lambda(O) = \lambda(\phi(O)) = \psi(O) = O. \qquad \square$$

Proposition 4.12. *Let E be an elliptic curve, and let Φ be a finite subgroup of E. Then there is a unique elliptic curve E' and a separable isogeny*

$$\phi : E \to E'$$

such that

$$\ker \phi = \Phi.$$

Remark 4.13.1. The elliptic curve whose existence is asserted in this corollary is often denoted by the quotient E/Φ. This clearly indicates the group structure, but there is no a priori reason why this group should correspond to the points on an elliptic curve. In fact, the quotient of any variety by a finite group of automorphisms is again a variety (cf. [Mum, §7]. The case of curves is done in (exer. 3.13).)

Remark 4.13.2. Suppose that E is defined over K, and that Φ is $G_{\bar{K}/K}$-invariant. (I.e. If $T \in \Phi$, then $T^\sigma \in \Phi$ for all $\sigma \in G_{\bar{K}/K}$.) Then it is actually possible to find an E' and a ϕ which are defined over K. (See exer. 3.13e.)

PROOF. As in (4.10b), each point $T \in \Phi$ gives rise to an automorphism τ_T^* of $\bar{K}(E)$. Let $\bar{K}(E)^\Phi$ be the subfield of $\bar{K}(E)$ fixed by every element of Φ. Then Galois theory says that $\bar{K}(E)$ is a Galois extension of $\bar{K}(E)^\Phi$ with Galois group isomorphic to Φ.

Now $\bar{K}(E)^\Phi$ is a field of transcendence degree 1 over $\bar{K}$, so from (II.2.4c) there is a unique curve $C/\bar{K}$ and a finite morphism

$$\phi : E \to C$$

such that

$$\phi^* \bar{K}(C) = \bar{K}(E)^\Phi.$$

Next we show that ϕ is unramified. Let $P \in E$ and $T \in \Phi$. Then for *every* function $f \in \bar{K}(C)$,

$$f(\phi(P + T)) = ((\tau_T^* \circ \phi^*)f)(P) = (\phi^*f)(P) = f(\phi(P)),$$

where the middle equality uses the fact that τ_T^* fixes every element of $\phi^* \bar{K}(C)$. It follows that $\phi(P + T) = \phi(P)$. Now let $Q \in C$, and choose any $P \in E$ with $\phi(P) = Q$. Then

$$\phi^{-1}(Q) \supset \{P + T : T \in \Phi\}.$$

But

$$\#\phi^{-1}(Q) \leqslant \deg \phi = \#\Phi,$$

with equality holding if and only if ϕ is unramified at Q (II.2.7). Since the points $P + T$ are distinct as T ranges over the elements of Φ, we conclude that ϕ is unramified at Q; and since Q was arbitrary, ϕ is unramified.

Now apply the Hurwitz genus formula (II.5.9) to ϕ. Since ϕ is unramified, the formula reads

$$2 \operatorname{genus}(E) - 2 = (\deg \phi)(2 \operatorname{genus}(C) - 2).$$

From this we conclude that C also has genus 1; so it becomes an elliptic curve, and ϕ becomes an isogeny, if we take $\phi(O)$ as the "zero point" on C. $\quad\square$

§5. The Invariant Differential

Let E/K be an elliptic curve given by the usual Weierstrass equation

$$y^2 + a_1 xy + a_3 y = x^3 + a_2 x^2 + a_4 x + a_6.$$

As we have seen (1.5), the differential

$$\omega = \frac{dx}{2y + a_1 x + a_3} \in \Omega_E$$

has neither zeros nor poles. We now justify its name of *invariant differential* by proving that it is invariant under translation.

Proposition 5.1. *With notation as above, for every* $Q \in E$,

$$\tau_Q^* \omega = \omega.$$

(*Here* τ_Q *is the translation-by-Q map* (4.7).)

PROOF. One can prove this proposition by a straightforward (but messy and unenlightening) calculation as follows. Write $x(P + Q)$ and $y(P + Q)$ out in terms of $x(P)$, $x(Q)$, $y(P)$, and $y(Q)$ using the addition formula (2.3c). Then use standard differentiation rules to calculate $dx(P + Q)$ as a rational function times $dx(P)$, treating $x(Q)$ as a constant. In this way one can directly verify that (for fixed Q).

$$\frac{dx(P + Q)}{2y(P + Q) + a_1 x(P + Q) + a_3} = \frac{dx(P)}{2y(P) + a_1 x(P) + a_3}.$$

We leave the details of this calculation to the reader, and instead give a more illuminating proof.

Since Ω_E is a 1-dimensional $\bar{K}(E)$-vector space (II.4.2), there is a function $a_Q \in \bar{K}(E)^*$, depending a priori on Q, so that

$$\tau_Q^* \omega = a_Q \omega.$$

(Note $a_Q \neq 0$ because τ_Q is an isomorphism.) Now

$$\text{div}(a_Q) = \text{div}(\tau_Q^* \omega) - \text{div}(\omega)$$

$$= \tau_Q^* \, \text{div}(\omega) - \text{div}(\omega)$$

$$= 0 \qquad \text{since div}(\omega) = 0 \text{ from (1.5)}.$$

Hence a_Q is a function on E with neither zeros nor poles, so by (II.1.2) it is constant, $a_Q \in \bar{K}^*$.

Next consider the map

$$f : E \to \mathbb{P}^1$$

$$Q \to [a_Q, 1].$$

From the calculation sketched above, even without doing it explicitly, it is clear that a_Q can be expressed as a rational function of $x(Q)$ and $y(Q)$. Hence f is a rational map from E to $\mathbb{P}^1$ which is not surjective. (It misses both $[0, 1]$ and $[1, 0]$.) From (II.2.1) and (II.2.3), we conclude that f is constant. Therefore

$$a_Q = a_O = 1 \qquad \text{for all } Q \in E,$$

which is the desired result.   □

Differential calculus is, in essence, a linearization tool. It will come as no surprise that the enormous utility of the invariant differential on an elliptic curve lies in its ability to linearize the otherwise quite complicated addition law on the curve.

Theorem 5.2. *Let E and E' be elliptic curves, let ω be an invariant differential on E, and let*

$$\phi, \psi : E' \to E$$

be two isogenies. Then

$$(\phi + \psi)^*\omega = \phi^*\omega + \psi^*\omega.$$

(*N.B. The two "plus signs" in this last equation respresent completely different operations. The first is addition in* $\text{Hom}(E', E)$*, which is essentially addition using the group law on E. The second is the more usual addition in the vector space of differentials* $\Omega_{E'}$*.*)

PROOF. If $\phi = [0]$ or $\psi = [0]$, the result is clear. If $\phi + \psi = [0]$, then using the fact that

$$\psi^* = (-\phi)^* = \phi^* \circ [-1]^*,$$

it suffices to check that

$$[-1]^*\omega = -\omega.$$

Since

$$[-1](x, y) = (x, -y - a_1 x - a_3),$$

we immediately obtain the desired result

$$[-1]^*\left(\frac{dx}{2y + a_1 x + a_3}\right) = \frac{dx}{2(-y - a_1 x - a_3) + a_1 x + a_3}$$

$$= -\frac{dx}{2y + a_1 x + a_3}.$$

We now assume that ϕ, ψ, and $\phi + \psi$ are all non-zero.

Let (x_1, y_1) and (x_2, y_2) be "independent" Weierstrass coordinates for E. By this we mean that they satisfy the given Weierstrass equation for E, but satisfy no other algebraic relation. (More formally, $([x_1, y_1, 1], [x_2, y_2, 1])$ gives coordinates for $E \times E$ sitting inside $\mathbb{P}^2 \times \mathbb{P}^2$. Alternatively, (x_1, y_1) and (x_2, y_2) are "independent generic points of E" in the sense of Weil (cf. [Ca 7]).)

Let

$$(x_3, y_3) = (x_1, y_1) + (x_2, y_2),$$

so x_3 and y_3 are the rational combinations of x_1, y_1, x_2, y_2 given by the addition formula on E (2.3c). Further, for any (x, y), let $\omega(x, y)$ denote the corresponding invariant differential,

$$\omega(x, y) = \frac{dx}{2y + a_1 x + a_3}.$$

Then using the addition formula (2.3c) and the standard rules for differentiation, we can express $\omega(x_3, y_3)$ in terms of $\omega(x_1, y_1)$ and $\omega(x_2, y_2)$. This yields

$$\omega(x_3, y_3) = f(x_1, y_1, x_2, y_2)\omega(x_1, y_1) + g(x_1, y_1, x_2, y_2)\omega(x_2, y_2),$$

where f and g are rational functions of the indicated variables. (In doing this calculation, remember that since x_i, y_i satisfy the given Weierstrass equation, the differentials dx_i and dy_i are related by

$$(2y_i + a_1 x_i + a_3)dy_i = (3x_i^2 + 2a_2 x_i + a_4 - a_1 y_i)\,dx_i.$$

In this way, $\omega(x_3, y_3)$ can be expressed as a $\bar{K}(x_1, y_1, x_2, y_2)$ linear combination of dx_1 and dx_2.)

We claim that f and g are both identically 1. Clearly this can be proven by an explicit calculation, a painful task that we leave for the reader. Instead, we use (5.1) to obtain the desired result. Suppose we assign fixed values to x_2 and y_2, say by choosing some $Q \in E$ and setting

$$x_2 = x(Q) \quad \text{and} \quad y_2 = y(Q).$$

Then

$$dx_2 = dx(Q) = 0, \qquad \text{so } \omega(x_2, y_2) = 0;$$

while from (5.1),

$$\omega(x_3, y_3) = \tau_Q^* \omega(x_1, y_1)$$
$$= \omega(x_1, y_1).$$

Substituting these in the above expression for $\omega(x_3, y_3)$, we find that

$$f(x_1, y_1, x(Q), y(Q)) \equiv 1$$

as a rational function in $\bar{K}(x_1, y_1)$. Further, this is true for *every* point $Q \in E$. It follows that f must be identically 1. Then reversing the roles of x_1, y_1 and x_2, y_2, we see that the same is true for g.

To recapitulate, we have shown that if

$$(x_3, y_3) = (x_1, y_1) + (x_2, y_2) \qquad (+ \text{ is addition on } E),$$

then

$$\omega(x_3, y_3) = \omega(x_1, y_1) + \omega(x_2, y_2) \qquad (+ \text{ is addition in } \Omega_E).$$

Now let (x', y') be Weierstrass coordinates on E', and set

$$(x_1, y_1) = \phi(x', y') \qquad (x_2, y_2) = \psi(x', y') \qquad (x_3, y_3) = (\phi + \psi)(x', y').$$

Substituting this above yields

$$(\omega \circ (\phi + \psi))(x', y') = (\omega \circ \phi)(x', y') + (\omega \circ \psi)(x', y'),$$

which says exactly that

$$(\phi + \psi)^*\omega = \phi^*\omega + \psi^*\omega. \qquad \square$$

Corollary 5.3. *Let ω be an invariant differential on an elliptic curve E. Let $m \in \mathbb{Z}$. Then*

$$[m]^*\omega = m\omega.$$

PROOF. The assertion is true for $m = 0$ (by definition) and $m = 1$ (clear). From (5.2) with $\phi = [m]$ and $\psi = [1]$ we obtain

$$[m + 1]^*\omega = [m]^*\omega + \omega.$$

The result now follows by (ascending and descending) induction. $\qquad \square$

As a first indication of the utility of the invariant differential, we give a new, less computational proof of part of (4.2a).

Corollary 5.4. *Let E/K be an elliptic curve, and let $m \in \mathbb{Z}$, $m \neq 0$. Assume either that char$(K) = 0$ or that m is prime to char(K). Then the multiplication-by-m map on E is a finite, separable endomorphism.*

PROOF. Let ω be an invariant differential on E. Then from (5.3),

$$[m]^*\omega = m\omega \neq 0,$$

so certainly $[m] \neq [0]$. Further, (II.4.2c) implies that $[m]$ is separable. $\qquad \square$

As a second application, we examine when a linear combination involving the Frobenius morphism is separable.

Corollary 5.5. *Let* char$(K) = p > 0$*, let E be defined over $\mathbb{F}_q$, let $\phi : E \to E$ be the qth-power Frobenius endomorphism (4.6), and let $m, n \in \mathbb{Z}$. Then the map*

$$m + n\phi : E \to E$$

is separable if and only if $p \nmid m$.
    In particular, the map $1 - \phi$ is separable.

PROOF. Let ω be an invariant differential on E. From (II.4.2c), a map $\psi : E \to E$ is inseparable if and only if $\psi^*\omega = 0$. We apply this to the map $\psi = m + n\phi$. Using (5.2) and (5.3), we compute

$$(m + n\phi)^*\omega = m\omega + n\phi^*\omega.$$

But $\phi^*\omega = 0$ because ϕ is inseparable (or by direct calculation, since

$\phi^* \, dx = d(x^q) = 0$), so

$$(m + n\phi)^* \omega = m\omega.$$

Now $m\omega = 0$ if and only if $p \mid m$, which gives the desired result.   □

§6. The Dual Isogeny

Let $\phi : E_1 \to E_2$ be a non-constant isogeny. Then ϕ induces a map (II.3.7)

$$\phi^* : \mathrm{Pic}^0(E_2) \to \mathrm{Pic}^0(E_1).$$

On the other hand, we have group isomorphisms (3.4)

$$\kappa_i : E_i \to \mathrm{Pic}^0(E_i)$$

$$P \to \text{class of } (P) - (O).$$

Hence we obtain a homomorphism going in the opposition direction to ϕ, namely the composition

$$E_2 \xrightarrow{\kappa_2} \mathrm{Pic}^0(E_2) \xrightarrow{\phi^*} \mathrm{Pic}^0(E_1) \xrightarrow{\kappa_1^{-1}} E_1.$$

As we will verify below, this map can be computed as follows. Let $Q \in E_2$ and choose any $P \in E_1$ satisfying $\phi(P) = Q$. Then

$$\kappa_1^{-1} \circ \phi^* \circ \kappa_2(Q) = [\deg \phi](P).$$

It is by no means clear that the homomorphism $\kappa_1^{-1} \circ \phi^* \circ \kappa_2$ is an isogeny; that is, given by a rational map. The process of finding a point P satisfying $\phi(P) = Q$ will involve taking roots of various polynomial equations. If ϕ is separable, one needs to check that applying $[\deg \phi]$ to P causes the conjugate roots to appear symmetrically. (That this is so is fairly clear if one explicitly writes out $\kappa_1^{-1} \circ \phi^* \circ \kappa_2$.) If ϕ is inseparable, this approach is more complicated. We now show that in all cases, there is an actual isogeny which can be computed in the manner described above.

Theorem 6.1. Let $\phi : E_1 \to E_2$ be a non-constant isogeny of degree m.
(a) There exists a unique isogeny

$$\hat{\phi} : E_2 \to E_1$$

satisfying

$$\hat{\phi} \circ \phi = [m].$$

(b) As a group homomorphism, $\hat{\phi}$ equals the composition

$$E_2 \to \mathrm{Div}^0(E_2) \xrightarrow{\phi^*} \mathrm{Div}^0(E_1) \xrightarrow{\text{sum}} E_1$$

$$Q \to (Q) - (O) \qquad \sum n_P(P) \to \sum [n_P]P.$$

PROOF. (a) First we show uniqueness. Suppose $\hat{\phi}$ and $\hat{\phi}'$ are two such isogenies. Then

$$(\hat{\phi} - \hat{\phi}') \circ \phi = [m] - [m] = [0].$$

Since ϕ is non-constant, it follows from (II.2.3) that $\hat{\phi} - \hat{\phi}'$ must be constant, so $\hat{\phi} = \hat{\phi}'$.

Next suppose that $\psi : E_2 \to E_3$ is another non-constant isogeny, say of degree n, and suppose that we know that $\hat{\phi}$ and $\hat{\psi}$ exist. Then

$$(\hat{\phi} \circ \hat{\psi}) \circ (\psi \circ \phi) = \hat{\phi} \circ [n] \circ \phi = [n] \circ \hat{\phi} \circ \phi = [nm].$$

Thus $\hat{\phi} \circ \hat{\psi}$ has the requisite property to be $\widehat{\psi \circ \phi}$. Hence using (II.2.12) to write an arbitrary isogeny ϕ as a compositon, it suffices to prove the existence of $\hat{\phi}$ when ϕ is either separable or the Frobenius morphism.

Case 1. ϕ is separable. Since ϕ has degree m, we have (4.10c)

$$\# \ker \phi = m;$$

so clearly

$$\ker \phi \subset \ker[m].$$

It now follows immediately from (4.11) that there is an isogeny

$$\hat{\phi} : E_2 \to E_1$$

satisfying

$$\hat{\phi} \circ \phi = [m].$$

Case 2. ϕ is a Frobenius morphism. If ϕ is the qth-power Frobenius morphism, and $q = p^e$, then clearly ϕ is the composition of the pth-power Frobenius morphism with itself e times. Hence it suffices to prove that $\hat{\phi}$ exists if ϕ is the pth-power Frobenius morphism, and so deg $\phi = p$ (II.2.11).

We look at the multiplication-by-p map on E. Let ω be an invariant differential. Then from (5.3) and the fact that char$(K) = p$,

$$[p]^*\omega = p\omega = 0.$$

Hence from (II.4.2c) we conclude that $[p]$ is not separable, so when $[p]$ is decomposed as some Frobenius morphism followed by a separable map (II.2.12), the Frobenius morphism does appear. In other words,

$$[p] = \psi \circ \phi^e$$

for some integer $e \geq 1$ and some separable isogeny ψ. Then we can take

$$\hat{\phi} = \psi \circ \phi^{e-1}.$$

(b) Let $Q \in E_2$. Then the image of Q under the indicated composition is

$$\text{sum}\{\phi^*((Q)-(O))\} = \sum_{P\in\phi^{-1}(Q)} [e_\phi(P)]P - \sum_{T\in\phi^{-1}(O)} [e_\phi(T)]T$$

$$\text{definition of } \phi^*$$

$$= [\deg_i \phi]\left(\sum_{P\in\phi^{-1}(Q)} P - \sum_{T\in\phi^{-1}(O)} T \right) \quad \text{from (4.10a)}$$

$$= [\deg_i \phi] \circ [\#\phi^{-1}(Q)]P \quad \text{for any } P\in\phi^{-1}(Q)$$

$$= [\deg \phi]P \quad \text{from (4.10a)}.$$

But by construction,

$$\hat{\phi}(Q) = \hat{\phi}\circ\phi(P) = [\deg \phi]P,$$

so the two maps are the same. □

Definition. Let $\phi : E_1 \to E_2$ be an isogeny. The *dual isogeny* to ϕ is the isogeny

$$\hat{\phi} : E_2 \to E_1$$

given by (6.1a). [This assumes $\phi \neq [0]$. If $\phi = [0]$, then we set $\hat{\phi} = [0]$.]

   The next theorem gives the basic properties of the dual isogeny. From these basic facts we will be able to deduce a number of very important corollaries, including a reasonably good description of the kernel of the "multiplication-by-m" map.

Theorem 6.2. *Let*

$$\phi : E_1 \to E_2$$

be an isogeny.
(a) *Let $m = \deg \phi$. Then*

$$\hat{\phi}\circ\phi = [m] \quad on \quad E_1;$$

$$\phi\circ\hat{\phi} = [m] \quad on \quad E_2.$$

(b) *Let $\lambda : E_2 \to E_3$ be another isogeny. Then*

$$\widehat{\lambda\circ\phi} = \hat{\phi}\circ\hat{\lambda}.$$

(c) *Let $\psi : E_1 \to E_2$ be another isogeny. Then*

$$\widehat{\phi+\psi} = \hat{\phi} + \hat{\psi}.$$

(d) *For all $m\in\mathbb{Z}$,*

$$\widehat{[m]} = [m] \quad and \quad \deg[m] = m^2.$$

(e) $$\deg \hat{\phi} = \deg \phi.$$

(f) $$\hat{\hat{\phi}} = \phi.$$

PROOF. If ϕ is constant, then the entire theorem is trivial; and similarly for (b) or (c) if λ or ψ is constant. We will thus assume that all isogenies are non-constant.

(a) The first statement is the defining property of $\hat{\phi}$. For the second, look at

$$(\phi \circ \hat{\phi}) \circ \phi = \phi \circ [m] = [m] \circ \phi.$$

Hence $\phi \circ \hat{\phi} = [m]$, since ϕ is not constant.

(b) Letting $n = \deg \lambda$, we have

$$(\hat{\phi} \circ \hat{\lambda}) \circ (\lambda \circ \phi) = \hat{\phi} \circ [n] \circ \phi = [n] \circ \hat{\phi} \circ \phi = [nm].$$

Hence from the uniqueness statement in (6.1a),

$$\hat{\phi} \circ \hat{\lambda} = \widehat{\lambda \circ \phi}.$$

(c) Let $x_1, y_1 \in K(E_1)$ and $x_2, y_2 \in K(E_2)$ be Weierstrass coordinates. We start by looking at E_2 considered as an elliptic curve defined over the field $K(E_1) = K(x_1, y_1)$. Then another way of saying that $\phi : E_1 \to E_2$ is an isogeny is to note that $\phi(x_1, y_1) \in E_2(K(x_1, y_1))$, and similarly for $(\phi + \psi)(x_1, y_1)$ and $\psi(x_1, y_1)$. Now consider the divisor

$$D = ((\phi + \psi)(x_1, y_1)) - (\phi(x_1, y_1)) - (\psi(x_1, y_1)) + (O) \in \mathrm{Div}_{K(x_1, y_1)}(E_2).$$

By definition of $\phi + \psi$, it sums to O, so by (3.5) it is linearly equivalent to 0. Thus there is a function

$$f \in K(x_1, y_1)(E_2) = K(x_1, y_1, x_2, y_2)$$

which, *when considered as a function of* x_2, y_2, has divisor D.

We now switch perspective, and look at f as a function of x_1, y_1; that is, f as a function on E_1 considered as a curve defined over $K(x_2, y_2)$. Suppose $P_1 \in E_1(\overline{K(x_2, y_2)})$ is a point satisfying $\phi(P_1) = (x_2, y_2)$. Then examining D, specifically the term $-(\phi(x_1, y_1))$, we see that f has a pole at P_1. (I.e. $f(x_1, y_1; x_2, y_2)$ will have a pole if x_1, y_1, x_2, y_2 satisfy $(x_2, y_2) = \phi(x_1, y_1)$.) Further,

$$\mathrm{ord}_{P_1}(f) = e_\phi(P_1).$$

Similarly, f has a pole at P_1 if $\psi(P_1) = (x_2, y_2)$, and a zero if $(\phi + \psi)(P_1) = (x_2, y_2)$. It follows that as a function of x_1, y_1, the divisor of f has the form

$$(\phi + \psi)^*((x_2, y_2)) - \phi^*((x_2, y_2)) - \psi^*((x_2, y_2)) + \sum n_i(P_i) \in \mathrm{Div}_{\overline{K(x_2, y_2)}}(E_1),$$

where the P_i's are in $E_1(\overline{K})$. [I.e. $\sum n_i(P_i) \in \mathrm{Div}_{\overline{K}}(E_1)$.] Since this is a divisor of a function, it sums to O, so using (6.1(b)) we conclude that

$$\widehat{(\phi + \psi)}(x_2, y_2) - \hat{\phi}(x_2, y_2) - \hat{\psi}(x_2, y_2)$$

does not depend on (x_2, y_2). [I.e. it is in $E_1(\overline{K})$.] Putting $(x_2, y_2) = O$ shows that it equals O, which completes the proof that

$$\widehat{\phi + \psi} = \hat{\phi} + \hat{\psi}.$$

(d) This is true for $m = 0$ (by definition) and $m = 1$ (clear). Then using (c) with $\phi = [m]$ and $\psi = [1]$, we find that

$$\widehat{[m + 1]} = \widehat{[m]} + \widehat{[1]};$$

so the first assertion holds for all m by (ascending and descending) induction.
   Now led $d = \deg[m]$ and look at multiplication by d.

$$
\begin{aligned}
[d] = \widehat{[m]} \circ [m] & \qquad \text{definition of dual isogeny} \\
= [m^2] & \qquad \text{since } \widehat{[m]} = [m].
\end{aligned}
$$

Since the endomorphism ring of an elliptic curve is a torsion-free $\mathbb{Z}$-module (4.2b), it follows that $d = m^2$.
(e) Let $m = \deg \phi$. Then using (d) and (a),

$$m^2 = \deg[m] = \deg(\phi \circ \hat\phi) = (\deg \phi)(\deg \hat\phi) = m(\deg \hat\phi).$$

Hence $m = \deg \hat\phi$.
(f) Again let $m = \deg \phi$. Then using (a), (b) and (d),

$$\hat\phi \circ \phi = [m] = \widehat{[m]} = \widehat{\hat\phi \circ \phi} = \hat\phi \circ \hat{\hat\phi}.$$

Therefore

$$\phi = \hat{\hat\phi}. \qquad\qquad \square$$

Definition. Let A be an abelian group. A function

$$d : A \to \mathbb{R}$$

is a *quadratic form* if

(i) $d(\alpha) = d(-\alpha)$ for all $\alpha \in A$; and
(ii) the pairing

$$A \times A \to \mathbb{R}$$

$$(\alpha, \beta) \to d(\alpha + \beta) - d(\alpha) - d(\beta)$$

is bilinear.

A quadratic form d is *positive definite* if

(iii) $d(\alpha) \geqslant 0$ \qquad for all $\alpha \in A$; and
(iv) $d(\alpha) = 0$ \qquad if and only if $\alpha = 0$.

Corollary 6.3. *Let E_1 and E_2 be elliptic curves. The degree map*

$$\deg : \operatorname{Hom}(E_1, E_2) \to \mathbb{Z}$$

is a positive definite quadratic form.

PROOF. Everything is clear except for the fact that the pairing

$$\langle \phi, \psi \rangle = \deg(\phi + \psi) - \deg(\phi) - \deg(\psi)$$

is bilinear. But using the injection

$$[\ \ ] : \mathbb{Z} \to \text{End}(E_1),$$

we have

$$[\langle \phi, \psi \rangle] = [\deg(\phi + \psi)] - [\deg(\phi)] - [\deg(\psi)]$$
$$= \widehat{(\phi + \psi)} \circ (\phi + \psi) - \hat{\phi} \circ \phi - \hat{\psi} \circ \psi$$
$$= \hat{\phi} \circ \psi + \hat{\psi} \circ \phi \qquad \text{from (6.2c).}$$

But again using (6.2c), we see that this last expression is linear in both ϕ and ψ.   □

Corollary 6.4. *Let E be an elliptic curve and $m \in \mathbb{Z}$, $m \neq 0$.*
(a) $\deg[m] = m^2$.
(b) If $\text{char}(K) = 0$ *or if m is prime to* $\text{char}(K)$, *then*

$$E[m] \cong (\mathbb{Z}/m\mathbb{Z}) \times (\mathbb{Z}/m\mathbb{Z}).$$

(c) If $\text{char}(K) = p$, *then either*

$$E[p^e] \cong \{O\} \qquad \text{for all } e = 1, 2, 3, \ldots; \text{ or}$$
$$E[p^e] \cong \mathbb{Z}/p^e\mathbb{Z} \qquad \text{for all } e = 1, 2, 3, \ldots.$$

(Recall that $E[m]$ is another notation for $\text{ker}[m]$, the set of points of E having order m.)

PROOF. (a) This was proven above (6.2d). We record it again here in order to point out that there are many other ways of proving this fact (e.g., exers. 3.7, 3.8, 3.11), and that the fundamental description of $E[m]$ given in (b) follows formally from (a).
(b) From the given conditions and the fact that $\deg[m] = m^2$, it follows that $[m]$ is a finite, separable map. Hence from (4.10c),

$$\# E[m] = \deg[m] = m^2.$$

Further, for every integer d dividing m, we similarly have

$$\# E[d] = d^2.$$

Writing the finite group $E[m]$ as a product of cyclic groups, one immediately sees that the only possibility is

$$E[m] = (\mathbb{Z}/m\mathbb{Z}) \times (\mathbb{Z}/m\mathbb{Z}).$$

(c) Let ϕ be the pth-power Frobenius morphism. Then

$$\# E[p^e] = \deg_s[p^e] \qquad (4.10a)$$
$$= (\deg_s(\hat{\phi} \circ \phi))^e \qquad (6.2a)$$
$$= (\deg_s \hat{\phi})^e \qquad (\text{II}.2.11b).$$

From (6.2e) and (II.2.11c),

$$\deg \hat{\phi} = \deg \phi = p,$$

so there are two cases. If $\hat{\phi}$ is inseparable, then $\deg_s \hat{\phi} = 1$, so

$$\# E[p^e] = 1 \qquad \text{for all } e.$$

Otherwise $\hat{\phi}$ is separable, so $\deg_s \hat{\phi} = p$ and

$$\# E[p^e] = p^e \qquad \text{for all } e.$$

This last is easily seen to imply that

$$E[p^e] = \mathbb{Z}/p^e\mathbb{Z}.$$

(For a more complete analysis of $E[p^e]$ in characteristic p, and its relation-ship to $\text{End}(E)$, see chapter V, §3.) $\square$

§7. The Tate Module

Let E/K be an elliptic curve and $m \geq 2$ an integer (prime to $\text{char}(K)$ if $\text{char}(K) > 0$.) As we have just seen (6.4b),

$$E[m] \cong (\mathbb{Z}/m\mathbb{Z}) \times (\mathbb{Z}/m\mathbb{Z}),$$

the isomorphism being one between abstract groups. However, the group $E[m]$ comes equipped with considerably more structure. Namely, each element of the Galois group $G_{\bar{K}/K}$ acts on $E[m]$, since if $[m]P = O$, then $[m](P^\sigma) = ([m]P)^\sigma = O$. We thus obtain a representation

$$G_{\bar{K}/K} \to \text{Aut}(E[m]) \cong GL_2(\mathbb{Z}/m\mathbb{Z}),$$

where the latter isomorphism involves choosing a basis for $E[m]$. Individually, for each m, these representations are not completely satisfactory, because it is generally easiest to deal with representations whose matrices have coefficients in a ring having characteristic 0. What we will do is to fit them together for varying m so as to achieve this end, the motivating example being the inverse limit construction of the ℓ-adic integers $\mathbb{Z}_\ell$ from the finite groups $\mathbb{Z}/\ell^n\mathbb{Z}$.

Definition. Let E be an elliptic curve and $\ell \in \mathbb{Z}$ a prime. The (ℓ-adic) *Tate module of E* is the group

$$T_\ell(E) = \varprojlim_n E[\ell^n],$$

the inverse limit being taken with respect to the natural maps

$$E[\ell^{n+1}] \overset{[\ell]}{\to} E[\ell^n].$$

Since each $E[\ell^n]$ is a $\mathbb{Z}/\ell^n\mathbb{Z}$-module, we see that the Tate module has a natural structure as a $\mathbb{Z}_\ell$-module. Note that since the multiplication-by-ℓ maps are surjective, the inverse limit topology on $T_\ell(E)$ is equivalent to the ℓ-adic topology it gains by being a $\mathbb{Z}_\ell$-module.

Proposition 7.1. *As a $\mathbb{Z}_\ell$-module, the Tate module has the following structure.*
(a) $T_\ell(E) \cong \mathbb{Z}_\ell \times \mathbb{Z}_\ell$ *if $\ell \neq \mathrm{char}(K)$.*
(b) $T_p(E) \cong \{0\}$ *or* $\mathbb{Z}_p$ *if $p = \mathrm{char}(K) > 0$.*

PROOF. This follows immediately from (6.4b, c). $\square$

Now the action of $G_{\bar{K}/K}$ on each $E[\ell^n]$ commutes with the multiplication-by-ℓ maps used to form the inverse limit, so $G_{\bar{K}/K}$ also acts on $T_\ell(E)$. Further, since the pro-finite group $G_{\bar{K}/K}$ acts continuously on each finite (discrete) group $E[\ell^n]$, the resulting action on $T_\ell(E)$ is also continuous.

Definition. *The ℓ-adic representation (of $G_{\bar{K}/K}$ on E),* denoted ρ_ℓ, *is the map*

$$\rho_\ell : G_{\bar{K}/K} \to \mathrm{Aut}(T_\ell(E))$$

giving the action of $G_{\bar{K}/K}$ on $T_\ell(E)$ as described above.

Convention. From here on, the number ℓ will always refer to a prime number distinct from the characteristic of K.

Remark 7.2. Notice that by choosing a $\mathbb{Z}_\ell$-basis for $T_\ell(E)$ we obtain a representation

$$G_{\bar{K}/K} \to GL_2(\mathbb{Z}_\ell);$$

and then the natural inclusion $\mathbb{Z}_\ell \subset \mathbb{Q}_\ell$ gives

$$G_{\bar{K}/K} \to GL_2(\mathbb{Q}_\ell).$$

In this way we obtain a 2-dimensional representation of $G_{\bar{K}/K}$ over a field of characteristic 0.

Remark 7.3. The above construction is analogous to the following one, which may be more familiar. Let

$$\boldsymbol{\mu}_{\ell^n} \subset \bar{K}^*$$

be the group of $(\ell^n)^{\mathrm{th}}$-roots-of-unity. Then raising to the ℓth-power gives maps

$$\boldsymbol{\mu}_{\ell^{n+1}} \xrightarrow{\ell} \boldsymbol{\mu}_{\ell^n},$$

and we can take the inverse limit as above to form the *Tate module of K*

$$T_\ell(\boldsymbol{\mu}) = \varprojlim_n \boldsymbol{\mu}_{\ell^n}.$$

As an abstract group,

$$T_\ell(\mu) \cong \mathbb{Z}_\ell.$$

Further, $G_{\overline{K}/K}$ acts on each μ_{ℓ^n}, so we obtain a 1-dimensional representation

$$G_{\overline{K}/K} \to \mathrm{Aut}(T_\ell(\mu)) \cong \mathbb{Z}_\ell^*.$$

For $K = \mathbb{Q}$, this cyclotomic representation is surjective, which is equivalent to the fact that the ℓ-power cyclotomic polynomials are all irreducible over $\mathbb{Q}$.

The Tate module is also a useful tool for studying isogenies. If

$$\phi : E_1 \to E_2$$

is an isogeny of elliptic curves, then ϕ gives maps

$$\phi : E_1[\ell^n] \to E_2[\ell^n],$$

and so it induces a ($\mathbb{Z}_\ell$-linear) map

$$\phi_\ell : T_\ell(E_1) \to T_\ell(E_2).$$

We thus obtain a homomorphism

$$\mathrm{Hom}(E_1, E_2) \to \mathrm{Hom}(T_\ell(E_1), T_\ell(E_2)).$$

(Notice if $E_1 = E_2 = E$, then the map

$$\mathrm{End}(E) \to \mathrm{End}(T_\ell(E))$$

is even a homomorphism of rings.) It is not hard to show that the above homomorphism is injective (see exer. 3.14), but to really analyze $\mathrm{Hom}(E_1, E_2)$ we will need the following stronger result.

Theorem 7.4. Let E_1 and E_2 be elliptic curves. Then the natural map

$$\mathrm{Hom}(E_1, E_2) \otimes \mathbb{Z}_\ell \to \mathrm{Hom}(T_\ell(E_1), T_\ell(E_2))$$

$$\phi \to \phi_\ell$$

is injective.

PROOF. We start by proving the following statement.

Let $M \subset \mathrm{Hom}(E_1, E_2)$ be a finitely generated subgroup, and let

(*) $M^{\mathrm{div}} = \{\phi \in \mathrm{Hom}(E_1, E_2) : [m] \circ \phi \in M \text{ for some integer } m \geqslant 1\}$.

Then M^{div} is also finitely generated.

To prove (*), we extend the degree mapping to the finite dimensional real vector space $M \otimes \mathbb{R}$, which we equip with the natural topology inherited from $\mathbb{R}$. Then the degree mapping is clearly continuous, so the set

$$U = \{\phi \in M \otimes \mathbb{R} : \deg \phi < 1\}$$

is an open neighborhood of 0. Further, since $\operatorname{Hom}(E_1, E_2)$ is a torsion-free $\mathbb{Z}$-module (4.2b), there is a natural inclusion

$$M^{\mathrm{div}} \subset M \otimes \mathbb{R};$$

and clearly

$$M^{\mathrm{div}} \cap U = \{0\},$$

since every non-zero isogeny has degree at least 1. Hence M^{div} is a discrete subgroup of the finite dimensional vector space $M \otimes \mathbb{R}$, so it is finitely generated.

We turn now to the proof of (7.4). Let $\phi \in \operatorname{Hom}(E_1, E_2) \otimes \mathbb{Z}_\ell$, and suppose that $\phi_\ell = 0$. Let

$$M \subset \operatorname{Hom}(E_1, E_2)$$

be a finitely generated subgroup so that $\phi \in M \otimes \mathbb{Z}_\ell$. Then with notation as above, M^{div} is finitely generated, so it is also free (since it is torsion-free (4.2b)). Let

$$\phi_1, \ldots, \phi_t \in \operatorname{Hom}(E_1, E_2)$$

be a basis for M^{div}, and write

$$\phi = \alpha_1 \phi_1 + \cdots + \alpha_t \phi_t \quad \text{with} \quad \alpha_i \in \mathbb{Z}_\ell.$$

Now choose $a_1, \ldots, a_t \in \mathbb{Z}$ so that

$$a_i \equiv \alpha_i \pmod{\ell^n}.$$

Then the fact that $\phi_\ell = 0$ implies that the isogeny

$$\psi = [a_1] \circ \phi_1 + \cdots + [a_t] \circ \phi_t \in \operatorname{Hom}(E_1, E_2)$$

annihilates $E_1[\ell^n]$. It follows from (4.11) that ψ factors through $[\ell^n]$, so there is an isogeny

$$\lambda \in \operatorname{Hom}(E_1, E_2) \quad \text{with} \quad \psi = [\ell^n] \circ \lambda.$$

Further, λ is in M^{div}, so there are integers $b_i \in \mathbb{Z}$ such that

$$\lambda = [b_1] \circ \phi_1 + \cdots + [b_t] \circ \phi_t.$$

Then, since the ϕ_i's form a $\mathbb{Z}$-basis of M^{div}, we have

$$a_i = \ell^n b_i,$$

hence

$$\alpha_i \equiv 0 \pmod{\ell^n}.$$

Since this holds for all n, it follows that all $\alpha_i = 0$, so $\phi = 0$. [N.B. The reason it is so important to use M^{div} is that it is essential that the $\mathbb{Z}$-basis used to express ϕ, ψ, and λ not depend on the choice of ℓ^n.] $\square$

Corollary 7.5. *Let E_1 and E_2 be elliptic curves. Then*

$$\operatorname{Hom}(E_1, E_2)$$

is a free $\mathbb{Z}$-module of rank at most 4.

PROOF. Since $\operatorname{Hom}(E_1, E_2)$ is torsion-free (4.2b), it follows that

$$\operatorname{rank}_{\mathbb{Z}} \operatorname{Hom}(E_1, E_2) = \operatorname{rank}_{\mathbb{Z}_\ell} \operatorname{Hom}(E_1, E_2) \otimes \mathbb{Z}_\ell,$$

in the sense that if one is finite, then they both are and they are equal. Next, from (7.4), we have the estimate

$$\operatorname{rank}_{\mathbb{Z}_\ell} \operatorname{Hom}(E_1, E_2) \otimes \mathbb{Z}_\ell \leqslant \operatorname{rank}_{\mathbb{Z}_\ell} \operatorname{Hom}(T_\ell(E_1), T_\ell(E_2)).$$

Finally, choosing $\mathbb{Z}_\ell$-bases for $T_\ell(E_1)$ and $T_\ell(E_2)$, we see from (7.1a) that

$$\operatorname{Hom}(T_\ell(E_1), T_\ell(E_2)) \cong M_2(\mathbb{Z}_\ell),$$

where $M_2(\mathbb{Z}_\ell)$ is the group of 2×2 matrices with $\mathbb{Z}_\ell$ coefficients. Since $M_2(\mathbb{Z}_\ell)$ has $\mathbb{Z}_\ell$-rank equal to 4, this gives the desired result. $\qquad\square$

Remark 7.6. By definition, an isogeny is defined over K if it commutes with the action of $G_{\bar{K}/K}$. Similarly, we can define

$$\operatorname{Hom}_K(T_\ell(E_1), T_\ell(E_2))$$

to be the group of $\mathbb{Z}_\ell$-linear maps from $T_\ell(E_1)$ to $T_\ell(E_2)$ which commute with the action of $G_{\bar{K}/K}$ as given by the ℓ-adic representation. Then we have a homomorphism

$$\operatorname{Hom}_K(E_1, E_2) \otimes \mathbb{Z}_\ell \to \operatorname{Hom}_K(T_\ell(E_1), T_\ell(E_2)),$$

which from (7.4) is injective. It turns out that in many cases this map is actually an isomorphism.

Theorem 7.7. *The natural map*

$$\operatorname{Hom}_K(E_1, E_2) \otimes \mathbb{Z}_\ell \to \operatorname{Hom}_K(T_\ell(E_1), T_\ell(E_2))$$

is an isomorphism if:

(a) ([Ta 7]) *K is a finite field;*
(b) ([Fa 1]) *K is a number field.*

The proofs, which make heavy use of abelian varieties of higher dimensions, are unfortunately beyond the scope of this book. Indeed, the methods used in proving (7.7b) include virtually all of the tools needed for Faltings' proof of the Mordell conjecture.

To understand what (7.7) says, one should think of the Tate module as a homology group, specifically as the first homology with $\mathbb{Z}_\ell$-coefficients. Then

(7.7) gives a characterization of when a map between homology groups comes from an actual geometric map.

Remark 7.8. Another natural question to ask is how large is the image $\rho_\ell(G_{\bar{K}/K})$ in $\text{Aut}(T_\ell(E))$. The following theorem of Serre provides an answer for number fields. We do not include the proof. (But see (IX.6.3) and exer. 9.7).

Theorem 7.9 (Serre). *Let K be a number field and E/K an elliptic curve without complex multiplication.*
(a) *$\rho_\ell(G_{\bar{K}/K})$ is of finite index in $\text{Aut}(T_\ell(E))$ for all primes ℓ.*
(b) *$\rho_\ell(G_{\bar{K}/K}) = \text{Aut}(T_\ell(E))$ for all but finitely many primes ℓ.*

PROOF. [Se 5] and [Se 6].   □

Remark 7.10. Let E/K be an elliptic curve. Then just as above, the elements of $\text{End}_K(E)$ commute with the elements of $G_{\bar{K}/K}$ in their action on $T_\ell(E)$. If

$$\text{End}_K(E) = \mathbb{Z},$$

this gives little information; but if E has complex multiplication, then this forces the action of $G_{\bar{K}/K}$ on $T_\ell(E)$ to be abelian (exer. 3.24). In particular, adjoining the coordinates of ℓ^n-torsion points to K leads to explicitly constructed abelian extensions, in much the same manner that abelian extensions of $\mathbb{Q}$ are obtained by adjoining roots of unity. (See appendix C §11 for a brief discussion.)

§8. The Weil Pairing

Let E/K be an elliptic curve. For this section we fix an integer $m \geq 2$, prime to $p = \text{char}(K)$ if $p > 0$. We will make frequent use of (3.5), which says that $\Sigma n_i(P_i)$ is the divisor of a function if and only if $\Sigma n_i = 0$ and $\Sigma[n_i]P_i = O$.

Let $T \in E[m]$. Then there is a function $f \in \bar{K}(E)$ such that

$$\text{div}(f) = m(T) - m(O).$$

Letting $T' \in E$ with $[m]T' = T$, there is similarly a function $g \in \bar{K}(E)$ satisfying

$$\text{div}(g) = [m]^*(T) - [m]^*(O) = \sum_{R \in E[m]} (T' + R) - (R).$$

(Note $\#E[m] = m^2$ (6.4b) and $[m^2]T' = O$.) One immediately verifies that the functions $f \circ [m]$ and g^m have the same divisor, so multiplying f by an element of $\bar{K}^*$, we may assume that

$$f \circ [m] = g^m.$$

Now suppose that $S \in E[m]$ is another m-torsion point ($S = T$ is allowed). Then for any point $X \in E$,

$$g(X + S)^m = f([m]X + [m]S) = f([m]X) = g(X)^m.$$

Hence we can define a pairing

$$e_m : E[m] \times E[m] \to \mu_m = m^{\text{th}} \text{ roots of unity}$$

by setting

$$e_m(S, T) = g(X + S)/g(X),$$

where $X \in E$ is any point such that $g(X + S)$ and $g(X)$ are both defined and non-zero. Note that although g is only defined up to multiplication by an element of $\bar{K}^*$, $e_m(S, T)$ does not depend on this choice. This pairing is called the *Weil e_m-pairing*. We begin by giving some of its basic properties.

Proposition 8.1. *The Weil e_m-pairing is:*

(a) *Bilinear:* $e_m(S_1 + S_2, T) = e_m(S_1, T)e_m(S_2, T)$

$$e_m(S, T_1 + T_2) = e_m(S, T_1)e_m(S, T_2);$$

(b) *Alternating:* $e_m(T, T) = 1,$ *so in particular* $e_m(S, T) = e_m(T, S)^{-1};$

(c) *Non-degenerate: If $e_m(S, T) = 1$ for all $S \in E[m]$, then $T = O$;*
(d) *Galois invariant: For all $\sigma \in G_{\bar{K}/K}$,*

$$e_m(S, T)^\sigma = e_m(S^\sigma, T^\sigma);$$

(e) *Compatible: If $S \in E[mm']$ and $T \in E[m]$, then*

$$e_{mm'}(S, T) = e_m([m']S, T).$$

PROOF. (a) Linearity in the first factor is easy.

$$e_m(S_1 + S_2, T) = \frac{g(X + S_1 + S_2)}{g(X + S_1)} \frac{g(X + S_1)}{g(X)} = e_m(S_2, T)e_m(S_1, T).$$

(Note how useful it is that in $e_m(S_2, T) = g(Y + S_2)/g(Y)$, we may choose any value for Y, such as $Y = X + S_1$.) For the second, let $f_1, f_2, f_3, g_1, g_2, g_3$ be functions as above for T_1, T_2, and $T_3 = T_1 + T_2$. Choose $h \in \bar{K}(E)$ with divisor

$$\text{div}(h) = (T_1 + T_2) - (T_1) - (T_2) + (O).$$

Then

$$\text{div}(f_3/f_1 f_2) = m \, \text{div}(h),$$

so

$$f_3 = cf_1 f_2 h^m \qquad \text{for some } c \in \bar{K}^*.$$

Compose with the multiplication-by-$[m]$ map, use $f_i \circ [m] = g_i^m$, and take m^{th}-roots to find

$$g_3 = c'g_1 g_2(h \circ [m]).$$

Now

$$e_m(S, T_1 + T_2) = \frac{g_3(X + S)}{g_3(X)} = \frac{g_1(X + S)g_2(X + S)h([m]X + [m]S)}{g_1(X)g_2(X)h([m]X)}$$

$$= e_m(S, T_1)e_m(S, T_2).$$

(b) From (a) we have

$$e_m(S + T, S + T) = e_m(S, S)e_m(S, T)e_m(T, S)e_m(T, T),$$

so it suffices to show that $e_m(T, T) = 1$ for all $T \in E[m]$. For any $P \in E$, recall that $\tau_P : E \to E$ denotes the translation-by-P map (4.7). Then

$$\operatorname{div}\left(\prod_{i=0}^{m-1} f \circ \tau_{[i]T}\right) = m \sum_{i=0}^{m-1} ([1 - i]T) - ([-i]T) = 0.$$

Hence $\prod_{i=0}^{m-1} f \circ \tau_{[i]T}$ is constant; and if we choose some $T' \in E$ with $[m]T' = T$, then $\prod_{i=0}^{m-1} g \circ \tau_{[i]T'}$ is also constant, because its mth-power is the above product of f's. Evaluating the product of g's at X and $X + T'$ yields

$$\prod_{i=0}^{m-1} g(X + [i]T') = \prod_{i=0}^{m-1} g(X + [i + 1]T').$$

Now cancelling like terms gives

$$g(X) = g(X + [m]T') = g(X + T),$$

so

$$e_m(T, T) = g(X + T)/g(X) = 1.$$

(c) If $e_m(S, T) = 1$ for all $S \in E[m]$, so $g(X + S) = g(X)$ for all $S \in E[m]$, then from (4.10b), $g = h \circ [m]$ for some function $h \in \bar{K}(E)$. But then

$$(h \circ [m])^m = g^m = f \circ [m],$$

so $f = h^m$. Hence

$$m \operatorname{div}(h) = \operatorname{div}(f) = m(T) - m(O),$$

so

$$\operatorname{div}(h) = (T) - (O).$$

Therefore $T = O$ (3.3).

(d) Let $\sigma \in G_{\bar{K}/K}$. If f, g are the functions for T as above, then clearly f^σ, g^σ are the corresponding functions for T^σ. Then

$$e_m(S^\sigma, T^\sigma) = \frac{g^\sigma(X^\sigma + S^\sigma)}{g^\sigma(X^\sigma)} = \left(\frac{g(X + S)}{g(X)}\right)^\sigma = e_m(S, T)^\sigma.$$

(e) Taking f, g as above, we have

$$\mathrm{div}(f^{m'}) = mm'(T) - mm'(O)$$

and

$$(g \circ [m'])^{mm'} = (f \circ [mm'])^{m'}.$$

Then from the definition of $e_{mm'}$ and e_m,

$$e_{mm'}(S, T) = \frac{g \circ [m'](X + S)}{g \circ [m'](X)} = \frac{g(Y + [m']S)}{g(Y)} = e_m([m']S, T). \qquad \square$$

The basic properties of the Weil pairing imply its surjectivity, as we now show.

Corollary 8.1.1. *There exist points S, $T \in E[m]$ such that $e_m(S, T)$ is a primitive m^{th}-root of unity. In particular, if $E[m] \subset E(K)$, then $\mu_m \subset K^*$.*

PROOF. The image of $e_m(S, T)$ as S and T range over $E[m]$ is a subgroup of μ_m, say equal to μ_d. It follows that for all S, $T \in E[m]$,

$$1 = e_m(S, T)^d = e_m([d]S, T).$$

The non-degeneracy of the e_m-pairing now implies that $[d]S = O$; and since S is arbitrary, (6.4) shows we must have $d = m$. Finally, if $E[m] \subset E(K)$, then from the Galois invariance of the e_m-pairing we see that $e_m(S, T) \in K^*$ for all S, $T \in E[m]$. Therefore $\mu_m \subset K^*$. $\qquad \square$

Recall that if E_1 and E_2 are elliptic curves and $\phi : E_1 \to E_2$ is an isogeny connecting them, then there is a dual isogeny $\hat{\phi} : E_2 \to E_1$ going in the other direction. The following proposition says that ϕ and $\hat{\phi}$ are dual (i.e. adjoint) with respect to the Weil pairing.

Proposition 8.2. *Let $S \in E_1[m]$, $T \in E_2[m]$, and $\phi : E_1 \to E_2$ an isogeny. Then*

$$e_m(S, \hat{\phi}(T)) = e_m(\phi(S), T).$$

PROOF. Let

$$\mathrm{div}(f) = m(T) - m(O) \quad \text{and} \quad f \circ [m] = g^m$$

be as above. Then

$$e_m(\phi S, T) = g(X + \phi S)/g(X).$$

Choose a function $h \in \bar{K}(E_1)$ so that

$$\phi^*((T)) - \phi^*((O)) = (\hat{\phi} T) - (O) + \mathrm{div}(h).$$

Such an h exists because, by (6.1b), $\hat{\phi} T$ is precisely the sum of the points of the divisor on the left-hand side of this equality. Now

$$\mathrm{div}\left(\frac{f \circ \phi}{h^m}\right) = \phi^* \,\mathrm{div}(f) - m\,\mathrm{div}(h)$$
$$= m(\hat\phi\, T) - m(O),$$

and

$$\left(\frac{g \circ \phi}{h \circ [m]}\right)^m = \frac{f \circ [m] \circ \phi}{(h \circ [m])^m} = \left(\frac{f \circ \phi}{h^m}\right) \circ [m].$$

Thus from the definition of the e_m-pairing,

$$e_m(S, \hat\phi\, T) = \frac{(g \circ \phi / h \circ [m])(X + S)}{(g \circ \phi / h \circ [m])(X)}$$

$$= \frac{g(\phi X + \phi S)}{g(\phi X)} \frac{h([m] X)}{h([m] X + [m] S)}$$

$$= e_m(\phi S, T). \qquad\qquad \square$$

Let ℓ be a prime number different from $\mathrm{char}(K)$. We would like to fit together the pairings

$$e_{\ell^n} : E[\ell^n] \times E[\ell^n] \to \boldsymbol{\mu}_{\ell^n}$$

for all $n = 1, 2, \ldots$ to give an ℓ-adic Weil pairing on the Tate module

$$e : T_\ell(E) \times T_\ell(E) \to T_\ell(\boldsymbol{\mu}).$$

Recall that the inverse limits for $T_\ell(E)$ and $T_\ell(\boldsymbol{\mu})$ are formed using the maps

$$E[\ell^{n+1}] \overset{[\ell]}{\to} E[\ell^n] \quad \text{and} \quad \boldsymbol{\mu}_{\ell^{n+1}} \overset{\ell}{\to} \boldsymbol{\mu}_{\ell^n}.$$

Thus to show that the e_{ℓ^n}-pairings are compatible with taking the inverse limit, we must show that for any $S, T \in E[\ell^{n+1}]$,

$$e_{\ell^{n+1}}(S, T)^\ell = e_{\ell^n}([\ell]S, [\ell]T).$$

But by linearity (8.1a),

$$e_{\ell^{n+1}}(S, T)^\ell = e_{\ell^{n+1}}(S, [\ell]T);$$

and then the desired result follows by applying (8.1e) to $(S, [\ell]T)$ with $m = \ell^n$ and $m' = \ell$. This proves that e is well-defined, and it inherits all of the properties from (8.1) and (8.2), which completes the proof of the following.

Proposition 8.3. *There exists a bilinear, alternating, non-degenerate, Galois invariant pairing*

$$e : T_\ell(E) \times T_\ell(E) \to T_\ell(\boldsymbol{\mu}).$$

Further, if $\phi : E_1 \to E_2$ is an isogeny, then ϕ and its dual isogeny $\hat\phi$ are adjoints for the pairing.

§9. The Endomorphism Ring

Let E/K be an elliptic curve. We are interested in characterizing which rings may occur as the endomorphism ring of E. So far, the following information has been collected:

(i) End(E) is a characteristic 0 integral domain of rank at most 4 over $\mathbb{Z}$ ((4.2c), (7.5));

(ii) End(E) possesses an anti-involution $\phi \to \hat{\phi}$ (6.2b, c, f);

(iii) For $\phi \in$ End(E), we have $\phi\hat{\phi} \in \mathbb{Z}$, $\phi\hat{\phi} \geqslant 0$, and $\phi\hat{\phi} = 0$ if and only if $\phi = 0$ ((6.2a), (6.3)).

It turns out that any ring satisfying (i)–(iii) is of a very special sort. After giving the relevant definitions, we will give the general classification of rings satisfying (i)–(iii), which may then be applied to the particular case of End(E).

Definition. Let $\mathscr{K}$ be a (not necessarily commutative) algebra, finitely generated over $\mathbb{Q}$. An *order* $\mathscr{R}$ of $\mathscr{K}$ is a subring of $\mathscr{K}$ which is finitely generated as $\mathbb{Z}$-module and which satisfies $\mathscr{R} \otimes \mathbb{Q} = \mathscr{K}$.

Example 9.1. Let $\mathscr{K}$ be a quadratic imaginary field and $\mathcal{O}$ its ring of integers. Then for each integer $f > 0$, the ring $\mathbb{Z} + f\mathcal{O}$ is an order of $\mathscr{K}$. (These are all the orders of $\mathscr{K}$. See exer. 3.20.)

Definition. A *quaternion algebra* is an algebra of the form

$$\mathscr{K} = \mathbb{Q} + \mathbb{Q}\alpha + \mathbb{Q}\beta + \mathbb{Q}\alpha\beta$$

with the multiplication rules

$$\alpha^2, \beta^2 \in \mathbb{Q}, \qquad \alpha^2 < 0, \qquad \beta^2 < 0, \qquad \beta\alpha = -\alpha\beta.$$

Remark 9.2. The quaternion algebras defined above are more properly called *definite quaternion algebras over* $\mathbb{Q}$. But since these are the only quaternion algebras that we will deal with in this book, we will generally drop the appellation "definite".

Theorem 9.3. *Let $\mathscr{R}$ be a ring of characteristic* 0 *with no zero divisors and having the following properties.*

(i) *$\mathscr{R}$ has rank at most 4 (as a $\mathbb{Z}$-module).*

(ii) *$\mathscr{R}$ possesses an anti-involution $\alpha \to \hat{\alpha}$. (I.e. $\widehat{\alpha + \beta} = \hat{\alpha} + \hat{\beta}, \widehat{\alpha\beta} = \hat{\beta}\hat{\alpha}, \hat{\hat{\alpha}} = \alpha$, and for $\alpha \in \mathbb{Z}, \hat{\alpha} = \alpha$.)*

(iii) *For $\alpha \in \mathscr{R}$, $\alpha\hat{\alpha}$ is a non-negative integer; and $\alpha\hat{\alpha} = 0$ if and only if $\alpha = 0$.*

Then $\mathscr{R}$ is one of the following three sorts of rings.

(a) *$\mathscr{R} \cong \mathbb{Z}$.*

(b) *$\mathscr{R}$ is an order in a quadratic imaginary extension of $\mathbb{Q}$.*

(c) *$\mathscr{R}$ is an order in a quaternion algebra over $\mathbb{Q}$.*

PROOF. Let $\mathcal{K} = \mathcal{R} \otimes \mathbb{Q}$. Since $\mathcal{R}$ is finitely generated (as a $\mathbb{Z}$-module), it suffices to show that either $\mathcal{K} = \mathbb{Q}$, $\mathcal{K}/\mathbb{Q}$ is a quadratic imaginary extension, or $\mathcal{K}/\mathbb{Q}$ is a quaternion algebra. We extend the anti-involution to $\mathcal{K}$, and define a (reduced) *norm* and *trace* from $\mathcal{K}$ to $\mathbb{Q}$ by

$$N\alpha = \alpha\hat{\alpha} \quad \text{and} \quad T\alpha = \alpha + \hat{\alpha}.$$

We make several observations about the trace. First, since

$$T\alpha = 1 + N\alpha - N(\alpha - 1),$$

$T\alpha$ is in $\mathbb{Q}$. Second, the trace is clearly $\mathbb{Q}$-linear. Third, if $\alpha \in \mathbb{Q}$, then $T\alpha = 2\alpha$. Finally, if $\alpha \in \mathcal{K}$ satisfies $T\alpha = 0$, then

$$0 = (\alpha - \alpha)(\alpha - \hat{\alpha}) = \alpha^2 - (T\alpha)\alpha + N\alpha = \alpha^2 + N\alpha,$$

so $\alpha^2 = -N\alpha$. Thus for elements with $T\alpha = 0$, either $\alpha = 0$, or else $\alpha^2 \in \mathbb{Q}$ and $\alpha^2 < 0$.

Now if $\mathcal{K} = \mathbb{Q}$, we are done. Otherwise we can choose some $\alpha \in \mathcal{K}$, $\alpha \notin \mathbb{Q}$. Replacing α by $\alpha - \frac{1}{2}T\alpha$, we may assume $T\alpha = 0$. Then from above $\alpha^2 < 0$, so $\mathbb{Q}(\alpha)$ is a quadratic imaginary field. If $\mathcal{K} = \mathbb{Q}(\alpha)$, we are again done.

Assume now $\mathcal{K} \neq \mathbb{Q}(\alpha)$, and choose $\beta \in \mathcal{K}$, $\beta \notin \mathbb{Q}(\alpha)$. As above, we may replace β by

$$\beta - \tfrac{1}{2}T\beta - \tfrac{1}{2}(T(\alpha\beta)/\alpha^2)\alpha.$$

Recalling that $T\alpha = 0$ and $\alpha^2 \in \mathbb{Q}^*$, one immediately verifies that $T\beta = T(\alpha\beta) = 0$. In particular, $\beta^2 < 0$. Further, writing

$$T\alpha = T\beta = T(\alpha\beta) = 0 \quad \text{as} \quad \alpha = -\hat{\alpha}, \; \beta = -\hat{\beta}, \; \alpha\beta = -\hat{\beta}\hat{\alpha},$$

we see by substituting the first two equalities into the third that

$$\alpha\beta = -\beta\alpha.$$

Hence

$$\mathbb{Q}[\alpha, \beta] = \mathbb{Q} + \mathbb{Q}\alpha + \mathbb{Q}\beta + \mathbb{Q}\alpha\beta$$

is a quaternion algebra. It remains to prove that $\mathbb{Q}[\alpha, \beta] = \mathcal{K}$. To do this, it suffices to show that $1, \alpha, \beta, \alpha\beta$ are $\mathbb{Q}$-linearly independent, since then $\mathbb{Q}[\alpha, \beta]$ and $\mathcal{K}$ will both have dimension 4 over $\mathbb{Q}$.

Suppose

$$w + x\alpha + y\beta + z\alpha\beta = 0$$

with $w, x, y, z \in \mathbb{Q}$ not all zero. Taking traces yields

$$2w = 0, \quad \text{so} \quad w = 0.$$

Then multiplying by α on the left and β on the right gives

$$(x\alpha^2)\beta + (y\beta^2)\alpha + z\alpha^2\beta^2 = 0,$$

contradicting the $\mathbb{Q}$-linear independence of $1, \alpha, \beta$. (Remember $\alpha^2, \beta^2 \in \mathbb{Q}^*$.) This completes the proof that $\mathcal{K} = \mathbb{Q}[\alpha, \beta]$. $\qquad\square$

Corollary 9.4. *The endomorphism ring of an elliptic curve is either* $\mathbb{Z}$, *an order in a quadratic imaginary field, or an order in a quaternion algebra.*

PROOF. As indicated above, we have proven all of the facts ((4.2b), (6.2), (6.3), (6.5)) needed to apply (9.3) to End(E). $\qquad\square$

It turns out that if char(K) = 0, then End(E) $\otimes$ $\mathbb{Q}$ cannot be a quaternion algebra. We will give an analytic proof of this later (VI.6.1b). (See also exer. 3.18b.) On the other hand, if K is a finite field, then End(E) is always larger than $\mathbb{Z}$ (V.3.1), and there are always elliptic curves (defined over $\bar{K}$) with End(E) non-commutative (V.4.1c). The complete description of End(E) can be found in Deuring's comprehensive article [De 1].

The following definition and result will be used in the exercises.

Definition. Let p be a prime (or ∞), and let $\mathbb{Q}_p$ be the completion of $\mathbb{Q}$ at p ($\mathbb{Q}_\infty = \mathbb{R}$). A quaternion algebra $\mathcal{K}$ is said to *split at p* if

$$\mathcal{K} \otimes_\mathbb{Q} \mathbb{Q}_p \cong M_2(\mathbb{Q}_p).$$

(Here M_2 is the algebra of 2×2 matrices.) Otherwise $\mathcal{K}$ is *ramified at p*. Define the *invariant of $\mathcal{K}$ at p* by

$$\text{inv}_p \mathcal{K} = \begin{cases} 0 & \text{if } \mathcal{K} \text{ splits at } p \\ \frac{1}{2} & \text{if } \mathcal{K} \text{ ramifies at } p. \end{cases}$$

Theorem 9.5. (a) *Let $\mathcal{K}$ be a quaternion algebra. Then* $\text{inv}_p(\mathcal{K}) = 0$ *for all but finitely many p, and*

$$\sum_p \text{inv}_p \mathcal{K} \in \mathbb{Z}.$$

(*Note that the sum includes* $p = \infty$.)
(b) *Two quaternion algebras $\mathcal{K}$ and $\mathcal{K}'$ are isomorphic (as $\mathbb{Q}$-algebras) if and only if* $\text{inv}_p(\mathcal{K}) = \text{inv}_p(\mathcal{K}')$ *for all p.*

PROOF. This is a very special case of the fact that the central simple algebras over a field K are classified by the Brauer group Br(K) = $H^2(G_{\bar{K}/K}, \bar{K}^*)$ ([Se 9, X §5]), and the fundamental exact sequence from class field theory ([Ta 3, §9.6])

$$0 \to \text{Br}(\mathbb{Q}) \to \bigoplus_p \text{Br}(\mathbb{Q}_p) \xrightarrow{\Sigma_p \text{inv}_p} \mathbb{Q}/\mathbb{Z} \to 0,$$

where

$$\text{Br}(\mathbb{Q}_p) \underset{\text{inv}_p}{\xrightarrow{\sim}} \begin{cases} \mathbb{Q}/\mathbb{Z} & p \neq \infty \\ \{0, \frac{1}{2}\} & p = \infty. \end{cases}$$

Quaternion algebras (definite and indefinite) correspond to elements of exact order 2 in Br($\mathbb{Q}$). $\qquad\square$

§10. The Automorphism Group

If an elliptic curve is given by a Weierstrass equation, it is in general a non-trivial matter to determine the exact structure of its endomorphism ring. For the automorphism group, however, the situation is much simpler.

Theorem 10.1. *Let E/K be an elliptic curve. Then its automorphism group* $\text{Aut}(E)$ *is a finite group of order dividing* 24. *More precisely, the order of* $\text{Aut}(E)$ *is given by the following list:*

   2 *if* $j(E) \neq 0, 1728$
   4 *if* $j(E) = 1728$ *and* $\text{char}(K) \neq 2, 3$
   6 *if* $j(E) = 0$ *and* $\text{char}(K) \neq 2, 3$
   12 *if* $j(E) = 0 = 1728$ *and* $\text{char}(K) = 3$
   24 *if* $j(E) = 0 = 1728$ *and* $\text{char}(K) = 2.$

PROOF. We restrict attention to the case $\text{char}(K) \neq 2, 3$ (see (1.3) and (A.1.2c)). Then E is given by an equation

$$E : y^2 = x^3 + Ax + B,$$

and every automorphism has the form

$$x = u^2 x' \qquad y = u^3 y'$$

for some $u \in \bar{K}^*$. Such a substitution will give an automorphism of E if and only if

$$u^{-4}A = A \quad \text{and} \quad u^{-6}B = B.$$

Hence if $AB \neq 0$ (so $j(E) \neq 0, 1728$), then the only possibilities are $u = \pm 1$; while if $B = 0$ ($j(E) = 1728$) or $A = 0$ ($j(E) = 0$), then u satisfies respectively $u^4 = 1$ or $u^6 = 1$, so $\text{Aut}(E)$ will be cyclic of order 4 or 6. $\square$

It is worth remarking that the proof of (10.1) actually gives the structure of $\text{Aut}(E)$ as a $G_{\bar{K}/K}$-module (at least for characteristic $\neq 2, 3$). We record this in the following corollary.

Corollary 10.2. *Let E/K be an elliptic curve over a field of characteristic $\neq 2$, 3, and let $n = 2$ (resp. 4, resp. 6) if $j(E) \neq 0$, 1728 (resp. $j(E) = 1728$, resp. $j(E) = 0$). Then as $G_{\bar{K}/K}$-modules,*

$$\text{Aut}(E) \cong \mu_n.$$

PROOF. In proving (10.1), we showed that the map

$$[\ \ ] : \mu_n \to \text{Aut}(E) \qquad [\zeta](x, y) = (\zeta^2 x, \zeta^3 y)$$

is an isomorphism of abstract groups. But this map clearly commutes with the action of $G_{\bar{K}/K}$, and so it is an isomorphism of $G_{\bar{K}/K}$-modules. $\square$

EXERCISES

3.1.   Show that the polynomials

$$x^4 - b_4 x^2 - 2b_6 x - b_8 \quad \text{and} \quad 4x^3 + b_2 x^2 + 2b_4 x + b_6$$

appearing in the duplication formula (2.3d) are relatively prime if and only if the discriminant Δ of the corresponding Weierstrass equation is non-zero.

3.2.   (a) Find a *triplication formula,* analogous to the duplication formula given in (2.3). (I.e. Express $x([3]P)$ as a rational function of $x(P)$ and $a_1, \ldots, a_6$.)
       (b) Use the result from (a) to show that if $\text{char}(K) \neq 3$, then E has a non-trivial point of order 3. Conclude that if $\gcd(m, 3) = 1$, then $[m] \neq [0]$. (*Warning:* This exercise probably requires a computer with a symbolic processor.)

3.3.   Assume $\text{char}(K) \neq 3$ and $A \in K^*$. Then the curve

$$E : X^3 + Y^3 = AZ^3$$

has genus 1 (exer. 2.7), so together with the point $O = [1, -1, 0]$ it becomes an elliptic curve.
       (a) Show that three points of E add to O if and only if they are collinear.
       (b) If $P = [X, Y, Z] \in E$, show that

$$-P = [Y, X, Z]$$

       and

$$[2]P = [-Y(X^3 + AZ^3), X(Y^3 + AZ^3), X^3 Z - Y^3 Z].$$

       (c) Develop an analogous formula for the sum of two distinct points.
       (d) Prove that E has j-invariant 0.

3.4.   Referring to example (2.4), express each of the points $P_2, P_4, P_5, P_6, P_7, P_8$ in the form $[m]P_1 + [n]P_3$ with $m, n \in \mathbb{Z}$.

3.5.   Let E/K be given by a singular Weierstrass equation.
       (a) Suppose that E has a node, and let the tangent lines at the node be $y = \alpha_i x + \beta_i$, $i = 1, 2$.
       (i) If $\alpha_1 \in K$, prove that $\alpha_2 \in K$ and

$$E_{ns}(K) \cong K^*.$$

       (ii) If $\alpha_1 \notin K$, prove that $L = K(\alpha_1, \alpha_2)$ is a quadratic extension of K. From (i), $E_{ns}(K) \subset E_{ns}(L) \cong L^*$. Show that

$$E_{ns}(K) \cong \{t \in L^* : N_{L/K}(t) = 1\}.$$

       (b) Suppose that E has a cusp. Prove that

$$E_{ns}(K) \cong K^+.$$

3.6.   Let C be a smooth curve of genus g, $P_0 \in C$, and $n \geq 2g + 1$ an integer. Let $\{f_0, f_1, \ldots, f_m\}$ be a basis for $\mathcal{L}(n(P_0))$ and

$$\phi = [f_0, \ldots, f_m] : C \to \mathbb{P}^m$$

the map determined by the f_i's.

(a)   Prove that the image $C' = \phi(C)$ is a curve in $\mathbb{P}^m$.
(b)   Prove that the map $\phi : C \to C'$ has degree 1.
(c)* Prove that C' is smooth, and so that $\phi : C \to C'$ is an isomorphism.

3.7. This exercise gives an elementary (highly computational) proof that the multiplication-by-m map has degree m^2. We will assume $\operatorname{char}(K) \neq 2, 3$, and take an elliptic curve

$$E : y^2 = x^3 + Ax + B.$$

Define *division polynomials* $\psi_m \in \mathbb{Z}[A, B, x, y]$ inductively as follows:

$$\psi_1 = 1, \qquad \psi_2 = 2y,$$

$$\psi_3 = 3x^4 + 6Ax^2 + 12Bx - A^2,$$

$$\psi_4 = 4y(x^6 + 5Ax^4 + 20Bx^3 - 5A^2x^2 - 4ABx - 8B^2 - A^3),$$

$$\psi_{2m+1} = \psi_{m+2}\psi_m^3 - \psi_{m-1}\psi_{m+1}^3 \qquad (m \geqslant 2),$$

$$2y\psi_{2m} = \psi_m(\psi_{m+2}\psi_{m-1}^2 - \psi_{m-2}\psi_{m+1}^2) \qquad (m \geqslant 3).$$

(One easily checks that the ψ_{2m}'s are polynomials.) Further define polynomials ϕ_m and ω_m by

$$\phi_m = x\psi_m^2 - \psi_{m+1}\psi_{m-1}$$

$$4y\omega_m = \psi_{m+2}\psi_{m-1}^2 - \psi_{m-2}\psi_{m+1}^2.$$

(a) Prove that ψ_m, ϕ_m, $y^{-1}\omega_m$ (for m odd) and $(2y)^{-1}\psi_m$, ϕ_m, ω_m (for m even) are polynomials in $\mathbb{Z}[A, B, x, y^2]$. Hence replacing y^2 by $x^3 + Ax + B$, we will treat them as polynomials in $\mathbb{Z}[A, B, x]$.
(b) As polynomials in x, show that

$$\phi_m(x) = x^{m^2} + \text{lower order terms},$$

$$\psi_m(x)^2 = m^2 x^{m^2-1} + \text{lower order terms}.$$

(c) If $\Delta = -16(4A^3 + 27B^2) \neq 0$, then $\phi_m(x)$ and $\psi_m(x)^2$ are relatively prime polynomials (in $K[x]$.)
(d) Again assume $\Delta \neq 0$, so E is an elliptic curve. Let $P = (x_0, y_0) \in E$. Then

$$[m]P = \left(\frac{\phi_m(P)}{\psi_m(P)^2}, \frac{\omega_m(P)}{\psi_m(P)^3} \right).$$

(e) The map $[m] : E \to E$ has degree m^2.

3.8. (a) Let $E/\mathbb{C}$ be an elliptic curve. We will later show (VI.5.1.1) that there is a lattice $L \subset \mathbb{C}$ and a complex analytic isomorphism of groups $\mathbb{C}/L \cong E(\mathbb{C})$. (N.B. This isomorphism is given by convergent power series, not by rational functions.) Assuming this, prove that

$$\deg[m] = m^2 \quad \text{and} \quad E[m] \cong \mathbb{Z}/m\mathbb{Z} \times \mathbb{Z}/m\mathbb{Z}.$$

(b) Let E/K be an elliptic curve with $\operatorname{char}(K) = 0$. Using (a), prove that $\deg[m] = m^2$. [*Hint*: If K can be embedded in $\mathbb{C}$, there is no problem. Reduce to this case.]

3.9. Let E/K be an elliptic curve given by a homogeneous Weierstrass equation $F(X_0, X_1, X_2) = 0$. (I.e. $x = X_0/X_2$ and $y = X_1/X_2$ are Weierstrass coordinate functions.) Let $P \in E$. Assume that $\mathrm{char}(K) \neq 2, 3$.

(a) Show that $[3]P = O$ if and only if the tangent line to E at P intersects E only at P.

(b) Show that $[3]P = O$ if and only if the Hessian matrix

$$((\partial^2 F/\partial X_i \partial X_j)(P))_{0 \leq i,j \leq 2}$$

has determinant 0.

(c) Show that $E[3]$ consists of 9 points.

3.10. Let E/K be an elliptic curve with Weierstrass coordinate functions x, y.

(a) Show that the map

$$\phi : E \to \mathbb{P}^3$$

$$\phi = [1, x, y, x^2]$$

maps E isomorphically onto the intersection of two quadric surfaces in $\mathbb{P}^3$. In particular, if $H \subset \mathbb{P}^3$ is a hyperplane, then $H \cap \phi(E)$ consists of 4 points (counted with appropriate multiplicity.)

(b) Show that $\phi(O) = [0, 0, 0, 1]$, and the hyperplane $\{T_0 = 0\}$ intersects $\phi(E)$ at the single point $\phi(O)$ with multiplicity 4.

(c) Let $P, Q, R \in E$. Prove $P + Q + R = O$ if and only if $\phi(P), \phi(Q), \phi(R), \phi(O)$ are coplanar.

(d) Let $P \in E$. Prove that $[4]P = O$ if and only if there exists a hyperplane $H \subset \mathbb{P}^3$ such that $H \cap \phi(E) = \{P\}$. Show that if $\mathrm{char}\, K \neq 2$, then there are exactly 16 such points.

(e) Assume $\mathrm{char}(K) \neq 2$. Show that after a linear change of variables (over $\bar{K}$), E has a model of the form

$$T_0^2 + T_2^2 = T_0 T_3$$

$$T_1^2 + \alpha T_2^2 = T_2 T_3.$$

For what value(s) of α is this model non-singular?

(f) Using the model in (e) and the addition law described by (c), derive formulas for $-P$, $P_1 + P_2$, and $[2]P$ analogous to those given in (2.3).

3.11. Generalize exercise 3.10 as follows. Let E/K be an elliptic curve, and choose a basis $f_1, \ldots, f_m$ for $\mathcal{L}(m(O))$. Then for $m \geq 3$, the map

$$\phi : E \to \mathbb{P}^{m-1}$$

$$\phi = [f_1, \ldots, f_m]$$

maps E isomorphically onto its image (exer. 3.6).

(a) Show that $\phi(E)$ is a curve of degree m. (I.e. The intersection of $\phi(E)$ and a hyperplane, counted with multiplicities, consists of m points.) [Hint: Find a hyperplane which intersects $\phi(E)$ at the single point $\phi(O)$, and show that it intersects with multiplicity m.]

(b) Let $P_1, \ldots, P_{m-1} \in E$. Prove that $P_1 + \cdots + P_{m-1} = O$ if and only if $\phi(P_1), \ldots, \phi(P_{m-1}), \phi(O)$ lie in a hyperplane. (Note that if some of the P_i's coincide, then we require the hyperplane to intersect $\phi(E)$ with correspondingly higher multiplicity.)

(c)* Let $P \in E$. Show that $[m]P = O$ if and only if there is a hyperplane $H \subset \mathbb{P}^{m-1}$ such that $H \cap \phi(E) = \{P\}$. If $\operatorname{char}(K) = 0$ or $\operatorname{char}(K) > m$, prove that there are exactly m^2 such points. Deduce that $\deg[m] = m^2$.

3.12. Let $m \geqslant 2$ be an integer, prime to $\operatorname{char}(K)$ if $\operatorname{char}(K) > 0$. Prove that the natural map

$$\operatorname{Aut}(E) \to \operatorname{Aut}(E[m])$$

is injective except for $m = 2$, when the kernel is ± 1. (Do not use (10.1).)

3.13. Generalize (4.12) as follows. Let $C/\bar{K}$ be a smooth curve, and let $\operatorname{Isom}(C)$ denote the group of isomorphisms from C to itself. (E.g. If C is an elliptic curve, then $\operatorname{Isom}(C)$ contains translation maps and $[\pm 1]$.) Let Φ be a finite subgroup of $\operatorname{Isom}(C)$.
   (a) Prove that there exists a unique smooth curve $C'/\bar{K}$ and a finite separable morphism $\phi : C \to C'$ such that $\phi^* \bar{K}(C') = \bar{K}(C)^\Phi$. (Here $\bar{K}(C)^\Phi$ denotes the subfield of $\bar{K}(C)$ fixed by Φ, where an element $\alpha \in \Phi$ acts on $\bar{K}(C)$ by $\alpha^* : \bar{K}(C) \to \bar{K}(C)$.)
   (b) Let $P \in C$. Prove that

$$e_\phi(P) = \#\{\alpha \in \Phi : \alpha P = P\}.$$

   (c) Prove that ϕ is unramified if and only if every non-trivial element of Φ has no fixed points.
   (d) Express the genus of C' in terms of the genus of C, $\#\Phi$, and the fixed points of the elements of Φ.
   (e)* Suppose that C is defined over K, and that Φ is $G_{\bar{K}/K}$-invariant. (I.e. If $\alpha \in \Phi$, then $\alpha^\sigma \in \Phi$ for all $\sigma \in G_{\bar{K}/K}$.) Prove that it is possible to find a C' so that C' and ϕ are defined over K. Further, show that C' is then unique up to isomorphism over K.

3.14. Use the non-degeneracy of the Weil pairing to give a quick proof that the map

$$\operatorname{Hom}(E_1, E_2) \to \operatorname{Hom}(T_\ell(E_1), T_\ell(E_2))$$

is injective. (Note this is not as strong as (7.4).)

3.15. Let $\phi : E_1 \to E_2$ be an isogeny of degree m, with m prime to $\operatorname{char}(K)$ if $\operatorname{char}(K) > 0$.
   (a) Mimic the construction in section 8 to construct a pairing

$$e_\phi : \ker \phi \times \ker \hat{\phi} \to \mu_m.$$

   (b) Prove that e_ϕ is bilinear, non-degenerate, and Galois invariant.
   (c) Prove that e_ϕ is compatible, in the sense that if $\psi : E_2 \to E_3$ is another isogeny, $P \in \ker(\psi \circ \phi)$, and $Q \in \ker(\hat{\phi})$, then

$$e_{\psi \circ \phi}(P, Q) = e_\psi(\phi P, Q).$$

3.16. *Alternative Definition of the Weil Pairing.* Let E be an elliptic curve. We define a pairing

$$\tilde{e}_m : E[m] \times E[m] \to \mu_m$$

as follows: Let $P, Q \in E[m]$, and choose divisors D_P, D_Q in $\operatorname{Div}^0(E)$ which add to P and Q respectively. (I.e. $\sigma(D_P) = P$ and $\sigma(D_Q) = Q$, where σ is as in (3.4a).) We further assume that D_P and D_Q are chosen with disjoint supports. Since P and Q have order m, there are functions $f_P, f_Q \in \bar{K}(E)$ such that

$$\text{div}(f_P) = mD_P \quad \text{and} \quad \text{div}(f_Q) = mD_Q.$$

Then we define

$$\tilde{e}_m(P, Q) = f_P(D_Q)/f_Q(D_P).$$

(See exer. 2.10 for the definition of the value of a function at a divisor.)

(a)   Prove that $\tilde{e}_m(P, Q)$ is well-defined.

(b)   Prove that $\tilde{e}_m(P, Q) \in \boldsymbol{\mu}_m$.

(c)* Prove that $\tilde{e}_m = e_m$, where e_m is the Weil pairing defined in section 8. [*Hint:* Use Weil reciprocity, exer. 2.11.]

3.17. Let $\mathscr{K}$ be a quaternion algebra. Show that $\mathscr{K}$ is ramified at ∞. [*Hint:* $M_2(\mathbb{R})$ contains zero-divisors.]

3.18. Let E/K be an elliptic curve, and assume that $\mathscr{K} = \text{End}(E) \otimes \mathbb{Q}$ is a quaternion algebra.

(a)   Prove that if $p \neq \infty$ and $p \neq \text{char}(K)$, then $\mathscr{K}$ splits at p. [*Hint:* Use (7.4).]

(b)   Prove that $\text{char}(K) > 0$. [*Hint:* Use exer. 3.17 and (9.5a).]

(c)   Prove that $\mathscr{K}$ is the unique quaternion algebra ramified at precisely ∞ and $\text{char}(K)$.

(d)* Prove that $\text{End}(E)$ is a maximal order in $\mathscr{K}$.

3.19. Let $\mathscr{K}$ be a quaternion algebra.

(a)   Show that $\mathscr{K} \otimes \bar{\mathbb{Q}} \cong M_2(\bar{\mathbb{Q}})$.

(b)   Show that $\mathscr{K} \otimes \mathscr{K} \cong M_4(\mathbb{Q})$. (This proves that $\mathscr{K}$ has order 2 in $\text{Br}(\mathbb{Q})$.) [*Hint:* First show that $\mathscr{K} \otimes \mathscr{K}$ is simple (i.e. has no two-sided ideals.) Then prove that the map

$$\mathscr{K} \otimes \mathscr{K} \to \text{End}(\mathscr{K}), \qquad a \otimes b \to (x \to ax\hat{b})$$

is an isomorphism.]

3.20. Let $\mathscr{K}$ be a quadratic imaginary field with ring of integers $\mathcal{O}$. Show that the orders of $\mathscr{K}$ are precisely the rings $\mathbb{Z} + f\mathcal{O}$ for integers $f > 0$. The integer f is called the *conductor* of the order.

3.21. Let $C/\bar{K}$ be a curve of genus 1. For any point $O \in C$, we can associate to the elliptic curve (C, O) its j-invariant $j(C, O)$. This exercise sketches a proof that the value $j(C, O)$ is independent of the choice of the basepoint O. Thus we can assign a *j-invariant* $j(C)$ to any curve C of genus 1. (We assume that $\text{char}(K) \neq 2$. The result is still true for $\text{char}(K) = 2$, but the method of proof must be modified and the ensuing algebra is more complicated.)

(a)   Choose a Legendre equation

$$y^2 = x(x - 1)(x - \lambda)$$

for the elliptic curve (C, O). Show that the map $x : C \to \mathbb{P}^1$ has degree 2 and is ramified exactly over the points $\{0, 1, \lambda, \infty\}$.

(b)   Let $O' \in C$ be another point, and choose a Legendre equation

$$w^2 = z(z - 1)(z - \mu)$$

for (C, O'). Let $\tau : C \to C$ be the translation-by-O' map on the elliptic curve (C, O). Show that there are constants $a \in \bar{K}$ and $b \in \bar{K}^*$ such that $\tau^*(z) = a + bx$. [*Hint*: Look at the divisor of $\tau^*(z)$.]

(c) Let $f : \mathbb{P}^1 \to \mathbb{P}^1$ be the map $f(t) = a + bt$. Prove the f maps the set $\{0, 1, \lambda\}$ bijectively to the set $\{0, 1, \mu\}$. [*Hint*: Compare the ramification of the maps $z \circ \tau$ and $f \circ x$.]

(d) Show that

$$\mu \in \{\lambda, 1/\lambda, 1 - \lambda, 1/(1 - \lambda), \lambda/(1 - \lambda), (\lambda - 1)/\lambda\}.$$

[*Hint*: Consider the six ways of matching $\{0, 1, \lambda\}$ with $\{0, 1, \mu\}$.]

(e) Deduce that $j(C, O) = j(C, O')$. [*Hint*: Show that the formula for $j(E_\lambda)$ in (1.7b) does not change if λ is replaced by any of the six expressions given in (d).]

3.22. Let C be a curve of genus 1 defined over K.

(a) Prove that $j(C) \in K$.

(b) Prove that C is an elliptic curve over K if and only if $C(K) \neq \varnothing$.

(c) Prove that C is always isomorphic (over $\bar{K}$) to an elliptic curve defined over K.

3.23. *Deuring Normal Form*. The following normal form for a Weierstrass equation is sometimes useful when dealing with elliptic curves over (algebraically closed) fields of arbitrary characteristic.

(a) Let E/K be an elliptic curve, and assume that either $\mathrm{char}(K) \neq 3$ or $j(E) \neq 0$. Prove that E has a Weierstrass equation *over* $\bar{K}$ of the form

$$E : y^2 + \alpha xy + y = x^3, \qquad \alpha \in \bar{K}.$$

(b) For the Weierstrass equation given in (a), show that $(0, 0) \in E[3]$.

(c) For what value(s) of α is the equation singular?

(d) Verify that

$$j(E) = \alpha^3(\alpha^3 - 24)^3/(\alpha^3 - 27).$$

3.24. Let E/K be an elliptic curve with complex multiplication *over* K (i.e. $\mathrm{End}_K(E)$ is strictly larger that $\mathbb{Z}$.) Prove that for all primes $\ell \neq \mathrm{char}(K)$, the action of $G_{\bar{K}/K}$ on the Tate module $T_\ell(E)$ is abelian. [*Hint*: Use the fact that the non-trivial endomorphisms in $\mathrm{End}_K(E)$ commute with the action of $G_{\bar{K}/K}$.]

CHAPTER IV

The Formal Group of an Elliptic Curve

Let E be an elliptic curve. In this chapter we start by studying an "infinitesimal" neighborhood of E centered at its origin O. In other words, we look at the local ring $K[E]_O$, and take the completion of this ring at its maximal ideal. This leads to a power series ring in one variable, say $K[\![z]\!]$, for some uniformizer z at O. We can then express the Weierstrass coordinates x and y as formal Laurent power series in z. Further, we can write down a power series $F(z_1, z_2) \in K[\![z_1, z_2]\!]$ which formally gives the group law on E. Such a power series, which might be described as a "group law without any group elements", is an example of a *formal group*. In the remainder of the chapter we study in some detail the principal properties of arbitrary (one-parameter) formal groups. The advantage of suppressing all mention of the elliptic curve which motivated this study in the first place is that working with formal power series tends to be fairly easy. Then, of course, having obtained results for arbitrary formal groups, we can apply them in particular to the formal group associated to our original elliptic curve.

§1. Expansion around O

In this section we investigate the structure of an elliptic curve and its addition law "close to the origin". To do this it is convenient to make a change of variables, so let

$$z = -\frac{x}{y} \quad \text{and} \quad w = -\frac{1}{y} \quad \left(\text{so } x = \frac{z}{w} \text{ and } y = -\frac{1}{w} \right).$$

The origin O on E is now the point $(z, w) = (0, 0)$, and z is a local uniformizer at O (i.e., z has a zero of order 1 at O.) The usual Weierstrass equation for E becomes

$$w = z^3 + a_1 zw + a_2 z^2 w + a_3 w^2 + a_4 zw^2 + a_6 w^3 \ (= f(z, w)).$$

The idea now is to substitute this equation into itself recursively so as to express w as a power series in z. Thus

$$
\begin{aligned}
w &= z^3 + (a_1 z + a_2 z^2)w + (a_3 + a_4 z)w^2 + a_6 w^3 \\
&= z^3 + (a_1 z + a_2 z^2)[z^3 + (a_1 z + a_2 z^2)w + (a_3 + a_4 z)w^2 + a_6 w^3] \\
&\quad + (a_3 + a_4 z)[z^3 + (a_1 z + a_2 z^2)w + (a_3 + a_4 z)w^2 + a_6 w^3]^2 \\
&\quad + a_6[z^3 + (a_1 z + a_2 z^2)w + (a_3 + a_4 z)w^2 + a_6 w^3]^3 \\
&\ \ \vdots \\
&= z^3 + a_1 z^4 + (a_1^2 + a_2)z^5 + (a_1^3 + 2a_1 a_2 + a_3)z^6 \\
&\quad + (a_1^4 + 3a_1^2 a_2 + 3a_1 a_3 + a_2^2 + a_4)z^7 + \cdots \\
&= z^3(1 + A_1 z + A_2 z^2 + \cdots),
\end{aligned}
$$

where $A_n \in \mathbb{Z}[a_1, \ldots, a_6]$ is a polynomial in the coefficients of E. Of course, we must show that this procedure actually converges to a power series $w(z) \in \mathbb{Z}[a_1, \ldots, a_6][[z]]$, and naturally we want the equality

$$w(z) = f(z, w(z))$$

to hold in the power series ring.

To more precisely describe the algorithm for producing $w(z)$, define a sequence of polynomials by

$$f_1(z, w) = f(z, w) \quad \text{and} \quad f_{m+1}(z, w) = f_m(z, f(z, w)).$$

Then we take

$$w(z) = \operatorname*{Lim}_{m \to \infty} f_m(z, 0)$$

provided this limit makes sense in $\mathbb{Z}[a_1, \ldots, a_6][[z]]$.

Proposition 1.1. (a) *The procedure described above gives a power series*

$$w(z) = z^3(1 + A_1 z + A_2 z^2 + \cdots) \in \mathbb{Z}[a_1, \ldots, a_6][[z]].$$

(b) *$w(z)$ is the unique power series satisfying*

$$w(z) = f(z, w(z)).$$

(c) *If $\mathbb{Z}[a_1, \ldots, a_6]$ is made into a graded ring by assigning weights $\mathrm{wt}(a_i) = i$, then A_n is a homogeneous polynomial of weight n.*

PROOF. Parts (a) and (b) are really special cases of Hensel's lemma, which we prove below (1.2). To prove the present proposition, use (1.2) with

$$R = \mathbb{Z}[a_1, \dots, a_6][[z]], \qquad I = (z),$$

$$F(w) = f(z, w) - w, \qquad a = 0, \qquad \alpha = -1.$$

Finally, to prove (c) we assign weights to z and w,

$$wt(z) = -1 \quad \text{and} \quad wt(w) = -3.$$

Then one sees that $f(z, w)$ is homogeneous of weight -3 in the graded ring $\mathbb{Z}[a_1, \dots, a_6, z, w]$, hence by an easy induction so is every $f_m(z, w)$. In particular,

$$f_m(z, 0) = z^3(1 + B_1 z + B_2 z^2 + \cdots + B_N z^N)$$

is homogeneous of weight -3, so each B_n is homogeneous of weight n in $\mathbb{Z}[a_1, \dots, a_6]$. Hence the A_n's have the same property, since $f_m(z, 0)$ converges to $w(z)$ as $m \to \infty$. $\square$

Lemma 1.2 (Hensel's Lemma). *Let R be a ring which is complete with respect to some ideal $I \subset R$, and let $F(w) \in R[w]$ be a polynomial. Suppose that $a \in R$ satisfies (for some integer $n \geq 1$)*

$$F(a) \in I^n \quad \text{and} \quad F'(a) \in R^*.$$

Then for any $\alpha \in R$ satisfying $\alpha \equiv F'(a) \pmod{I}$, the sequence

$$w_0 = a \qquad w_{m+1} = w_m - F(w_m)/\alpha$$

converges to an element $b \in R$ satisfying

$$F(b) = 0 \quad \text{and} \quad b \equiv a \pmod{I^n}.$$

If R is integral domain, then these conditions determine b uniquely.
(We remark that Hensel's lemma is usually proven for complete local rings, and generally one uses Newton's iteration $w_{m+1} = w_m - F(w_m)/F'(w_m)$. For this reason, we include a quick proof of (1.2).)

PROOF. To ease notation, we replace $F(w)$ by $F(w + a)/\alpha$, so we are now dealing with the recursion

$$w_0 = 0, \qquad F(0) \in I^n, \qquad F'(0) \equiv 1 \pmod{I}, \qquad w_{m+1} = w_m - F(w_m).$$

Since $F(0) \in I^n$, it is clear that if $w_m \in I^n$, then $w_m - F(w_m)$ is also in I^n. It follows that

$$w_m \in I^n \qquad \text{for all } m \geq 0.$$

We now show by induction that

$$w_m \equiv w_{m+1} \pmod{I^{m+n}} \qquad \text{for all } m \geq 0.$$

For $m = 0$, this just says $F(0) \equiv 0 \pmod{I^n}$, which is one of our initial assumptions. Assume now that this congruence is true for all integers less than m. Let X and Y be new variables, and factor

$$F(X) - F(Y) = (X - Y)(F'(0) + XG(X, Y) + YH(X, Y))$$

with polynomials $G, H \in R[X, Y]$. Then

$$w_{m+1} - w_m = (w_m - F(w_m)) - (w_{m-1} - F(w_{m-1}))$$
$$= (w_m - w_{m-1}) - (F(w_m) - F(w_{m-1}))$$
$$= (w_m - w_{m-1})[1 - F'(0) - w_m G(w_m, w_{m-1})$$
$$- w_{m-1} H(w_m, w_{m-1})] \in I^{m+n}.$$

Here the last line follows from the induction hypothesis and the fact that $F'(0) \equiv 1 \pmod{I}$ and $w_m, w_{m-1} \in I^n$. This proves that $w_m - w_{m+1} \in I^{m+n}$ for all $m \geqslant 0$.

Since R is complete with respect to I, it follows that the sequence w_m converges to an element $b \in R$; and since every $w_m \in I^n$, $b \in I^n$ also. Further, taking the limit of the relation $w_{m+1} = w_m - F(w_m)$ as $m \to \infty$ yields $b = b - F(b)$, so $F(b) = 0$.

Finally, to show uniqueness (under the assumption that R is an integral domain), suppose that also $c \in I^n$ and $F(c) = 0$. Then

$$0 = F(b) - F(c) = (b - c)(F'(0) + bG(b, c) + cH(b, c)).$$

If $b \neq c$, then $F'(0) + bG(b, c) + cH(b, c) = 0$. But $bG(b, c) + cH(b, c) \in I$, so it would follow that $F'(0) \in I$. This contradiction shows that $b = c$. □

Using the power series $w(z)$ from (1.1), we find *Laurent series* for x and y,

$$x(z) = \frac{z}{w(z)} = \frac{1}{z^2} - \frac{a_1}{z} - a_2 - a_3 z - (a_4 + a_1 a_3)z^2 - \cdots$$

$$y(z) = \frac{-1}{w(z)} = -\frac{1}{z^3} + \frac{a_1}{z^2} + \frac{a_2}{z} + a_3 + (a_4 + a_1 a_3)z + \cdots.$$

Similarly the invariant differential has an expansion

$$\omega(z) = (1 + a_1 z + (a_1^2 + a_2)z^2 + (a_1^3 + 2a_1 a_2 + 2a_3)z^3$$
$$+ (a_1^4 + 3a_1^2 a_2 + 6a_1 a_3 + a_2^2 + 2a_4)z^4 + \cdots) dz.$$

We note that the series $x(z)$, $y(z)$, and $\omega(z)$ have coefficients in $\mathbb{Z}[a_1, \ldots, a_6]$. This is clear for $x(z)$ and $y(z)$; while for $\omega(z)$ it follows from the two expressions

$$\frac{\omega(z)}{dz} = \frac{dx(z)/dz}{2y + a_1 x + a_3} = \frac{-2z^{-3} + \cdots}{-2z^{-3} + \cdots} \in \mathbb{Z}[\tfrac{1}{2}, a_1, \ldots, a_6][\![z]\!]$$

$$\frac{\omega(z)}{dz} = \frac{dy(z)/dz}{3x^2 + 2a_2 x + a_4 - a_1 y} = \frac{3z^{-4} + \cdots}{3z^{-4} + \cdots} \in \mathbb{Z}[\tfrac{1}{3}, a_1, \ldots, a_6][\![z]\!],$$

which show that any denominator is simultaneously a power of 2 and a power of 3.

Now the pair $(x(z), y(z))$ provides a "formal solution" to the Weierstrass equation

$$E: y^2 + a_1 xy + a_3 y = x^3 + a_2 x^2 + a_4 x + a_6;$$

that is, a solution in the quotient field of the ring of formal power series. If E is defined over a field K, we might try to produce points of E by taking $z \in K$ and looking at $(x(z), y(z))$. In general, there is no obvious way to attach a meaning to an infinite series such as $x(z)$. But if K is a complete local field with ring of integers R and maximal ideal $\mathcal{M}$, and if the coefficients satisfy $a_i \in R$, and if $z \in \mathcal{M}$, then the power series $x(z)$ and $y(z)$ will converge to give a point of $E(K)$. This gives an injection (the inverse is $z = -x(z)/y(z)$)

$$\mathcal{M} \to E(K),$$

and it is easy to characterize the image as those (x, y) with $xy^{-1} \in \mathcal{M}$. This map will be a key tool when we study elliptic curves over local fields in chapter VII.

Returning now to formal power series, we look for the power series formally giving the addition law on E. Thus let z_1, z_2 be independent indeterminates, and let $w_i = w(z_i)$ for $i = 1, 2$. In the (z, w)-plane, the line connecting (z_1, w_1) to (z_2, w_2) has slope

$$\lambda = \lambda(z_1, z_2) = \frac{w_2 - w_1}{z_2 - z_1} = \sum_{n=3}^{\infty} A_{n-3} \frac{z_2^n - z_1^n}{z_2 - z_1} \in \mathbb{Z}[a_1, \ldots, a_6][\![z_1, z_2]\!].$$

Note that λ has no constant or linear term. The A_n's come from (1.1a). Letting

$$v = v(z_1, z_2) = w_1 - \lambda z_1 \in \mathbb{Z}[a_1, \ldots, a_6][\![z_1, z_2]\!],$$

the connecting line has equation $w = \lambda z + v$. Substituting this into the Weierstrass equation gives a cubic in z, two of whose roots are z_1 and z_2. Looking at the quadratic term, we see that the third root (say z_3) can be expressed as a power series in z_1 and z_2,

$$z_3 = z_3(z_1, z_2)$$

$$= -z_1 - z_2 + \frac{a_1 \lambda + a_3 \lambda^2 - a_2 v - 2a_4 \lambda v - 3a_6 \lambda^2 v}{1 + a_2 \lambda + a_4 \lambda^2 + a_6 \lambda^3}$$

$$\in \mathbb{Z}[a_1, \ldots, a_6][\![z_1, z_2]\!].$$

For the group law on E, the points $(z_1, w_1), (z_2, w_2), (z_3, w_3)$ add up to zero. Thus to add the first two, we need the formula for the inverse. In the (x, y)-plane, the inverse of (x, y) is $(x, -y - a_1 x - a_3)$. Hence the inverse of (z, w) will have z-coordinate (remember $z = -x/y$)

$$i(z) = \frac{x(z)}{y(z) + a_1 x(z) + a_3} = \frac{z^{-2} - a_1 z^{-1} - \cdots}{-z^{-3} + 2a_1 z^{-2} + \cdots} \in \mathbb{Z}[a_1, \ldots, a_6][\![z]\!].$$

This gives the formal addition law

$$F(z_1, z_2) = i(z_3(z_1, z_2))$$

$$= z_1 + z_2 - a_1 z_1 z_2 - a_2(z_1^2 z_2 + z_1 z_2^2)$$

$$- (2a_3 z_1^3 z_2 - (a_1 a_2 - 3a_3)z_1^2 z_2^2 + 2a_3 z_1 z_2^3) + \cdots$$

$$\in \mathbb{Z}[a_1, \ldots, a_6]\llbracket z_1, z_2 \rrbracket.$$

From the corresponding properties for E we deduce that $F(z_1, z_2)$ satisfies

$$F(z_1, z_2) = F(z_2, z_1) \qquad \text{(commutativity)}$$

$$F(z_1, F(z_2, z)) = F(F(z_1, z_2), z) \qquad \text{(associativity)}$$

$$F(z, i(z)) = 0 \qquad \text{(inverse)}.$$

The power series $F(z_1, z_2)$ might be described as "a group law without any group elements". Such objects are called *formal groups*. We could now continue with the study of the particular formal group coming from our elliptic curve, but since it is little more difficult to analyze arbitrary (one-parameter) formal groups, and in fact the abstraction tends to clarify the situation, we will take the latter approach. The reader should, however, keep the example of an elliptic curve in mind when reading the rest of this chapter.

§2. Formal Groups

Let R be a ring.

Definition. A (*one-parameter commutative*) *formal group* $\mathcal{F}$ *defined over* R is a power series $F(X, Y) \in R\llbracket X, Y \rrbracket$ satisfying:

(a) $F(X, Y) = X + Y + \text{(terms of degree} \geq 2)$.
(b) $F(X, F(Y, Z)) = F(F(X, Y), Z)$ (associativity).
(c) $F(X, Y) = F(Y, X)$ (commutativity).
(d) There is a unique power series $i(T) \in R\llbracket T \rrbracket$ such that $F(T, i(T)) = 0$ (inverse).
(e) $F(X, 0) = X$ and $F(0, Y) = Y$.

We call $F(X, Y)$ the *formal group law of* $\mathcal{F}$.

Remark 2.1. It is in fact easy to show that (a) and (b) imply (d) and (e) (exer. 4.1). It is also true that (a) and (b) imply (c) provided that R has no torsion nilpotents (see exer. 4.2b), but we will only prove this below if $\text{char}(R) = 0$.

Definition. Let $(\mathcal{F}, F)$ and $(\mathcal{G}, G)$ be formal groups defined over R. A *homomorphism from* $\mathcal{F}$ *to* $\mathcal{G}$ *defined over* R is a power series (with no constant term) $f(T) \in R\llbracket T \rrbracket$ satisfying

$$f(F(X, Y)) = G(f(X), f(Y)).$$

$\mathscr{F}$ and $\mathscr{G}$ are *isomorphic over R* if there are homomorphisms $f : \mathscr{F} \to \mathscr{G}$ and $g : \mathscr{G} \to \mathscr{F}$ defined over R with

$$f(g(T)) = g(f(T)) = T.$$

Example 2.2.1. The *formal additive group*, denoted $\hat{\mathbb{G}}_a$, is given by

$$F(X, Y) = X + Y.$$

Example 2.2.2. The *formal multiplicative group*, denoted $\hat{\mathbb{G}}_m$, is given by

$$F(X, Y) = X + Y + XY = (1 + X)(1 + Y) - 1.$$

Example 2.2.3. Let E be an elliptic curve given by a Weierstrass equation with coefficients in R. The *formal group associated to E*, denoted $\hat{E}$, is given by the power series $F(z_1, z_2)$ described in section 1.

Example 2.2.4. Let $(\mathscr{F}, F)$ be a formal group. We can define homomorphisms

$$[m] : \mathscr{F} \to \mathscr{F}$$

inductively for $m \in \mathbb{Z}$ by

$$[0](T) = 0 \qquad [m + 1](T) = F([m](T), T)$$
$$[m - 1](T) = F([m](T), i(T)).$$

One easily checks (by induction) that $[m]$ is a homomorphism. We call $[m]$ the *multiplication-by-m map*. The following elementary proposition, which explains when $[m]$ is invertible, will be of great importance. (The progression is $(2.3) \Rightarrow (3.2b) \Rightarrow (VII.3.1)$, and the latter provides a key fact for the proof of the weak Mordell–Weil theorem (VIII.1.1).)

Proposition 2.3. *Let $\mathscr{F}$ be a formal group over R, and let $m \in \mathbb{Z}$.*

(a) $[m](T) = mT + $ *(higher order terms).*
(b) *If $m \in R^*$, then $[m] : \mathscr{F} \to \mathscr{F}$ is an isomorphism.*

PROOF. (a) For $m \geqslant 0$ this is a trivial induction using the recursive definition of $[m]$ and the fact that $F(X, Y) = X + Y + \cdots$. Then, from

$$0 = F(T, i(T)) = T + i(T) + \cdots,$$

we see that $i(T) = -T + \cdots$; and now the downward induction for $m < 0$ is also clear.

(b) This follows from (a) and the following lemma, which we will have occasion to use several times. $\square$

Lemma 2.4. *Let $a \in R^*$ and $f(T) \in R[\![T]\!]$ a power series starting*

$$f(T) = aT + \cdots.$$

Then there is a unique power series $g(T) \in R[\![T]\!]$ such that $f(g(T)) = T$. It further satisfies $g(f(T)) = T$.

PROOF. We construct a sequence of polynomials $g_n(T) \in R[T]$ satisfying

$$f(g_n(T)) \equiv T \ (\mathrm{mod} \ T^{n+1}) \quad \text{and} \quad g_{n+1}(T) \equiv g_n(T) \ (\mathrm{mod} \ T^{n+1}).$$

Then $g(T) = \operatorname{Lim} g_n(T)$ exists and clearly satisfies $f(g(T)) = T$.

To start the induction, let $g_1(T) = a^{-1}T$. Now suppose $g_{n-1}(T)$ has been constructed. We look for $\lambda \in R$ so that

$$g_n(T) = g_{n-1}(T) + \lambda T^n$$

has the desired property. We compute

$$
\begin{aligned}
f(g_n(T)) &= f(g_{n-1}(T) + \lambda T^n) \\
&\equiv f(g_{n-1}(T)) + a\lambda T^n \ (\mathrm{mod} \ T^{n+1}) \\
&\equiv T + bT^n + a\lambda T^n \ (\mathrm{mod} \ T^{n+1})
\end{aligned}
$$

for some $b \in R$ by the induction hypothesis. It thus suffices to take $\lambda = -b/a$, which is in R because $a \in R^*$. This shows that $g(T)$ exists.

Next, applying g to $f(g(T)) = T$ gives $g(f(g(T))) = g(T)$. This is an identity in the power-series ring $R[\![g(T)]\!]$, so $g(f(T)) = T$. Finally, if $f(h(T)) = T$, then

$$g(T) = g(f(h(T))) = (g \circ f)(h(T)) = h(T),$$

which shows that $g(T)$ is unique. $\qquad\qquad\square$

§3. Groups Associated to Formal Groups

In general a formal group is merely a group operation, with no actual underlying group. But if the ring R is local and complete, and if the variables are assigned values in the maximal ideal of R, then the power series giving the formal group will converge. In this section we give some basic facts about the resulting group. The following notation will be used:

R a complete local ring
$\mathcal{M}$ the maximal ideal of R
k the residue field $R/\mathcal{M}$
$\mathscr{F}$ a formal group defined over R, with formal group law $F(X, Y)$.

Definition. The *group associated to* $\mathscr{F}/R$, denoted $\mathscr{F}(\mathcal{M})$, is the set $\mathcal{M}$ with the group operations

$$x \oplus_{\mathscr{F}} y = F(x, y) \quad \text{(addition)} \qquad \text{for } x, y \in \mathcal{M},$$

$$\ominus_{\mathscr{F}} x = i(x) \quad \text{(inverse)} \qquad \text{for } x \in \mathcal{M}.$$

Similarly, for $n \geq 1$, $\mathscr{F}(\mathscr{M}^n)$ is the subgroup of $\mathscr{F}(\mathscr{M})$ consisting of the set $\mathscr{M}^n$.

Since R is complete, the power series $F(x, y)$ and $i(x)$ converge in R for $x, y \in \mathscr{M}$; and then the axioms for a formal group immediately imply that $\mathscr{F}(\mathscr{M})$ is a group and $\mathscr{F}(\mathscr{M}^n)$ a subgroup.

Example 3.1.1. The additive group $\hat{\mathbb{G}}_a(\mathscr{M})$ is just $\mathscr{M}$ with its usual addition law. Notice the exact sequence (of additive groups)

$$0 \rightarrow \hat{\mathbb{G}}_a(\mathscr{M}) \rightarrow R \rightarrow k \rightarrow 0.$$

Example 3.1.2. The multiplicative group $\hat{\mathbb{G}}_m(\mathscr{M})$ is the group of 1-units (i.e. $1 + \mathscr{M}$) with its usual multiplication. Notice we again have an exact sequence

$$0 \rightarrow \hat{\mathbb{G}}_m(\mathscr{M}) \rightarrow R^* \rightarrow k^* \rightarrow 0.$$

Example 3.1.3. Let $\hat{E}$ be the formal group associated to an elliptic curve E/K (2.2.3), where K is the quotient field of R. As we noted in section 1, the power series $x(z)$ and $y(z)$ give a map

$$\mathscr{M} \rightarrow E(K)$$

$$z \rightarrow (x(z), y(z)).$$

From the way the power series for $\hat{E}$ was defined, this map gives a homomorphism of $\hat{E}(\mathscr{M})$ to $E(K)$. As we will see in chapter VII, there is often an exact sequence

$$0 \rightarrow \hat{E}(\mathscr{M}) \rightarrow E(K) \rightarrow \tilde{E}(k) \rightarrow 0,$$

where $\tilde{E}$ is a certain elliptic curve defined over the residue field k. In this way the study of $E(K)$ is reduced to the study of the formal group $\hat{E}$ and the study of an elliptic curve over a smaller (so hopefully simpler) field.

Proposition 3.2. (a) *For each $n \geq 1$, the map*

$$\mathscr{F}(\mathscr{M}^n)/\mathscr{F}(\mathscr{M}^{n+1}) \rightarrow \mathscr{M}^n/\mathscr{M}^{n+1}$$

induced by the identity map on sets is an isomorphism of groups.
(b) *Let p be the characteristic of k ($p = 0$ is allowed). Then every torsion element of $\mathscr{F}(\mathscr{M})$ has order a power of p.* (See section 6 for a more precise description.)

PROOF. (a) Since the underlying sets are the same, it suffices to show that the map is a homomorphism. But for $x, y \in \mathscr{M}^n$,

$$x \oplus_{\mathscr{F}} y = F(x, y) = x + y + \cdots$$

$$\equiv x + y \pmod{\mathscr{M}^{2n}}.$$

(b) We give two proofs of this important fact. Multiplying an arbitrary torsion element by an appropriate power of p, it suffices to prove that there are no non-zero torsion elements of order prime to p. Thus let $m \geq 1$ be prime to

p (arbitrary if $p = 0$) and $x \in \mathscr{F}(\mathscr{M})$ an element with $[m](x) = 0$. We must show $x = 0$.

First, since m is prime to p, we see that $m \notin \mathscr{M}$. Hence from (2.3b), $[m]$ is an isomorphism of the formal group $\mathscr{F}/R$ to itself, so it induces an isomorphism

$$[m] : \mathscr{F}(\mathscr{M}) \xrightarrow{\sim} \mathscr{F}(\mathscr{M}).$$

In particular, it has trivial kernel, so $x = 0$.

For the second proof, we assume that R is Noetherian. We show inductively that $x \in \mathscr{M}^n$ for all $n \geq 1$, which implies $x = 0$ from Krull's theorem ([A–M, Corollary 10.20]). By assumption, $x \in \mathscr{M}$. Suppose $x \in \mathscr{M}^n$. Look at the image $\bar{x}$ of x in $\mathscr{F}(\mathscr{M}^n)/\mathscr{F}(\mathscr{M}^{n+1})$. On the one hand, $\bar{x}$ has order dividing m. On the other hand, $\mathscr{F}(\mathscr{M}^n)/\mathscr{F}(\mathscr{M}^{n+1})$ has only p-torsion, since from (a) it is isomorphic to the k vector space $\mathscr{M}^n/\mathscr{M}^{n+1}$. Hence $\bar{x} = 0$, so $x \in \mathscr{M}^{n+1}$ as desired. $\square$

§4. The Invariant Differential

We return to the study of a formal group $\mathscr{F}$ defined over an arbitrary ring R. In such a formal setting, a differential form is simply an expression $P(T) \, dT$ with $P(T) \in R[\![T]\!]$. Of particular interest are those differential forms which respect the group structure of $\mathscr{F}$.

Definition. An *invariant differential* on $\mathscr{F}/R$ is a differential form

$$\omega(T) = P(T) \, dT \in R[\![T]\!] \, dT$$

satisfying

$$\omega \circ F(T, S) = \omega(T).$$

[In other words, satisfying

$$P(F(T, S))F_X(T, S) = P(T),$$

where $F_X(X, Y)$ is the partial derivative of F with respect to the first variable.] An invariant differential as above is said to be *normalized* if $P(0) = 1$.

Example 4.1.1. On the additive group $\hat{\mathbb{G}}_a$, an invariant differential is $\omega = dT$.

Example 4.1.2. On the multiplicative group $\hat{\mathbb{G}}_m$, an invariant differential is

$$\omega = (1 + T)^{-1} \, dT = (1 - T + T^2 - \cdots) \, dT.$$

Proposition 4.2. *Let $\mathscr{F}/R$ be a formal group. There exists a unique normalized invariant differential on $\mathscr{F}/R$, given by the formula*

$$\omega = F_X(0, T)^{-1} \, dT.$$

Every invariant differential on $\mathscr{F}/R$ is of the form $a\omega$ for some $a \in R$.

PROOF. Suppose $P(T)\,dT$ is an invariant differential on $\mathscr{F}/R$. Thus

$$P(F(T, S))F_X(T, S) = P(T).$$

Putting $T = 0$ (remember $F(0, S) = S$) gives

$$P(S)F_X(0, S) = P(0).$$

Since $F_X(0, S) = 1 + \cdots$, we see that $P(T)$ is determined by $P(0)$, and every possible invariant differential is of the form $a\omega$ with $a \in R$ and

$$\omega(T) = F_X(0, T)^{-1}\,dT.$$

Since this ω is normalized, it only remains to show that it is invariant.

Thus we must show that

$$F_X(0, F(T, S))^{-1}F_X(T, S) = F_X(0, T)^{-1}.$$

To prove this, differentiate the associative law

$$F(U, F(T, S)) = F(F(U, T), S)$$

with respect to U to obtain (chain rule!)

$$F_X(U, F(T, S)) = F_X(F(U, T), S)F_X(U, T).$$

Now putting $U = 0$ (note $F(0, T) = T$) yields

$$F_X(0, F(T, S)) = F_X(T, S)F_X(0, T),$$

which is the desired result. $\square$

Corollary 4.3. *Let $\mathscr{F}$, $\mathscr{G}/R$ be formal groups with normalized invariant differentials $\omega_{\mathscr{F}}$, $\omega_{\mathscr{G}}$. Let $f : \mathscr{F} \to \mathscr{G}$ be a homomorphism. Then*

$$\omega_{\mathscr{G}} \circ f = f'(0)\omega_{\mathscr{F}}.$$

(Here $f'(T)$ is the *formal derivative* of the power series, obtained by differentiating $f(T)$ term by term.)

PROOF. Let $F(X, Y)$, $G(X, Y)$ be the formal group laws for $\mathscr{F}$ and $\mathscr{G}$. We verify that $\omega_{\mathscr{G}} \circ f$ is an invariant differential on $\mathscr{F}$:

$$\omega_{\mathscr{G}} \circ f(F(T, S)) = \omega_{\mathscr{G}}(G(f(T), f(S))) \text{ since } f \text{ is a homomorphism}$$

$$= \omega_{\mathscr{G}} \circ f(T) \text{ since } \omega_{\mathscr{G}} \text{ is invariant for } \mathscr{G}.$$

Hence from (4.2), $\omega_{\mathscr{G}} \circ f$ equals $a\omega_{\mathscr{F}}$. Comparing initial terms gives $a = f'(0)$.

$\square$

Corollary 4.4. *Let $\mathscr{F}/R$ be a formal group and $p \in \mathbb{Z}$ a prime. Then there are power series $f(T), g(T) \in R[\![T]\!]$ with $f(0) = g(0) = 0$ such that*

$$[p](T) = pf(T) + g(T^p).$$

PROOF. Let $\omega(T)$ be the normalized invariant differential on $\mathscr{F}$. From (2.3a) we have $[p]'(0) = p$, so (4.3) implies that

$$p\omega(T) = \omega \circ [p](T) = (1 + \cdots)[p]'(T)\,dT.$$

Since the series $(1 + \cdots)$ is invertible in $R[\![T]\!]$, it follows that $[p]'(T) \in pR[\![T]\!]$; hence every term aT^n in the series $[p](T)$ satisfies either $a \in pR$ or $p \mid n$.   □

Example 4.5. Let $\hat{E}$ be the formal group associated to an elliptic curve (2.2.3). Then in terms of the coefficients of a Weierstrass equation for E, one finds

$$[2](T) = 2\{T - a_2 T^3 + \cdots\} + \{-a_1 T^2 + (a_1 a_2 - 7a_3)T^4 + \cdots\},$$

$$[3](T) = 3\{T - a_1 T^2 + (4a_1 a_2 - 13a_3)T^4 + \cdots\} + \{(a_1^2 - 8a_2)T^3 + \cdots\}.$$

§5. The Formal Logarithm

By integrating an invariant differential, one might hope to obtain a homomorphism to the additive group. Unfortunately, integration tends to introduce denominators, but at least in characteristic 0 we can proceed fairly well.

Definition. Let R be a ring of characteristic 0, $K = R \otimes \mathbb{Q}$, and $\mathscr{F}/R$ a formal group. Let

$$\omega(T) = (1 + c_1 T + c_2 T^2 + c_3 T^3 + \cdots)\,dT$$

be the normalized invariant differential on $\mathscr{F}/R$. The *formal logarithm* of $\mathscr{F}/R$ is the power series

$$\log_{\mathscr{F}}(T) = \int \omega(T) = T + \frac{c_1}{2}T^2 + \frac{c_2}{3}T^3 + \cdots \in K[\![T]\!].$$

The *formal exponential* of $\mathscr{F}/R$ is the unique power series $\exp_{\mathscr{F}}(T) \in K[\![T]\!]$ satisfying

$$\log_{\mathscr{F}} \circ \exp_{\mathscr{F}}(T) = \exp_{\mathscr{F}} \circ \log_{\mathscr{F}}(T) = T.$$

(Note $\exp_{\mathscr{F}}$ exists and is unique from (2.4).)

Example 5.1. The formal group law and invariant differential of the formal multiplicative group $\mathscr{F} = \hat{\mathbb{G}}_m$ are

$$F_{\mathscr{F}}(X, Y) = X + Y + XY \quad \text{and} \quad \omega_{\mathscr{F}}(T) = (1 + T)^{-1}\,dT.$$

Thus its formal logarithm and exponential are given by

$$\log_{\mathscr{F}}(T) = \int (1 + T)^{-1}\,dT = \sum_{n=1}^{\infty} (-1)^{n-1} T^n / n$$

and

$$\exp_{\mathscr{F}}(T) = \sum_{n=1}^{\infty} T^n/n!.$$

(Remember that the "identity" is at $T = 0$, so in terms of the usual series these series are $\log(1 + T)$ and $e^T - 1$.)

Proposition 5.2. *Let $\mathscr{F}/R$ be a formal group with* $\text{char}(R) = 0$. *Then*

$$\log_{\mathscr{F}} : \mathscr{F} \to \hat{\mathbb{G}}_a$$

is an isomorphism of formal groups over $K = R \otimes \mathbb{Q}$. *(N.B. Due to the denominators in* $\log_{\mathscr{F}}$, *it is not in general an isomorphism over R.)*

PROOF. Let $\omega(T)$ be the normalized invariant differential on $\mathscr{F}/R$. Thus

$$\omega(F(T, S)) = \omega(T).$$

Integrating this with respect to T gives

$$\log_{\mathscr{F}} F(T, S) = \log_{\mathscr{F}}(T) + f(S)$$

for some "constant of integration" $f(S) \in K[\![S]\!]$. Taking $T = 0$ shows that $f(S) = \log_{\mathscr{F}}(S)$, which proves that $\log_{\mathscr{F}}$ is indeed a homomorphism. Its inverse is $\exp_{\mathscr{F}}$, so $\log_{\mathscr{F}}$ is an isomorphism. $\qquad\square$

Application 5.3. Suppose R is a ring of characteristic 0 and $F(X, Y) \in R[\![X, Y]\!]$ is a power series satisfying

$$F(X, F(Y, Z)) = F(F(X, Y), Z), \qquad F(X, 0) = X, \qquad F(0, Y) = Y.$$

We note that in constructing the invariant differential, formal logarithm, and formal exponential, and in proving their basic properties, we used only these three facts about $F(X, Y)$. Thus letting $K = R \otimes \mathbb{Q}$, we have shown the existence of power series $\log(T), \exp(T) \in K[\![T]\!]$ such that

$$F(X, Y) = \exp(\log(X) + \log(Y)).$$

In particular, we see that $F(X, Y) = F(Y, X)$. In other words, every one-parameter formal group in characteristic 0 is automatically commutative. (For a more precise statement, see exer. 4.2b.)

For certain applications it is useful to have a bound for the denominators appearing in log and exp. For the former, it is clear from the definition, while for the latter we use the following calculation.

Lemma 5.4. *Let R be a ring with* $\text{char}(R) = 0$, *and let*

$$f(T) = \sum_{n=1}^{\infty} \frac{a_n}{n!} T^n$$

be a power series with $a_n \in R$ and $a_1 \in R^*$. Then the unique power series satisfy-

ing $f(g(T)) = T$ (cf. 2.4) can be written

$$g(T) = \sum_{n=1}^{\infty} \frac{b_n}{n!} T^n$$

with $b_n \in R$.

PROOF. Differentiating $f(g(T)) = T$ gives

$$f'(g(T))g'(T) = 1;$$

so evaluating at $T = 0$ shows that

$$b_1 = g'(0) = 1/f'(0) = 1/a_1 \in R^*.$$

Differentiating again yields

$$f'(g(T))g''(T) + f''(g(T))g'(T)^2 = 0.$$

Now repeated differentiation will show that for every $n \geqslant 2$, $f'(g(T))g^{(n)}(T)$ can be expressed as a polynomial (with integer coefficients) in the variables $f^{(i)}(g(T))$, $1 \leqslant i \leqslant n$, and $g^{(j)}(T)$, $1 \leqslant j \leqslant n - 1$. Hence evaluating at $T = 0$ expresses $a_1 b_n$ as a polynomial in $a_1, \ldots, a_n, b_1, \ldots, b_{n-1}$. Since $a_1, b_1 \in R^*$, an easy induction now shows that every $b_n \in R$.   □

Proposition 5.5. *Let R be a ring with char$(R) = 0$, and let $\mathscr{F}/R$ be a formal group. Then*

$$\log_{\mathscr{F}}(T) = \sum_{n=1}^{\infty} \frac{a_n}{n} T^n \quad \text{and} \quad \exp_{\mathscr{F}}(T) = \sum_{n=1}^{\infty} \frac{b_n}{n!} T^n$$

with $a_n, b_n \in R$ and $a_1 = b_1 = 1$.

PROOF. The expression for $\log_{\mathscr{F}}$ follows directly from the definition, and then the above lemma (5.4) shows that $\exp_{\mathscr{F}}$ has the desired form.   □

§6. Formal Groups over Discrete Valuation Rings

Let R be a complete local ring with maximal ideal $\mathscr{M}$, and let $\mathscr{F}/R$ be a formal group. As we have seen (3.2b), the associated group $\mathscr{F}(\mathscr{M})$ has no torsion of order prime to $p = \text{char}(R/\mathscr{M})$. We now analyze more closely the p-primary torsion for the case of discrete valuation rings.

Theorem 6.1. *Let R be a discrete valuation ring which is complete with respect to its maximal ideal $\mathscr{M}$, let $p = \text{char}(R/\mathscr{M})$, and let v be the valuation on R. Let $\mathscr{F}/R$ be a formal group, and suppose that $x \in \mathscr{F}(\mathscr{M})$ has exact order p^n for some $n \geqslant 1$. (I.e. $[p^n](x) = 0$ and $[p^{n-1}](x) \neq 0$.) Then*

$$v(x) \leqslant \frac{v(p)}{p^n - p^{n-1}}.$$

PROOF. The statement is automatic (and uninteresting) if $\text{char}(R) \neq 0$ or $p = 0$, since then $v(p) = \infty$. We assume this is not the case. From (4.4), we know that

$$[p](T) = pf(T) + g(T^p);$$

and from (2.3a), $f(T) = T + \cdots$. We prove the theorem by induction on n.

Suppose $x \neq 0$ and $[p](x) = 0$. Thus

$$0 = pf(x) + g(x^p).$$

Since R is a discrete valuation ring, the only way that the leading term of $pf(x)$ can be eliminated is to have

$$v(px) \geqslant v(x^p).$$

Hence

$$v(p) \geqslant (p - 1)v(x),$$

which proves the theorem for $n = 1$.

Now assume that the theorem is true for n, and let $x \in \mathscr{F}(\mathscr{M})$ have exact order p^{n+1}. Then

$$v([p](x)) = v(pf(x) + g(x^p))$$
$$\geqslant \min\{v(px), v(x^p)\}.$$

But $[p](x)$ has exact order p^n, so by the induction hypothesis

$$v(p)/(p^n - p^{n-1}) \geqslant v([p](x)).$$

Therefore

$$v(p)/(p^n - p^{n-1}) \geqslant \min\{v(px), v(x^p)\}.$$

But since $v(x) > 0$ and $n \geqslant 1$, it certainly is not possible to have

$$v(p)/(p^n - p^{n-1}) \geqslant v(px).$$

We conclude that

$$v(p)/(p^n - p^{n-1}) \geqslant v(x^p) = pv(x),$$

which is exactly the desired result.   □

Example 6.1.1. Let $\mathscr{F}$ be a formal group defined over $\mathbb{Z}_p$, the ring of p-adic integers. If $p \geqslant 3$, then (6.1) says that $\mathscr{F}(p\mathbb{Z}_p)$ has no torsion at all; and even for $p = 2$ it has at most elements of order 2. The same holds for the ring of integers in any finite *unramified* extension of $\mathbb{Q}_p$. For a general finite extension, the determining factor is the ramification degree (which equals $v(p)$ if one takes a normalized valuation.)

Next we show that $\mathscr{F}(\mathscr{M})$ has a large piece that looks like the additive group. The idea is to use the formal logarithm to define the map, but the presence of denominators means that convergence is no longer automatic. The following two lemmas will thus be useful.

Lemma 6.2. *Let v be a valuation and $p \in \mathbb{Z}$ a prime with $0 < v(p) < \infty$. Then for all integers $n \geq 1$,*

$$v(n!) \leq \frac{(n-1)v(p)}{p-1}.$$

PROOF. We compute

$$v(n!) = \sum_{i=1}^{\infty} \left[\frac{n}{p^i}\right] v(p) \leq \sum_{i=1}^{[\log_p n]} \frac{nv(p)}{p^i}$$

$$= \frac{nv(p)}{p-1}(1 - p^{-[\log_p n]}) \leq \frac{(n-1)v(p)}{p-1}. \qquad \square$$

Lemma 6.3. *Let R be a ring of characteristic 0, complete with respect to a discrete valuation v, and let $p \in \mathbb{Z}$ be a prime with $v(p) > 0$.*
(a) *Let $f(T)$ be a power series of the form*

$$f(T) = \sum_{n=1}^{\infty} \frac{a_n}{n} T^n \quad \text{with} \quad a_n \in R.$$

If $x \in R$ satisfies $v(x) > 0$, then the series $f(x)$ converges in R.
(b) *Let $g(T)$ be a power series of the form*

$$g(T) = \sum_{n=1}^{\infty} \frac{b_n}{n!} T^n \quad \text{with} \quad b_n \in R \quad \text{and} \quad b_1 \in R^*.$$

If $x \in R$ satisfies $v(x) > v(p)/(p-1)$, then the series $g(x)$ converges in R, and

$$v(g(x)) = v(x).$$

PROOF. (a) For a general term of $f(x)$, we have

$$v(a_n x^n/n) \geq nv(x) - v(n) \qquad \text{since } a_n \in R$$

$$\geq nv(x) - (\log_p n)v(p);$$

and this last expression goes to ∞ as n goes to ∞. Since v is non-archimedean, $f(x)$ converges.
(b) For a general term of the series $g(x)$, we have

$$v(b_n x^n/n!) \geq nv(x) - v(n!) \qquad \text{since } b_n \in R$$

$$\geq nv(x) - (n-1)v(p)/(p-1) \qquad \text{from (6.2)}$$

$$= v(x) + (n-1)\left\{v(x) - \frac{v(p)}{p-1}\right\}.$$

Hence from the initial assumption on $v(x)$, we have

$$v(b_n x^n/n!) \to \infty \qquad \text{as } n \to \infty, \text{ and}$$

$$v(b_n x^n/n!) > v(x) \qquad \text{for } n \geqslant 2.$$

Since v is non-archimedean, the former implies that $g(x)$ converges, and the latter shows that the leading term predominates. (Note $v(b_1 x) = v(x)$.) $\qquad \square$

Theorem 6.4. *Let K be a field of characteristic 0, complete with respect to a normalized discrete valuation v (i.e. $v(K^*) = \mathbb{Z}$), R the valuation ring of K, $\mathcal{M}$ the maximal ideal of R, and $p \in \mathbb{Z}$ a prime with $v(p) > 0$. Let $\mathscr{F}/R$ be a formal group.*
(a) *The formal logarithm induces a homomorphism*

$$\log_{\mathscr{F}} : \mathscr{F}(\mathcal{M}) \to K \qquad \text{(taken additively)}.$$

(b) *Let $r > v(p)/(p-1)$ be an integer. Then the formal logarithm induces an isomorphism*

$$\log_{\mathscr{F}} : \mathscr{F}(\mathcal{M}^r) \xrightarrow{\sim} \hat{\mathbb{G}}_a(\mathcal{M}^r).$$

PROOF. (a) Since

$$\log_{\mathscr{F}} F(X, Y) = \log_{\mathscr{F}} X + \log_{\mathscr{F}} Y$$

as power series (5.2), it suffices to prove that $\log_{\mathscr{F}}(x)$ converges for $x \in \mathcal{M}$. This follows from (5.5) and (6.3a).
(b) Similarly, since $\log_{\mathscr{F}}$ and $\exp_{\mathscr{F}}$ give inverse homomorphisms as power series (5.2), it suffices to show that for $x \in \mathcal{M}^r$, both $\log_{\mathscr{F}}(x)$ and $\exp_{\mathscr{F}}(x)$ converge and are in $\mathcal{M}^r$. This follows immediately from (5.5) and (6.3b). (Note that since v is normalized, $x \in \mathcal{M}^r$ is equivalent to $v(x) \geqslant r$.) $\qquad \square$

Remark 6.5. If $r > v(p)/p - 1$, then (6.4) implies that $\mathscr{F}(\mathcal{M}^r)$ is torsion free, since $\hat{\mathbb{G}}_a(\mathcal{M}^r)$ certainly is. We thus recover the $n = 1$ case of (6.1).

§7. Formal Groups in Characteristic p

For this section we let R be a ring of characteristic $p > 0$.

Definition. Let $\mathscr{F}, \mathscr{G}/R$ be formal groups and $f : \mathscr{F} \to \mathscr{G}$ a homomorphism defined over R. The *height of f*, denoted $ht(f)$, is the largest integer h such that

$$f(T) = g(T^{p^h})$$

for some power series $g(T) \in R[\![T]\!]$. (If $f = 0$, then $ht(f) = \infty$.) The *height of $\mathscr{F}$*, denoted $ht(\mathscr{F})$, is the height of the multiplication by p map $[p] : \mathscr{F} \to \mathscr{F}$.

Example 7.1. If $m \geqslant 1$ is prime to p, then $ht([m]) = 0$, since $[m](T) = mT + \cdots$ (2.3a). On the other hand, (4.4) implies that $ht([p]) \geqslant 1$, so the height of a formal group is always a positive integer.

Proposition 7.2. Let $\mathscr{F}, \mathscr{G}/R$ be formal groups and $f: \mathscr{F} \to \mathscr{G}$ a homomorphism defined over R.
(a) If $f'(0) = 0$, then $f(T) = f_1(T^p)$ for some $f_1(T) \in R[\![T]\!]$.
(b) Write $f(T) = g(T^{p^h})$ with $h = ht(f)$. Then $g'(0) \neq 0$.

PROOF. (a) Let $\omega_{\mathscr{F}}$ and $\omega_{\mathscr{G}}$ be the normalized invariant differentials on $\mathscr{F}$ and $\mathscr{G}$. Then

$$0 = f'(0)\omega_{\mathscr{F}}(T) \qquad\qquad \text{since } f'(0) = 0$$

$$= \omega_{\mathscr{G}}(f(T)) \qquad\qquad \text{from (4.3)}$$

$$= (1 + \cdots)f'(T)\,dT.$$

Hence $f'(T) = 0$, so $f(T) = f_1(T^p)$.
(b) Let $q = p^h$, and if $F(X, Y) = \Sigma a_{ij}X^i Y^j$ is the power series for $\mathscr{F}$, let $\mathscr{F}^{(q)}$ denote the formal group with group law $F^{(q)}(X, Y) = \Sigma a_{ij}^q X^i Y^j$. One easily checks that since char$(R) = p$, $\mathscr{F}^{(q)}$ is a formal group. We now show that g is a homomorphism from $\mathscr{F}^{(q)}$ to $\mathscr{G}$.

$$g(F^{(q)}(X, Y)) = g(F(S, T)^q) \qquad \text{writing } S^q = X, T^q = Y$$

$$= f(F(S, T))$$

$$= G(f(S), f(T)) \qquad \text{since } f \text{ is a homomorphism}$$

$$= G(g(S^q), g(T^q))$$

$$= G(g(X), g(Y)).$$

Hence if $g'(0) = 0$, then from (a) we would have $g(T) = g_1(T^p)$. This would mean that

$$f(T) = g(T^{p^h}) = g_1(T^{p^{h+1}}),$$

contradicting the fact that $h = ht(f)$. Therefore $g'(0) \neq 0$.   □

Next we show that the height behaves well under composition.

Proposition 7.3. Let $\mathscr{F}, \mathscr{G}, \mathscr{H}/R$ be formal groups and

$$\mathscr{F} \xrightarrow{f} \mathscr{G} \xrightarrow{g} \mathscr{H}$$

a chain of homomorphisms. Then

$$ht(g \circ f) = ht(f) + ht(g).$$

PROOF. Write

$$f(T) = f_1(T^{p^{ht(f)}}) \quad \text{and} \quad g(T) = g_1(T^{p^{ht(g)}}).$$

Then

$$g \circ f(T) = g_1(f_1(T^{p^{ht(f)}})^{p^{ht(g)}}) = g_1(\tilde{f}_1(T^{p^{ht(f)+ht(g)}})),$$

where $\tilde{f}_1$ is obtained from f_1 by raising each coefficient to the $p^{ht(g)}$ power. Since g_1 and f_1 have non-zero linear terms (7.2b), it follows that

$$ht(g \circ f) = ht(f) + ht(g). \qquad \square$$

Finally, we return to the study of elliptic curves, and relate the inseparable degree of an isogeny to the height of the corresponding map on the formal groups.

Theorem 7.4. *Let K be a field of characteristic $p > 0$, $E_1, E_2/K$ elliptic curves, and $\phi: E_1 \to E_2$ a non-zero isogeny defined over K. Further let $f: \hat{E}_1 \to \hat{E}_2$ be the homomorphism of formal groups induced by ϕ. Then*

$$\deg_i(\phi) = p^{ht(f)}.$$

Corollary 7.5. *Let E/K be an elliptic curve. Then*

$$ht(\hat{E}) = 1 \text{ or } 2.$$

PROOF. We start with two special cases.

Case 1. ϕ is the p^r-power Frobenius map. Then $\deg_i \phi = p^r$ (II.2.11), while $f(T) = T^{p^r}$, so $ht(f) = r$.

Case 2. ϕ is separable. Let ω be an invariant differential on E_2/K, and let $\omega(T)$ be the corresponding differential on the formal group $\hat{E}_2$. Since ϕ is separable, we have $\phi^* \omega \neq 0$ (II.4.2c), so using (4.3),

$$\omega \circ f(T) = f'(0)\omega(T) \neq 0.$$

Hence $f'(0) \neq 0$, so $ht(f) = 0$.

Now from (II.2.12) every isogeny is the composition of a Frobenius map and a separable map. The theorem now follows from the above two cases and the fact that inseparable degrees multiply and heights add (7.3) under composition.

The corollary is immediate on applying the theorem with $\phi = [p]$, since the map $[p]$ has degree p^2 (III.6.4a). $\qquad \square$

EXERCISES

4.1.   Let $F(X, Y) \in R[\![X, Y]\!]$ be a power series satisfying

$$F(X, Y) = X + Y + \cdots \quad \text{and} \quad F(X, F(Y, Z)) = F(F(X, Y), Z).$$

   (a) Show that there is a unique power series $i(T) \in R[\![T]\!]$ satisfying $F(T, i(T)) = 0$. Show that $i(T)$ also satisfies $F(i(T), T) = 0$.
   (b) Show that $F(X, 0) = X$ and $F(0, Y) = Y$.

4.2. (a) Let $R = \mathbb{F}_p[\varepsilon]/(\varepsilon^2)$. Show that

$$F(X, Y) = X + Y + \varepsilon X Y^p$$

defines a "non-commutative formal group". (I.e. F satisfies all the properties of a formal group law except $F(X, Y) = F(Y, X)$.)

(b) Let R be a ring. Show that there exists a non-commutative formal group defined over R if and only if there is an $\varepsilon \in R$, $\varepsilon \neq 0$, and integers $m, n \geq 1$ such that $m\varepsilon = \varepsilon^n = 0$.

4.3. Let R be the ring of integers in a finite extension of $\mathbb{Z}_p$ and let $\mathscr{F}/R$ be a formal group.

(a) Show that for every $x \in \mathscr{F}(\mathscr{M})$,

$$\underset{n \to \infty}{\text{Limit}} \, [p^n](x) = 0.$$

(b) Show that for every $\alpha \in \mathbb{Z}_p$ there exists a unique homomorphism $[\alpha] : \mathscr{F} \to \mathscr{F}$ with

$$[\alpha](T) = \alpha T + \cdots \in R[\![T]\!].$$

4.4. Let R and $\mathscr{F}/R$ be as in (exer. 4.3), and let h be the height of the formal group over $R/\mathscr{M}$ obtained by reducing modulo $\mathscr{M}$ the coefficients of the formal group law for $\mathscr{F}$. Show that there is a finite extension R' of R with maximal ideal $\mathscr{M}'$ such that the p-torsion in $\mathscr{F}(\mathscr{M}')$ is isomorphic to $(\mathbb{Z}/p\mathbb{Z})^h$. [*Hint:* Use the p-adic version of the Weierstrass preparation theorem [La 8, Ch. 5, Thm. 11.2].] This provides an alternative proof of (7.5).

4.5. Let E be the elliptic curve $y^2 = x^3 + Ax$.

(a) Let $w(z) = \Sigma A_n z^n$ be the power series for E described in section 1. Prove that

$$A_n = 0 \quad \text{unless} \quad n \equiv 3 \, (\text{mod } 4).$$

(b) Let $F(X, Y) = \Sigma F_n(X, Y)$ be the formal group law for E, where $F_n(X, Y)$ is a homogeneous polynomial of degree n. Prove that

$$F_n = 0 \quad \text{unless} \quad n \equiv 1 \, (\text{mod } 4).$$

(c) Prove the analogous statements for the curve $y^2 = x^3 + A$.

4.6. Using notation as in (6.1), let $k = R/\mathscr{M}$, and let h be the height of the formal group $\tilde{\mathscr{F}}/k$ obtained by reducing the coefficients of the formal group law $F(X, Y)$ modulo $\mathscr{M}$. Prove that if $x \in \mathscr{F}(\mathscr{M})$ has exact order p^{n+1}, then

$$v(x) \leq \left[\frac{v(p)}{p^{hn}(p^h - 1)} \right].$$

(Since every formal group has height $h \geq 1$, this strengthens (6.1).)

Chapter V

Elliptic Curves over Finite Fields

In this chapter we study elliptic curves defined over a finite field. The most important arithmetic quantity associated with such a curve is its number of rational points. We start by proving a theorem of Hasse which says that if K is a field with q elements, and E/K is an elliptic curve, then $E(K)$ contains approximately q points, with an error of no more than $2\sqrt{q}$. Following Weil, we then reinterpret and extend this result in terms of a certain generating function, the zeta-function of the curve. In the final two sections we study in some detail the endomorphism ring of an elliptic curve defined over a finite field, and in particular give the relationship between $\mathrm{End}(E)$ and the existence of non-trivial p-torsion points. The notation for chapter V is:

K a perfect field of characteristic $p > 0$

q a power of p

§1. Number of Rational Points

Let K be a finite field with q elements and let E/K be an elliptic curve. We wish to estimate how many points there are in $E(K)$; or equivalently, one more than the number of solutions to the equation

$$E : y^2 + a_1 xy + a_3 y = x^3 + a_2 x^2 + a_4 x + a_6$$

with $(x, y) \in K^2$. Since each value of x yields at most two values of y, a trivial upper bound is $2q + 1$. But since a "randomly chosen" quadratic equation has a 50% chance of being solvable in K, one would expect the right order of magnitude to be q. The following theorem, conjectured by E. Artin in his thesis and proved by Hasse in the 1930's, shows that this heuristic reasoning is correct.

Theorem 1.1. *Let E/K be an elliptic curve defined over the field with q elements. Then*

$$|\#E(K) - q - 1| \leqslant 2\sqrt{q}.$$

PROOF. Choose a Weierstrass equation for E with coefficients in K, and let

$$\phi : E \to E$$

$$(x, y) \to (x^q, y^q)$$

be the qth-power Frobenius morphism (III.4.6). Since the Galois group $G_{\bar{K}/K}$ is (topologically) generated by the qth-power map on $\bar{K}$, we see that for a point $P \in E(\bar{K})$,

$$P \in E(K) \qquad \text{if and only if} \qquad \phi(P) = P.$$

Thus

$$E(K) = \ker(1 - \phi),$$

so

$$\#E(K) = \#\ker(1 - \phi)$$

$$= \deg(1 - \phi) \qquad \text{(III.5.5 and III.4.10c).}$$

(Note the importance of knowing that the map $1 - \phi$ is separable.) Since the degree map on $\text{End}(E)$ is a positive definite quadratic form (III.6.3), and $\deg \phi = q$ (II.2.11c), the following version of the Cauchy–Schwarz inequality gives the desired result. $\qquad\qquad\qquad\qquad\qquad\qquad\qquad\qquad\qquad\qquad\quad\square$

Lemma 1.2. *Let A be an abelian group and*

$$d : A \to \mathbb{Z}$$

a positive definite quadratic form. Then for all $\psi, \phi \in A$,

$$|d(\psi - \phi) - d(\phi) - d(\psi)| \leqslant 2\sqrt{d(\phi)\, d(\psi)}.$$

PROOF. For $\psi, \phi \in A$, let

$$L(\psi, \phi) = d(\psi - \phi) - d(\phi) - d(\psi).$$

By definition of quadratic form, L is bilinear. Since d is positive definite, we have for all $m, n \in \mathbb{Z}$,

$$0 \leqslant d(m\psi - n\phi) = m^2 d(\psi) + mnL(\psi, \phi) + n^2 d(\phi).$$

In particular, taking

$$m = -L(\psi, \phi) \quad \text{and} \quad n = 2d(\psi)$$

yields

$$0 \leqslant d(\psi) \, [4d(\psi) \, d(\phi) - L(\psi, \phi)^2].$$

This gives the desired result provided $\psi \neq 0$, while for $\psi = 0$ the original inequality is trivial. $\square$

Application 1.3. Let $K = \mathbb{F}_q$ be a finite field with q odd. One can use Hasse's result to estimate the value of certain character sums on K. Thus let

$$f(x) = ax^3 + bx^2 + cx + d \in K[x]$$

be a cubic polynomial with distinct roots (in $\bar{K}$), and let

$$\chi : K^* \to \{\pm 1\}$$

be the unique non-trivial character of order 2. (I.e. $\chi(t) = 1$ if and only if t is a square in K^*.) Extend χ to K by setting $\chi(0) = 0$. We wish to use χ to count the K-rational points on the elliptic curve

$$E : y^2 = f(x).$$

Each $x \in K$ will yield 0 (respectively 1 or 2) point(s) $(x, y) \in E(K)$ if $f(x)$ is a non-square (respectively zero or a non-zero square) in K. Thus in terms of χ we find (remember the point at infinity)

$$\# E(K) = 1 + \sum_{x \in K} (\chi(f(x)) + 1)$$

$$= 1 + q + \sum_{x \in K} \chi(f(x)).$$

Comparing this with (1.1), we have proven

Corollary 1.4. *With notation as above,*

$$\left| \sum_{x \in K} \chi(f(x)) \right| \leqslant 2\sqrt{q}.$$

Notice that the sum consists of q terms, each ± 1. Thus (1.4) says that as x runs through K, the values of a cubic polynomial $f(x)$ tend to be equally distributed between squares and non-squares.

§2. The Weil Conjectures

In 1949, André Weil made a series of very general conjectures concerning the number of points on varieties defined over finite fields. In this section we will state Weil's conjectures and prove them for elliptic curves.

Let K be a field with q elements; and for each integer $n \geqslant 1$, let K_n be the extension of K of degree n, so $\# K_n = q^n$. Let V/K be a projective variety, so

V is the set of zeros

$$f_1(x_0, \ldots, x_N) = \cdots f_m(x_0, \ldots, x_N) = 0$$

of a collection of homogeneous polynomials with coefficients in K. Then $V(K_n)$ is the set of points of V with coordinates in K_n. We code the number of such points into a generating function.

Definition. The *zeta function of V/K* is the power series

$$Z(V/K; T) = \exp\left(\sum_{n=1}^{\infty} (\# V(K_n)) \frac{T^n}{n} \right).$$

(Here if $F(T) \in \mathbb{Q}[[T]]$ is a power series with no constant term, then $\exp(F(T))$ is the power series $\Sigma_{i=0}^{\infty} F(T)^i/i!$.) As usual, if we know $Z(V/K; T)$, then we can recover the numbers $\# V(K_n)$ by the formula

$$\# V(K_n) = \frac{1}{(n-1)!} \frac{d^n}{dT^n} \log Z(V/K; T) \Big|_{T=0}.$$

The reason for defining $Z(V/K; T)$ in this way, rather than using the more natural series $\Sigma(\# V(K_n)) T^n$, will become apparent below.

Example 2.1. Let $V = \mathbb{P}^N$. Then a point of $V(K_n)$ is given by homogeneous coordinates $[x_0, \ldots, x_N]$ with $x_i \in K_n$ not all zero. Two sets of coordinates give the same point if they differ by multiplication by an element of K_n^*. Hence

$$\# V(K_n) = \frac{q^{n(N+1)} - 1}{q^n - 1} = \sum_{i=0}^{N} q^{ni},$$

so

$$\log Z(V/K; T) = \sum_{n=1}^{\infty} \left(\sum_{i=0}^{N} q^{ni} \right) \frac{T^n}{n}$$

$$= \sum_{i=0}^{N} -\log(1 - q^i T).$$

Thus

$$Z(\mathbb{P}^N/K; T) = \frac{1}{(1 - T)(1 - qT) \cdots (1 - q^N T)}.$$

Notice that in this case the zeta function is actually in $\mathbb{Q}(T)$. In general, if there are numbers $\alpha_1, \ldots, \alpha_r \in \mathbb{C}$ such that

$$\# V(K_n) = \alpha_1^n \pm \cdots \pm \alpha_r^n \qquad \text{for all } n = 1, 2, \ldots,$$

then $Z(V/K; T)$ will be a rational function.

Theorem 2.2 (Weil Conjectures). *Let K be a field with q elements and V/K a smooth projective variety of dimension n.*

(a) *Rationality*

$$Z(V/K; T) \in \mathbb{Q}(T).$$

(b) *Functional Equation*
   There is an integer ε (*the Euler characteristic of* V) *so that*

$$Z(V/K; 1/q^n T) = \pm q^{n\varepsilon/2} \, T^\varepsilon \, Z(V/K; T).$$

(c) *Riemann Hypothesis*
   There is a factorization

$$Z(V/K; T) = \frac{P_1(T) \cdots P_{2n-1}(T)}{P_0(T)P_2(T) \cdots P_{2n}(T)}$$

with each $P_i(T) \in \mathbb{Z}[T]$. *Further* $P_0(T) = 1 - T$, $P_{2n}(T) = 1 - q^n T$, *and for each* $1 \leqslant i \leqslant 2n - 1$, $P_i(T)$ *factors* (*over* $\mathbb{C}$) *as*

$$P_i(T) = \prod_j (1 - \alpha_{ij} T) \quad \text{with} \quad |\alpha_{ij}| = q^{i/2}.$$

This conjecture was proposed by Weil [We 3] in 1949, and proven by him for curves and abelian varieties. The rationality of the zeta function in general was established by Dwork [Dw] in 1960 using techniques of p-adic functional analysis. Soon thereafter the ℓ-adic cohomology theory developed by M. Artin, Grothendieck, and others gave another proof of the rationality and the functional equation. Then in 1973 Deligne ([Del]) proved the Riemann hypothesis. For a nice overview of Deligne's proof, see [Ka].

We now prove the Weil conjectures for elliptic curves. Let ℓ be a prime different from char(K). Recall that we have a representation (III §7)

$$\text{End}(E) \to \text{End}(T_\ell(E))$$

$$\psi \to \psi_\ell.$$

If we choose a $\mathbb{Z}_\ell$-basis for $T_\ell(E)$, then we can write ψ_ℓ as a 2×2 matrix, and in particular can compute

$$\det(\psi_\ell), \, \text{tr}(\psi_\ell) \in \mathbb{Z}_\ell.$$

Of course, the determinant and trace do not depend on the choice of basis.

Proposition 2.3. *Let* $\psi \in \text{End}(E)$. *Then*

$$\det(\psi_\ell) = \deg(\psi) \quad \text{and} \quad \text{tr}(\psi_\ell) = 1 + \deg(\psi) - \deg(1 - \psi).$$

In particular, $\det(\psi_\ell)$ *and* $\text{tr}(\psi_\ell)$ *are in* $\mathbb{Z}$ *and are independent of* ℓ.

PROOF. Let v_1, v_2 be a $\mathbb{Z}_\ell$-basis for $T_\ell(E)$, and write the matrix of ψ_ℓ for this basis as

$$\psi_\ell = \begin{pmatrix} a & b \\ c & d \end{pmatrix}.$$

Recall there is a non-degenerate, bilinear, alternating pairing (III.8.3)

$$e : T_\ell(E) \times T_\ell(E) \to T_\ell(\mu).$$

We compute

$$\begin{aligned}
e(v_1, v_2)^{\deg \psi} &= e([\deg \psi]v_1, v_2) \\
&= e(\hat{\psi}_\ell \psi_\ell v_1, v_2) \quad\quad \text{(III.6.1a)} \\
&= e(\psi_\ell v_1, \psi_\ell v_2) \quad\quad \text{(III.8.3 and III.6.2f)} \\
&= e(av_1 + cv_2, bv_1 + dv_2) \\
&= e(v_1, v_2)^{ad-bc} \\
&= e(v_1, v_2)^{\det \psi_\ell}.
\end{aligned}$$

Since e is non-degenerate, we conclude that $\deg \psi = \det \psi_\ell$. Finally, for any 2×2 matrix A, a trivial calculation yields

$$\text{tr}(A) = 1 + \det A - \det(1 - A). \quad\quad\quad\quad \square$$

Now let

$$\phi : E \to E$$

be the qth-power Frobenius endomorphism, so as we saw in section 1,

$$\# E(K) = \deg(1 - \phi) \quad\quad \text{(III.5.5 and III.4.10c)}.$$

Similarly, for each integer $n \geqslant 1$, ϕ^n is the $(q^n)^{\text{th}}$-power Frobenius endomorphism, so

$$\# E(K_n) = \deg(1 - \phi^n).$$

From (2.3), the characteristic polynomial of ϕ_ℓ has coefficients in $\mathbb{Z}$, so we can factor it over $\mathbb{C}$ as (say)

$$\det(T - \phi_\ell) = T^2 - \text{tr}(\phi_\ell)T + \det(\phi_\ell) = (T - \alpha)(T - \beta).$$

Further, since for every rational number $m/n \in \mathbb{Q}$,

$$\det((m/n) - \phi_\ell) = \det(m - n\phi_\ell)/n^2 = \deg(m - n\phi)/n^2 \geqslant 0,$$

it follows that the quadratic polynomial $\det(T - \phi_\ell)$ has complex conjugate roots or a double root. Thus $|\alpha| = |\beta|$, so from

$$\alpha\beta = \det \phi_\ell = \deg \phi = q,$$

we conclude that

$$|\alpha| = |\beta| = \sqrt{q}.$$

Finally we note that the characteristic polynomial of ϕ_ℓ^n is given by

$$\det(T - \phi_\ell^n) = (T - \alpha^n)(T - \beta^n).$$

(To compute this, we may put ϕ_ℓ in Jordan normal form, so it is upper triangular with α and β on the diagonal.) In particular,

$$\# E(K_n) = \deg(1 - \phi^n)$$

$$= \det(1 - \phi_\ell^n) \qquad \text{from (2.3)}$$

$$= 1 - \alpha^n - \beta^n + q^n,$$

where $\alpha, \beta \in \mathbb{C}$ are complex conjugates of absolute value $\sqrt{q}$. From this expression it is easy to verify the Weil conjectures for elliptic curves as follows.

Theorem 2.4. *Let K be a field with q elements and E/K an elliptic curve. Then there is an $a \in \mathbb{Z}$ so that*

$$Z(E/K; T) = \frac{1 - aT + qT^2}{(1 - T)(1 - qT)}.$$

Further

$$Z(E/K; 1/qT) = Z(E/K; T), \text{ and}$$

$$1 - aT + qT^2 = (1 - \alpha T)(1 - \beta T) \quad \text{with} \quad |\alpha| = |\beta| = \sqrt{q}.$$

PROOF. We compute

$$\log Z(E/K; T) = \sum_{n=1}^{\infty} (\# E(K_n)) T^n/n \qquad \text{definition}$$

$$= \sum_{n=1}^{\infty} (1 - \alpha^n - \beta^n + q^n) T^n/n \qquad \text{from above}$$

$$= -\log(1 - T) + \log(1 - \alpha T) + \log(1 - \beta T) - \log(1 - qT).$$

Hence

$$Z(E/K; T) = \frac{(1 - \alpha T)(1 - \beta T)}{(1 - T)(1 - qT)},$$

which has the desired form, since from above α and β are complex conjugates of absolute value $\sqrt{q}$, and

$$a = \alpha + \beta = \text{tr}(\phi_\ell) = 1 + q - \deg(1 - \phi) \in \mathbb{Z}.$$

The functional equation is immediate (with $\varepsilon = 0$).   □

Remark 2.5. To see why (2.2c) is called the Riemann hypothesis, we make a change of variable and let $T = q^{-s}$. Thus for an elliptic curve we define

$$\zeta_{E/K}(s) = Z(E/K; q^{-s}) = \frac{1 - aq^{-s} + q^{1-2s}}{(1 - q^{-s})(1 - q^{1-s})}$$

Now the functional equation reads

$$\zeta_{E/K}(1-s) = \zeta_{E/K}(s),$$

which certainly looks familiar. Further, the Riemann hypothesis for $Z(E/K; T)$ proved above says that if $\zeta_{E/K}(s) = 0$, then $|q^s| = \sqrt{q}$, so $\mathrm{Re}(s) = \frac{1}{2}$.

§3. The Endomorphism Ring

Let K be a field of characteristic p, and let E/K be an elliptic curve. We have seen (III.6.4) that there are two possibilities for the group of p-torsion points $E[p]$, namely 0 and $\mathbb{Z}/p\mathbb{Z}$. Similarly, there are several possibilities for the endomorphism ring $\mathrm{End}(E)$ (III §9). The next result shows that the seemingly unrelated values of $E[p]$ and $\mathrm{End}(E)$ are in fact far from independent.

Theorem 3.1 ([De 1]). *Let K be a (perfect) field of characteristic p and E/K an elliptic curve. For each integer $r \geq 1$, let*

$$\phi_r : E \to E^{(p^r)} \quad and \quad \hat{\phi}_r : E^{(p^r)} \to E$$

be the p^r-power Frobenius map and its dual.
(a) *The following are equivalent.*

  (i) *$E[p^r] = 0$ for one (all) $r \geq 1$.*
  (ii) *$\hat{\phi}_r$ is (purely) inseparable for one (all) $r \geq 1$.*
  (iii) *The map $[p] : E \to E$ is purely inseparable and $j(E) \in \mathbb{F}_{p^2}$.*
  (iv) *$\mathrm{End}(E)$ is an order in a quaternion algebra. (Note $\mathrm{End}(E)$ means $\mathrm{End}_{\bar{K}}(E)$.)*
  (v) *The formal group $\hat{E}/K$ associated to E has height 2. (cf. IV, §7.)*

(b) *If the equivalent conditions in (a) do not hold, then*

$$E[p^r] = \mathbb{Z}/p^r\mathbb{Z} \qquad for\ all\ r \geq 1,$$

and the formal group $\hat{E}/K$ has height 1. Further, if $j(E) \in \bar{\mathbb{F}}_p$, then $\mathrm{End}(E)$ is an order in a quadratic imaginary field. (For $j(E)$ transcendental over $\mathbb{F}_p$, see exer. 5.8.)

Definition. If E has the properties given by (3.1a), then we say that E is *supersingular*, or that E has *Hasse invariant 0*. Otherwise we say that E is *ordinary*, or that E has *Hasse invariant 1*.

Remark 3.2.1. There are yet further characterizations of supersingular elliptic curves which are quite important in various applications. See [Har IV §4] for a description in terms of sheaf cohomology, and [La 3, app. 2 §5] for one involving residues of differentials.

Remark 3.2.2. Do not confuse the notions of singularity and supersingularity.

By definition a supersingular elliptic curve is an elliptic curve, so in particular it is a non-singular (i.e. smooth) curve.

PROOF OF (3.1). For notational convenience, we let $\phi = \phi_1$.
(a) Since the Frobenius map is purely inseparable (II.2.11b), we have

$$\deg_s(\hat{\phi}_r) = \deg_s[p^r] = (\deg_s[p])^r = (\deg_s \hat{\phi})^r.$$

Combining this with (III.4.10a) yields

$$\# E[p^r] = \deg_s(\hat{\phi}_r) = \deg_s(\hat{\phi})^r,$$

from which the equivalence of (i) and (ii) follows immediately.
    Next, from (IV.7.4) and the fact that ϕ is purely inseparable, we have

$$\deg_i \hat{\phi} = (\deg_i[p])/p = p^{ht(\hat{E})-1}.$$

Since $\hat{\phi}$ has degree p, this shows that (ii) and (v) are equivalent.
    We now prove (ii) $\Rightarrow$ (iii) $\Rightarrow$ (iv) $\Rightarrow$ (ii).
(ii) $\Rightarrow$ (iii). From (ii), it is immediate that $[p] = \hat{\phi} \circ \phi$ is purely inseparable, so we must show that $j(E) \in \mathbb{F}_{p^2}$. We apply (II.2.12) to the map $\hat{\phi} : E^{(p)} \to E$. Since $\hat{\phi}$ is purely inseparable by assumption, it follows from (II.2.12) and comparison of degrees that $\hat{\phi}$ factors as

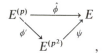

where ϕ' is the pth-power Frobenius map on $E^{(p)}$ and ψ has degree 1. But then ψ is an isomorphism (II.2.4.1), so

$$j(E) = j(E^{(p^2)}) = j(E)^{p^2}   \text{(cf. III.4.6).}$$

Hence $j(E) \in \mathbb{F}_{p^2}$.
(iii) $\Rightarrow$ (iv). Suppose End(E) is not an order in a quaternion algebra. We proceed to derive a contradiction. From (III.9.4) we see that

$$\mathscr{K} = \text{End}(E) \otimes \mathbb{Q}$$

is a number field (either $\mathbb{Q}$ or quadratic imaginary over $\mathbb{Q}$).
    Let E' be any elliptic curve isogenous to E, say $\psi : E \to E'$. Since $\psi \circ [p] = [p] \circ \psi$ and $[p]$ is purely inseparable on E, comparing inseparability degrees shows $[p]$ is also purely inseparable on E'. Hence

$$\# E'[p] = \deg_s[p] = 1,$$

so from our already proven (i) $\Rightarrow$ (ii) $\Rightarrow$ (iii), $j(E') \in \mathbb{F}_{p^2}$. This gives the crucial fact that up to isomorphism, there are only finitely many elliptic curves isogenous to E.
    Now choose a prime $\ell \in \mathbb{Z}$, $\ell \neq p$, so that ℓ remains prime in the rings

End(E') for every E' isogenous to E. (Since there are only finitely many possible End(E')'s, and each is a subring of $\mathcal{K}$, it is easy to find such an ℓ. See exer. 5.5.) From (III.6.4b),

$$E[\ell^i] \cong \mathbb{Z}/\ell^i\mathbb{Z} \times \mathbb{Z}/\ell^i\mathbb{Z},$$

so we can find a sequence of subgroups

$$\Phi_1 \subset \Phi_2 \subset \cdots \subset E \quad \text{with} \quad \Phi_i \cong \mathbb{Z}/\ell^i\mathbb{Z}.$$

Let $E_i = E/\Phi_i$ be the quotient of E by Φ_i (III.4.12), so there is an isogeny $E \to E_i$ with kernel Φ_i. From above, there are only finitely many distinct E_i's, so we can choose integers $m, n > 0$ such that E_{m+n} and E_m are isomorphic. Composing this isomorphism with the natural projection from E_m to E_{m+n}, we produce an endomorphism of E_m,

$$\lambda : E_m \xrightarrow{\text{proj}} E_{m+n} \cong E_m.$$

Note that the kernel of λ is *cyclic* of order ℓ^n. (I.e. $\ker(\lambda) \cong \Phi_{m+n}/\Phi_m$.) But ℓ is prime in the ring End(E_m), so just by comparing degrees we must have $\lambda = u \circ [\ell^{n/2}]$ for some $u \in \text{Aut}(E_m)$. (Also n must be even.) But the kernel of $[\ell^{n/2}]$ is not cyclic for any $n > 0$. This contradiction proves the desired result.

(iv) $\Rightarrow$ (ii). Suppose that (ii) is false, so $\hat{\phi}_r$ is separable for all $r \geq 1$. We proceed to prove that End(E) is commutative, which will contradict (iv) and so give the desired result.

First we show that the natural map

$$\text{End}(E) \to \text{End}(T_p(E))$$

is injective. Suppose that $\psi \in \text{End}(E)$ goes to 0. Then from the definition of $T_p(E)$ we have $\psi(E[p^r]) = 0$ for all $r \geq 1$. Since $[p^r] = \phi_r \circ \hat{\phi}_r$ and ϕ_r is surjective (II.2.3), it follows that

$$\phi_r(\ker \psi) \supset \ker \hat{\phi}_r, \quad \text{and so} \quad \#\ker \psi \geq \#\ker \hat{\phi}_r \quad \text{for all } r \geq 1.$$

On the other hand, we know that

$$\#\ker \hat{\phi}_r = \deg \hat{\phi}_r \quad \text{from (III.4.10c), since } \hat{\phi}_r \text{ is separable,}$$

$$\deg \hat{\phi}_r = \deg \phi_r \quad \text{from (III.6.2e), and}$$

$$\deg \phi_r = p^r \quad \text{from (II.2.11c).}$$

Therefore $\#\ker \psi \geq p^r$ for all $r \geq 1$, which means that $\psi = 0$.

Next, from (III.7.1b), we know that $T_p(E)$ is either 0 or $\mathbb{Z}_p$. But $T_p(E)/pT_p(E) \cong E[p]$, and by assumption $E[p] \neq 0$, so $T_p(E) \cong \mathbb{Z}_p$. Now combining this fact with the injection proven above, we have

$$\text{End}(E) \hookrightarrow \text{End}(T_p(E)) \cong \text{End}(\mathbb{Z}_p) \cong \mathbb{Z}_p.$$

Therefore End(E) is commutative.

(b) From (III.6.4c), $E[p^r]$ is 0 or $\mathbb{Z}/p^r\mathbb{Z}$ for every $r \geq 1$. Hence if condition (i) of (a) is false, then we must have

$$E[p^r] \cong \mathbb{Z}/p^r\mathbb{Z} \qquad \text{for all } r \geqslant 1.$$

Further, since (v) is assumed not to hold, (IV.7.5) implies that $\hat{E}/K$ has height 1.

Suppose now that $j(E) \in \overline{\mathbb{F}}_p$ and E does not satisfy the conditions in (a). We can find an elliptic curve E', defined over a finite field K, which is isomorphic to E (III.1.4bc). Let $\#K = p^r$, so ϕ_r is an endomorphism of E'. Suppose that

$$\phi_r \in \mathbb{Z} \subset \text{End}(E').$$

Then comparing degrees, it would follow that

$$\phi_r = [\pm p^{r/2}]$$

(and necessarily r is even.) But then by (III.4.10) and (II.2.11b),

$$\#E'[p^{r/2}] = \deg_s \phi_r = 1,$$

contradicting the assumption that (i) does not hold. Therefore $\phi_r \notin \mathbb{Z}$, so $\text{End}(E')$ is strictly larger than $\mathbb{Z}$. By assumption, it is not an order in a quaternion algebra, so from (III.9.4) the only remaining possibility is an order in a quadratic imaginary field. Since $\text{End}(E') \cong \text{End}(E)$, this completes the proof.   □

§4. Calculating the Hasse Invariant

From (3.1a) we see that up to isomorphism, there are only finitely many elliptic curves with Hasse invariant 0, since each has j-invariant in $\mathbb{F}_{p^2}$. For $p = 2$, one can easily check (exer. 5.7) that the only supersingular elliptic curve is

$$E : y^2 + y = x^3.$$

For $p > 2$, the following theorem gives a simple criterion for determining whether an elliptic curve is supersingular.

Theorem 4.1. *Let K be finite field of characteristic $p > 2$.*
(a) *Let E/K be an elliptic curve with Weierstrass equation*

$$E : y^2 = f(x),$$

where $f(x) \in K[x]$ is a cubic polynomial with distinct roots (in $\bar{K}$). Then E is supersingular if and only if the coefficient of x^{p-1} in $f(x)^{(p-1)/2}$ is zero.
(b) *Let $m = (p-1)/2$, and define a polynomial*

$$H_p(t) = \sum_{i=0}^{m} \binom{m}{i}^2 t^i.$$

Let $\lambda \in \bar{K}$, $\lambda \neq 0$, 1. Then the elliptic curve

$$E : y^2 = x(x - 1)(x - \lambda)$$

is supersingular if and only if $H_p(\lambda) = 0$.
(c) *The polynomial $H_p(t)$ has distinct roots in $\bar{K}$. Up to isomorphism, there are exactly*

$$[p/12] + \varepsilon_p$$

supersingular elliptic curves in characteristic p, where $\varepsilon_3 = 1$, and for $p \geqslant 5$,

$$\varepsilon_p = 0, 1, 1, 2 \quad if \quad p \equiv 1, 5, 7, 11 \pmod{12}.$$

Remark 4.1.1. The results of this theorem (and more) are mostly in [De 1]. Our proof of (a) follows [Man 1], and (c) is from [Ig]. For a beautiful generalization to curves of higher genus, see [Man 1].

PROOF. (a) Let $q = \#K$, let

$$\chi : K^* \to \{\pm 1\}$$

be the unique non-trivial character of order 2, and extend χ to K be setting $\chi(0) = 0$. As we have seen (1.3), χ can be used to count the number of points of E,

$$\#E(K) = 1 + q + \sum_{x \in K} \chi(f(x)).$$

Since K^* is cyclic of order $q - 1$, for any $z \in K$ we have

$$\chi(z) = z^{(q-1)/2} \qquad \text{in } K.$$

Hence

$$\#E(K) = 1 + \sum_{x \in K} f(x)^{(q-1)/2} \qquad \text{in } K.$$

But again from the cyclic nature of K^*, we have the easy result

$$\sum_{x \in K} x^i = \begin{cases} -1 & \text{if } q - 1 | i \\ 0 & \text{if } q - 1 \nmid i. \end{cases}$$

Since $f(x)$ has degree 3, if we multiply out $f(x)^{(q-1)/2}$ and sum over $x \in K$, the only non-zero term comes from x^{q-1}. Hence if we let

$$A_q = \text{coefficient of } x^{q-1} \text{ in } f(x)^{(q-1)/2},$$

then

$$\#E(K) = 1 - A_q.$$

(Note this equality is taking place in K, so it is actually only a formula for $\#E(K)$ modulo p.)

On the other hand, letting $\phi : E \to E$ be the qth-power Frobenius endomorphism, we have (cf. §2)

$$\# E(K) = \deg(1 - \phi)$$
$$= 1 - a + q,$$

where

$$a = 1 - \deg(1 - \phi) + \deg \phi.$$

(I.e. $[a] = \phi + \hat{\phi}$.) Comparing these two expressions for $\# E(K)$, we see that

$$a = A_q \qquad \text{(as an element of } K\text{)}.$$

Since a is an integer, this shows that

$$A_q = 0 \Leftrightarrow a \equiv 0 \,(\text{mod } p).$$

But $\hat{\phi} = [a] - \phi$, so

$$a \equiv 0 \,(\text{mod } p) \Leftrightarrow \hat{\phi} \text{ is inseparable} \qquad (\text{III.5.5})$$
$$\Leftrightarrow E \text{ is supersingular} \qquad (3.1a(ii)).$$

This proves that

$$A_q = 0 \Leftrightarrow E \text{ is supersingular}.$$

It remains to show that $A_q = 0$ if and only if $A_p = 0$. Writing

$$f(x)^{(p^{r+1}-1)/2} = f(x)^{(p^r-1)/2}(f(x)^{(p-1)/2})^{p^r}$$

and equating coefficients (remember f is a cubic) yields

$$A_{p^{r+1}} = A_{p^r} A_p^{p^r}.$$

This easily gives the desired result by induction on r.

(b) This is a special case of (a). We need the coefficient of x^{p-1} in $[x(x-1)(x-\lambda)]^m$, so the coefficient of x^m in $(x-1)^m(x-\lambda)^m$. That coefficient is

$$\sum_{i=0}^{m} \binom{m}{i} (-\lambda)^i \binom{m}{m-i} (-1)^{m-i},$$

which differs from $H_p(\lambda)$ by a factor of $(-1)^m$.

(c) Let $\mathscr{D}$ be the differential operator

$$\mathscr{D} = 4t(1-t)\frac{d^2}{dt^2} + 4(1-2t)\frac{d}{dt} - 1.$$

Then by direct calculation and rearranging terms, one finds (remember $m = (p-1)/2$)

$$\mathscr{D}H_p(t) = p \sum_{i=0}^{m} (p-2-4i)\binom{m}{i}^2 t^i.$$

In particular, since $\text{char}(K) = p$,

$$\mathscr{D}H_p(t) = 0 \quad \text{in} \quad K[t].$$

Hence the only possible multiple roots of $H_p(t)$ in $\bar{K}$ are $t = 0$ and $t = 1$. But

$$H_p(0) = 1 \quad \text{and} \quad H_p(1) = \binom{p-1}{m} \equiv (-1)^m \pmod{p},$$

so the roots of $H_p(t)$ are indeed distinct; and each root λ gives an elliptic curve

$$E_\lambda : y^2 = x(x-1)(x-\lambda).$$

Now for $p = 3$, $H_p(t) = 1 + t$, so there is exactly one supersingular elliptic curve. It has j-invariant $j(-1) = 1728 = 0$. We assume now that $p \geqslant 5$. Recall that the association $\lambda \to j(\lambda) = j(E_\lambda)$ is exactly six-to-one except for $j = 0$ and 1728, where it is two-to-one and three-to-one respectively (III.1.7). Further, if $H_p(\lambda) = 0$, then for every λ' satisfying $j(\lambda') = j(\lambda)$ we must have $H_p(\lambda') = 0$; since $E_{\lambda'} \cong E_\lambda$, and the roots of $H_p(t)$ give every λ for which E_λ is supersingular. Let $\varepsilon_p(j) = 1$ if the elliptic curve with j-invariant j is supersingular over $\mathbb{F}_p$, and $\varepsilon_p(j) = 0$ if it is ordinary. Then using the fact that $H_p(t)$ has distinct roots, the above considerations imply that the number of supersingular elliptic curves in characteristic $p \geqslant 5$ is

$$\frac{1}{6}\left(\frac{p-1}{2} - 2\varepsilon_p(0) - 3\varepsilon_p(1728)\right) + \varepsilon_p(0) + \varepsilon_p(1728)$$

$$= \frac{p-1}{12} + \frac{2}{3}\varepsilon_p(0) + \frac{1}{2}\varepsilon_p(1728).$$

As we will compute directly below (4.4, 4.5), $\varepsilon_p(0)$ is 0 or 1 according as $p \equiv 1$ or 2 (mod 3), and $\varepsilon_p(1728)$ is 0 or 1 according as $p \equiv 1$ or 3 (mod 4). Taking the four possibilities for p (mod 12) gives the desired result.   □

Remark 4.2. The differential operator $\mathscr{D}$ which we used to prove (4.1c) probably seems rather mysterious. This operator is called the *Picard–Fuchs differential operator* for the Legendre equation

$$y^2 = x(x-1)(x-t).$$

It arises quite naturally when one looks at the Legendre equation as defining a family of elliptic curves parametrized by a complex variable t (i.e. an elliptic surface over $\mathbb{P}^1$). For a nice informal discussion of this connection, see [Cle, §2.10].

Example 4.3. For $p = 11$,

$$H_p(t) = t^5 + 3t^4 + t^3 + t^2 + 3t + 1$$

$$= (t^2 - t + 1)(t + 1)(t - 2)(t + 5) \pmod{11}.$$

The supersingular j-invariants in characteristic 11 are $j = 0$ and $j = 1 = 1728$.

Example 4.4. For which primes $p \geqslant 5$ is the elliptic curve

$$E : y^2 = x^3 + 1$$

supersingular? Notice this curve has $j(E) = 0$. From the criterion of (4.1a), we must compute the coefficient of x^{p-1} in $(x^3 + 1)^{(p-1)/2}$. If $p \equiv 2(3)$, then there is no x^{p-1} term, so E is supersingular; while if $p \equiv 1(3)$, then the coefficient is $\binom{(p-1)/2}{(p-1)/3}$, which is non-zero modulo p, so in this case E is ordinary.

Example 4.5. Similarly we compute for which primes $p \geqslant 3$ the $j = 1728$ elliptic curve

$$E : y^2 = x^3 + x$$

is supersingular. This is determined by the coefficient of $x^{(p-1)/2}$ in $(x^2 + 1)^{(p-1)/2}$, which equals 0 if $p \equiv 3(4)$ and $\binom{(p-1)/2}{(p-1)/4}$ if $p \equiv 1(4)$. Hence E is supersingular if $p \equiv 3(4)$ and ordinary if $p \equiv 1(4)$.

The above examples might suggest that for a given Weierstrass equation with coefficients in $\mathbb{Z}$, the resulting elliptic curve is supersingular in characteristic p for half of the primes. This is in fact true *provided* the elliptic curve has complex multiplication over $\overline{\mathbb{Q}}$, as the $j = 0$ and $j = 1728$ curves do. There is a more precise result, due to Deuring, which we will not give. The next example shows that for elliptic curves without complex multiplication, such supersingular primes seem to be quite rare.

Example 4.6. Let E be given by the equation

$$E : y^2 + y = x^3 - x^2 - 10x - 20,$$

so $j(E) = -2^{12}\,31^3/11^5$. Then either by using the criterion of (4.1a) directly, or else using (exer. 5.10) and $[B-K$, table 3$]$, one finds that the only primes $p < 100$ for which E is supersingular in characteristic p are $p = 2, 19, 29$. (D. H. Lehmer has calculated that there are exactly 27 primes $p < 31500$ for which this E is supersingular.)

It is always true that there are infinitely many primes for which E is ordinary (exer. 5.11); and if E does not have complex multiplication, then Serre has shown that the set of supersingular primes for E has density 0 ([Se 3]). There is a more precise conjecture, due to Lang and Trotter [L–T], which says that for such E,

$$\#\{p < x : E/\mathbb{F}_p \text{ is supersingular}\} \sim c\sqrt{x}/\log x$$

as $x \to \infty$, where $c > 0$ is a constant depending on E. Elkies has shown that every elliptic curve $E/\mathbb{Q}$ has infinitely many primes for which $E/\mathbb{F}_p$ is supersingular (*Invent. Math.* 89, 1987).

EXERCISES

5.1. Verify the Weil conjectures for $V = \mathbb{P}^N$.

5.2. Let K be a finite field, V/K a smooth projective variety of dimension n, and ε the Euler characteristic of V (cf. 2.2b). Show that up to ± 1, the function

$$q^{-\varepsilon s/2} Z(V/K; q^{-s})$$

is invariant under the substitution $s \to n - s$.

5.3. Show that for any square matrix A,

$$\exp\left(\sum_{n=1}^{\infty} (\text{trace } A^n) T^n/n\right) = 1/\det(1 - AT).$$

5.4. Let K be a finite field and $E, E'/K$ elliptic curves.
   (a) If E and E' are isogenous, show that

$$\# E(K) = \# E'(K).$$

   (b) Prove the converse. [*Hint*: Use (III.7.7a).]

5.5. Let $\mathscr{K}/\mathbb{Q}$ be a quadratic imaginary field and let $\mathscr{R}_1, \ldots, \mathscr{R}_n$ be orders in $\mathscr{K}$. Show that there is a prime $\ell \in \mathbb{Z}$ so that $\ell\mathscr{R}_i$ is a prime ideal of $\mathscr{R}_i$ for each $i = 1, 2, \ldots, n$.

5.6. Let $E/\mathbb{F}_q$ be an elliptic curve.
   (a) Prove that $E(\mathbb{F}_q) \cong \mathbb{Z}/m\mathbb{Z} \times \mathbb{Z}/mn\mathbb{Z}$ for some integers $m, n \geqslant 1$ with $\gcd(m, q) = 1$.
   (b) With notation as in (a), prove that $q \equiv 1 \pmod{m}$.
   (c) Suppose that $q = p \geqslant 5$ is prime and that E is supersingular. Prove that $m = 1$ or 2. If $p \equiv 1 \pmod 4$, prove that $m = 1$.

5.7. Show that the only supersingular elliptic curve in characteristic 2 is the curve with j-invariant 0.

5.8. If char $K = p$ and E/K is an elliptic curve with $j(E) \notin \overline{\mathbb{F}}_p$, show that $\text{End}(E) = \mathbb{Z}$. [*Hints*: From (III.9.4) it suffices to show that $\text{End}(E)$ is not an order in a quadratic imaginary field. Now mimic the proof of (3.1a, (iii) $\Rightarrow$ (iv)).]

5.9. Prove the following "mass formula" of Eichler and Deuring:

$$\sum_{\substack{E/\overline{\mathbb{F}}_p \\ \text{supersingular}}} \frac{1}{|\text{Aut } E|} = \frac{p-1}{24}.$$

5.10. Let $E/\mathbb{F}_q$ be an elliptic curve, and $\phi : E \to E$ the qth-power Frobenius endomorphism. Let $p = \text{char}(\mathbb{F}_q)$.
   (a) Prove that E is supersingular if and only if

$$\text{tr}(\phi) \equiv 0 \pmod p.$$

   (Here the trace of ϕ is computed in $\text{End}(T_\ell(E))$ for any prime $\ell \neq p$.)
   (b) Suppose now that $q = p \neq 2, 3$. Prove that E is supersingular if and only if

$$\# E(\mathbb{F}_p) = p + 1.$$

5.11. Let E be an elliptic curve defined over $\mathbb{Q}$, and fix a Weierstrass equation for E with coefficients in $\mathbb{Z}$. Show that there are infinitely many primes $p \in \mathbb{Z}$ so that the reduced curve $E/\mathbb{F}_p$ has Hasse invariant 1. [*Hint*: Fix a prime ℓ, look at those primes p which split completely in the field $\mathbb{Q}(E[\ell])$ obtained by adjoining the coordinates of all ℓ-torsion points of E to $\mathbb{Q}$, and use exer. 5.10.]

5.12. Prove that for every prime $p \geqslant 3$, the elliptic curve

$$E : y^2 = x^3 + x$$

satisfies

$$\# E(\mathbb{F}_p) \equiv 0 \pmod 4.$$

CHAPTER VI

Elliptic Curves over $\mathbb{C}$

Evaluation of the integral giving arc-length on a circle, namely $\int 1/\sqrt{1-x^2}\,dx$, leads to an (inverse) trigonometric function. The analogous problem for the arc-length of an ellipse yields an integral which is not computable in terms of so-called "elementary" functions. Due to the indeterminacy in the sign of the square root, the study of such integrals over $\mathbb{C}$ leads one to look at the Riemann surface on which they are most naturally defined. For the ellipse, this Riemann surface turns out to be the set of complex points on an elliptic curve E. We thus begin our study of elliptic curves over $\mathbb{C}$ by studying certain *elliptic integrals*, which are line integrals on $E(\mathbb{C})$. (In fact, the reason that elliptic curves are so named is because they are the Riemann surfaces associated to the integrals for the arc-length of ellipses. In terms of their geometry, ellipses and elliptic curves actually have little in common, the former having genus 0 and the latter genus 1.)

This study of elliptic integrals leads to questions which are fairly difficult to answer if one restricts attention to integrals. But, as with the more familiar circular functions, it is much easier to develop a theory of the corresponding inverse functions. (Thus trigonometry is not generally built up around the function $\int 1/\sqrt{1-x^2}\,dx$, but rather its inverse $\sin(x)$.) In sections 2 and 3 we give the rudiments of this theory of *elliptic functions*, which are those meromorphic functions having two $\mathbb{R}$-linearly independent periods. We then relate this theory back to our original study of elliptic integrals, and use the relationship to make various deductions about elliptic curves over $\mathbb{C}$. In the final section we amplify on the remark that the study of elliptic curves over $\mathbb{C}$ essentially encompasses the theory of elliptic curves over arbitrary algebraically closed fields of characteristic 0.

The analytic theory of elliptic functions and integrals is a beautiful, but vast, body of knowledge. The contents of this chapter represent a very modest beginning in the study of that theory. Further, we have restricted ourselves to the function theory of a single elliptic curve. There is another sort of function theory which is quite important, namely the theory of *modular functions,* in which one studies functions whose domain is the set of all elliptic curves over $\mathbb{C}$. (See C §12 for a brief discussion and a list of references.) We do not touch on the subject of modular functions in this chapter.

§1. Elliptic Integrals

Let E be an elliptic curve defined over $\mathbb{C}$. Since char$(\mathbb{C}) = 0$ and $\mathbb{C}$ is algebraically closed, there is a Weierstrass equation for E in Legendre form (III.1.7),

$$E : y^2 = x(x - 1)(x - \lambda).$$

Then the natural map

$$E(\mathbb{C}) \to \mathbb{P}^1$$

$$(x, y) \to x$$

is a double cover ramified over precisely the four points $0, 1, \lambda, \infty \in \mathbb{P}^1(\mathbb{C})$.

Recall (III.1.5) that $\omega = dx/y$ is a holomorphic differential form on E. Suppose that we try to define a map

$$E(\mathbb{C}) \xrightarrow{?} \mathbb{C}$$

$$P \to \int_O^P \omega,$$

where the integral is along some path connecting O to P. Of course, this map may not be well-defined. To see why, let $P = (x, y)$, and look at what is happening in $\mathbb{P}^1$.

We are attempting to compute the complex line integral

$$\int_\infty^x \frac{dt}{\sqrt{t(t - 1)(t - \lambda)}}.$$

The problem is that this integral is not path-independent, because the square-

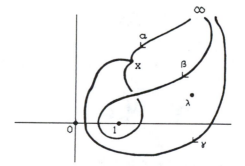

Figure 6.1

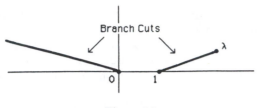

Figure 6.2

root is not single valued. Thus in Figure 6.1, the integrals $\int_\alpha \omega, \int_\beta \omega, \int_\gamma \omega$ are not equal.

In order to make the integral well-defined, it is necessary to make branch cuts. For example, the integral will be path-independent on the complement of the branch cuts illustrated in Figure 6.2, because in this region it is possible to define a single-valued branch of $\sqrt{t(t-1)(t-\lambda)}$. More generally, since the square-root is double-valued, we should take two copies of $\mathbb{P}^1(\mathbb{C})$, make the indicated branch cuts (Figure 6.3), and glue them together along the branch cuts to form a Riemann surface (Figure 6.4). (Note that $\mathbb{P}^1(\mathbb{C}) = \mathbb{C} \cup \{\infty\}$ is topologically nothing more than a 2-sphere.) As is readily seen, the resulting Riemann surface is a torus. It is on this surface that one should really study the integral $\int dt/\sqrt{t(t-1)(t-\lambda)}$; and in fact, elliptic curves first arose when people began to study such integrals. (The very reason that they

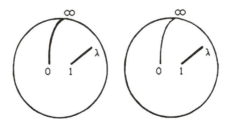

Figure 6.3

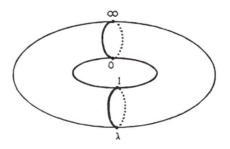

Figure 6.4

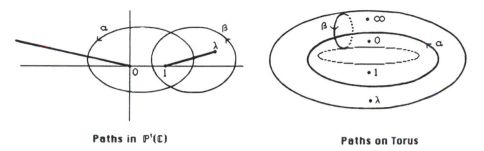

Paths in $\mathbb{P}^1(\mathbb{C})$ **Paths on Torus**

Figure 6.5

are called elliptic curves is because such "elliptic integrals" arise when one attempts to calculate the arc-length of an ellipse (exer. 6.13b).)

Returning now to our hypothetical map

$$E(\mathbb{C}) \to \mathbb{C}$$

$$P \to \int_O^P \omega,$$

it is seen that the indeterminacy comes from integrating across branch cuts in $\mathbb{P}^1$ (or around non-contractible loops on the torus). Figure 6.5 illustrates two closed paths α and β for which the integrals $\int_\alpha \omega$ and $\int_\beta \omega$ may be non-zero. We thus obtain two complex numbers, which are called *periods* of E,

$$\omega_1 = \int_\alpha \omega \quad \text{and} \quad \omega_2 = \int_\beta \omega.$$

Notice that the paths α and β generate the first homology of the torus. Thus any two paths from O to P differ by something homologous to $n_1 \alpha + n_2 \beta$ for some $n_1, n_2 \in \mathbb{Z}$, so the integral $\int_O^P \omega$ is well-defined up to addition of a number of the form $n_1 \omega_1 + n_2 \omega_2$. Let

$$\Lambda = \{n_1 \omega_1 + n_2 \omega_2 : n_1, n_2 \in \mathbb{Z}\}.$$

We have thus shown that there is a well-defined map

$$F : E(\mathbb{C}) \to \mathbb{C}/\Lambda$$

$$P \to \int_O^P \omega \ (\text{mod } \Lambda).$$

Further, using the translation invariance of ω (III.5.1), we can easily verify that F is a homomorphism. (The group law on $\mathbb{C}/\Lambda$ being induced by addition on $\mathbb{C}$.) Thus

$$\int_O^{P+Q} \omega \equiv \int_O^P \omega + \int_P^{P+Q} \omega \equiv \int_O^P \omega + \int_O^Q \tau_P^* \omega \equiv \int_O^P \omega + \int_O^Q \omega \ (\text{mod } \Lambda).$$

Now the quotient space $\mathbb{C}/\Lambda$ will be a Riemann surface (i.e. a one-dimensional complex manifold) if and only if Λ is a lattice; that is, if and only if the periods ω_1 and ω_2 which generate Λ are linearly independent over $\mathbb{R}$. This turns out to be the case; and further, F gives a complex analytic isomorphism from $E(\mathbb{C})$ to $\mathbb{C}/\Lambda$. However, rather than proving these facts here, we will instead turn to the study of the space $\mathbb{C}/\Lambda$ for a given lattice Λ. In section 3 we will construct the inverse to the mapping F, and show that $\mathbb{C}/\Lambda$ is analytically isomorphic to $E_\Lambda(\mathbb{C})$ for a certain elliptic curve $E_\Lambda/\mathbb{C}$. The uniformization theorem (5.1) then says that every elliptic curve $E/\mathbb{C}$ is isomorphic to some E_Λ, from which we will be able to deduce (5.2) that the periods of $E/\mathbb{C}$ are $\mathbb{R}$-linearly independent and that F is a complex analytic isomorphism. (For a direct proof of the independence, which uses only Stokes' theorem in $\mathbb{R}^2$, see [Cle, §2.9].)

§2. Elliptic Functions

Let $\Lambda \subset \mathbb{C}$ be a *lattice*; that is, Λ is a discrete subgroup of $\mathbb{C}$ which contains an $\mathbb{R}$-basis for $\mathbb{C}$. In this section we will study meromorphic functions on the quotient space $\mathbb{C}/\Lambda$; or equivalently, meromorphic functions on $\mathbb{C}$ which are periodic with respect to the lattice Λ.

Definition. An *elliptic function* (*relative to the lattice* Λ) is a meromorphic function $f(z)$ on $\mathbb{C}$ which satisfies

$$f(z + \omega) = f(z) \qquad \text{for all } \omega \in \Lambda, z \in \mathbb{C}.$$

The set of all such functions is denoted $\mathbb{C}(\Lambda)$. $\mathbb{C}(\Lambda)$ is clearly a field.

Definition. A *fundamental parallelogram* for Λ is a set of the form

$$D = \{a + t_1\omega_1 + t_2\omega_2 : 0 \leqslant t_1, t_2 < 1\},$$

where $a \in \mathbb{C}$ and ω_1, ω_2 are a basis for Λ. Thus the map of sets $D \to \mathbb{C}/\Lambda$ is bijective. We denote the closure of D in $\mathbb{C}$ by $\bar{D}$. (A lattice and three different fundamental parallelograms are illustrated in Figure 6.6.)

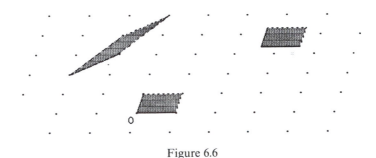

Figure 6.6

Proposition 2.1. *An elliptic function with no poles (or no zeros) is constant.*

PROOF. Suppose that $f(z) \in \mathbb{C}(\Lambda)$ is holomorphic. Let D be a fundamental parallelogram for Λ. Then the periodicity of f implies that

$$\sup_{z \in \mathbb{C}} |f(z)| = \sup_{z \in \bar{D}} |f(z)|.$$

But f is continuous and $\bar{D}$ is compact, so $|f(z)|$ is bounded on $\bar{D}$, hence it is bounded on all of $\mathbb{C}$. Therefore, by Liouville's theorem ([Ahl, ch. 4, §2.3]), f is constant. Finally, if f has no zeros, look at $1/f$. $\qquad\square$

Let f be an elliptic function, and let $w \in \mathbb{C}$. Then, as for any meromorphic function, we can define

$$\mathrm{ord}_w(f) = \text{order of vanishing of } f \text{ at } w, \text{ and}$$

$$\mathrm{res}_w(f) = \text{residue of } f \text{ at } w$$

(cf. [Ahl, ch. 4, §3.2, §5.1]). However, since f is elliptic, we see that the order and residue of f remain the same if w is replaced by $w + \omega$ for any $\omega \in \Lambda$. This prompts the following convention.

Notation. By $\Sigma_{w \in \mathbb{C}/\Lambda}$ we mean a sum over $w \in D$, where D is a fundamental parallelogram for Λ. (By implication, the resulting sum is independent of the choice of D.)

Notice that (2.1) is the complex analogue of (II.1.2), which says that an algebraic function without poles is constant. The next theorem and corollary continue this theme by proving for $\mathbb{C}/\Lambda$ results analogous to parts of (II.3.1) and (III.3.5).

Theorem 2.2. *Let $f \in \mathbb{C}(\Lambda)$.*

(a) $\displaystyle\sum_{w \in \mathbb{C}/\Lambda} \mathrm{res}_w(f) = 0.$

(b) $\displaystyle\sum_{w \in \mathbb{C}/\Lambda} \mathrm{ord}_w(f) = 0.$

(c) $\displaystyle\sum_{w \in \mathbb{C}/\Lambda} \mathrm{ord}_w(f) w \in \Lambda.$

PROOF. Let D be a fundamental parallelogram for Λ such that $f(z)$ has no poles or zeros on the boundary ∂D of D. All three parts of the theorem are simple applications of the residue theorem [Ahl, ch. 4, thm. 19] applied to appropriately chosen functions on D.

(a) By the residue theorem,

$$\sum_{w \in \mathbb{C}/\Lambda} \mathrm{res}_w(f) = \frac{1}{2\pi i} \int_{\partial D} f(z)\, dz.$$

Now the periodicity of f implies that the integrals along the opposite sides of the parallelogram cancel, so the total integral around the boundary of D is zero.

(b) The periodicity of $f(z)$ implies that $f'(z)$ is also periodic, so applying (a) to the elliptic function $f'(z)/f(z)$ gives

$$\sum_{w \in \mathbb{C}/\Lambda} \mathrm{ord}_w(f) = \frac{1}{2\pi i} \int_{\partial D} \frac{f'(z)}{f(z)} \, dz = 0.$$

(c) We apply the residue theorem to the function $zf'(z)/f(z)$

$$\sum_{w \in \mathbb{C}/\Lambda} \mathrm{ord}_w(f)w = \frac{1}{2\pi i} \int_{\partial D} zf'(z)/f(z) \, dz$$

$$= \frac{1}{2\pi i} \left(\int_a^{a+\omega_1} + \int_{a+\omega_1}^{a+\omega_1+\omega_2} + \int_{a+\omega_1+\omega_2}^{a+\omega_2} + \int_{a+\omega_2}^{a} \right) zf'(z)/f(z) \, dz.$$

Now in the second (respectively third) integral make the change of variable $z \to z - \omega_1$ (respectively $z - \omega_2$). Then using the periodicity of f'/f yields

$$\sum_{w \in \mathbb{C}/\Lambda} \mathrm{ord}_w(f)w = -\frac{\omega_2}{2\pi i} \int_a^{a+\omega_1} \frac{f'(z)}{f(z)} \, dz + \frac{\omega_1}{2\pi i} \int_a^{a+\omega_2} \frac{f'(z)}{f(z)} \, dz.$$

But for any meromorphic function $g(z)$, the integral

$$\frac{1}{2\pi i} \int_a^b \frac{g'(z)}{g(z)} \, dz$$

is the winding number around 0 of the path

$$[0, 1] \to \mathbb{C}, \qquad t \to g((1 - t)a + tb);$$

and in particular, if $g(a) = g(b)$, then the integral is an integer. Hence the periodicity of $f'(z)/f(z)$ implies that $\sum \mathrm{ord}_w(f)w$ has the desired form. $\square$

Definition. The *order* of an elliptic function is its number of poles (counted with multiplicity) in any fundamental parallelogram. (Note that from (2.2b), the order is also equal to the number of zeros.)

Corollary 2.3. *A non-constant elliptic function has order at least 2.*

PROOF. If $f(z)$ has a single simple pole, then from (2.2a) the residue at that pole is 0, so f is actually holomorphic. Now apply (2.1). $\square$

We now define the *divisor group* $\mathrm{Div}(\mathbb{C}/\Lambda)$ to be the group of formal linear combinations $\sum_{w \in \mathbb{C}/\Lambda} n_w(w)$ with $n_w \in \mathbb{Z}$ and $n_w = 0$ for all but finitely many w. Then for $D = \sum n_w(w) \in \mathrm{Div}(\mathbb{C}/\Lambda)$, we define

$$\deg D = \text{degree of } D = \sum n_w \quad \text{and} \quad \mathrm{Div}^0(\mathbb{C}/\Lambda) = \{D \in \mathrm{Div}(\mathbb{C}/\Lambda) : \deg D = 0\}.$$

From (2.2b), for any $f \in \mathbb{C}(\Lambda)^*$ we can define a divisor $\mathrm{div}(f) \in \mathrm{Div}^0(\mathbb{C}/\Lambda)$ by

$$\mathrm{div}(f) = \sum_{w \in \mathbb{C}/\Lambda} \mathrm{ord}_w(f)(w).$$

Clearly the map $\mathrm{div}: \mathbb{C}(\Lambda)^* \to \mathrm{Div}^0(\mathbb{C}/\Lambda)$ is a homomorphism, since each ord_w is a valuation. Finally, we define a *summation map*

$$\mathit{sum}: \mathrm{Div}^0(\mathbb{C}/\Lambda) \to \mathbb{C}/\Lambda \qquad \mathit{sum}(\textstyle\sum n_w(w)) = \sum n_w w \ (\mathrm{mod}\ \Lambda).$$

The following exact sequence encompasses our main results on $\mathbb{C}/\Lambda$, as well as one fact (3.4) to be proven in the next section.

Theorem 2.4. *The sequence*

$$1 \to \mathbb{C}^* \to \mathbb{C}(\Lambda)^* \xrightarrow{\mathrm{div}} \mathrm{Div}^0(\mathbb{C}/\Lambda) \xrightarrow{\mathrm{sum}} \mathbb{C}/\Lambda \to 0$$

is exact.

PROOF. Exactness on the left is clear, and on the right follows from $\mathit{sum}((w) - (0)) = w$. Exactness at $\mathbb{C}(\Lambda)^*$ is (2.1), and exactness at $\mathrm{Div}^0(\mathbb{C}/\Lambda)$ is (2.2c) and (3.4). $\qquad\square$

§3. Construction of Elliptic Functions

In order to show that the results of section 2 are not vacuous, we must construct some non-constant elliptic functions. By (2.3), any such function will have order at least 2. Following Weierstrass, we look for a function with a pole of order 2 at $z = 0$.

Definition. Let $\Lambda \subset \mathbb{C}$ be a lattice. The *Weierstrass $\wp$-function (relative to Λ)* is defined by the series

$$\wp(z; \Lambda) = \frac{1}{z^2} + \sum_{\substack{\omega \in \Lambda \\ \omega \neq 0}} \frac{1}{(z-\omega)^2} - \frac{1}{\omega^2}.$$

The *Eisenstein series of weight $2k$ (for Λ)* is the series

$$G_{2k}(\Lambda) = \sum_{\substack{\omega \in \Lambda \\ \omega \neq 0}} \omega^{-2k}.$$

(For notational convenience, we write $\wp(z)$ and G_{2k} if the lattice Λ has been fixed.)

Theorem 3.1. *Let $\Lambda \subset \mathbb{C}$ be a lattice.*
(a) *The Eisenstein series G_{2k} for Λ is absolutely convergent for all $k > 1$.*
(b) *The series defining the Weierstrass $\wp$-function converges absolutely and*

uniformly on every compact subset of $\mathbb{C} - \Lambda$. It defines a meromorphic function on $\mathbb{C}$ having a double pole with residue 0 at each lattice point and no other poles.
(c) The Weierstrass $\wp$-function is an even elliptic function.

PROOF. (a) Since Λ is discrete in $\mathbb{C}$, one easily checks that there is a constant $c = c(\Lambda)$ so that for all $N \geqslant 1$, the number of lattice points in an annulus satisfies

$$\#\{\omega \in \Lambda : N \leqslant |\omega| < N + 1\} < cN.$$

(See exer. 6.2.) Hence

$$\sum_{\substack{\omega \in \Lambda \\ |\omega| \geqslant 1}} \frac{1}{|\omega|^{2k}} \leqslant \sum_{N=1}^{\infty} \frac{\#\{\omega \in \Lambda : N \leqslant |\omega| < N + 1\}}{N^{2k}} < \sum_{N=1}^{\infty} \frac{c}{N^{2k-1}} < \infty.$$

(b) If $|\omega| > 2|z|$, then

$$\left| \frac{1}{(z - \omega)^2} - \frac{1}{\omega^2} \right| = \left| \frac{z(2\omega - z)}{\omega^2(z - \omega)^2} \right| \leqslant \frac{10|z|}{|\omega|^3}.$$

Hence from (a) we see that the series for $\wp(z)$ is absolutely convergent for $z \in \mathbb{C} - \Lambda$, and uniformly convergent on every compact subset of $\mathbb{C} - \Lambda$. Therefore it defines a holomorphic function on $\mathbb{C} - \Lambda$; and from the series expansion it is clear that $\wp(z)$ has a double pole with residue 0 at each point of Λ.

(c) Clearly $\wp(z) = \wp(-z)$. (Replace ω by $-\omega$ in the sum.) Since the series for $\wp$ is uniformly convergent, we can compute its derivative $\wp'(z)$ by termwise differentiation:

$$\wp'(z) = -2 \sum_{\omega \in \Lambda} \frac{1}{(z - \omega)^3}.$$

From this expression it is clear that $\wp'$ is an elliptic function, so integrating yields

$$\wp(z + \omega) = \wp(z) + c(\omega) \qquad \text{for all } z \in \mathbb{C} - \Lambda,$$

where $c(\omega) \in \mathbb{C}$ is independent of z. Now let $z = -\omega/2$ and use the evenness of $\wp(z)$ to conclude that $c(\omega) = 0$. $\square$

Next we show that every elliptic function can be expressed in terms of the Weierstrass $\wp$-function and its derivative. (This is the analogue of (III.3.1.1).)

Theorem 3.2. *Let Λ be a lattice. Then*

$$\mathbb{C}(\Lambda) = \mathbb{C}(\wp(z), \wp'(z)).$$

(I.e. Every elliptic function is a rational combination of $\wp$ and $\wp'$.)

PROOF. Let $f(z) \in \mathbb{C}(\Lambda)$. Writing

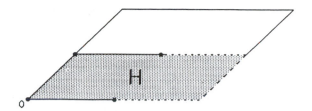

Figure 6.7

$$f(z) = \frac{1}{2}[f(z) + f(-z)] + \frac{1}{2}[f(z) - f(-z)],$$

we see that it suffices to prove the theorem for odd and even functions. But if f is odd, then $\wp'f$ is even, so we are reduced to the case that f is even.

Now if f is even, we have

$$\mathrm{ord}_w f = \mathrm{ord}_{-w} f$$

for every $w \in \mathbb{C}$. Further, we claim that if $2w \in \Lambda$, then $\mathrm{ord}_w f$ is even. To see this, differentiate $f(z) = f(-z)$ repeatedly to obtain

$$f^{(i)}(z) = (-1)^i f^{(i)}(-z).$$

Hence if $2w \in \Lambda$, so $f^{(i)}(w) = f^{(i)}(-w)$, then this implies that $f^{(i)}(w) = 0$ for all odd i, so $\mathrm{ord}_w f$ must be even.

Now let D be a fundamental parallelogram for Λ, and let H be "half" of D, so that H is a fundamental domain for $(\mathbb{C}/\Lambda)/\{\pm 1\}$. (I.e.

$$\mathbb{C} = (H + \Lambda) \cup (-H + \Lambda).$$

See Figure 6.7.) The above considerations imply that the divisor of $f(z)$ has the form

$$\sum_{w \in H} n_w[(w) + (-w)]$$

for certain *integers* n_w. (Note if $2w \in \Lambda$, we are using the fact that $\mathrm{ord}_w f$ is even.)

Next consider the function

$$g(z) = \prod_{w \in H - 0} [\wp(z) - \wp(w)]^{n_w}.$$

Since the divisor of $\wp(z) - \wp(w)$ is $(w) + (-w) - 2(0)$, we see that f and g have exactly the same zeros and poles except possibly at $w = 0$. But then (2.2b) implies that they have the same order at 0, also. Therefore $f(z)/g(z)$ is a holomorphic elliptic function, hence is constant (2.1). This proves that $f(z) = cg(z) \in \mathbb{C}(\wp(z), \wp'(z))$.   □

In order to prove the converse to (2.2), it is convenient to introduce a "theta function" for Λ.

Definition. The *Weierstrass σ-function (relative to* Λ*)* is the function defined by the product

$$\sigma(z) = \sigma(z; \Lambda) = z \prod_{\substack{\omega \in \Lambda \\ \omega \neq 0}} \left(1 - \frac{z}{\omega}\right) e^{(z/\omega) + (z/\omega)^2/2}.$$

The following lemma gives the basic facts that we will need concerning $\sigma(z)$. For a further description, see exers. 6.3, 6.4.

Lemma 3.3. (a) *The infinite product for* $\sigma(z)$ *defines a holomorphic function on all of* $\mathbb{C}$. *It has simple zeros at each* $z \in \Lambda$, *and no other zeros.*

(b) $$\frac{d^2}{dz^2} \log \sigma(z) = -\wp(z) \qquad \text{for all } z \in \mathbb{C} - \Lambda.$$

(c) *For any* $\omega \in \Lambda$ *there are constants* $a, b \in \mathbb{C}$ *such that*

$$\sigma(z + \omega) = e^{az+b} \sigma(z) \qquad \text{for all } z \in \mathbb{C}.$$

PROOF. (a) That the infinite product is absolutely and uniformly convergent on $\mathbb{C}$ follows from (3.1a) and standard facts about convergence of infinite products ([Ahl, ch. 5, §2.3]). The location and order of the zeros is clear by inspection.

(b) From (a) we can differentiate

$$\log \sigma(z) = \log z + \sum \left\{ \log\left(1 - \frac{z}{\omega}\right) - \frac{z}{\omega} - \frac{1}{2}\left(\frac{z}{\omega}\right)^2 \right\}$$

term by term. Its second derivative is, up to sign, exactly the series defining $\wp(z)$.

(c) From (3.1c), $\wp(z + \omega) = \wp(z)$. Now integrate twice and use (b) to obtain

$$\log \sigma(z + \omega) = \log \sigma(z) + az + b$$

for constants of integration $a, b \in \mathbb{C}$.   □

Proposition 3.4. *Let* $n_1, \ldots, n_r \in \mathbb{Z}$ *and* $z_1, \ldots, z_r \in \mathbb{C}$ *satisfy*

$$\sum n_i = 0 \quad \text{and} \quad \sum n_i z_i \in \Lambda.$$

Then there exists an elliptic function $f(z) \in \mathbb{C}(\Lambda)$ *satisfying*

$$\text{div}(f) = \sum n_i(z_i).$$

More precisely, if we normalize so that $\sum n_i z_i = 0$, *then*

$$f(z) = \prod \sigma(z - z_i)^{n_i}.$$

PROOF. Let $\lambda = \Sigma n_i z_i \in \Lambda$. Replacing $n_1(z_1) + \cdots + n_r(z_r)$ by $n_1(z_1) + \cdots + n_r(z_r) + (0) - (\lambda)$, we may assume that $\Sigma n_i z_i = 0$. Then (3.3a) implies that

$$f(z) = \prod \sigma(z - z_i)^{n_i}$$

has the correct zeros and poles; while (3.3c) allows us to compute (for any $\omega \in \Lambda$)

$$f(z + \omega)/f(z) = \prod e^{(a(z-z_i)+b)n_i}$$
$$= e^{(az+b)\Sigma n_i} e^{-a\Sigma n_i z_i}$$
$$= 1.$$

Therefore $f(z) \in \mathbb{C}(\Lambda)$.   □

We next derive the Laurent series expansion for $\wp(z)$ about $z = 0$, from which we will deduce the fundamental *algebraic* relation satisfied by $\wp(z)$ and $\wp'(z)$.

Theorem 3.5. (a) *The Laurent series for* $\wp(z)$ *about* $z = 0$ *is given by*

$$\wp(z) = z^{-2} + \sum_{k=1}^{\infty} (2k + 1)G_{2k+2} z^{2k}.$$

(b) *For all* $z \in \mathbb{C}$ *with* $z \notin \Lambda$,

$$\wp'(z)^2 = 4\wp(z)^3 - 60G_4 \wp(z) - 140G_6.$$

PROOF. (a) Provided $|z| < |\omega|$, we have

$$(z - \omega)^{-2} - \omega^{-2} = \omega^{-2}[(1 - z/\omega)^{-2} - 1]$$
$$= \sum_{n=1}^{\infty} (n + 1)z^n/\omega^{n+2}.$$

Substituting this into the series for $\wp(z)$ and reversing the order of summation gives the desired result.

(b) We write out the first few terms in various Laurent expansions:

$$\wp'(z)^2 = 4z^{-6} - 24G_4 z^{-2} - 80G_6 + \cdots$$
$$\wp(z)^3 = z^{-6} + 9G_4 z^{-2} + 15G_6 + \cdots$$
$$\wp(z) = z^{-2} + 3G_4 z^2 + \cdots.$$

Comparing these, we see that the function

$$f(z) = \wp'(z)^2 - 4\wp(z)^3 + 60G_4 \wp(z) + 140G_6$$

is holomorphic around $z = 0$ and vanishes at $z = 0$. But it is also elliptic relative to Λ, and from (3.1b) it is holomorphic away from Λ, hence it is a holomorphic elliptic function. From (2.1), we conclude that $f(z)$ is identically zero.   □

Remark 3.5.1. It is standard notation to set

$$g_2 = g_2(\Lambda) = 60G_4 \quad \text{and} \quad g_3 = g_3(\Lambda) = 140G_6.$$

Then the algebraic relation between $\wp(z)$ and $\wp'(z)$ reads

$$\wp'(z)^2 = 4\wp(z)^3 - g_2\wp(z) - g_3.$$

Let $E/\mathbb{C}$ be an elliptic curve. Since the group law $E \times E \to E$ is given by everywhere locally defined rational functions (III.3.6), we see in particular that $E = E(\mathbb{C})$ is a *complex Lie group*. (I.e. It is a complex manifold with a group law given locally by complex analytic functions.) Similarly, if $\Lambda \subset \mathbb{C}$ is a lattice, then $\mathbb{C}/\Lambda$ with its natural addition is a complex Lie group. The next proposition shows that $\mathbb{C}/\Lambda$ is always complex analytically isomorphic to an elliptic curve.

Proposition 3.6. *Let g_2 and g_3 be the quantities associated to a lattice $\Lambda \subset \mathbb{C}$.*
(a) *The polynomial*

$$f(x) = 4x^3 - g_2 x - g_3$$

has distinct roots. Its discriminant

$$\Delta(\Lambda) = g_2^3 - 27g_3^2$$

is not zero.
(b) *Let $E/\mathbb{C}$ be the curve*

$$E : y^2 = 4x^3 - g_2 x - g_3,$$

which is an elliptic curve from (a). *Then the map*

$$\phi : \mathbb{C}/\Lambda \to E \subset \mathbb{P}^2(\mathbb{C})$$

$$z \to [\wp(z), \wp'(z), 1]$$

is a complex analytic isomorphism of complex Lie groups. (I.e. It is an isomorphism of Riemann surfaces which is a group homomorphism.)

PROOF. (a) Let $\{\omega_1, \omega_2\}$ be a basis for Λ, and let $\omega_3 = \omega_1 + \omega_2$. Then since $\wp'(z)$ is an odd elliptic function, we see that

$$\wp'(\omega_i/2) = -\wp'(-\omega_i/2) = -\wp'(\omega_i/2),$$

so $\wp'(\omega_i/2) = 0$. Hence from (3.5b), $f(x)$ vanishes at each $x = \wp(\omega_i/2)$, so it suffices to show that these three values are distinct.

The function $\wp(z) - \wp(\omega_i/2)$ is even, hence has at least a double zero at $z = \omega_i/2$. But since it has order 2, these are the only zeros (in an appropriate fundamental parallelogram). Therefore $\wp(\omega_j/2) \neq \wp(\omega_i/2)$ for $j \neq i$.
(b) The image of ϕ is contained in E from (3.5b). To see that ϕ is surjective, let $(x, y) \in E$. Then $\wp(z) - x$ is a non-constant elliptic function, so from (2.1) it has a zero, say $z = a$. It follows that $\wp'(a)^2 = y^2$, so replacing a by $-a$ if necessary, we obtain $\wp'(a) = y$. Then $\phi(a) = (x, y)$.

Next suppose that $\phi(z_1) = \phi(z_2)$. Assume first that $2z_1 \notin \Lambda$. Then the function $\wp(z) - \wp(z_1)$ has order 2 and zeros z_1, $-z_1$, z_2. It follows that

$z_2 \equiv \pm z_1 \pmod{\Lambda}$; and now

$$\wp'(z_1) = \wp'(z_2) = \wp'(\pm z_1) = \pm \wp'(z_1)$$

implies that $z_2 \equiv z_1 \pmod{\Lambda}$. [Note $\wp'(z_1) \neq 0$ from the proof of (a).] Now if instead $2z_1 \in \Lambda$, then $\wp(z) - \wp(z_1)$ has a double zero at z_1, and vanishes at z_2, so again we conclude that $z_2 \equiv z_1 \bmod \Lambda$. This proves that ϕ is injective.

Next, in order to show that ϕ is an analytic isomorphism, we compute its effect on the cotangent space. At every point of E, dx/y is a non-vanishing holomorphic differential. Since

$$\phi^*(dx/y) = d\wp(z)/\wp'(z) = dz$$

is similarly non-vanishing and holomorphic at every point of $\mathbb{C}/\Lambda$, we see that ϕ is a local isomorphism. But ϕ is bijective from above, so this implies that it is a global isomorphism.

Finally, to see that ϕ is a group homomorphism, let $z_1, z_2 \in \mathbb{C}$. From (3.4), there is a function $f(z) \in \mathbb{C}(\Lambda)$ with divisor

$$\mathrm{div}(f) = (z_1 + z_2) - (z_1) - (z_2) + (0).$$

Using (3.2), we can write $f(z) = F(\wp(z), \wp'(z))$ for some *rational* function $F(X, Y) \in \mathbb{C}(X, Y)$; and then considering $F(x, y) \in \mathbb{C}(x, y) = \mathbb{C}(E)$, we have

$$\mathrm{div}(F) = (\phi(z_1 + z_2)) - (\phi(z_1)) - (\phi(z_2)) + (\phi(0)).$$

It follows from (III.3.5) that

$$\phi(z_1 + z_2) = \phi(z_1) + \phi(z_2). \qquad \square$$

§4. Maps—Analytic and Algebraic

In this section we investigate complex analytic maps between complex tori. It turns out that they all have a particularly simple form; and, somewhat more surprisingly, the maps which they induce on the corresponding elliptic curves via (3.6b) are actually isogenies (i.e. given by rational functions).

Thus let Λ_1 and Λ_2 be lattices in $\mathbb{C}$. If $\alpha \in \mathbb{C}$ has the property that $\alpha\Lambda_1 \subset \Lambda_2$, then scalar multiplication by α

$$\phi_\alpha : \mathbb{C}/\Lambda_1 \to \mathbb{C}/\Lambda_2 \qquad \phi_\alpha(z) = \alpha z \pmod{\Lambda_2}$$

is clearly a holomorphic homomorphism. We now show that these are essentially the only holomorphic maps.

Theorem 4.1. (a) *With notation as above, the association*

$$\{\alpha \in \mathbb{C} : \alpha\Lambda_1 \subset \Lambda_2\} \to \{\text{holomorphic maps } \phi : \mathbb{C}/\Lambda_1 \to \mathbb{C}/\Lambda_2 \text{ with } \phi(0) = 0\}$$

$$\alpha \to \phi_\alpha$$

is a bijection.

(b) *Let E_1 and E_2 be the elliptic curves corresponding to the lattices Λ_1 and Λ_2 as in (3.6b). Then the natural inclusion*

$$\{\text{isogenies } \phi : E_1 \to E_2\} \to \{\text{holomorphic maps } \phi : \mathbb{C}/\Lambda_1 \to \mathbb{C}/\Lambda_2 \text{ with } \phi(0) = 0\}$$

is a bijection.

PROOF. (a) If $\phi_\alpha = \phi_\beta$, then for all $z \in \mathbb{C}$, $\alpha z \equiv \beta z \pmod{\Lambda_2}$. Hence the map $z \to (\alpha - \beta)z$ sends $\mathbb{C}$ to Λ_2; and since Λ_2 is discrete, this map must be constant. Therefore $\alpha = \beta$.

Next let $\phi : \mathbb{C}/\Lambda_1 \to \mathbb{C}/\Lambda_2$ be a holomorphic map with $\phi(0) = 0$. Then, since $\mathbb{C}$ is simply connected, we can lift ϕ to a holomorphic map $f : \mathbb{C} \to \mathbb{C}$ with $f(0) = 0$ so that the following diagram commutes:

$$
\begin{array}{ccc}
\mathbb{C} & \overset{f}{\to} & \mathbb{C} \\
\downarrow & & \downarrow \\
\mathbb{C}/\Lambda_1 & \overset{\phi}{\to} & \mathbb{C}/\Lambda_2.
\end{array}
$$

Now for any $\omega \in \Lambda_1$, $f(z + \omega) \equiv f(z) \pmod{\Lambda_2}$ for all $z \in \mathbb{C}$. Again using the discreteness of Λ_2, we see that $f(z + \omega) - f(z)$ must be independent of z. Thus

$$f'(z + \omega) = f'(z) \qquad \text{for all } z \in \mathbb{C} \text{ and all } \omega \in \Lambda_1.$$

This says that $f'(z)$ is a holomorphic elliptic function, so from (2.1) it is constant. Therefore $f(z) = \alpha z + \gamma$ for some $\alpha, \gamma \in \mathbb{C}$. Now $f(0) = 0$ implies that $\gamma = 0$, and $f(\Lambda_1) \subset \Lambda_2$ implies $\alpha\Lambda_1 \subset \Lambda_2$, so $\phi = \phi_\alpha$.

(b) First note that since an isogeny is given locally by *everywhere defined* rational functions (i.e. it is a morphism), the map induced on the corresponding complex tori will be holomorphic. Thus our association

$$\text{Hom}(E_1, E_2) \to \text{Holom. Map}(\mathbb{C}/\Lambda_1, \mathbb{C}/\Lambda_2)$$

is well-defined; and it is clearly injective.

We now prove surjectivity. From (a), it suffices to consider a map of the form ϕ_α, where $\alpha \in \mathbb{C}^*$ satisfies $\alpha\Lambda_1 \subset \Lambda_2$. The induced map on Weierstrass equations is given by

$$E_1 \to E_2$$

$$[\wp(z, \Lambda_1), \wp'(z, \Lambda_2), 1] \to [\wp(\alpha z, \Lambda_2), \wp'(\alpha z, \Lambda_2), 1],$$

so we must show that $\wp(\alpha z, \Lambda_2)$ and $\wp'(\alpha z, \Lambda_2)$ can be expressed as rational functions of $\wp(z, \Lambda_1)$ and $\wp'(z, \Lambda_1)$. But using the fact that $\alpha\Lambda_1 \subset \Lambda_2$, we see that for any $\omega \in \Lambda_1$,

$$\wp(\alpha(z + \omega), \Lambda_2) = \wp(\alpha z + \alpha\omega, \Lambda_2) = \wp(\alpha z, \Lambda_2);$$

and similarly for $\wp'(\alpha z, \Lambda_2)$. Thus $\wp(\alpha z, \Lambda_2)$ and $\wp'(\alpha z, \Lambda_2)$ are in $\mathbb{C}(\Lambda_1)$. But $\mathbb{C}(\Lambda_1) = \mathbb{C}(\wp(z, \Lambda_1), \wp'(z, \Lambda_1))$ from (3.2), which gives the desired result. □

Corollary 4.1.1. *Let $E_1/\mathbb{C}$ and $E_2/\mathbb{C}$ be elliptic curves corresponding to lattices Λ_1 and Λ_2 as in (3.6b). Then E_1 and E_2 are isomorphic (over $\mathbb{C}$) if and only if Λ_1 and Λ_2 are homothetic. (I.e. $\Lambda_1 = \alpha\Lambda_2$ for some $\alpha \in \mathbb{C}^*$.)*

PROOF. Clear from (4.1).   □

Remark 4.2. Since the maps ϕ_α are clearly homomorphisms, (4.1) implies that every complex analytic map from $E_1(\mathbb{C})$ to $E_2(\mathbb{C})$ taking O to O is necessarily a homomorphism. This is the analogue of (III.4.8), which says that every isogeny of elliptic curves is a homomorphism.

§5. Uniformization

The uniformization theorem for elliptic curves says that every elliptic curve over $\mathbb{C}$ is parametrized by elliptic functions. The most natural proof of this fact uses the theory of modular functions; that is, functions on the set of lattices in $\mathbb{C}$. (For example, $g_2(\Lambda)$ and $g_3(\Lambda)$ are modular functions.) The proof is not difficult, but would take us rather far afield, so we will be content to merely state the result here and use it to make various deductions.

Theorem 5.1. *Uniformization Theorem. Let $A, B \in \mathbb{C}$ satisfy $A^3 - 27B^2 \neq 0$. Then there exists a unique lattice $\Lambda \subset \mathbb{C}$ such that $g_2(\Lambda) = A$ and $g_3(\Lambda) = B$.*

PROOF. See [Ap, Thm. 2.9], [Rob, I.3.13], [Shi 1, §4.2], or [Se 7, VII Prop. 5].
                                                                                    □

Corollary 5.1.1. *Let $E/\mathbb{C}$ be an elliptic curve. Then there exists a lattice $\Lambda \subset \mathbb{C}$, unique up to homothety, and a complex analytic isomorphism*

$$\phi : \mathbb{C}/\Lambda \to E(\mathbb{C}) \qquad \phi(z) = [\wp(z, \Lambda), \wp'(z, \Lambda), 1]$$

of complex Lie groups.

PROOF. The existence is immediate from (3.6b) and (5.1), and the uniqueness is (4.1.1).   □

We are now in a position to prove the results left undone in section 1.

Proposition 5.2. *Let $E/\mathbb{C}$ be an elliptic curve with Weierstrass coordinate functions x, y.*
(a) *Let α and β be paths on $E(\mathbb{C})$ giving a basis for $H_1(E, \mathbb{Z})$. Then the periods*

$$\omega_1 = \int_\alpha dx/y \quad and \quad \omega_2 = \int_\beta dx/y$$

are $\mathbb{R}$-linearly independent.

(b) *Let $\Lambda \subset \mathbb{C}$ be the lattice generated by ω_1 and ω_2. Then the map*

$$F : E(\mathbb{C}) \to \mathbb{C}/\Lambda \qquad F(P) = \int_O^P dx/y \,(\mathrm{mod}\ \Lambda)$$

is a complex analytic isomorphism of Lie groups. Its inverse is the map given in (5.1.1).

PROOF. (a) From (5.1.1) there exists some lattice Λ_1 such that the map

$$\phi_1 : \mathbb{C}/\Lambda_1 \to E(\mathbb{C}) \qquad \phi_1(z) = [\wp(z, \Lambda_1), \wp'(z, \Lambda_1), 1]$$

is a complex analytic isomorphism. It follows that $\phi_1^{-1} \circ \alpha$ and $\phi_1^{-1} \circ \beta$ are a basis for $H_1(\mathbb{C}/\Lambda_1, \mathbb{Z})$. Note further that $H_1(\mathbb{C}/\Lambda_1, \mathbb{Z})$ is isomorphic to Λ_1 via the map $\gamma \to \int_\gamma dz$; while the differential dx/y on E pulls back to

$$\phi_1^*(dx/y) = d\wp(z)/\wp'(z) = dz$$

on $\mathbb{C}/\Lambda_1$. Therefore the periods

$$\omega_1 = \int_\alpha dx/y = \int_{\phi_1^{-1} \circ \alpha} dz \quad \text{and} \quad \omega_2 = \int_\beta dx/y = \int_{\phi_1^{-1} \circ \beta} dz$$

are a basis for Λ_1, so in particular they are $\mathbb{R}$-linearly independent.

(b) We have just shown above that the lattice Λ_1 corresponding to E in (5.1.1) is precisely the lattice Λ generated by the periods of E. The composition $F \circ \phi$ thus gives an analytic map

$$F \circ \phi : \mathbb{C}/\Lambda \to \mathbb{C}/\Lambda \qquad F \circ \phi(z) = \int_O^{(\wp(z), \wp'(z))} dx/y.$$

Since

$$F^*(dz) = dx/y \quad \text{and} \quad \phi^*(dx/y) = d\wp(z)/\wp'(z) = dz,$$

we see that

$$(F \circ \phi)^* \, dz = dz.$$

On the other hand, (4.1a) says that any analytic map $\mathbb{C}/\Lambda \to \mathbb{C}/\Lambda$ has the form $\psi_a(z) = az$ for some $a \in \mathbb{C}^*$. Since $\psi_a^*(dz) = adz$, we see that $F \circ \phi(z) = z$. (I.e. $F \circ \phi$ is the identity map.) But we already know (3.6b) that ϕ is an analytic isomorphism; and so $F = \phi^{-1}$ is, too. $\qquad\qquad\square$

Much of the preceding material can be summarized as an equivalence of certain categories.

Theorem 5.3. *The following categories are equivalent.*

(a) *Objects*: *Elliptic curves over $\mathbb{C}$.*
   Maps: *Isogenies.*

(b) *Objects*: *Elliptic curves over* $\mathbb{C}$.
   Maps: *Complex analytic maps taking O to O.*
(c) *Objects*: *Lattices* $\Lambda \subset \mathbb{C}$, *up to homothety*.
   Maps: *Map* $(\Lambda_1, \Lambda_2) = \{\alpha \in \mathbb{C} : \alpha\Lambda_1 \subset \Lambda_2\}$.

PROOF. The one-to-one correspondence between elliptic curves over $\mathbb{C}$ and lattices (modulo homothety) follows from (3.6b), (5.1.1), and (5.2). The match-up of the maps in (a), (b), (c) is precisely the content of (4.1). □

Remark 5.3.1. The equivalence of (a) and (b) in (5.3) is a very special case of a general principle (GAGA [Se 1]), which says (among other things) that any complex analytic map between projective varieties over $\mathbb{C}$ is necessarily given by rational functions. (For an introductory discussion, see [Har, app. B].)

We now use the uniformization theorem (really (5.1.1)) to make some general deductions about elliptic curves over $\mathbb{C}$. It is worth remarking that even without knowing (5.1.1), everything that we are about to prove would at least apply to those elliptic curves which occur in (3.6b). The uniformization theorem merely says that this class of curves includes every elliptic curve over $\mathbb{C}$.

Proposition 5.4. *Let* $E/\mathbb{C}$ *be an elliptic curve and* $m \geqslant 1$ *an integer.*
(a) *As abstract groups,*

$$E[m] \cong \mathbb{Z}/m\mathbb{Z} \times \mathbb{Z}/m\mathbb{Z}.$$

(b) *The multiplication-by-m map* $[m] : E \to E$ *has degree* m^2.

PROOF. (a) From (5.1.1), $E(\mathbb{C})$ is isomorphic to $\mathbb{C}/\Lambda$ for some lattice $\Lambda \subset \mathbb{C}$. Hence

$$E[m] \cong (\mathbb{C}/\Lambda)[m] \cong \frac{1}{m}\Lambda/\Lambda \cong (\mathbb{Z}/m\mathbb{Z})^2.$$

(b) Since char $\mathbb{C} = 0$ and $[m]$ is unramified, the degree of $[m]$ is just the number of points in $E[m] = [m]^{-1}\{O\}$. □

Let $E/\mathbb{C}$ be an elliptic curve. Notice that (4.1) allows us to identify $\text{End}(E)$ with a certain subring of $\mathbb{C}$. Thus if $E(\mathbb{C}) \cong \mathbb{C}/\Lambda$ as in (5.1.1), then

$$\text{End}(E) \cong \{\alpha \in \mathbb{C} : \alpha\Lambda \subseteq \Lambda\}.$$

Since Λ is unique up to homothety (4.1.1), this ring is independent of Λ. We now use this description of $\text{End}(E)$ to completely characterize the possible endomorphism rings which can occur. We recall the following definition from (III §9).

Definition. Let $\mathscr{K}$ be a number field. An *order* $\mathscr{R}$ of $\mathscr{K}$ is a subring of $\mathscr{K}$ which is finitely generated as a $\mathbb{Z}$-module and satisfies $\mathscr{R} \otimes \mathbb{Q} = \mathscr{K}$.

Theorem 5.5. *Let $E/\mathbb{C}$ be an elliptic curve, and let ω_1, ω_2 be generators for the lattice Λ associated to E by (5.1.1). Then either*

(i) $\mathrm{End}(E) = \mathbb{Z}$; *or*
(ii) $\mathbb{Q}(\omega_1/\omega_2)$ *is a quadratic imaginary extension of $\mathbb{Q}$, and $\mathrm{End}(E)$ is isomorphic to an order in $\mathbb{Q}(\omega_1/\omega_2)$.*

PROOF. Let $\tau = \omega_1/\omega_2$. Since Λ is homothetic to $\mathbb{Z} + \mathbb{Z}\tau$ (multiply by $1/\omega_2$), we may replace Λ by $\mathbb{Z} + \mathbb{Z}\tau$. Let

$$\mathscr{R} = \{\alpha \in \mathbb{C} : \alpha\Lambda \subset \Lambda\},$$

so $\mathscr{R} \cong \mathrm{End}(E)$ from (4.1). Then for any $\alpha \in \mathscr{R}$, there are integers a, b, c, d such that

$$\alpha = a + b\tau \quad \text{and} \quad \alpha\tau = c + d\tau.$$

Eliminating τ yields

$$\alpha^2 - (a + d)\alpha + ad - bc = 0.$$

Thus $\mathscr{R}$ is an integral extension of $\mathbb{Z}$.

Now suppose that $\mathscr{R} \neq \mathbb{Z}$, and choose $\alpha \in \mathscr{R}$ with $\alpha \notin \mathbb{Z}$. Then with notation as above, $b \neq 0$, so eliminating α gives a non-trivial equation

$$b\tau^2 - (a - d)\tau - c = 0.$$

Therefore $\mathbb{Q}(\tau)$ is a quadratic imaginary (since $\tau \notin \mathbb{R}$) extension of $\mathbb{Q}$. Finally, since $\mathscr{R} \subset \mathbb{Q}(\tau)$ and $\mathscr{R}$ is integral over $\mathbb{Z}$, it follows that $\mathscr{R}$ is an order of $\mathbb{Q}(\tau)$. $\qquad \square$

§6. The Lefschetz Principle

The *Lefschetz principle* says roughly that algebraic geometry over an arbitrary algebraically closed field of characteristic 0 is "the same" as algebraic geometry over $\mathbb{C}$. One can, of course, make this precise by formulating an equivalence of suitably defined categories; but we will be content here to give a more informal presentation.

The first observation to make is that if the given field K can be embedded in $\mathbb{C}$, then everything proceeds smoothly. For example, if $K \subset \mathbb{C}$ is any field and if E/K is an elliptic curve, then the fact that $[m] : E \to E$ is an algebraic map (i.e. given by rational functions) implies that $E[m] \subset E(\bar{K}) \subset E(\mathbb{C})$. Hence using (5.4), we obtain a proof that

$$E[m] = E(\bar{K})[m] = E(\mathbb{C})[m] \cong (\mathbb{Z}/m\mathbb{Z})^2.$$

Note that the embedding $K \subset \mathbb{C}$ need not be topological (assuming K has a topology in the first place.) It does not matter that we may have used the topology of $\mathbb{C}$ to reach our conclusions (such as in the analytic isomorphism $E(\mathbb{C}) \cong \mathbb{C}/\Lambda$), as long as our hypotheses and conclusions are purely algebraic.

The second observation is that theorems in algebraic geometry generally deal with finite (or at worst, countable) sets. For example, any variety is defined by a finite set of polynomial equations (Hilbert basis theorem), and each equation has finitely many coefficients. Similarly, an algebraic map between varieties is given by a finite set of polynomials, each having a finite number of coefficients. Now suppose that $\{V_1, V_2, \ldots\}$ is a finite (or countable) set of varieties defined over some field K of characteristic 0, and suppose that $\{\phi_1, \phi_2, \ldots\}$ is a finite (or countable) set of rational maps (defined over K) between various of the V_i's. Let $K_0 \subset K$ be the field generated over $\mathbb{Q}$ by all of the coefficients of all of the polynomials defining all of the V_i's and all of the ϕ_j's. Then $\mathrm{trdeg}(K_0/\mathbb{Q})$ clearly has cardinality at most that of the natural numbers, so we can embed $K_0 \subset \mathbb{C}$ (Zorn's lemma). Now from the above discussion concerning subfields of $\mathbb{C}$, we will be able to reduce most algebro-geometric questions concerning the V_i's and ϕ_j's to the corresponding question over $\mathbb{C}$, where we may be able to profitably employ techniques from complex analysis and differential geometry.

To illustrate the procedure outlined above, we prove the following.

Theorem 6.1. *Let K be a field of characteristic 0 and E/K an elliptic curve.*
(a) *Let $m \geq 1$ be an integer. Then*

$$E[m] \cong \mathbb{Z}/m\mathbb{Z} \times \mathbb{Z}/m\mathbb{Z}.$$

(b) *The endomorphism ring of E is either $\mathbb{Z}$ or an order in a quadratic imaginary extension of $\mathbb{Q}$. (Compare with (III.9.4).)*

PROOF. (a) This is immediate from (5.4) and the Lefschetz principle.
(b) Here we can apply the Lefschetz principle to (5.5), once we note that $\mathrm{End}(E)$ is countably (in fact finitely) generated from (III.7.5). Alternatively, even without (III.7.5), we can argue as follows. If $\mathrm{End}(E)$ is neither $\mathbb{Z}$ nor quadratic imaginary, then it contains a finitely generated subring with the same property. Now applying the Lefschetz principle to the maps in this subring will contradict (5.5). □

EXERCISES

6.1. Let $\Lambda = \mathbb{Z}\omega_1 + \mathbb{Z}\omega_2$ be a lattice, and let $\theta(z)$ be an entire function (i.e. holomorphic on all of $\mathbb{C}$.) Suppose that there are constants $a_1, a_2 \in \mathbb{C}$ such that

$$\theta(z + \omega_1) = a_1 \theta(z) \quad \text{and} \quad \theta(z + \omega_2) = a_2 \theta(z) \qquad \text{for all } z \in \mathbb{C}.$$

Prove that

$$\theta(z) = be^{cz} \qquad \text{for some } b, c \in \mathbb{C}.$$

6.2. Let $\Lambda \subset \mathbb{C}$ be a lattice.

(a) Prove that every fundamental parallelogram for Λ has the same area. Denote this area by $A(\Lambda)$.

(b) Prove that as $R \to \infty$,

$$\#\{\omega \in \Lambda : |\omega| \leqslant R\} = A(\Lambda)^{-1}\pi R^2 + O(R).$$

(The big-O constant depends on Λ, of course.)

(c) Prove that there is a constant $c = c(\Lambda)$ such that for all $R > 0$,

$$\#\{\omega \in \Lambda : R \leqslant |\omega| < R + 1\} < cR.$$

6.3. (a) Prove that for all $z, a \in \mathbb{C}$,

$$\wp(z) - \wp(a) = -\frac{\sigma(z+a)\sigma(z-a)}{\sigma(z)^2\sigma(a)^2}.$$

[*Hint*: Compare zeros and poles.]

(b) Prove that

$$\wp'(z) = -\frac{\sigma(2z)}{\sigma(z)^4}.$$

(c) Prove that for every integer n, the function $\sigma(nz)/\sigma(z)^{n^2}$ is in $\mathbb{C}(\Lambda)$.

(d) More precisely, prove that

$$(-1)^{n-1}\{1!2!\ldots(n-1)!\}\frac{\sigma(nz)}{\sigma(z)^{n^2}} = \det(\wp^{(i+j-1)}(z))_{1 \leqslant i, j \leqslant n-1}.$$

6.4. Define the *Weierstrass ζ-function* $\zeta(z)$ (not to be confused with the Riemann ζ-function) by the series

$$\zeta(z) = \frac{1}{z} + \sum_{\substack{\omega \in \Lambda \\ \omega \neq 0}} \left\{\frac{1}{z-\omega} + \frac{1}{\omega} + \frac{z}{\omega^2}\right\}.$$

(a) Prove that

$$\frac{d}{dz}\log\sigma(z) = \zeta(z) \quad \text{and} \quad \frac{d}{dz}\zeta(z) = -\wp(z).$$

(b) Prove that

$$\zeta(-z) = -\zeta(z);$$

and that for all $\omega \in \Lambda$,

$$\zeta(z + \omega) = \zeta(z) + \eta(\omega),$$

where $\eta(\omega)$ is independent of z. If $\omega \notin 2\Lambda$, show $\eta(\omega) = 2\zeta(\omega/2)$.

(c) Prove that the map $\eta : \Lambda \to \mathbb{C}$ given in (b) is linear.

(d) Write $\Lambda = \mathbb{Z}\omega_1 + \mathbb{Z}\omega_2$ with $\text{Im}(\omega_1/\omega_2) > 0$. Prove the *Legendre relation*

$$\omega_1\eta(\omega_2) - \omega_2\eta(\omega_1) = 2\pi i.$$

[*Hint*: Integrate $\zeta(z)$ around a fundamental parallelogram.] $\eta(\omega_1)$ and $\eta(\omega_2)$ are called *quasi-periods*.

(e) Prove that

$$\sigma(z + \omega) = \pm e^{\eta(\omega)(z + \omega/2)}\sigma(z),$$

where the sign is positive or negative according to $\omega \in 2\Lambda$ or $\omega \notin 2\Lambda$ respectively.

(f) Extend $\eta : \Lambda \to \mathbb{C}$ to an $\mathbb{R}$-*linear* map $\eta : \mathbb{C} \to \mathbb{C}$ by identifying $\Lambda \otimes_{\mathbb{Z}} \mathbb{R}$ with $\mathbb{C}$. Let

$$G(z) = e^{-z\eta(z)/2}\sigma(z).$$

Prove that for all $\omega \in \Lambda$,

$$|G(z + \omega)| = |G(z)|.$$

Hence $|G(z)|$ defines a real analytic function $\mathbb{C}/\Lambda \to \mathbb{R}$.

6.5. Verify the following indefinite integrals.
(a) $\int \wp(z)^2 \, dz = \frac{1}{6}\wp'(z) + \frac{1}{12}g_2 z + C.$
(b) $\int \wp(z)^3 \, dz = \frac{1}{120}\wp'''(z) - \frac{3}{20}g_2\zeta(z) + \frac{1}{10}g_3 z + C.$

6.6. For a lattice $\Lambda \subset \mathbb{C}$, let $g_2(\Lambda)$ and $g_3(\Lambda)$ be as in (3.5.1), and define

$$\Delta(\Lambda) = g_2(\Lambda)^3 - 27g_3(\Lambda)^2 \quad \text{and} \quad j(\Lambda) = 1728g_2(\Lambda)^3/\Delta(\Lambda).$$

(a) Let $\alpha \in \mathbb{C}^*$. Prove that

$$g_2(\alpha\Lambda) = \alpha^{-4}g_2(\Lambda) \qquad g_3(\alpha\Lambda) = \alpha^{-6}g_3(\Lambda);$$

and so

$$\Delta(\alpha\Lambda) = \alpha^{-12}\Delta(\Lambda) \qquad j(\alpha\Lambda) = j(\Lambda).$$

(b) Prove that $j(\Lambda_1) = j(\Lambda_2)$ if and only if there is an $\alpha \in \mathbb{C}^*$ such that $\alpha\Lambda_1 = \Lambda_2$.
(c) Prove that

$$j(\mathbb{Z} + \mathbb{Z}i) = 1728 \quad \text{and} \quad j(\mathbb{Z} + \mathbb{Z}e^{\pi i/3}) = 0.$$

6.7. *Elliptic curves over* $\mathbb{R}$. Let $E/\mathbb{C}$ be an elliptic curve corresponding to a lattice $\Lambda \subset \mathbb{C}$.
(a) Prove that E can be defined over $\mathbb{R}$ if and only if there is an $\alpha \in \mathbb{C}^*$ such that $\alpha\Lambda$ is mapping to itself by complex conjugation. [*Hint*: First show that $\overline{j(\Lambda)} = j(\bar{\Lambda})$.]
(b) Suppose E is defined over $\mathbb{R}$, and that we have chosen a lattice Λ for E which is invariant under complex conjugation. Prove that $\Delta(\Lambda) \in \mathbb{R}$; and that $E(\mathbb{R})$ is connected if and only if $\Delta(\Lambda) < 0$.
(c) Let $E/\mathbb{C}$ have a Legendre equation

$$E : y^2 = x(x - 1)(x - \lambda).$$

Prove that $\lambda \in \mathbb{R}$ if and only if E can be defined over $\mathbb{R}$ and $E[2] \subset E(\mathbb{R})$.
(d) If E is defined over $\mathbb{R}$ and $E[2] \subset E(\mathbb{R})$, prove that there is a lattice for E which is rectangular (i.e. of the form $\mathbb{Z}\omega_1 + \mathbb{Z}\omega_2 i$ with $\omega_1, \omega_2 \in \mathbb{R}$).

6.8. Let $\mathscr{K}/\mathbb{Q}$ be a quadratic imaginary field, $\mathscr{R}$ the ring of integers of $\mathscr{K}$, and $h_{\mathscr{R}}$ the class number of $\mathscr{R}$. Prove that up to isomorphism, there are exactly $h_{\mathscr{R}}$ elliptic curves $E/\mathbb{C}$ with $\text{End}(E) \cong \mathscr{R}$. If E is such a curve, conclude that $j(E)$ is an algebraic number satisfying $[\mathscr{K}(j(E)) : \mathscr{K}] \leq h_{\mathscr{R}}$. (In fact, $\mathscr{K}(j(E))$ is the Hilbert class field of $\mathscr{K}$. See (C §11) and the references listed there.)

6.9. Let $E_1/\mathbb{C}$ and $E_2/\mathbb{C}$ be elliptic curves, and assume that E_1 has complex multiplication. Prove that E_1 is isogenous to E_2 if and only if

$$\text{End}(E_1) \otimes \mathbb{Q} \cong \text{End}(E_2) \otimes \mathbb{Q}.$$

6.10. Let $\phi : E_1 \to E_2$ be an isogeny of elliptic curves over $\mathbb{C}$, and let $\alpha \in \mathbb{C}^*$ correspond to ϕ via the equivalence in (4.1). (I.e. $E_i \cong \mathbb{C}/\Lambda_i$ and $\alpha\Lambda_1 \subset \Lambda_2$.) Let $m = \deg(\phi)$. Prove that $m\alpha^{-1}$ corresponds to the dual isogeny $\hat{\phi} : E_2 \to E_1$. If $\Lambda_1 = \Lambda_2$, prove that $\bar{\alpha}$ corresponds to $\hat{\phi}$.

Elliptic Integrals. The following exercises (6.11–6.13) develop a minute portion of the classical theory of elliptic integrals.

6.11. Let $E/\mathbb{C}$ be an elliptic curve given by a Legendre equation

$$E : Y^2 = X(X - 1)(X - \lambda).$$

(a) Prove that there is a $k \in \mathbb{C} - \{0, \pm1\}$ such that E has an equation of the form

$$E : y^2 = (1 - x^2)(1 - k^2 x^2).$$

   [*Hint*: Let $X = (ax + b)/(cx + d)$ and $Y = ey/(cx + d)^2$ for appropriate $a, b, c, d, e \in \mathbb{C}$.]

(b) For a given value of λ, find all possible values of k. Conversely, given k, find all values of λ.

(c) Express the j-invariant $j(E)$ in terms of k.

(d) Suppose $\lambda \in \mathbb{R}$. (See exer. 6.7.) Show that k may be chosen so as to satisfy $0 < k < 1$.

6.12. *Complete Elliptic Integrals.* Let E be an elliptic curve given by an equation

$$E : y^2 = (1 - x^2)(1 - k^2 x^2).$$

To simplify matters, assume that $0 < k < 1$. (See exer. 6.11d.) Define *complete elliptic integrals to the modulus k* by

$$K(k) = \int_0^1 \frac{1}{y} dx = \int_0^1 \frac{1}{(1 - x^2)^{1/2}(1 - k^2 x^2)^{1/2}} dx \qquad First\ Kind$$

$$T(k) = \int_0^1 \frac{y}{1 - x^2} dx = \int_0^1 \left(\frac{1 - k^2 x^2}{1 - x^2}\right)^{1/2} dx.$$

(a) Make appropriate branch cuts, and show that the lattice for E is generated by the periods

$$4\int_0^1 \{(1 - x^2)(1 - k^2 x^2)\}^{-1/2} dx \quad \text{and} \quad 2i \int_1^{1/k} \{(x^2 - 1)(1 - k^2 x^2)\}^{-1/2} dx.$$

(b) Let k' be the *complementary modulus* defined by $k^2 + k'^2 = 1$, $0 < k' < 1$. Prove that

$$\int_1^{1/k} \{(x^2 - 1)(1 - k^2 x^2)\}^{-1/2} dx = \int_0^1 \{(1 - X^2)(1 - k'^2 X^2)\}^{-1/2} dx.$$

[*Hint*: Let $x = (1 - k'^2 X^2)^{-1/2}$.] Conclude that the period lattice for the elliptic curve $E/\mathbb{C}$ is generated by $4K(k)$ and $2iK(k')$.

(c) Prove the transformation formulas

$$K\left(\frac{2\sqrt{k}}{1+k}\right) = (1+k)K(k) \quad \text{and} \quad K\left(\frac{1-k}{1+k}\right) = \frac{1+k}{2}K(k').$$

[*Hint*: For the former, use the substitution $x = (k+1)X/(1+kX^2)$.]

6.13. (a) Show that the complete elliptic integrals defined above may also be written as

$$K(k) = \int_0^{\pi/2} (1 - k^2 \sin^2 \theta)^{-1/2} \, d\theta,$$

$$T(k) = \int_0^{\pi/2} (1 - k^2 \sin^2 \theta)^{1/2} \, d\theta.$$

(b) Prove that the arclength of the ellipse

$$x^2/a^2 + y^2/b^2 = 1$$

is given by the complete elliptic integral

$$4aT(\sqrt{1 - (b/a)^2})$$

(We assume $a \geqslant b > 0$.)

(c) Prove that the arclength of the *lemniscate*

$$r^2 = \cos 2\theta$$

is given by the complete elliptic integral $2\sqrt{2}K(1/\sqrt{2})$. Show that it also equals $4\int_0^1 (1 - x^4)^{-1/2} \, dx$. (Thus the arclength of the lemniscate resembles the arclength of the unit circle, namely $2\pi = 4\int_0^1 (1 - x^2)^{-1/2} \, dx$.)

6.14. *The Arithmetic-Geometric Mean*. For $a, b \in \mathbb{R}$ with $a \geqslant b > 0$, we define two sequences $\{a_n\}$ and $\{b_n\}$ by

$$a_0 = a \qquad b_0 = b$$

$$a_{n+1} = \tfrac{1}{2}(a_n + b_n) \qquad b_{n+1} = \sqrt{a_n b_n}.$$

(a) Prove that

$$0 \leqslant a_{n+1} - b_{n+1} \leqslant \tfrac{1}{2}(a_n - b_n).$$

Deduce that the limit

$$M(a, b) = \text{Lim } a_n = \text{Lim } b_n$$

exists. $M(a, b)$ is called the *arithmetic-geometric mean of a and b*.

(b) Prove that

$$M(a, b) = M(a_1, b_1) = M(a_2, b_2) = \cdots,$$

and

$$M(ca, cb) = cM(a, b) \qquad \text{for } c > 0.$$

(c) Define the integral $I(a, b)$ by

$$I(a, b) = \int_0^{\pi/2} (a^2 \cos^2 \theta + b^2 \sin^2 \theta)^{-1/2} \, d\theta.$$

Show that $I(a, b)$ is related to certain complete elliptic integrals by the formulas

$$I(a, b) = a^{-1} K \left(\frac{2\sqrt{k}}{1 + k} \right) \quad \text{and} \quad I(a_1, b_1) = a_1^{-1} K(k).$$

$$\left[\text{Hint: Take } k = \frac{a - b}{a + b}. \right]$$

(d) Prove that

$$M(a, b) I(a, b) = \pi/2.$$

[*Hint:* Use (c) and (exer. 6.12c) to prove that $I(a, b) = I(a_1, b_1)$. Then calculate Lim $I(a_n, b_n)$.] Combining (c) and (d), note that complete elliptic integrals of the first kind (for $0 < k < 1$) may be computed in terms of the arithmetic-geometric mean.

(e) Prove that the rate of convergence predicted by (a), namely $a_n - b_n \leqslant 2^{-n}(a - b)$, is far slower than the reality. More precisely, use (b) to show that it suffices to compute $M(a, b)$ in the case that $b \geqslant 1$; and under this assumption, prove that

$$a_{n+m} - b_{n+m} \leqslant 8 \left(\frac{a_n - b_n}{8} \right)^{2^m} \quad \text{for all } m, n \geqslant 0.$$

In particular, since eventually $a_n - b_n < 1$, the sequences $\{a_n\}$ and $\{b_n\}$ eventually converge doubly exponentially.

(f) Show that

$$\int_0^1 (1 - z^4)^{-1/2} \, dz = \pi/2M(\sqrt{2}, 1),$$

and use this equality to calculate the value of the complete elliptic integral. (It was the observation that these two numbers, calculated independently, agree to eleven decimal places which led Gauss to initiate his extensive study of the arithmetic-geometric mean. For a fascinating account of this subject, see [Cox].)

CHAPTER VII

Elliptic Curves over Local Fields

In this chapter we study the group of rational points on an elliptic curve defined over a field which is complete with respect to a discrete valuation. We start with some basic facts concerning Weierstrass equations and "reduction modulo π". This enables us to break our problem up into several pieces; and then by examining each piece individually, we will be able to deduce a great deal about the group of rational points as a whole. Unless explicitly stated otherwise, we will use the following notation.

K	a local field, complete with respect to a discrete valuation v
R	the ring of integers of $K = \{x \in K : v(x) \geqslant 0\}$
R^*	the unit group of $R = \{x \in K : v(x) = 0\}$
$\mathcal{M}$	the maximal ideal of $R = \{x \in K : v(x) > 0\}$
π	a uniformizer for R (i.e. $\mathcal{M} = \pi R$)
k	the residue field of $R = R/\mathcal{M}$.

We will further assume that v is normalized so that $v(\pi) = 1$. Note that by convention, $v(0) = \infty$ is assigned a value larger than every real number. Finally, in keeping with our general policy, we will assume that both K and k are perfect fields.

§1. Minimal Weierstrass Equations

Let E/K be an elliptic curve, and let

$$y^2 + a_1 xy + a_3 y = x^3 + a_2 x^2 + a_4 x + a_6$$

be a Weierstrass equation for E/K. Since replacing (x, y) by $(u^{-2}x, u^{-3}y)$ causes each a_i to become $u^i a_i$, if we choose u divisible by a large power of π, then we can find a Weierstrass equation with all coefficients $a_i \in R$. Then the discriminant Δ satisfies $v(\Delta) \geqslant 0$; and since v is discrete, we can look for an equation with $v(\Delta)$ as small as possible.

Definition. Let E/K be an elliptic curve. A Weierstrass equation as above is called a *minimal (Weierstrass) equation for E at v* if $v(\Delta)$ is minimized subject to the condition $a_1, a_2, a_3, a_4, a_6 \in R$. This value of $v(\Delta)$ is the *valuation of the minimal discriminant of E at v*.

Remark 1.1. How can one tell if a given Weierstrass equation is minimal? First, by definition, all of the a_i's must be in R, so in particular the discriminant Δ is in R. If the equation is not minimal, then there is a coordinate change giving a new equation with discriminant $\Delta' = u^{-12}\Delta \in R$ (cf. III.1.2). Thus $v(\Delta)$ can only be changed by multiples of 12, so we conclude:

If $a_i \in R$ and $v(\Delta) < 12$, then the equation is minimal.

Similarly, since $c_4' = u^{-4}c_4$ and $c_6' = u^{-6}c_6$, we have:

If $a_i \in R$ and $v(c_4) < 4$ (or $v(c_6) < 6$), then the equation is minimal.

If $\operatorname{char}(k) \neq 2, 3$, then the converse holds, namely minimality implies either $v(\Delta) < 12$ or $v(c_4) < 4$. (See exer. 7.1.) For arbitrary K there is an algorithm of Tate ([Ta 6]) which will determine if a given equation is minimal.

Example 1.2. Let p be a prime and consider the Weierstrass equation

$$E : y^2 + xy + y = x^3 + x^2 + 22x - 9$$

over the field $\mathbb{Q}_p$. This equation has discriminant $\Delta = -2^{15}5^2$ and $c_4 = -5 \cdot 211$. Hence using the above criteria (1.1), this is a minimal Weierstrass equation at p for every prime $p \in \mathbb{Z}$.

Proposition 1.3. (a) *Every elliptic curve E/K has a minimal Weierstrass equation.*
(b) *A minimal Weierstrass equation is unique up to a change of coordinates*

$$x = u^2 x' + r \qquad y = u^3 y' + u^2 sx' + t$$

with $u \in R^*$ and $r, s, t \in R$.
(c) *The invariant differential*

$$\omega = dx/(2y + a_1 x + a_3)$$

associated with a minimal Weierstrass equation is unique up to multiplication by an element of R^*.
(d) *Conversely, if one starts with any Weierstrass equation with coefficients $a_i \in R$, then any change of coordinates*

$$x = u^2 x' + r \qquad y = u^3 y' + u^2 sx' + t$$

used to produce a minimal Weierstrass equation satisfies $u, r, s, t \in R$.

PROOF. (a) One can easily find some Weierstrass equation with all $a_i \in R$, and among such there is a minimal $v(\Delta)$ since v is discrete.

(b) We know (III.3.1b) that any Weierstrass equation for E/K is unique up to the indicated change of coordinates with $u \in K^*$ and $r, s, t \in K$. Now suppose the given equation and the new equation are both minimal. We use the transformation formulas (III.1.2). From the definition of minimality, we have $v(\Delta) = v(\Delta')$. But $u^{12}\Delta' = \Delta$, so $u \in R^*$. From the transformation for b_6 (respectively b_8) we see that $4r^3$ (respectively $3r^4$) is in R, hence $r \in R$. Now the transformation for a_2 gives $s \in R$, and that for a_6 gives $t \in R$.

(c) Clear from (b), since $\omega' = u\omega$.

(d) Since $u^{12}\Delta' = \Delta$ and $v(\Delta') \leqslant v(\Delta)$ (because the new equation is to be minimal), we see that $v(u) \geqslant 0$, so $u \in R$. Now the proof in (b) can be repeated to show that $r, s, t \in R$.   □

§2. Reduction Modulo π

We next look at the operation of "reduction modulo π", which we denote by a tilde. Thus, for example, the natural reduction map $R \to k = R/\pi R$ is denoted $t \to \tilde{t}$. Now having chosen a minimal Weierstrass equation for E/K, we can reduce its coefficients modulo π to obtain a (possibly singular) curve over k, namely

$$\tilde{E} : y^2 + \tilde{a}_1 xy + \tilde{a}_3 y = x^3 + \tilde{a}_2 x^2 + \tilde{a}_4 x + \tilde{a}_6.$$

The curve $\tilde{E}/k$ is called the *reduction of E modulo π*. From (1.3b), since we started with a minimal equation for E, the equation for $\tilde{E}$ is unique up to the standard change of coordinates (III.3.1b) for Weierstrass equations *over* k.

Next let $P \in E(K)$. We can find homogeneous coordinates $P = [x_0, y_0, z_0]$ with $x_0, y_0, z_0 \in R$ and at least one of x_0, y_0, z_0 in R^*. Then the reduced point $\tilde{P} = [\tilde{x}_0, \tilde{y}_0, \tilde{z}_0]$ is in $\tilde{E}(k)$. This gives a *reduction map*

$$E(K) \to \tilde{E}(k)$$

$$P \to \tilde{P}.$$

(More generally, one can similarly define a *reduction map*

$$\mathbb{P}^n(K) \to \mathbb{P}^n(k).$$

The above map is just its restriction to $E(K) \subset \mathbb{P}^2(K)$.)

Now the curve $\tilde{E}/k$ may or may not be singular (more on this later), but recall (III.2.5) that in any case its set of non-singular points, denoted $\tilde{E}_{ns}(k)$, forms a group. We define two subsets of $E(K)$ as follows:

$$E_0(K) = \{P \in E(K) : \tilde{P} \in \tilde{E}_{ns}(k)\};$$

$$E_1(K) = \{P \in E(K) : \tilde{P} = \tilde{O}\}.$$

In words, $E_0(K)$ is the set of points with *non-singular reduction*, and $E_1(K)$ is the *kernel of reduction*. From (1.3b), they do not depend on which minimal Weierstrass equation we choose.

Proposition 2.1. *There is an exact sequence of abelian groups*

$$0 \to E_1(K) \to E_0(K) \to \tilde{E}_{ns}(k) \to 0,$$

where the right-hand map is reduction modulo π.

PROOF. The group laws on $E(K)$ and $\tilde{E}_{ns}(k)$ are defined by taking the intersection of the curve with lines in $\mathbb{P}^2$. Since the reduction map $\mathbb{P}^2(K) \to \mathbb{P}^2(k)$ takes lines to lines, it follows that $E_0(K)$ is a group, and that the map $E_0(K) \to \tilde{E}_{ns}(k)$ is a homomorphism. Exactness at the left and center now comes directly from the definition of $E_1(K)$.

It remains to show that the reduction map is surjective. This will follow from Hensel's lemma and the completeness of K. Thus let

$$f(x, y) = y^2 + a_1 xy + a_3 y - x^3 - a_2 x^2 - a_4 x - a_6 = 0$$

be a minimal Weierstrass equation, $\tilde{f}(x, y)$ the corresponding polynomial with coefficients reduced modulo π, and $\tilde{P} = (\alpha, \beta) \in \tilde{E}_{ns}(k)$ any point. Since $\tilde{P}$ is a non-singular point of $\tilde{E}$, we know that either

$$\frac{\partial \tilde{f}}{\partial x}(\tilde{P}) \neq 0 \quad \text{or} \quad \frac{\partial \tilde{f}}{\partial y}(\tilde{P}) \neq 0,$$

say the former. (The other case is entirely similar.) Choose any $y_0 \in R$ with $\tilde{y}_0 = \beta$, and look at the equation

$$f(x, y_0) = 0.$$

When reduced modulo π, this equation has α as a *simple* root, since $(\partial \tilde{f}/\partial x)(\alpha, \tilde{y}_0) \neq 0$. Hence by Hensel's lemma ([La 2, Ch. II, Prop. 2]), the root α can be lifted to an $x_0 \in R$ such that $\tilde{x}_0 = \alpha$ and $f(x_0, y_0) = 0$. Then the point $P = (x_0, y_0) \in E_0(K)$ reduces to $\tilde{P}$. $\qquad\square$

Note that if $v(\Delta) = 0$, so $\tilde{\Delta} \neq 0$, then $\tilde{E}$ is non-singular, $\tilde{E}_{ns} = \tilde{E}$, and so $E_0(K) = E(K)$. In this case (2.1) says that $E(K)$ is built up from two pieces, $E_1(K)$ and $\tilde{E}(k)$. Now $\tilde{E}(k)$ is the set of points on an elliptic curve defined over a smaller field; and we will often consider the case where k is a finite field, a situation analyzed in some detail in chapter V.

On the other hand, the following proposition shows that $E_1(K)$ is also an object with which we are already familiar.

Proposition 2.2. *Let E/K be given by a minimal Weierstrass equation, let $\hat{E}/R$ be the formal group associated to E (IV.2.2.3), and let $w(z) \in R\llbracket z \rrbracket$ be the power series from (IV.1.1). Then the map*

$$\hat{E}(\mathcal{M}) \to E_1(K)$$

$$z \to \left(\frac{z}{w(z)}, \ -\frac{1}{w(z)} \right)$$

is an isomorphism. (We understand that $z = 0$ goes to O. For the definition of the group $\hat{E}(\mathcal{M})$, see (IV §3).)

PROOF. From (IV.1.1b) the point $(z/w(z), -1/w(z))$, when considered as a pair of power series, satisfies the Weierstrass equation for E. Since $w(z) = z^3(1 + \cdots) \in R[\![z]\!]$, it follows that $w(z)$ converges for any $z \in \mathcal{M}$. Hence $(z/w(z), -1/w(z))$ is in $E(K)$ for $z \in \mathcal{M}$, and since $v(-1/w(z)) = -3v(z)$, it is even in $E_1(K)$. Thus we have a well-defined map

$$\hat{E}(\mathcal{M}) \to E_1(K)$$

$$z \to (z/w(z), -1/w(z)).$$

Further, in deriving the power series giving the group law on $\hat{E}$, we simply used the group law on E (in the (z, w)-plane) and then replaced w by $w(z)$. Therefore the map is a group homomorphism. Since $w(z) = 0$ only for $z = 0$, it is injective, so it remains to show that the image is all of $E_1(K)$.

Let $(x, y) \in E_1(K)$. Since (x, y) reduces modulo π to the point at infinity on $\tilde{E}(k)$, we see that $v(x) < 0$ and $v(y) < 0$. But then from the Weierstrass equation $y^2 + \cdots = x^3 + \cdots$, we must have

$$3v(x) = 2v(y) = -6r$$

for some integer $r \geqslant 1$. Hence $x/y \in \mathcal{M}$, so the map

$$E_1(K) \to \hat{E}(\mathcal{M})$$

$$(x, y) \to -x/y$$

is well-defined. Again because the group law on $\hat{E}(\mathcal{M})$ is defined by using the group law on E, this map is a homomorphism; and it is clearly injective. Hence we have two injections

$$\hat{E}(\mathcal{M}) \to E_1(K) \to \hat{E}(\mathcal{M})$$

whose composition is the identity, so they are isomorphisms. □

§3. Points of Finite Order

In this section we analyze the points of finite order in the group $E(K)$. Although we will prove a stronger result below (3.4), we start with the following easy proposition, which will provide a crucial ingredient in the proof of the weak Mordell–Weil theorem (VIII.1.1).

Proposition 3.1. *Let E/K be an elliptic curve and $m \geqslant 1$ an integer relatively prime to* char(k).
(a) *The subgroup $E_1(K)$ has no non-trivial points of order m.*
(b) *If the reduced curve $\tilde{E}/k$ is non-singular, then the reduction map*

$$E(K)[m] \rightarrow \tilde{E}(k)$$

is injective. (Here $E(K)[m]$ denotes the set of points of order m in $E(K)$.)

PROOF. From (2.1) we have an exact sequence

$$0 \rightarrow E_1(K) \rightarrow E_0(K) \rightarrow \tilde{E}_{ns}(k) \rightarrow 0.$$

But from (2.2), $E_1(K) \cong \hat{E}(\mathcal{M})$, where $\hat{E}$ is the formal group associated to E; and from our general result on formal groups (IV.3.2b), $\hat{E}(\mathcal{M})$ has no non-trivial elements of order m. This proves (a). Now if $\tilde{E}$ is non-singular, then $E_0(K) = E(K)$ and $\tilde{E}_{ns}(k) = \tilde{E}(k)$, so the m-torsion in $E(K)$ injects into $\tilde{E}(k)$, which proves (b). $\square$

Application 3.2. The above proposition (3.1) generally provides the quickest method for finding the torsion subgroup of an elliptic curve defined over a number field. Thus let K be a number field and K_v the completion of K at some discrete valuation v. Then clearly $E(K)$ injects into $E(K_v)$, so by applying (3.1) for several differents v's, one can obtain information about the torsion in $E(K)$. We illustrate with several examples over $\mathbb{Q}$.

Example 3.3.1. Let $E/\mathbb{Q}$ be the elliptic curve

$$E : y^2 + y = x^3 - x + 1.$$

Its discriminant $\Delta = -611 = -13 \cdot 47$, so $\tilde{E}$(modulo 2) is non-singular. One easily checks that $\tilde{E}(\mathbb{F}_2) = \{O\}$ and $E(\mathbb{Q})[2] = \{O\}$, hence from (3.1) we conclude that $E(\mathbb{Q})$ has no non-zero torsion points.

Example 3.3.2. Let $E/\mathbb{Q}$ be the elliptic curve

$$E : y^2 = x^3 + 3.$$

Its discriminant is $\Delta = -3^5 2^4$, so $\tilde{E}$(modulo p) is non-singular for every $p \geqslant 5$. One easily checks that

$$\#\tilde{E}(\mathbb{F}_5) = 6 \quad \text{and} \quad \#\tilde{E}(\mathbb{F}_7) = 13.$$

Hence $E(\mathbb{Q})$ can have no non-trivial torsion. In particular, the point $(1,2) \in E(\mathbb{Q})$ has infinite order, and so $E(\mathbb{Q})$ is an infinite set, two facts which are by no means obvious. (For the complete analysis of the rational torsion points on the curves $y^2 = x^3 + D$ with $D \in \mathbb{Z}$, see [Fue] or exer. 10.19.)

Example 3.3.3. Let $E/\mathbb{Q}$ be the elliptic curve

$$E : y^2 = x^3 + x,$$

whose discriminant is $\Delta = -64$. The point $(0,0) \in E(\mathbb{Q})$ is a point of order 2. We compute

$$\# E(\mathbb{F}_3) = 4 \qquad \# E(\mathbb{F}_5) = 4 \qquad \# E(\mathbb{F}_7) = 8.$$

As can easily be checked (exer. 5.12), $\# E(\mathbb{F}_p)$ is divisible by 4 for every $p \geqslant 5$. But suppose we look at the actual groups,

$$E(\mathbb{F}_3) = \{O, (0,0), (2,1), (2,2)\},$$

$$E(\mathbb{F}_5) = \{O, (0,0), (2,0), (3,0)\}.$$

Now a point of E has order 2 if and only if its y-coordinate is zero. Hence

$$E(\mathbb{F}_3) \cong \mathbb{Z}/4\mathbb{Z} \quad \text{and} \quad E(\mathbb{F}_5) \cong (\mathbb{Z}/2\mathbb{Z})^2,$$

so $(0,0)$ is the only torsion point in $E(\mathbb{Q})$.

The next result, due to Cassels, gives a precise bound for the denominator of a torsion point. Following Katz–Lang ([La 5, Thm. III.3.7]), we give a proof based on general facts concerning formal groups. For an exposition of Cassel's original proof, which involves a careful analysis of division polynomials, see [Ca 1, Thm. 17.2] or [La 5, Thm. III.1.5].

Theorem 3.4. *Assume* $\text{char}(K) = 0$ *and* $p = \text{char}(k) > 0$. *Let* E/K *be an elliptic curve given by a Weierstrass equation*

$$E : y^2 + a_1 xy + a_3 y = x^3 + a_2 x^2 + a_4 x + a_6$$

with all $a_i \in R$. *(N.B. The equation need not be minimal.) Let* $P \in E(K)$ *be a point of exact order* $m \geqslant 2$.
(a) *If* m *is not a power of* p, *then* $x(P), y(P) \in R$.
(b) *If* $m = p^n$, *then*

$$\pi^{2r} x(P), \pi^{3r} y(P) \in R \quad \text{with} \quad r = \left[\frac{v(p)}{p^n - p^{n-1}} \right].$$

(Here [ ] *is greatest integer.)*

PROOF. If $x(P) \in R$ there is nothing to prove, so we assume $v(x(P)) < 0$. If the equation for E is not minimal, and (x', y') are coordinates for a minimal equation, then from (1.3d) we see that

$$v(x(P)) \geqslant v(x'(P)) \quad \text{and} \quad v(y(P)) \geqslant v(y'(P)).$$

It thus suffices to prove the theorem for a minimal Weierstrass equation.

Since $v(x(P)) < 0$, we see from the Weierstrass equation that

$$3v(x(P)) = 2v(y(P)) = -6s \qquad \text{for some integer } s \geqslant 1.$$

Further, P is in $E_1(K)$, the kernel of the reduction map, so under the isomorphism of (2.2) it corresponds to the element $-x(P)/y(P)$ in the formal group $\hat{E}(\mathcal{M})$. But from (IV.3.2b), $\hat{E}(\mathcal{M})$ contains no torsion of order prime to p, which proves (a).

To prove (b) we use (IV.6.1). Since $-x(P)/y(P)$ has exact order p^n in $\hat{E}(\mathcal{M})$, it follows from (IV.6.1) that

$$s = v(-x(P)/y(P)) \leqslant v(p)/(p^n - p^{n-1}).$$

Since $\pi^{2s} x(P)$ and $\pi^{3s} y(P)$ are in R, this gives the desired result.   □

Application 3.5. Let $E/\mathbb{Q}$ be an elliptic curve given by a Weierstrass equation having coefficients in $\mathbb{Z}$. Let $P \in E(\mathbb{Q})$ be a point of exact order m. By embedding $E(\mathbb{Q})$ into $E(\mathbb{Q}_p)$ for various primes p, we deduce integrality conditions on the coordinates of P. Thus if m is not a prime power, then (3.4a) implies $x(P), y(P) \in \mathbb{Z}$. But even if $m = p^n$ for some prime p corresponding to a normalized valuation v, we have

$$[v(p)/(p^n - p^{n-1})] = [1/(p^n - p^{n-1})] = 0$$

unless $p = 2$ and $n = 1$. We conclude that $x(P), y(P) \in \mathbb{Z}$ for every torsion point $P \in E(\mathbb{Q})$ of exact order $m \geqslant 3$. This is best possible, as the example

$$E : y^2 + xy = x^3 + 4x + 1 \qquad (-1/4, 1/8) \in E(\mathbb{Q})\ [2]$$

shows. For a further discussion of torsion points over number fields, see (VIII §7).

§4. The Action of Inertia

In this section we will reinterpret the injectivity of torsion (3.1b) in terms of the action of Galois. We set the following notation:

K^{nr} the maximal unramified extension of K,
I_v the inertia subgroup of $G_{\bar{K}/K}$.

Since the unramified extensions of K correspond to the extensions of the residue field k, $G_{\bar{K}/K}$ has a decomposition

$$1 \to G_{\bar{K}/K^{nr}} \to G_{\bar{K}/K} \to G_{K^{nr}/K} \to 1$$
$$\|\qquad\qquad\qquad\qquad\|$$
$$I_v \qquad\qquad\qquad G_{\bar{k}/k}$$

In words, the inertia group I_v is the set of elements of $G_{\bar{K}/K}$ which act trivially on the residue field $\bar{k}$. (For these basic facts about local fields, see e.g. [Frö 1, §7] or [La 2, Ch. I, II]. Remember that K and k are both assumed to be perfect.)

Definition. Let Σ be a set on which $G_{\bar{K}/K}$ acts. We say that Σ is *unramified at v* if the action of I_v on Σ is trivial.

Recall that if E/K is an elliptic curve, then we have seen (III §7) that $G_{\bar{K}/K}$ acts on the torsion subgroups $E[m]$ and the Tate modules $T_\ell(E)$ of E.

Proposition 4.1. *Let E/K be an elliptic curve, and suppose that the reduced curve $\tilde{E}/k$ is non-singular.*
(a) *Let $m \geq 1$ be an integer relatively prime to char(k) (i.e. $v(m) = 0$). Then $E[m]$ is unramified at v.*
(b) *Let $\ell \neq$ char(k) be a prime. Then $T_\ell(E)$ is unramified at v.*

PROOF. (a) Take a finite extension K'/K so that $E[m] \subset E(K')$, and let

$$R' = \text{ring of integers of } K'$$

$$\mathscr{M}' = \text{maximal ideal of } R'$$

$$k' = \text{residue field of } R' = R'/\mathscr{M}'$$

$$v' = \text{valuation on } K'.$$

By assumption, if we take a minimal Weierstrass equation for E at v, then its discriminant Δ satisfies $v(\Delta) = 0$ (since $\tilde{E}/k$ is non-singular.) But v' restricted to K is just a multiple of v, so $v'(\Delta) = 0$. Hence the Weierstrass equation is also minimal at v', and $\tilde{E}/k'$ is non-singular. Now (3.1b) implies that the reduction map

$$E[m] \to \tilde{E}(k')$$

is injective.

Let $\sigma \in I_v$ and $P \in E[m]$. We must show that $P^\sigma = P$. From the definition of the inertia group, σ acts trivially on $\tilde{E}(k')$, so

$$\widetilde{P^\sigma - P} = \tilde{P}^\sigma - \tilde{P} = \tilde{O}.$$

But $P^\sigma - P$ is clearly in $E[m]$, so from the injectivity proven above we conclude $P^\sigma - P = O$.
(b) This follows immediately from (a) and the definition $T_\ell(E) = \varprojlim E[\ell^n]$. $\qquad\square$

There is a converse to this proposition, known as the criterion of Néron–Ogg–Shafarevich, which characterizes when $\tilde{E}/k$ is non-singular in terms of the action of the inertia group on torsion points. We will return to this in section 7, after first studying the reduced curve $\tilde{E}$ more closely.

§5. Good and Bad Reduction

Let E/K be an elliptic curve. Then from our general knowledge of Weierstrass equations (III.1.4), the reduced curve $\tilde{E}$ is one of three types. We classify E according to these possibilities.

Definition. Let E/K be an elliptic curve, and let $\tilde{E}$ be the reduced curve for a *minimal* Weierstrass equation.
(a) E has *good* (or *stable*) *reduction* over K if $\tilde{E}$ is non-singular.

(b) *E* has *multiplicative* (or *semi-stable*) *reduction* over *K* if $\tilde{E}$ has a node.
(c) *E* has *additive* (or *unstable*) *reduction* over *K* if $\tilde{E}$ has a cusp.
In cases (b) and (c), *E* is naturally said to have *bad reduction*. If *E* has multiplicative reduction, then the reduction is said to be *split* (respectively *non-split*) if the slopes of the tangent lines at the node are in *k* (respectively not in *k*).

It is quite easy to read off the reduction type of an elliptic curve from a minimal Weierstrass equation.

Proposition 5.1. *Let E/K be an elliptic curve with minimal Weierstrass equation*

$$y^2 + a_1 xy + a_3 y = x^3 + a_2 x^2 + a_4 x + a_6.$$

Let Δ *be the discriminant of this equation and* c_4 *the usual combination of the* a_i*'s (cf. III §1).*
(a) E has good reduction if and only if $v(\Delta) = 0$ *(i.e.* $\Delta \in R^*$*). In this case* $\tilde{E}/k$ *is an elliptic curve.*
(b) E has multiplicative reduction if and only if $v(\Delta) > 0$ *and* $v(c_4) = 0$ *(i.e.* $\Delta \in \mathcal{M}$ *and* $c_4 \in R^*$*). In this case* $\tilde{E}_{ns}$ *is the multiplicative group,*

$$\tilde{E}_{ns}(\bar{k}) \cong \bar{k}^*.$$

(c) E has additive reduction if and only if $v(\Delta) > 0$ *and* $v(c_4) > 0$ *(i.e.* $\Delta, c_4 \in \mathcal{M}$*). In this case* $\tilde{E}_{ns}$ *is the additive group,*

$$\tilde{E}_{ns}(\bar{k}) \cong \bar{k}^+.$$

PROOF. The type of reduction for *E* follows from (III.1.4) applied to the reduced Weierstrass equation $\tilde{E}$ over the field *k*. Then the group $\tilde{E}_{ns}(\bar{k})$ is given by (III.2.5). $\qquad\square$

Example 5.2. Let $p \geqslant 5$ be a prime. Then the elliptic curve

$$E_1 : y^2 = x^3 + px^2 + 1$$

has good reduction over $\mathbb{Q}_p$, while

$$E_2 : y^2 = x^3 + x^2 + p$$

has (split) multiplicative reduction over $\mathbb{Q}_p$, and

$$E_3 : y^2 = x^3 + p$$

has additive reduction over $\mathbb{Q}_p$. Notice that E_3 has good reduction over $\mathbb{Q}_p(\sqrt[6]{p})$, since the given equation is then not minimal. (Make the substitution $x = \sqrt[3]{p}\, x'$, $y = \sqrt[2]{p}\, y'$.) On the other hand, E_2 still has multiplicative reduction over any extension of $\mathbb{Q}_p$. This is in fact true in general; after extending the ground field, additive reduction turns either multiplicative or good, while the latter two do not change. (See (5.4) below.) This suggests the origins of the terms "stable", "semi-stable", "unstable", although they do have quite precise definitions in terms of the stability of points on moduli space. (For a high-powered account of the general theory, see [M–F].)

Even if an elliptic curve E/K has bad reduction, it is often useful to know whether it attains good reduction over some extension of K. We give this property a name.

Definition. Let E/K be an elliptic curve. E has *potential good reduction* over K if there is a finite extension K'/K so that E has good reduction over K'.

Example 5.3. If K is a finite extension of $\mathbb{Q}_p$, and if E/K has complex multiplication, then E has potential good reduction. (See exer. 7.10.)

The next result explains how reduction type behaves under field extension, and the one immediately following provides a useful characterization of when an elliptic curve has potential good reduction.

Proposition 5.4 (Semi-stable reduction theorem). *Let E/K be an elliptic curve.*
(a) *Let K'/K be an unramified extension. Then the reduction type of E over K (i.e. good, multiplicative, or additive) is the same as the reduction type of E over K'.*
(b) *Let K'/K be any finite extension. If E has either good or multiplicative reduction over K, then it has the same type of reduction over K'.*
(c) *There exists a finite extension K'/K so that E has either good or (split) multiplicative reduction over K'.*

Proposition 5.5. *Let E/K be an elliptic curve. Then E has potential good reduction if and only if its j-invariant is integral (i.e. if $j(E) \in R$).*

PROOF OF (5.4). (a) For arbitrary K this follows from Tate's algorithm [Ta 6]. We will assume $\operatorname{char}(k) \geqslant 5$, so E has a minimal Weierstrass equation over K of the form

$$E : y^2 = x^3 + Ax + B.$$

Let R' be the ring of integers in K', v' the valuation on K' extending v, and

$$x = (u')^2 x' \qquad y = (u')^3 y'$$

a change of coordinates producing a minimal equation for E over K'. Since K'/K is unramified, we can find $u \in K$ with $(u/u') \in (R')^*$. Then the substitution

$$x = u^2 x' \qquad y = u^3 y'$$

also gives a minimal equation for E/K', since

$$v'(u^{-12}\Delta) = v'((u')^{-12}\Delta).$$

But this new equation has coefficients in R, so by the minimality of the original equation over K, we have $v(u) = 0$. Hence the original equation is also minimal over K'. Since $v(\Delta) = v'(\Delta)$ and $v(c_4) = v'(c_4)$, using (5.1) we see that E has the same reduction type over K and K'.

(b) Take a minimal Weierstrass equation for E over K, with corresponding quantities Δ and c_4. Let R' be the ring of integers in K', v' the valuation on K' extending v,

$$x = u^2 x' + r \qquad y = u^3 y' + su^2 x' + t$$

a change of coordinates giving a minimal Weierstrass equation for E over K'. For this new equation the associated Δ' and c_4' satisfy

$$0 \leqslant v'(\Delta') = v'(u^{-12}\Delta) \quad \text{and} \quad 0 \leqslant v'(c_4') = v'(u^{-4}c_4).$$

From (1.3d) we also have $u \in R'$, hence

$$0 \leqslant v'(u) \leqslant \min\{\tfrac{1}{12}v'(\Delta), \tfrac{1}{4}v'(c_4)\}.$$

But for good (resp. multiplicative) reduction we have $v(\Delta) = 0$ (resp. $v(c_4) = 0$) (5.1a, b), so in both cases $v'(u) = 0$. Hence

$$v'(\Delta') = v'(\Delta) \quad \text{and} \quad v'(c_4') = v'(c_4),$$

so again using the characterization in (5.1), E has good (resp. multiplicative) reduction over K'.

(c) We assume char$(k) \neq 2$, and extend K so that E has a Weierstrass equation in Legendre normal form (III.1.7)

$$E : y^2 = x(x - 1)(x - \lambda), \qquad \lambda \neq 0, 1.$$

(For char$(k) = 2$, see (A.1.4a).) For this equation,

$$c_4 = 16(\lambda^2 - \lambda + 1) \quad \text{and} \quad \Delta = 16\lambda^2(\lambda - 1)^2.$$

We consider three cases.

Case 1. $\lambda \in R$, $\lambda \not\equiv 0, 1 \pmod{\mathcal{M}}$. Then $\Delta \in R^*$, so the given equation has good reduction.

Case 2. $\lambda \in R$, $\lambda \equiv 0$ or $1 \pmod{\mathcal{M}}$. Then $\Delta \in \mathcal{M}$ and $c_4 \in R^*$, so the given equation has (split) multiplicative reduction.

Case 3. $\lambda \notin R$. Choose the integer $r \geqslant 1$ so that $\pi^r \lambda \in R^*$. Then the substitution $x = \pi^{-r}x'$, $y = \pi^{-3r/2}y'$ (where we replace K by $K(\pi^{1/2})$ if necessary) gives a Weierstrass equation

$$(y')^2 = x'(x' - \pi^r)(x' - \pi^r\lambda)$$

for E with integral coefficients, $\Delta' \in \mathcal{M}$, and $c_4' \in R^*$, so E has (split) multiplicative reduction. $\square$

PROOF OF (5.5). As above, we assume char$(k) \neq 2$ and extend K so that E has a Weierstrass equation in Legendre form (III.1.7)

$$E : y^2 = x(x - 1)(x - \lambda), \qquad \lambda \neq 0, 1.$$

(For char$(k) = 2$, see (A.1.4b).) By assumption, $j = j(E) \in R$; and λ is related to j by

$$(1 - \lambda(1 - \lambda))^3 - j\lambda^2(1 - \lambda)^2 = 0.$$

From this equation and the integrality of j it is immediate that

$$\lambda \in R \quad \text{and} \quad \lambda \not\equiv 0 \text{ or } 1 \pmod{\mathcal{M}},$$

so the given Legendre equation has integral coefficients and good reduction.

Conversely, suppose E has potential good reduction. Let K'/K be a finite extension so that E has good reduction over K', let R' be the ring of integers of K', and let Δ' and c_4' be the quantities associated to a minimal Weierstrass equation for E over K'. Since E has good reduction over K', we have $\Delta' \in (R')^*$, and hence

$$j(E) = (c_4')^3/\Delta' \in R'.$$

But $j(E) \in K$, since E is defined over K, hence $j(E) \in R$. $\qquad \square$

§6. The Group E/E_0

Recall that the group $E_0(K)$ consists of those points of $E(K)$ whose reduction to $\tilde{E}(k)$ is not a singular point. Further, from (2.1), $E_0(K)$ is made up of two pieces that we have analyzed fairly closely, namely $\tilde{E}_{ns}(k)$ and the formal group $E_1(K) \cong \hat{E}(\mathcal{M})$. We are left to study the remaining piece, the quotient $E(K)/E_0(K)$.

The most important fact about this quotient is that it is finite. As the theorem given below indicates, one can actually say quite a bit more. Unfortunately, a direct proof, working explicitly with Weierstrass equations, is quite lengthy. Since even the simplifying assumption char$(k) \geqslant 5$ leads to a long case-by-case proof, we will not give one here (but see exer. 7.7). If the residue field k is finite, then the mere finiteness of $E(K)/E_0(K)$ can be proven by an easy compactness argument (exer. 7.6).

Theorem 6.1 (Kodaira, Néron). *Let E/K be an elliptic curve. If E has split multiplicative reduction over K, then $E(K)/E_0(K)$ is a cyclic group of order $v(\Delta) = -v(j)$. In all other cases, $E(K)/E_0(K)$ is a finite group of order at most* 4.

Corollary 6.2. *The subgroup $E_0(K)$ is of finite index in $E(K)$.*

PROOF. The finiteness of $E(K)/E_0(K)$ follows from the existence of the Néron model, which is a group scheme over $\mathrm{Spec}(R)$ whose generic fiber is E/K. The specific description of $E(K)/E_0(K)$ comes from the complete classification of the possible special fibers of a Néron model. One can also give an elementary (but lengthy) proof by doing explicit computations using Weierstrass equations. See (C §15) for a further discussion   □

Our most important application of (6.2) will be in the proof of the criterion of Néron–Ogg–Shafarevich, which we give in the next section. Another interesting application is the following.

Proposition 6.3. *Let K be a finite extension of $\mathbb{Q}_p$ (so $\mathrm{char}(K) = 0$ and k is a finite field). Then $E(K)$ contains a subgroup of finite index which is isomorphic to R^+ (i.e. taken additively).*

PROOF. From (6.2), $E(K)/E_0(K)$ is finite; and from (2.1), $E_0(K)/E_1(K)$ is isomorphic to $\tilde{E}_{ns}(k)$, which is finite since k is finite. Hence it suffices to prove that $E_1(K)$ has a subgroup of finite index isomorphic to R^+. Now $E_1(K)$ is isomorphic to the formal group $\hat{E}(\mathcal{M})$ (2.2). Further, from (IV.3.2a), $\hat{E}(\mathcal{M})$ has a filtration

$$\hat{E}(\mathcal{M}) \supset \hat{E}(\mathcal{M}^2) \supset \hat{E}(\mathcal{M}^3) \supset \cdots;$$

and each quotient $E(\mathcal{M}^i)/E(\mathcal{M}^{i+1})$ is isomorphic to $\mathcal{M}^i/\mathcal{M}^{i+1}$, which is also finite since k is finite. Finally, for an appropriate r (IV.6.4b), the formal logarithm map provides an isomorphism

$$\hat{E}(\mathcal{M}^r) \xrightarrow{\sim} \mathcal{M}^r = \pi^r R \qquad \text{(taken additively)},$$

which gives the desired result.   □

§7. The Criterion of Néron–Ogg–Shafarevich

If an elliptic curve E/K has good reduction, and $m \geq 1$ is an integer prime to $\mathrm{char}(k)$, then we have seen that the torsion subgroup $E[m]$ is unramified (4.1). Various partial converses were proven by Néron, Ogg, and Shafarevich, and these were vastly generalized by Serre and Tate. We follow the exposition in [S–T].

Theorem 7.1 (Criterion of Néron–Ogg–Shafarevich). *Let E/K be an elliptic curve. The following are equivalent.*

(a) *E has good reduction over K.*
(b) *$E[m]$ is unramified at v for all integers $m \geq 1$ relatively prime to $\mathrm{char}(k)$.*
(c) *The Tate module $T_\ell(E)$ is unramified at v for some (all) primes ℓ with $\ell \neq \mathrm{char}(k)$.*

(d) $E[m]$ is unramified at v for infinitely many integers $m \geq 1$ relatively prime to char(k).

PROOF. We have already proven (a) $\Rightarrow$ (b) (4.1), and clearly (b) $\Rightarrow$ (c) $\Rightarrow$ (d). (Note that $T_\ell(E)$ being unramified is the same as $E[\ell^n]$ being unramified for all $n \geq 1$.) It remains to prove that (d) implies (a).

Assume (d) holds. Let K^{nr} be the maximal unramified extension of K. Choose an integer m satisfying

(i) m is relatively prime to char(k);
(ii) $m > \# E(K^{nr})/E_0(K^{nr})$;
(iii) $E[m]$ is unramified at v.

Such an m exists, since we are assuming (d), and $E(K^{nr})/E_0(K^{nr})$ is finite from (6.2).

Now consider the two exact sequences

$$0 \to E_0(K^{nr}) \to E(K^{nr}) \to E(K^{nr})/E_0(K^{nr}) \to 0$$

$$0 \to E_1(K^{nr}) \to E_0(K^{nr}) \to \tilde{E}_{ns}(\bar{k}) \to 0.$$

(Note $\bar{k}$ is the residue field of the ring of integers in K^{nr}.) Since $E[m] \subset E(K^{nr})$, we see that $E(K^{nr})$ has a subgroup isomorphic to $(\mathbb{Z}/m\mathbb{Z})^2$. But from (ii), $E(K^{nr})/E_0(K^{nr})$ has order strictly less than m. It follows from the first exact sequence that we can find a prime ℓ dividing m so that $E_0(K^{nr})$ contains a subgroup $(\mathbb{Z}/\ell\mathbb{Z})^2$. Now look at the second exact sequence. From (3.1a), $E_1(K^{nr})$ has no non-trivial ℓ-torsion, so we conclude that $\tilde{E}_{ns}(\bar{k})$ has a subgroup isomorphic to $(\mathbb{Z}/\ell\mathbb{Z})^2$.

Now suppose that E has bad reduction over K^{nr}. If the reduction is multiplicative, then from (5.1b),

$$\tilde{E}_{ns}(\bar{k}) = (\bar{k})^*;$$

but then the ℓ-torsion in $\tilde{E}_{ns}(\bar{k})$ would be $\mathbb{Z}/\ell\mathbb{Z}$. Hence this type of reduction cannot occur. Similarly, if E has additive reduction over K^{nr}, then from (5.1c),

$$\tilde{E}_{ns}(\bar{k}) = \bar{k} \qquad \text{(taken additively)},$$

which has no ℓ-torsion at all. This eliminates multiplicative and additive reduction as possibilities, so all that remains is for E to have good reduction over K^{nr}. Finally, since K^{nr}/K is unramified, we conclude (5.4a) that E has good reduction over K. $\square$

Corollary 7.2. Let $E_1, E_2/K$ be elliptic curves which are isogenous over K. Then either they both have good reduction over K, or neither one does.

PROOF. Let $\phi : E_1 \to E_2$ be a non-zero isogeny defined over K, and let $m \geq 2$ be an integer relatively prime to both char(k) and deg ϕ. Then the induced map

$$\phi : E_1[m] \to E_2[m]$$

is an isomorphism of $G_{\bar{K}/K}$-modules, so in particular either both are unramified at v, or neither one is. Now use (7.1, a $\Leftrightarrow$ d).   □

Another immediate corollary of (7.1) is a criterion, in terms of the action of inertia, for when an elliptic curve has potential good reduction.

Corollary 7.3. *Let E/K be an elliptic curve. Then E has potential good reduction if and only if the inertia group I_v acts on the Tate module $T_\ell(E)$ through a finite quotient for some (all) prime(s) $\ell \neq \mathrm{char}(k)$.*

PROOF. Suppose E has potential good reduction. Then there is a finite extension K'/K so that E has good reduction over K'. Extending K', we may assume K'/K is Galois. Let v' be the valuation on K' and $I_{v'}$ the inertia group of K'. From (7.1), $I_{v'}$ acts trivially on $T_\ell(E)$ for any $\ell \neq \mathrm{char}(k)$. Hence the action of I_v on $T_\ell(E)$ factors through the finite quotient $I_v/I_{v'}$. This proves one implication.

Assume now that for some $\ell \neq \mathrm{char}(k)$, I_v acts on $T_\ell(E)$ through a finite quotient, say I_v/J. Then the fixed field of J, which we denote $\bar{K}^J$, is a finite extension of $K^{nr} = \bar{K}^{I_v}$. Hence we can find a finite extension K'/K so that $\bar{K}^J$ is the compositum

$$\bar{K}^J = K' K^{nr}.$$

Then the inertia group of K' is equal to J, and by assumption J acts trivially on $T_\ell(E)$. Now (7.1) implies that E has good reduction over K'.   □

EXERCISES

7.1. Assume that $\mathrm{char}(k) \neq 2,3$.
  (a) Let E/K be an elliptic curve given by a Weierstrass equation with coefficients $a_i \in R$. Prove that the equation is minimal if and only if either $v(\Delta) < 12$ or $v(c_4) < 4$.
  (b) Let E/K be given by a minimal Weierstrass equation of the form

$$E : y^2 = x^3 + Ax + B.$$

  Prove that E has
    (i) good reduction $\Leftrightarrow 4A^3 + 27B^2 \in R^*$;
    (ii) multiplicative reduction $\Leftrightarrow 4A^3 + 27B^2 \in \mathcal{M}$ and $AB \in R^*$;
    (iii) additive reduction $\Leftrightarrow A, B \in \mathcal{M}$.

7.2. Let E/K be an elliptic curve with j-invariant $j(E) \in R$. Prove that the minimal discriminant Δ of E satisfies

$$v(\Delta) < 12 + 12v(2) + 6v(3).$$

7.3. Describe all Weierstrass equations

$$E : y^2 + a_1 xy + a_3 y = x^3 + a_2 x^2 + a_4 x + a_6$$

with $a_i \in \mathbb{Z}$ and $\Delta \neq 0$ for which $E(\mathbb{Q})$ contains a torsion point P satisfying $x(P) \notin \mathbb{Z}$. [*Hint*: cf. (3.5).]

7.4. Let E/K be an elliptic curve given by a minimal Weierstrass equation, and define subsets of $E(K)$ by

$$E_n(K) = \{P \in E(K) : v(x(P)) \leqslant -2n\} \cup \{O\}.$$

(a) Prove that each $E_n(K)$ is a subgroup of $E(K)$.
(b) Prove that for $n \geqslant 1$,

$$E_n(K)/E_{n+1}(K) \cong k^+.$$

7.5. Show that the following elliptic curves have good reduction over the indicated field by writing down a minimal Weierstrass equation over that field.
   (a) $E : y^2 = x^3 + x$ $\mathbb{Q}_2(\pi, \zeta), \pi^8 = 2, \zeta^4 = -1$.
   (b) $E : y^2 + y = x^3$ $\mathbb{Q}_3(\pi, \zeta), \pi^4 = 3, \zeta^3 = -1$.
   (c) $E : y^2 = x^3 + x^2 - 3x - 2$ $\mathbb{Q}_5(\pi), \pi^4 = 5$.

7.6. Assume that K is locally compact for the topology induced by the discrete valuation v. (This is equivalent to the assumption that k is finite, cf. [Ca 8, §7].) The following steps provide a proof of (6.2) for such fields.
   (a) Use v to define a topology on $\mathbb{P}^N(K)$, and show that $\mathbb{P}^N(K)$ is compact for this topology.
   (b) Let E/K be an elliptic curve and $E(K) \subset \mathbb{P}^2(K)$ the inclusion coming from a minimal Weierstrass equation. Prove that with the induced topology, $E(K)$ is compact; and that the translation map $\tau_P : E(K) \to E(K)$ is continuous for any $P \in E(K)$.
   (c) Prove that $E_0(K)$ is an open subset of $E(K)$. (It is also a closed subset!)
   (d) Prove that $E(K)/E_0(K)$ is finite.

7.7. The following examples illustrate some special cases of (6.1). We assume throughout that $\mathrm{char}(k) \neq 2, 3$. Let E/K be an elliptic curve given by a Weierstrass equation

$$E : y^2 = x^3 + Ax + B.$$

   (a) If $v(A) \geqslant 1$ and $v(B) = 1$, then $E(K) = E_0(K)$.
   (b) If $v(A) = 1$ and $v(B) \geqslant 2$, then $E(K)/E_0(K) \cong \mathbb{Z}/2\mathbb{Z}$.
       [*Hint*: If $P, Q \notin E_0(K)$, use the addition formula to show that $P + Q \in E_0(K)$.]
   (c) If $v(A) \geqslant 2$ and $v(B) = 2$, then $E(K)/E_0(K)$ is either 0 or $\mathbb{Z}/3\mathbb{Z}$.

7.8. Let E/K be an elliptic curve and m an integer relatively prime to $\mathrm{char}(k)$. Prove that

$$E_0(K^{nr})/mE_0(K^{nr}) = 0.$$

7.9. Let E/K be an elliptic curve with potential good reduction, let $m \geqslant 3$ be an integer relatively prime to $\mathrm{char}(k)$, and let $K(E[m])$ be the field obtained by adjoining to K the coordinates of the points of $E[m]$.
   (a) Prove that the inertia group of $K(E[m])/K$ is independent of m. [*Hint*: For each prime $\ell \neq \mathrm{char}(k)$, let $\ell' = \ell$ if $\ell \geqslant 3$ and $\ell' = 4$ if $\ell = 2$. Show that $\rho_\ell(I_v)$ has trivial intersection with the kernel of the map

$$\mathrm{Aut}(T_\ell(E)) \to \mathrm{Aut}(T_\ell(E)/\ell' T_\ell(E)) \cong GL_2(\mathbb{Z}/\ell'\mathbb{Z}).$$

Characterize the inertia group of $K(E[m])/K$ in terms of the kernels of the various ρ_ℓ's.]

(b) Prove that $K(E[m])/K$ is unramified if and only if E has good reduction at v.

(c) Prove that $K(E[m])/K$ is tamely ramified if char$(k) > 3$.

7.10. Let K be a finite extension of $\mathbb{Q}_p$, R the ring of integers of K, and E/K an elliptic curve with complex multiplication. Prove that $j(E) \in R$. [*Hint:* Use the description of the maximal abelian extension K^{ab} of K provided by local class field theory to prove that the action of $G_{K^{ab}/K}$ on $T_\ell(E)$ factors through a finite quotient. Then apply (exer. 3.24), (7.3), and (5.5).]

7.11. Use (exer. 3.23) to prove (5.4c) and (5.5) in characteristic 2.

7.12. Let $[K : \mathbb{Q}_p] = 2$, let E/K be an elliptic curve given by a Weierstrass equation with coefficients in R, and let $P \in E(K)$ be a point of exact order $m \geqslant 2$ such that $x(P) \notin R$ (i.e. $v(x(P)) < 0$).

(a) Prove that $p = 2$ or 3 and that $m = 2, 3$, or 4. Give examples to show that each value of m is possible.

(b) Suppose that the reduced curve $\tilde{E}/k$ is supersingular. Prove that $p = m = 2$.

CHAPTER VIII

Elliptic Curves over Global Fields

Let K be a number field and E/K an elliptic curve. Our main goal in this chapter is to prove the following result.

Mordell–Weil Theorem. *The group $E(K)$ is finitely generated.*

The proof of this theorem consists of two quite distinct parts, the so-called "weak Mordell–Weil theorem" (§1) and the "infinite descent" using height functions (§3, 5, 6). We also give a separate proof of the descent step in the simplest case (§4), where the general theory of height functions can be replaced by explicit polynomial calculations.

From the Mordell–Weil theorem we see that the *Mordell–Weil group $E(K)$* has the form

$$E(K) \cong E_{\text{tors}}(K) \times \mathbb{Z}^r,$$

where the *torsion subgroup $E_{\text{tors}}(K)$* is finite and the *rank r* of $E(K)$ is a non-negative integer. For any given elliptic curve, it is possible to describe quite precisely the torsion subgroup (§7). The rank is much more difficult to compute, and in general there is no known procedure which is guaranteed to yield an answer. We will return to this question in more detail in chapter X.

The following notation will be used for the next three chapters.

K	a number field
M_K	a complete set of inequivalent absolute values on K
M_K^∞	the archimedean absolute values in M_K
M_K^0	the non-archimedean absolute values in M_K
$v(x)$	$= -\log|x|_v$ for absolute values $v \in M_K$
ord_v	normalized valuation for $v \in M_K^0$ (i.e. $\text{ord}_v(K^*) = \mathbb{Z}$)

R	the ring of integers of $K = \{x \in K : v(x) \geqslant 0 \text{ for all } v \in M_K^0\}$
R^*	the unit group of $R = \{x \in K : v(x) = 0 \text{ for all } v \in M_K^0\}$
K_v	the completion of K at v for $v \in M_K$
$R_v, \mathcal{M}_v, k_v$	the ring of integers, maximal ideal, and residue field associated to K_v for $v \in M_K^0$.

Finally, in those situations where it is important to have the absolute values in M_K coherently normalized, such as the theory of height functions, we will always adopt the "standard normalization" as described in section 5.

§1. The Weak Mordell–Weil Theorem

Our goal in this section is to prove the following result.

Theorem 1.1 (Weak Mordell–Weil Theorem). *Let K be a number field, E/K an elliptic curve, and $m \geqslant 2$ an integer. Then*

$$E(K)/mE(K)$$

is a finite group.

For the rest of this section, E/K and m will be as in the statement of (1.1). We start with the following reduction lemma.

Lemma 1.1.1. *Let L/K be a finite Galois extension. If $E(L)/mE(L)$ is finite, then $E(K)/mE(K)$ is also finite.*

PROOF. Let Φ be the kernel of the natural map $E(K)/mE(K) \to E(L)/mE(L)$. Thus

$$\Phi = (E(K) \cap mE(L))/mE(K),$$

so for each P (mod $mE(K)$) in Φ, we can choose a point $Q_P \in E(L)$ with $[m]Q_P = P$. (Q_P need not be unique, of course.) Having done this, we define a map of sets (which is *not* in general a group homomorphism)

$$\lambda_P : G_{L/K} \to E[m], \qquad \lambda_P(\sigma) = Q_P^\sigma - Q_P.$$

(Notice that $Q_P^\sigma - Q_P$ is in $E[m]$, since

$$[m](Q_P^\sigma - Q_P) = ([m]Q_P)^\sigma - [m]Q_P = P^\sigma - P = 0.$$

The map λ_P is actually a 1-cocycle; see section 2.)
    Suppose now that $\lambda_P = \lambda_{P'}$ for two points $P, P' \in E(K) \cap mE(L)$. Then

$$(Q_P - Q_{P'})^\sigma = Q_P - Q_{P'} \qquad \text{for all } \sigma \in G_{L/K},$$

so $Q_P - Q_{P'} \in E(K)$. Therefore

$$P - P' = [m]Q_P - [m]Q_{P'} \in mE(K), \qquad \text{so } P \equiv P' \ (\text{mod } mE(K)).$$

This proves that the association

$$\Phi \to \text{Map}(G_{L/K}, E[m]), \qquad P \to \lambda_P,$$

is one-to-one. But $G_{L/K}$ and $E[m]$ are finite sets, so there are only a finite number of maps between them. Therefore Φ is finite.

Finally, the exact sequence

$$0 \to \Phi \to E(K)/mE(K) \to E(L)/mE(L)$$

nests $E(K)/mE(K)$ between two finite groups, so it too is finite.   □

In view of (1.1.1), it suffices to prove the weak Mordell–Weil theorem (1.1) under the additional assumption that

$$E[m] \subset E(K).$$

For the remainder of this section we will assume, without further comment, that this inclusion is true.

The next step is to translate the putative finiteness of $E(K)/mE(K)$ into a statement about a certain field extension of K. For this purpose, we use the following tool.

Definition. The *Kummer pairing*

$$\kappa : E(K) \times G_{\bar{K}/K} \to E[m]$$

is defined as follows. Let $P \in E(K)$, and choose any $Q \in E(\bar{K})$ satisfying $[m]Q = P$. Then

$$\kappa(P, \sigma) = Q^\sigma - Q.$$

Proposition 1.2. (a) *The Kummer pairing is well-defined.*
(b) *The Kummer pairing is bilinear.*
(c) *The kernel of the Kummer pairing on the left is $mE(K)$.*
(d) *The kernel of the Kummer pairing on the right is $G_{\bar{K}/L}$, where*

$$L = K([m]^{-1}E(K))$$

is the compositum of all fields $K(Q)$ as Q ranges over the points of $E(\bar{K})$ satisfying $[m]Q \in E(K)$.

Hence the Kummer pairing induces a perfect bilinear pairing

$$E(K)/mE(K) \times G_{L/K} \to E[m],$$

where L is the field given in (d).

PROOF. Most of this proposition follows immediately from basic facts concerning group cohomology. (See section 2.) We will give a direct proof here.

(a) We must show that $\kappa(P, \sigma)$ is in $E[m]$ and does not depend on the choice of Q. For the former,

$$[m]\kappa(P, \sigma) = [m]Q^\sigma - [m]Q = P^\sigma - P = 0,$$

since $P \in E(K)$ and σ fixes K. For the latter, note that any other choice has the form $Q + T$ for some $T \in E[m]$. Then

$$(Q + T)^\sigma - (Q + T) = Q^\sigma + T^\sigma - Q - T = Q^\sigma - Q,$$

because by assumption $E[m] \subset E(K)$, so σ fixes T.

(b) The linearity in P is obvious. For the other side, let $\sigma, \tau \in G_{\bar{K}/K}$. Then

$$\kappa(P, \sigma\tau) = Q^{\sigma\tau} - Q = (Q^\sigma - Q)^\tau + Q^\tau - Q = \kappa(P, \sigma)^\tau + \kappa(P, \tau).$$

But $\kappa(P, \sigma) \in E[m]$ is contained in $E(K)$, so it is fixed by τ.

(c) Suppose $P \in mE(K)$, say $P = [m]Q$ with $Q \in E(K)$. Then any $\sigma \in G_{\bar{K}/K}$ fixes Q, so

$$\kappa(P, \sigma) = Q^\sigma - Q = 0.$$

Conversely, suppose $\kappa(P, \sigma) = 0$ for all $\sigma \in G_{\bar{K}/K}$. Thus choosing $Q \in E(\bar{K})$ with $[m]Q = P$, we have

$$Q^\sigma = Q \qquad \text{for all } \sigma \in G_{\bar{K}/K}.$$

Therefore $Q \in E(K)$, so $P = [m]Q \in mE(K)$.

(d) Suppose $\sigma \in G_{\bar{K}/L}$. Then

$$\kappa(P, \sigma) = Q^\sigma - Q = 0,$$

since $Q \in E(L)$ from the definition of L. Conversely, suppose $\sigma \in G_{\bar{K}/K}$ and $\kappa(P, \sigma) = 0$ for all $P \in E(K)$. Then for every $Q \in E(\bar{K})$ satisfying $[m]Q \in E(K)$,

$$0 = \kappa([m]Q, \sigma) = Q^\sigma - Q.$$

But L is the compositum of $K(Q)$ over all such Q, so σ fixes L. Hence $\sigma \in G_{\bar{K}/L}$.

Finally, the last statement of (1.2) is clear from what precedes it, once we note that L/K is Galois because $G_{\bar{K}/K}$ takes $[m]^{-1}E(K)$ to itself. (Alternatively, from (d), $G_{\bar{K}/L}$ is the kernel of the homomorphism

$$G_{\bar{K}/K} \to \text{Hom}(E(K), E[m]), \qquad \sigma \to \kappa(\cdot, \sigma),$$

so it is a normal subgroup.) $\square$

Using (1.2), we see that the finiteness of $E(K)/mE(K)$ is equivalent to the finiteness of the extension L/K. The next step is to analyze this extension. Our main tool will be (VII.3.1), which we restate after making appropriate definitions.

Definition. Let K be a number field and E/K an elliptic curve. Let $v \in M_K$ be a discrete valuation (i.e. $v \in M_K^0$). Then E is said to have *good* (respectively *bad*) *reduction at* v if E has good (respectively bad) reduction when considered over the completion K_v (cf. VII §5). Taking a minimal Weierstrass equation for E over K_v, we denote the reduced curve over the residue field by $\tilde{E}_v/k_v$. [N.B. It may not be possible to choose a single Weierstrass equation for E over K which is simultaneously minimal for all K_v. However, this can be done if $K = \mathbb{Q}$. For further details, see section 8.]

Remark 1.3. Take any Weierstrass equation for E/K,

$$E : y^2 + a_1 xy + a_3 y = x^3 + a_2 x^2 + a_4 x + a_6,$$

say with discriminant Δ. Then for all but finitely many $v \in M_K^0$, we have

$$v(a_i) \geqslant 0 \qquad \text{for } i = 1, \ldots, 6 \qquad \text{and } v(\Delta) = 0.$$

Now for such v, the given equation is already a minimal Weierstrass equation, and the reduced curve $\tilde{E}_v/k_v$ is non-singular. This shows that E *has good reduction at v for all but finitely many* $v \in M_K^0$.

Proposition 1.4 (restatement of VII.3.1b). *Let* $v \in M_K^0$, *and suppose that* $v(m) = 0$ *and E has good reduction at v. Then the reduction map*

$$E(K)[m] \to \tilde{E}_v(k_v)$$

is injective.

We are now ready to analyze the extension L/K.

Proposition 1.5. *Let*

$$L = K([m]^{-1} E(K))$$

be the field defined in (1.2d).
(a) L/K *is an abelian extension of exponent m. (I.e. $G_{L/K}$ is abelian and every element has order dividing m.)*
(b) *Let*

$$S = \{v \in M_K^0 : E \text{ has bad reduction at } v\} \cup \{v \in M_K^0 : v(m) \neq 0\} \cup M_K^\infty.$$

Then L/K is unramified outside S. (I.e. If $v \in M_K$ and $v \notin S$, then L/K is unramified at v.)

PROOF. (a) This follows immediately from (1.2), which implies that there is an injection

$$G_{L/K} \to \operatorname{Hom}(E(K), E[m])$$

$$\sigma \to \kappa(\cdot, \sigma).$$

(b) Let $v \in M_K$ with $v \notin S$, let $Q \in E(\bar{K})$ satisfy $[m]Q \in E(K)$, and let $K' = K(Q)$. It suffices to show that K'/K is unramified at v, since L is the compositum of all such K'. Let $v' \in M_{K'}$ be a place of K' lying above v, and let $k'_{v'}/k_v$ be the corresponding extension of residue fields. Since E has good reduction at v

(remember $v \notin S$), it certainly has good reduction at v' (take the same Weierstrass equation). Thus we have the usual reduction map

$$E(K') \to \tilde{E}_{v'}(k'_{v'}),$$

which we denote as usual by a tilde.

Now let $I_{v'/v} \subset G_{K'/K}$ be the inertia group for v'/v, and let $\sigma \in I_{v'/v}$. By definition of inertia, σ acts trivially on $\tilde{E}_{v'}(k'_{v'})$, so

$$\widetilde{Q^\sigma - Q} = \tilde{Q}^\sigma - \tilde{Q} = \tilde{O}.$$

On the other hand,

$$[m](Q^\sigma - Q) = ([m]Q)^\sigma - [m]Q = O,$$

since $[m]Q \in E(K)$. Thus $Q^\sigma - Q$ is a point of order m which is in the kernel of the "reduction modulo v" map. It follows from (1.4) that

$$Q^\sigma - Q = O.$$

This proves that Q is fixed by every element of the inertia group $I_{v'/v}$, hence $K' = K(Q)$ is unramified over K at v'. Since this holds for every v' over v, and for every $v \notin S$, we have proven that K'/K is unramified outside S. $\qquad\square$

To complete the proof of the weak Mordell–Weil theorem, all that remains is to show that any field extension L/K satisfying the conditions of (1.5) is necessarily a finite extension. The proof of this fact relies on the two fundamental finiteness theorems of algebraic number theory, namely the finiteness of the ideal class group and the finite generation of the group of S-units.

Proposition 1.6. *Let K be a number field, $S \subset M_K$ a finite set of places containing M_K^∞, and $m \geqslant 2$ an integer. Let L/K be the maximal abelian extension of K having exponent m which is unramified outside of S. Then L/K is a finite extension.*

PROOF. Suppose the proposition were true for some finite extension K' of K, where S' is the set of places of K' lying over S. Then LK'/K', being abelian of exponent m unramified outside S', would be finite; and so L/K would also be finite. It thus suffices to prove the proposition under the assumption that K contains the mth-roots of unity μ_m.

Similarly, we may increase the set S, since this only has the effect of making L larger. Using the fact that the class number of K is finite, we can thus add a finite number of elements to S so that the *ring of S-integers*

$$R_S = \{a \in K : v(a) \geqslant 0 \text{ for all } v \in M_K, v \notin S\}$$

is a principal ideal domain. We may also enlarge S so that $v(m) = 0$ for all $v \notin S$.

Now the main theorem of Kummer theory says that if a field (of characteristic 0) contains μ_m, then its maximal abelian extension of exponent m is obtained by adjoining m^{th}-roots. (See any basic text on field theory, for example [Bi §2] or [Ar, theorem 25]; or do exer. 8.4.) Thus L is the largest subfield of

$$K(\sqrt[m]{a} : a \in K)$$

which is unramified outside S.

Let $v \in M_K$, $v \notin S$. Looking at the equation

$$X^m - a = 0$$

over the local field K_v, and remembering that $v(m) = 0$, it is clear that $K_v(\sqrt[m]{a})/K_v$ is unramified if and only if

$$\text{ord}_v(a) \equiv 0 \pmod{m}.$$

(Recall ord_v is the normalized valuation associated to v.) Now when adjoining m^{th}-roots, it is only necessary to take one representative for each class in $K^*/(K^*)^m$. We conclude that

$$L = K(\sqrt[m]{a} : a \in T_S),$$

where

$$T_S = \{a \in K^*/(K^*)^m : \text{ord}_v(a) \equiv 0 \ (m) \text{ for all } v \in M_K, v \notin S\}.$$

To finish the proof, it thus suffices to show that the set T_S is finite.

Consider the natural map

$$R_S^* \to T_S.$$

We claim that it is surjective. To see this, suppose $a \in K^*$ represents an element of T_S. Then the ideal aR_S is the mth-power of an ideal in R_S, since the prime ideals of R_S correspond to the valuations $v \notin S$. Since R_S is a principal ideal domain, there is a $b \in K^*$ so that $aR_S = b^m R_S$, Hence there is a $u \in R_S^*$ so that

$$a = ub^m.$$

Then a and u give the same element of T_S, so R_S^* surjects onto T_S. Now the kernel of this map certainly contains $(R_S^*)^m$, so we have a surjection

$$R_S^*/(R_S^*)^m \twoheadrightarrow T_S.$$

(It is actually an isomorphism.) But Dirichlet's S-unit theorem [La 2, V §1] says that R_S^* is finitely generated, so this proves that T_S is finite, and thereby completes the proof of the proposition. $\square$

The three propositions proven above may now be combined to give our main result.

PROOF OF THE WEAK MORDELL–WEIL THEOREM (1.1). Let $L = K([m]^{-1}E(K))$ be the field defined in (1.2d). Since $E[m]$ is finite, the perfect pairing given in (2.1) shows that $E(K)/mE(K)$ is finite if and only if $G_{L/K}$ is finite. Now (1.5) shows that L has certain properties, and (1.6) shows that any extension of K with those properties is a finite extension, which gives the desired result. (Note that the set S of (1.5b) is a finite set; cf. (1.3).)   □

Remark 1.7. The heart of the proof of the weak Mordell–Weil theorem lies in the assertion that the field $L = K([m]^{-1}E(K))$ is a finite extension of K. We proved this by first showing (1.5) that it is abelian, of exponent m, and unramified outside a certain finite set $S \subset M_K$. The desired result then followed from the basic Kummer theory of fields given in the proof of (1.6). It is worth pointing out that instead of (1.6), we could have used the more general theorem of Minkowski which asserts that there are only finitely many extensions of K of bounded degree which are unramified outside of S. To apply this in the present instance, note that for any $Q \in [m]^{-1}E(K)$, the field $K(Q)$ has degree at most m^2 over K. (The $G_{\bar{K}/K}$ conjugates of Q all have the form $Q + T$ for some $T \in E[m]$.) It follows from Minkowski's theorem that as Q ranges over $[m]^{-1}E(K)$, there are only finitely many possibilities for the fields $K(Q)$. Hence their compositum $K([m]^{-1}E(K))$ is a finite extension of K.

Remark on Effectivity

Let E/K be an elliptic curve with $E[m] \subset E(K)$, let $S \subset M_K$ be the usual set of bad places for E/K (as in (1.5b)), and let L/K be the maximal abelian extension of K having exponent m which is unramified outside S. Then from (1.2) and (1.5), the Kummer pairing induces an injection

$$E(K)/mE(K) \to \mathrm{Hom}(G_{L/K}, E[m]).$$

Now it is possible to make the proof of (1.6) completely explicit, and so exactly determine the finite group $G_{L/K}$ (see exer. 8.1). Thus one can describe all of the elements of the group $\mathrm{Hom}(G_{L/K}, E[m])$, and the crucial question becomes that of determining which of these elements come from points of $E(K)/mE(K)$. It is this last question for which there is at present no known effective procedure for answering. We will examine this problem in more detail in chapter X. There we will exhibit a smaller group into which $E(K)/mE(K)$ injects, and see what can be said about the cokernel. Let us also note that this is the only point at which the Mordell–Weil theorem is ineffective; if one can produce generators for $E(K)/mE(K)$, then one can find generators for $E(K)$. (See (3.2) and exer. 8.18.)

§2. The Kummer Pairing via Cohomology

In this section we reinterpret the Kummer pairing of §1 in terms of group cohomology. The methods used here will not be used again until chapter X, and may be omitted by the reader wishing to proceed directly to the proof of the Mordell–Weil theorem. For the basic facts on group cohomology which we will use, see appendix B and/or the references listed there.

We start with the short exact sequence of $G_{\bar{K}/K}$-modules,

$$0 \to E[m] \to E(\bar{K}) \overset{[m]}{\to} E(\bar{K}) \to 0,$$

where $m \geq 2$ is a fixed integer. Taking $G_{\bar{K}/K}$ cohomology yields a long exact sequence which starts

$$
\begin{aligned}
0 \to \quad E(K)[m] \quad &\to \quad E(K) \quad \overset{[m]}{\to} \quad E(K) \\
\overset{\delta}{\to} H^1(G_{\bar{K}/K}, E[m]) &\to H^1(G_{\bar{K}/K}, E(\bar{K})) \overset{[m]}{\to} H^1(G_{\bar{K}/K}, E(\bar{K})).
\end{aligned}
$$

Now from the middle of this long exact sequence we can extract the following short exact sequence, which we call the *Kummer sequence for E/K*:

$$0 \to \frac{E(K)}{mE(K)} \overset{\delta}{\to} H^1(G_{\bar{K}/K}, E[m]) \to H^1(G_{\bar{K}/K}, E(\bar{K}))[m] \to 0.$$

(As usual, for any abelian group A, $A[m]$ denotes the m-torsion subgroup of A.)

From general principles, the connecting homomorphism δ is computed as follows. Let $P \in E(K)$ and choose some $Q \in E(\bar{K})$ satisfying $[m]Q = P$. Then a 1-cocycle representing $\delta(P)$ is given by

$$c : G_{\bar{K}/K} \to E[m]$$

$$c_\sigma = Q^\sigma - Q.$$

But this is exactly the Kummer pairing defined in §1,

$$c_\sigma = \kappa(P, \sigma).$$

(This assumes we use the same Q for both sides, of course.)

Now suppose that $E[m]$ is contained in $E(K)$. Then

$$H^1(G_{\bar{K}/K}, E[m]) = \mathrm{Hom}(G_{\bar{K}/K}, E[m]),$$

so in this case we have an injective homomorphism given by

$$E(K)/mE(K) \hookrightarrow \mathrm{Hom}(G_{\bar{K}/K}, E[m])$$

$$P \to \kappa(P, \cdot).$$

This provides an alternative proof of (1.2abc).

Similarly, we can use the inflation-restriction sequence (B.2.4) to obtain a quick proof of reduction lemma (1.1.1). Thus if L/K is a finite Galois extension (say with $E[m] \subset E(L)$), then we have a commutative diagram

$$
\begin{array}{ccccccc}
0 & \to & \Phi & \to & E(K)/mE(K) & \to & E(L)/mE(L) \\
 & & \downarrow & & \downarrow & & \downarrow \\
0 & \to & H^1(G_{L/K}, E[m]) & \overset{\mathrm{inf}}{\to} & H^1(G_{\bar{K}/K}, E[m]) & \overset{\mathrm{res}}{\to} & H^1(G_{\bar{L}/L}, E[m]).
\end{array}
$$

Since $G_{L/K}$ and $E[m]$ are finite groups, the cohomology group $H^1(G_{L/K}, E[m])$

is finite, so Φ is finite also. (The map $\lambda_P : G_{L/K} \to E[m]$ defined in the proof of (1.1.1) is a cocycle whose cohomology class is precisely the image of $P \in \Phi$ in $H^1(G_{L/K}, E[m])$.)

Returning now to the general case, we reinterpret (1.5b) in terms of cohomology.

Definition. Let M be a $G_{\bar{K}/K}$-module, $v \in M_K^0$ a discrete valuation, and $I_v \subset G_{\bar{K}/K}$ the inertia group for v. A cohomology class $\xi \in H^r(G_{\bar{K}/K}, M)$ is said to be *unramified at v* if it is trivial in $H^r(I_v, M)$.

Proposition 2.1. *Let*

$$S = \{v \in M_K^0 : E \text{ has bad reduction at } v\} \cup \{v \in M_K^0 : v(m) \neq 0\} \cup M_K^\infty.$$

Then the image of $E(K)$ in $H^1(G_{\bar{K}/K}, E[m])$ under the connecting homomorphism δ consists of cohomology classes which are unramified at every $v \in M_K$, $v \notin S$.

PROOF. Let $P \in E(K)$, and as above let

$$c_\sigma = Q^\sigma - Q$$

be a cocycle representing $\delta(P)$, where $[m]Q = P$. Then from (1.5b), the field $K(Q)$ is unramified over v. (Note that the proof of (1.5b) did not use the assumption that $E[m]$ is contained in $E(K)$.) Hence I_v acts trivially on Q, so for all $\sigma \in I_v$, $c_\sigma = 0$. $\qquad\qquad\square$

The Kummer Sequence for Fields

The exact sequences derived above are analogous to the usual ones related to Kummer theory for a field. To make the analogy clear, we briefly recall the relevant facts. Corresponding to the multiplication-by-m sequence for E used above is the exact sequence of $G_{\bar{K}/K}$-modules

$$1 \to \mu_m \to \bar{K}^* \overset{m}{\to} \bar{K}^* \to 1,$$

where the map denoted m is raising to the mth-power. Taking $G_{\bar{K}/K}$ cohomology yields a long exact sequence, from which we extract

$$1 \to K^*/(K^*)^m \overset{\delta}{\to} H^1(G_{\bar{K}/K}, \mu_m) \to H^1(G_{\bar{K}/K}, \bar{K}^*).$$

Now Hilbert's famous "theorem 90" (B.2.5) asserts that

$$H^1(G_{\bar{K}/K}, \bar{K}^*) = 0,$$

so the connecting homomorphism δ is an isomorphism. This is in marked contrast to the situation for elliptic curves, where the non-triviality of $H^1(G_{\bar{K}/K}, E(\bar{K}))$ provides much added complication. (See chapter X.) Collecting the above facts, and using an explicit computation of the connecting homomorphism, we have the following.

Proposition 2.2. *There is an isomorphism*

$$\delta : K^*/(K^*)^m \xrightarrow{\sim} H^1(G_{\bar{K}/K}, \mathbf{\mu}_m)$$

given by

$$\delta(a) = \text{cohomology class of } \{\sigma \to \alpha^\sigma/\alpha\},$$

where $\alpha \in \bar{K}^*$ *satisfies* $\alpha^m = a$.

§3. The Descent Procedure

Our main goal in this chapter is to prove that $E(K)$, the group of rational points on an elliptic curve, is finitely generated. So far, we know (1.1) that the quotient group $E(K)/mE(K)$ is finite. It is easy to see that this is not enough. For example, $\mathbb{R}/m\mathbb{R} = 0$ for every integer $m \geq 1$, but $\mathbb{R}$ is certainly not finitely generated. Similarly, if $E/\mathbb{Q}_p$ is an elliptic curve, then (VII.6.3) says that $E(\mathbb{Q}_p)$ has a subgroup of finite index isomorphic to the additive group $\mathbb{Z}_p$. Hence $E(\mathbb{Q}_p)/mE(\mathbb{Q}_p)$ is finite and $E(\mathbb{Q}_p)$ is not finitely generated.

An examination of these two examples shows that the problem occurs because of the large number of elements in the group which are divisible by m. The idea used to finish the proof of the Mordell–Weil theorem is to show that on an elliptic curve over a number field, the multiplication by m map tends to increase the "size" of a point; and that there are only finitely many points with small "size". This will bound how high a power of m can divide a point, and so eliminate problems such as in the above examples. Of course, all of this is very vague until we explain what is meant by the "size" of a point.

In this section we will axiomatize the situation and describe the type of size (or height) function needed to prove that an abelian group is finitely generated. Then in the next section we will define such a function on an elliptic curve in the simplest case, and use explicit formulas to prove that it has the desired properties. This will suffice to prove a special case of the Mordell–Weil theorem (4.1). After that, we will turn back to the general case and develop the theory of height functions in sufficient generality to both prove the Mordell–Weil theorem (6.7) and be useful for future applications.

Proposition 3.1 (Descent theorem). *Let A be an abelian group. Suppose there is a "height" function*

$$h : A \to \mathbb{R}$$

with the following three properties:

(i) *Let $Q \in A$. There is a constant C_1, depending on A and Q, so that for all $P \in A$,*

$$h(P + Q) \leqslant 2h(P) + C_1.$$

(ii) *There is an integer $m \geqslant 2$ and a constant C_2, depending on A, so that for all $P \in A$,*

$$h(mP) \geqslant m^2 h(P) - C_2.$$

(iii) *For every constant C_3,*

$$\{P \in A : h(P) \leqslant C_3\}$$

is a finite set.

Suppose further that for the integer m in (ii), the quotient group A/mA is finite. Then A is finitely generated.

PROOF. Choose elements $Q_1, \ldots, Q_r \in A$ to represent the finitely many cosets in A/mA. Now let $P \in A$. The idea is to show that the difference between P and an appropriate linear combination of $Q_1, \ldots, Q_r$ is a multiple of a point whose height is less than a constant which is *independent of P*. Then $Q_1, \ldots, Q_r$ and the finitely many points with height less than this constant will generate A.

Write

$$P = mP_1 + Q_{i_1} \qquad \text{for some } 1 \leqslant i_1 \leqslant r.$$

Continuing in this fashion,

$$P_1 = mP_2 + Q_{i_2},$$

$$\vdots$$

$$P_{n-1} = mP_n + Q_{i_n}.$$

Now for any j, we have

$$h(P_j) \leqslant \frac{1}{m^2}[h(mP_j) + C_2] \qquad \text{from (ii)}$$

$$= \frac{1}{m^2}[h(P_{j-1} - Q_{i_j}) + C_2]$$

$$\leqslant \frac{1}{m^2}[2h(P_{j-1}) + C_1' + C_2] \qquad \text{from (i)},$$

where we take C_1' to be the maximum of the constants from (i) for $Q = -Q_i$, $1 \leqslant i \leqslant r$. Note that C_1' and C_2 do not depend on P.

Now use the above inequality repeatedly, starting from P_n and working back to P. This yields

$$h(P_n) \leqslant \left(\frac{2}{m^2}\right)^n h(P) + \left[\frac{1}{m^2} + \frac{2}{m^4} + \frac{4}{m^6} + \cdots + \frac{2^{n-1}}{m^{2n}}\right](C_1' + C_2)$$

$$< \left(\frac{2}{m^2}\right)^n h(P) + \frac{C_1' + C_2}{m^2 - 2}$$

$$\leqslant 2^{-n} h(P) + (C_1' + C_2)/2 \qquad \text{since } m \geqslant 2.$$

It follows that by taking n sufficiently large, we will have (say)

$$h(P_n) \leqslant 1 + (C_1' + C_2)/2.$$

Since (from above)

$$P = m^n P_n + \sum_{j=1}^{n} m^{j-1} Q_{i_j},$$

it follows that every $P \in A$ is a linear combination of the points in the set

$$\{Q_1, \ldots, Q_r\} \cup \{Q \in A : h(Q) \leqslant 1 + (C_1' + C_2)/2\}.$$

From (iii), this is a finite set, which proves that A is finitely generated. $\qquad\square$

Remark 3.2. What is needed to make the descent theorem effective; that is, to allow us to find generators for the group A? First, we must be able to calculate the constants $C_1 = C_1(Q_i)$ for each of the elements $Q_1, \ldots, Q_r \in A$ representing the cosets of A/mA. Second, we must be able to calculate the constant C_2. Third, for any constant C_3, we must be able to determine the elements in the finite set $\{P \in A : h(P) \leqslant C_3\}$. The reader may check (exer. 8.18) that for the height functions which we will define on elliptic curves (§4, 5, 6), all of these constants are effectively computable *provided* we can find elements of $E(K)$ which generate the finite group $E(K)/mE(K)$. Unfortunately, at present there is no known procedure which is guaranteed to give generators for $E(K)/mE(K)$. We will return to this question in chapter X.

§4. The Mordell–Weil Theorem over $\mathbb{Q}$

In this section we will prove the following special case of the Mordell–Weil theorem.

Theorem 4.1. *Let $E/\mathbb{Q}$ be an elliptic curve. Then the group $E(\mathbb{Q})$ is finitely generated.*

We will, of course, soon be ready to prove the general case (6.7). But it seems worthwhile to give the proof of (4.1) first, since in this case the necessary height computations using explicit formulas are not too cumbersome.

Fix a Weierstrass equation for $E/\mathbb{Q}$ of the form

$$E : y^2 = x^3 + Ax + B$$

with $A, B \in \mathbb{Z}$. From (1.1) we know that $E(\mathbb{Q})/2E(\mathbb{Q})$ is finite, so to use the descent theorem (3.1), we need to define a height function on $E(\mathbb{Q})$.

Definition. Let $t \in \mathbb{Q}$ and write $t = p/q$ as a fraction in lowest terms. The *height* of t, denoted $H(t)$, is defined by

$$H(t) = \max\{|p|, |q|\}.$$

Definition. The *height on $E(\mathbb{Q})$* (relative to the given Weierstrass equation) is the function

$$h_x : E(\mathbb{Q}) \to \mathbb{R}$$

$$h_x(P) = \begin{cases} \log H(x(P)) & \text{if } P \neq O \\ 0 & \text{if } P = O. \end{cases}$$

Notice $h_x(P)$ is always non-negative.

The following lemma gives us the necessary information about this height function

Lemma 4.2. (a) *Let $P_0 \in E(\mathbb{Q})$. There is a constant C_1, depending on P_0, A, B, so that for all $P \in E(\mathbb{Q})$,*

$$h_x(P + P_0) \leqslant 2h_x(P) + C_1.$$

(b) *There is a constant C_2, depending on A, B, so that for all $P \in E(\mathbb{Q})$,*

$$h_x([2]P) \geqslant 4h_x(P) - C_2.$$

(c) *For every constant C_3, the set*

$$\{P \in E(\mathbb{Q}) : h_x(P) \leqslant C_3\}$$

is finite.

PROOF. (a) Taking $C_1 > \max\{h_x(P_0), h_x([2]P_0)\}$, we may assume $P_0 \neq O$ *and* $P \neq O, \pm P_0$. Then writing

$$P = (x, y) = \left(\frac{a}{d^2}, \frac{b}{d^3}\right) \qquad P_0 = (x_0, y_0) = \left(\frac{a_0}{d_0^2}, \frac{b_0}{d_0^3}\right)$$

(where the indicated fractions are in lowest terms), the addition formula (III.2.3d) reads

$$x(P + P_0) = \left(\frac{y - y_0}{x - x_0}\right)^2 - x - x_0.$$

Now multiplying this out and using that P and P_0 satisfy the Weierstrass equation yields

$$x(P + P_0) = \frac{(xx_0 + A)(x + x_0) + 2B - 2yy_0}{(x - x_0)^2}$$

$$= \frac{(aa_0 + Ad^2d_0^2)(ad_0^2 + a_0d^2) + 2Bd^4d_0^4 - 2bdb_0d_0}{(ad_0^2 - a_0d^2)^2}$$

In computing the height of a rational number, cancellation between numerator and denominator can only decrease the height, so we find by an easy estimation that

$$H(x(P + P_0)) \leqslant C_1' \max\{|a|^2, |d|^4, |bd|\},$$

where C_1' has a simple expression in terms of A, B, a_0, b_0, d_0. Since $H(x(P)) = \max\{|a|, |d|^2\}$, this is exactly what we want except for the presence of the $|bd|$. But since P is on the curve,

$$b^2 = a^3 + Aad^4 + Bd^6,$$

so

$$|b| \leqslant C_1'' \max\{|a|^{3/2}, |d|^3\}.$$

Using this above yields

$$H(x(P + P_0)) \leqslant C_1 \max\{|a|^2, |d|^4\} = C_1 H(x(P))^2,$$

and now taking logarithms gives the desired result.

(b) By choosing $C_2 \geqslant 4h_x(T)$ for each of the points $T \in E(\mathbb{Q})[2]$, we may assume that $[2]P \neq O$. Then writing $P = (x, y)$, the duplication formula (III.2.3d) reads

$$x([2]P) = \frac{x^4 - 2Ax^2 - 8Bx + A^2}{4x^3 + 4Ax + 4B}.$$

It is convenient to define homogeneous polynomials

$$F(X, Z) = X^4 - 2AX^2Z^2 - 8BXZ^3 + A^2Z^4,$$

$$G(X, Z) = 4X^3Z + 4AXZ^3 + 4BZ^4.$$

Then if we write $x = x(P) = a/b$ as a fraction in lowest terms, $x([2]P)$ can be written as a quotient of integers

$$x([2]P) = F(a, b)/G(a, b).$$

However, in contrast to (a), we are looking for a lower bound for $H(x([2]P))$, so it will be important to bound how much cancellation can occur between numerator and denominator.

The idea is to use the fact that $F(X, 1)$ and $G(X, 1)$ are relatively prime polynomials, so they generate the unit ideal in $\mathbb{Q}[X]$. This implies that identities of the following sort exist.

Sublemma 4.3. *Let* $\Delta = 4A^3 + 27B^2$,

$$F(X, Z) = X^4 - 2AX^2Z^2 - 8BXZ^3 + A^2Z^4,$$

$$G(X, Z) = 4X^3Z + 4AXZ^3 + 4BZ^4,$$

$$f_1(X, Z) = 12X^2Z + 16AZ^3,$$

$$g_1(X, Z) = 3X^3 - 5AXZ^2 - 27BZ^3,$$

$$f_2(X, Z) = 4(4A^3 + 27B^2)X^3 - 4A^2BX^2Z$$
$$\qquad\qquad + 4A(3A^3 + 22B^2)XZ^2 + 12B(A^3 + 8B^2)Z^3,$$

$$g_2(X, Z) = A^2BX^3 + A(5A^3 + 32B^2)X^2Z$$
$$\qquad\qquad + 2B(13A^3 + 96B^2)XZ^2 - 3A^2(A^3 + 8B^2)Z^3.$$

Then the following identities hold in $\mathbb{Q}[X, Z]$:

$$f_1(X, Z)F(X, Z) - g_1(X, Z)G(X, Z) = 4\Delta Z^7$$

$$f_2(X, Z)F(X, Z) + g_2(X, Z)G(X, Z) = 4\Delta X^7.$$

PROOF. Since $F(X, Z)$ and $G(X, Z)$ are relatively prime homogeneous polynomials (provided $\Delta \neq 0$), it is clear a priori that identities of this sort will exist. To check the validity of the two given identities is at worst a tedious calculation, which we leave for the reader. (To actually find the polynomials f_1, g_1, f_2, g_2, one can use the Euclidean algorithm or the theory of resultants.)
$\qquad\qquad\qquad\qquad\qquad\qquad\qquad\qquad\qquad\qquad\qquad\qquad\qquad\qquad$ $\square$

We return to the proof of (4.2b). Let

$$\delta = \gcd(F(a, b), G(a, b))$$

be the cancellation in our fraction for $x([2]P)$. From the equations

$$f_1(a, b)F(a, b) - g_1(a, b)G(a, b) = 4\Delta b^7$$

$$f_2(a, b)F(a, b) + g_2(a, b)G(a, b) = 4\Delta a^7,$$

we see that δ divides 4Δ. Hence we obtain the bound

$$|\delta| \leqslant |4\Delta|,$$

and so

$$H(x([2]P)) \geqslant \max\{|F(a, b)|, |G(a, b)|\}/|4\Delta|.$$

On the other hand, the same identities give the estimates

$$|4\Delta b^7| \leqslant 2 \max\{|f_1(a, b)|, |g_1(a, b)|\}\max\{|F(a, b)|, |G(a, b)|\},$$

$$|4\Delta a^7| \leqslant 2 \max\{|f_2(a, b)|, |g_2(a, b)|\}\max\{|F(a, b)|, |G(a, b)|\}.$$

Now looking at the expressions for f_1, f_2, g_1, and g_2 in (4.3), we have

$$\max\{|f_1(a, b)|, |g_1(a, b)|, |f_2(a, b)|, |g_2(a, b)|\} \leqslant C \max\{|a|^3, |b|^3\},$$

where C is a constant depending on A and B. Combining the last three inequalities yields

$$\max\{|4\Delta a^7|, |4\Delta b^7|\} \leqslant 2C \max\{|a|^3, |b|^3\} \max\{|F(a, b)|, |G(a, b)|\},$$

and so cancelling $\max\{|a|^3, |b|^3\}$ gives

$$\max\{|F(a, b)|, |G(a, b)|\}/|4\Delta| \geqslant (2C)^{-1} \max\{|a|, |b|\}.^4$$

Since $\max\{|a|, |b|\} = H(x(P))$, this gives the desired estimate

$$H(x([2]P)) \geqslant (2C)^{-1} H(x(P)).^4$$

(c) For any constant C, the set

$$\{t \in \mathbb{Q} : H(t) \leqslant C\}$$

is clearly finite. (It certainly has fewer than $(2C + 1)^2$ elements.) But given any value for x, there are at most two values of y for which (x, y) is a point of E. Therefore

$$\{P \in E(\mathbb{Q}) : h_x(P) \leqslant C_3\}$$

is also a finite set. $\square$

Proving (4.1) is now just a matter of fitting together what we have already proven.

PROOF OF (4.1). From (1.1), $E(\mathbb{Q})/2E(\mathbb{Q})$ is finite. Now (4.2) says that the height function

$$h_x : E(\mathbb{Q}) \to \mathbb{R}$$

satisfies the conditions necessary to apply the descent theorem (3.1) (with $m = 2$). The conclusion from (3.1) is that $E(\mathbb{Q})$ is finitely generated. $\square$

§5. Heights on Projective Space

In order to use the descent theorem (3.1) to prove the Mordell–Weil theorem in general, it is necessary to define a height function on the K-rational points of an elliptic curve. It is possible to proceed in an ad hoc manner using explicit equations, as in the last section; but rather than do this, we will instead develop the general theory of height functions, from which will follow all of the necessary properties plus considerably more. Since our elliptic curves are given as subsets of projective space, in this section we will study a certain height function defined on all of projective space; and then in the next section we will examine its properties when restricted to the points of an elliptic curve.

Example 5.1. Suppose $P \in \mathbb{P}^N(\mathbb{Q})$. Since $\mathbb{Z}$ is a principal ideal domain, we can find homogeneous coordinates for P, say

$$P = [x_0, \ldots, x_N],$$

which satisfy

$$x_0, \ldots, x_N \in \mathbb{Z} \quad \text{and} \quad \gcd(x_0, \ldots, x_N) = 1.$$

Then a natural measure of the *height of P* would be

$$H(P) = \max\{|x_0|, \ldots, |x_N|\}.$$

Notice that with this definition, it is clear that for any constant C, the set

$$\{P \in \mathbb{P}^N(\mathbb{Q}) : H(P) \leqslant C\}$$

is a finite set. (It has fewer than $(2C + 1)^{N+1}$ elements.) This is the sort of finiteness property needed to apply the descent theorem (3.1).

Now in trying to directly generalize (5.1) to arbitrary number fields, one runs into difficulty when the ring of integers is not a principal ideal domain. We thus take a somewhat different approach, for which purpose we now specify more precisely how the absolute values in M_K are to be normalized.

Definition. The *set of standard absolute values on* $\mathbb{Q}$, which we again denote by $M_\mathbb{Q}$, consists of the following:

(i) $M_\mathbb{Q}$ contains one archimedean absolute value, given by

$$|x|_\infty = \text{usual absolute value} = \max\{x, -x\}.$$

(ii) For each prime $p \in \mathbb{Z}$, $M_\mathbb{Q}$ contains one non-archimedean (p-adic) absolute value, given by

$$\left| p^n \frac{a}{b} \right|_p = p^{-n} \quad \text{for } a, b \in \mathbb{Z}, \quad \gcd(p, ab) = 1.$$

The *set of standard absolute values on* K, denoted M_K, consists of all absolute values on K whose restriction to $\mathbb{Q}$ is one of the absolute values in $M_\mathbb{Q}$.

Definition. For $v \in M_K$, the *local degree at* v, denoted n_v, is given by

$$n_v = [K_v : \mathbb{Q}_v].$$

(Here K_v and $\mathbb{Q}_v$ denote, as usual, the completion of the indicated field with respect to the absolute value v.)

With these definitions, we can state the two basic facts from algebraic number theory which will be needed.

Extension Formula 5.2. *Let $L/K/\mathbb{Q}$ be a tower of number fields, and $v \in M_K$. Then*

$$\sum_{\substack{w \in M_L \\ w|v}} n_w = [L : K] n_v.$$

(*Here $w|v$ means that w equals v when restricted to K.*)

Product Formula 5.3. *Let* $x \in K^*$. *Then*

$$\prod_{v \in M_K} |x|_v^{n_v} = 1.$$

For proofs of these two formulas, see [La 2, II §1 and V §1].
We are now ready to define the height of a point in projective space.

Definition. Let $P \in \mathbb{P}^N(K)$ be a point with homogeneous coordinates

$$P = [x_0, \ldots, x_N], \qquad x_i \in K.$$

The *height of P (relative to K)* is defined by

$$H_K(P) = \prod_{v \in M_K} \max\{|x_0|_v, \ldots, |x_N|_v\}^{n_v}.$$

Proposition 5.4. *Let* $P \in \mathbb{P}^N(K)$.
(a) *The height* $H_K(P)$ *does not depend on the choice of homogeneous coordinates for P.*

(b) $$H_K(P) \geqslant 1.$$

(c) *Let* L/K *be a finite extension. Then*

$$H_L(P) = H_K(P)^{[L:K]}.$$

PROOF. (a) Any other choice of homogeneous coordinates for P has the form $[\lambda x_0, \ldots, \lambda x_N]$ for some $\lambda \in K^*$. Then using the product formula (5.3), we have

$$\prod_{v \in M_K} \max_i \{|\lambda x_i|_v\}^{n_v} = \prod_{v \in M_K} |\lambda|_v^{n_v} \max_i \{|x_i|_v\}^{n_v} = \prod_{v \in M_K} \max_i \{|x_i|_v\}^{n_v}.$$

(b) For any point in projective space, one can find homogeneous coordinates so that one of the coordinates is 1. Then every factor in the product defining $H_K(P)$ is at least 1.
(c) We compute

$$
\begin{aligned}
H_L(P) &= \prod_{w \in M_L} \max\{|x_i|_w\}^{n_w} \\
&= \prod_{v \in M_K} \prod_{\substack{w \in M_L \\ w|v}} \max\{|x_i|_v\}^{n_w} \qquad \text{since } x_i \in K \\
&= \prod_{v \in M_K} \max\{|x_i|_v\}^{[L:K]n_v} \qquad \text{from (5.2)} \\
&= H_K(P)^{[L:K]}. \qquad\qquad\qquad \square
\end{aligned}
$$

Remark 5.5. If $K = \mathbb{Q}$, then $H_{\mathbb{Q}}$ agrees with the more intuitive height function given in (5.1). Thus let $P \in \mathbb{P}^N(\mathbb{Q})$, and choose homogeneous coordinates $[x_0, \ldots, x_N]$ for P so that $x_i \in \mathbb{Z}$ and $\gcd(x_0, \ldots, x_N) = 1$. Then for every non-archimedean absolute value $v \in M_{\mathbb{Q}}$, we have $|x_i|_v \leqslant 1$ for all i and $|x_i|_v = 1$ for at least one i. Hence in the product for $H_{\mathbb{Q}}(P)$, only the term for the archimedean absolute value contributes, so

$$H_{\mathbb{Q}}(P) = \max\{|x_0|_\infty, \ldots, |x_N|_\infty\}.$$

In particular, it follows that

$$\{P \in \mathbb{P}^N(\mathbb{Q}): H_{\mathbb{Q}}(P) \leqslant C\}$$

is a finite set for any constant C. One of our goals is to extend this result to H_K, and we will actually prove something even stronger (5.11).

It is sometimes easier to use a height function which is not relative to a given field. In view of (5.4c), the following definition makes sense.

Definition. Let $P \in \mathbb{P}^N(\bar{\mathbb{Q}})$. The (*absolute*) *height of* P, denoted $H(P)$, is defined as follows. Choose any field K such that $P \in \mathbb{P}^N(K)$. Then

$$H(P) = H_K(P)^{1/[K:\mathbb{Q}]} \qquad \text{(positive root)}.$$

We now investigate how the height changes under mappings between projective spaces. We recall the following definition (cf. I.3.3).

Definition. A *morphism of degree d* between projective spaces is a map

$$F: \mathbb{P}^N \to \mathbb{P}^M$$

$$F(P) = [f_0(P), \ldots, f_M(P)],$$

where $f_0, \ldots, f_M \in \bar{\mathbb{Q}}[X_0, \ldots, X_N]$ are homogeneous polynomials of degree d with no common zero in $\bar{\mathbb{Q}}$ other than $X_0 = \cdots = X_N = 0$. If F can be written with polynomials f_i having coefficients in K, then F is said to be *defined over* K.

Theorem 5.6. *Let*

$$F: \mathbb{P}^N \to \mathbb{P}^M$$

be a morphism of degree d. Then there are positive constants C_1 and C_2, depending on F, so that for all points $P \in \mathbb{P}^N(\bar{\mathbb{Q}})$,

$$C_1 H(P)^d \leqslant H(F(P)) \leqslant C_2 H(P)^d.$$

PROOF. Write $F = [f_0, \ldots, f_M]$ with homogeneous polynomials f_i, and let $P = [x_0, \ldots, x_N] \in \mathbb{P}^N(\bar{\mathbb{Q}})$. Choose some number field K containing $x_0, \ldots, x_N$ and all of the coefficients of all of the f_i's. Then for each $v \in M_K$, let

$$|P|_v = \max_{0 \leqslant i \leqslant N} \{|x_i|_v\}, \qquad |F(P)|_v = \max_{0 \leqslant j \leqslant M} \{|f_j(P)|_v\},$$

and

$$|F|_v = \max\{|a|_v : a \text{ is a coefficient of some } f_i\}.$$

Then from the definition of height,

$$H_K(P) = \prod_{v \in M_K} |P|_v^{n_v} \quad \text{and} \quad H_K(F(P)) = \prod_{v \in M_K} |F(P)|_v^{n_v},$$

so it makes sense to define

$$H_K(F) = \prod_{v \in M_K} |F|_v^{n_v}.$$

(I.e. $H_K(F) = H_K([a_0, a_1, \dots])$, where the a_j's are the coefficients of the f_i's.) Finally, we let $C_1, C_2, \dots$ denote constants which depend only on M, N and d, and set

$$\varepsilon(v) = \begin{cases} 1 & \text{if } v \in M_K^{\infty} \\ 0 & \text{if } v \in M_K^0. \end{cases}$$

(To illustrate the utility of $\varepsilon(v)$, we note that the triangle inequality can be concisely written as

$$|t_1 + \cdots + t_n|_v \leqslant n^{\varepsilon(v)} \max\{|t_1|_v, \dots, |t_n|_v\}$$

for all $v \in M_K$, both archimedean and non-archimedean.)

Having set notation, we turn to the proof of (5.6). The upper bound is relatively easy. Let $v \in M_K$. The triangle inequality yields

$$|f_i(P)|_v \leqslant C_1^{\varepsilon(v)} |F|_v |P|_v^d,$$

since f_i is homogeneous of degree d. Here C_1 could equal the number of terms in f_i, which is at most $\binom{N+d}{N}$ (i.e. this is the number of monomials of degree d in $N + 1$ variables). Since this holds for each i, we find

$$|F(P)|_v \leqslant C_1^{\varepsilon(v)} |F|_v |P|_v^d.$$

Now raise to the n_v-power, multiply over all $v \in M_K$, and take the $[K : \mathbb{Q}]^{\text{th}}$-root. This yields the desired upper bound

$$H(F(P)) \leqslant C_1 H(F) H(P)^d.$$

(Note that

$$\sum_{v \in M_K} \varepsilon(v) n_v = \sum_{v \in M_K^{\infty}} n_v = [K : \mathbb{Q}] \qquad \text{from (5.2).)}$$

It is worth mentioning that in proving this upper bound, we did not use the fact that the f_i's have no common non-trivial zero. But for the lower bound we will certainly need this fact, since otherwise there are easy counter-examples (see exer. 8.10).

Thus we now assume that the set

$$\{Q \in \mathbb{A}^{N+1}(\overline{\mathbb{Q}}) : f_0(Q) = \cdots = f_M(Q) = 0\}$$

consists of the single point $(0, \dots, 0)$. It follows from the Nullstellensatz ([Har, I.3A]) that the ideal generated by $f_0, \dots, f_M$ in $\overline{\mathbb{Q}}[X_0, \dots, X_N]$ contains some power of each of $X_0, \dots, X_N$, since each X_i also vanishes at

$(0, \ldots, 0)$. Thus for an appropriate integer $e \geqslant 1$, there are polynomials $g_{ij} \in \overline{\mathbb{Q}}[X_0, \ldots, X_N]$ such that

$$X_i^e = \sum_{j=0}^{M} g_{ij} f_j \qquad \text{for each } 0 \leqslant i \leqslant N.$$

Replacing K by a finite extension, we may assume that each $g_{ij} \in K[X_0, \ldots, X_N]$. Further, by discarding all terms except those which are homogeneous of degree e, we may assume that each g_{ij} is homogeneous of degree $e - d$. Let us set the further reasonable notation

$$|G|_v = \max\{|b|_v : b \text{ is a coefficient of some } g_{ij}\}$$

$$H_K(G) = \prod_{v \in M_K} |G|_v^{n_v}.$$

(We note that e and $H_K(G)$ may be bounded in terms of M, N, d, and $H_K(F)$, although to give a good bound is not at all an easy task. See (5.7) for a discussion. For our purposes it is enough to note that e and $H_K(G)$ do not depend on the point P.)

Recalling that $P = [x_0, \ldots, x_N]$, the equations described above imply that for each i,

$$|x_i|_v^e = \left| \sum_{j=0}^{M} g_{ij}(P) f_j(P) \right|_v$$

$$\leqslant C_2^{\varepsilon(v)} \max_{0 \leqslant j \leqslant M} \{|g_{ij}(P) f_j(P)|_v\}.$$

Now taking the maximum over i gives

$$|P|_v^e \leqslant C_2^{\varepsilon(v)} \max_{\substack{0 \leqslant j \leqslant M \\ 0 \leqslant i \leqslant N}} \{|g_{ij}(P)|_v\} |F(P)|_v.$$

But since each g_{ij} has degree $e - d$, the usual application of the triangle inequality yields

$$|g_{ij}(P)|_v \leqslant C_3^{\varepsilon(v)} |G|_v |P|_v^{e-d}.$$

(Here C_3 may also depend on e; but as mentioned above, e may be bounded in terms of M, N, and d.) Substituting this in above and multiplying through by $|P|_v^{d-e}$ gives

$$|P|_v^d \leqslant C_4^{\varepsilon(v)} |G|_v |F(P)|_v;$$

and now the usual raising to the n_v-power, multiplying over $v \in M_K$, and taking the $[K : \mathbb{Q}]^{\text{th}}$-root yields the desired lower bound.   □

Remark 5.7. As indicated during the proof of (5.6), in the inequality

$$C_1 H(P)^d \leqslant H(F(P)),$$

the dependence of C_1 on F is not at all straightforward. Precisely, C_1 can be given in terms of the coefficients of certain polynomials whose existence is guaranteed by the Nullstellensatz. Now the Nullstellensatz can be made completely mechanical by the use of elimination theory, but using this method directly leads to a very poor estimate. For an explicit version of the Nullstellensatz where an effort has been made to give good estimates for the coefficients, see [M–W].

We also record the special case of (5.6) corresponding to an automorphism of $\mathbb{P}^N$.

Corollary 5.8. Let $A \in GL_{N+1}(\overline{\mathbb{Q}})$, so matrix multiplication by A induces an automorphism $A : \mathbb{P}^N \to \mathbb{P}^N$. Then there are constants C_1 and C_2, depending on the entries of the matrix A, so that for all $P \in \mathbb{P}^N(\overline{\mathbb{Q}})$,

$$C_1 H(P) \leqslant H(AP) \leqslant C_2 H(P).$$

PROOF. This is (5.6) for a morphism of degree 1.   □

We next investigate the relationship between the height of the coefficients of a polynomial and the height of its roots.

Notation. For $x \in \overline{\mathbb{Q}}$, let

$$H(x) = H([x, 1]).$$

Similarly, if $x \in K$, then

$$H_K(x) = H_K([x, 1]).$$

Theorem 5.9. *Let*

$$f(T) = a_0 T^d + a_1 T^{d-1} + \cdots + a_d = a_0(T - \alpha_1) \cdots (T - \alpha_d) \in \overline{\mathbb{Q}}[T]$$

be a polynomial of degree d (i.e. $a_0 \neq 0$). Then

$$2^{-d} \prod_{j=1}^{d} H(\alpha_j) \leqslant H([a_0, \ldots, a_d]) \leqslant 2^{d-1} \prod_{j=1}^{d} H(\alpha_j).$$

PROOF. First note that the inequality to be proven remains unchanged if $f(T)$ is replaced by $(1/a_0)f(T)$. It thus suffices to prove the result under the assumption that $a_0 = 1$.

Let $K = \mathbb{Q}(\alpha_1, \ldots, \alpha_d)$, and for $v \in M_K$, set

$$\varepsilon(v) = \begin{cases} 2 & \text{if } v \in M_K^\infty \\ 1 & \text{if } v \in M_K^0. \end{cases}$$

(Note this notation differs from that used in the proof of (5.6). In the present instance, the triangle inequality reads

$$|x + y|_v \leqslant \varepsilon(v) \max\{|x|_v, |y|_v\} \qquad \text{for } v \in M_K, x, y \in K.$$

Of course, if $v \in M_K^0$ and $|x|_v \neq |y|_v$, then it is an equality.) We will now prove that

$$\varepsilon(v)^{-d} \prod_{j=1}^{d} \max\{|\alpha_j|_v, 1\} \leqslant \max_{0 \leqslant i \leqslant d} \{|a_i|_v\} \leqslant \varepsilon(v)^{d-1} \prod_{j=1}^{d} \max\{|\alpha_j|_v, 1\}.$$

Once this is done, raising to the n_v-power, multiplying over $v \in M_K$, and taking $[K : \mathbb{Q}]^{\text{th}}$-roots gives the desired result.

The proof is by induction on $d = \deg(f)$. For $d = 1$, $f(T) = T - \alpha_1$, so the inequality is clear. Assume now that we know the result for all polynomials (with roots in K) of degree $d - 1$. Choose an index k so that

$$|\alpha_k|_v \geqslant |\alpha_j|_v \qquad \text{for all } 0 \leqslant j \leqslant d,$$

and define a polynomial

$$g(T) = (T - \alpha_1) \cdots (T - \alpha_{k-1})(T - \alpha_{k+1}) \cdots (T - \alpha_d)$$
$$= b_0 T^{d-1} + b_1 T^{d-2} + \cdots + b_{d-1}.$$

Thus $f(T) = (T - \alpha_k)g(T)$, so comparing coefficients yields

$$a_i = b_i - \alpha_k b_{i-1}.$$

(This holds in the entire range $0 \leqslant i \leqslant d$ if we set $b_{-1} = b_d = 0$.)

We now prove the upper bound stated above.

$$\max_{0 \leqslant i \leqslant d} \{|a_i|_v\} = \max_{0 \leqslant i \leqslant d} \{|b_i - \alpha_k b_{i-1}|_v\}$$
$$\leqslant \varepsilon(v) \max_{0 \leqslant i \leqslant d} \{|b_i|_v, |\alpha_k b_{i-1}|_v\} \qquad \text{triangle inequality}$$
$$\leqslant \varepsilon(v) \max_{0 \leqslant i \leqslant d} \{|b_i|_v\} \max\{|\alpha_k|_v, 1\}$$
$$\leqslant \varepsilon(v)^{d-1} \prod_{j=1}^{d} \max\{|\alpha_j|_v, 1\} \qquad \begin{array}{l} \text{induction hypothesis} \\ \text{applied to } g. \end{array}$$

Next, to prove the lower bound, we consider two cases. First, if $|\alpha_k|_v \leqslant \varepsilon(v)$, then by the choice of the index k,

$$\prod_{j=1}^{d} \max\{|\alpha_j|_v, 1\} \leqslant \max\{|\alpha_k|_v, 1\}^d \leqslant \varepsilon(v)^d,$$

so the result is clear. (Remember $a_0 = 1$.) Next, suppose that $|\alpha_k|_v > \varepsilon(v)$. Then

$$\max_{0 \leqslant i \leqslant d} \{|a_i|_v\} = \max_{0 \leqslant i \leqslant d} \{|b_i - \alpha_k b_{i-1}|_v\}$$
$$\geqslant \varepsilon(v)^{-1} \max_{0 \leqslant i \leqslant d-1} \{|b_i|_v\} \max\{|\alpha_k|_v, 1\}.$$

Here the last line is an equality for $v \in M_K^0$, while for $v \in M_K^\infty$ we are using the calculation

$$\max_{0 \leqslant i \leqslant d} \{|b_i - \alpha_k b_{i-1}|_v\} \geqslant (|\alpha_k|_v - 1) \max_{0 \leqslant i \leqslant d-1} \{|b_i|_v\}$$

$$> \varepsilon(v)^{-1}|\alpha_k|_v \max_{0 \leqslant i \leqslant d-1} \{|b_i|_v\}$$

$$\text{since } |\alpha_k|_v > \varepsilon(v) = 2.$$

Now applying the induction hypothesis to g gives the desired lower bound, which completes the proof of (5.9). ☐

Our first application of (5.9) will be to show that there are only finitely many points of bounded height in projective space. To do this, we will need to know that the action of Galois does not affect the height of a point.

Lemma 5.10. *Let* $P \in \mathbb{P}^N(\overline{\mathbb{Q}})$ *and* $\sigma \in G_{\overline{\mathbb{Q}}/\mathbb{Q}}$. *Then*

$$H(P^\sigma) = H(P).$$

PROOF. Let $K/\mathbb{Q}$ be a field with $P \in \mathbb{P}^N(K)$. σ gives an isomorphism $\sigma: K \overset{\sim}{\to} K^\sigma$, and it likewise identifies the sets of absolute values,

$$\sigma: M_K \overset{1-1}{\to} M_{K^\sigma}$$

$$v \to v^\sigma.$$

(I.e. For $x \in K$ and $v \in M_K$, $|x^\sigma|_{v^\sigma} = |x|_v$.) Clearly σ also gives an isomorphism $K_v \overset{\sim}{\to} K_{v^\sigma}^\sigma$, so $n_v = n_{v^\sigma}$. We now compute

$$H_{K^\sigma}(P^\sigma) = \prod_{w \in M_{K^\sigma}} \max\{|x_i^\sigma|_w\}^{n_w}$$

$$= \prod_{v \in M_K} \max\{|x_i^\sigma|_{v^\sigma}\}^{n_{v^\sigma}}$$

$$= \prod_{v \in M_K} \max\{|x_i|_v\}^{n_v}$$

$$= H_K(P).$$

Since $[K:\mathbb{Q}] = [K^\sigma:\mathbb{Q}]$, this is the desired result. ☐

Theorem 5.11. *Let C and d be constants. Then the set*

$$\{P \in \mathbb{P}^N(\overline{\mathbb{Q}}): H(P) \leqslant C \quad and \quad [\mathbb{Q}(P):\mathbb{Q}] \leqslant d\}$$

contains only finitely many points. In particular, for any number field K,

$$\{P \in \mathbb{P}^N(K): H_K(P) \leqslant C\}$$

is a finite set. (Recall (I §2) that $\mathbb{Q}(P)$ is the minimal field of definition for P.)

PROOF. Let $P \in \mathbb{P}^N(\overline{\mathbb{Q}})$. Take homogeneous coordinates for P, say

$$P = [x_0, \ldots, x_N],$$

with some $x_j = 1$. Then $\mathbb{Q}(P) = \mathbb{Q}(x_0, \ldots, x_N)$, and we have the easy estimate

$$H_{\mathbb{Q}(P)}(P) = \prod_{v \in M_{\mathbb{Q}(P)}} \max_{0 \leqslant i \leqslant N} \{|x_i|_v\}^{n_v}$$

$$\geqslant \max_{0 \leqslant i \leqslant N} \left(\prod_{v \in M_{\mathbb{Q}(P)}} \max\{|x_i|_v, 1\}^{n_v} \right)$$

$$= \max_{0 \leqslant i \leqslant N} H_{\mathbb{Q}(P)}(x_i).$$

Thus if $H(P) \leqslant C$ and $[\mathbb{Q}(P) : \mathbb{Q}] \leqslant d$, then

$$\max_{0 \leqslant i \leqslant N} H(x_i) \leqslant C \quad \text{and} \quad \max_{0 \leqslant i \leqslant N} [\mathbb{Q}(x_i) : \mathbb{Q}] \leqslant d.$$

It thus suffices to prove that the set

$$\{x \in \bar{\mathbb{Q}} : H(x) \leqslant C \quad \text{and} \quad [\mathbb{Q}(x) : \mathbb{Q}] \leqslant d\}$$

is finite. (I.e. We have reduced to the case $N = 1$.)

Suppose $x \in \bar{\mathbb{Q}}$ is in this set, and let $e = [\mathbb{Q}(x) : \mathbb{Q}]$, so $e \leqslant d$. Further let $x = x_1, x_2, \ldots, x_e$ be the conjugates of x (in $\bar{\mathbb{Q}}$), so the minimal polynomial of x over $\mathbb{Q}$ is

$$f_x(T) = (T - x_1) \cdots (T - x_e) = T^e + a_1 T^{e-1} + \cdots + a_e \in \mathbb{Q}[T].$$

Now

$$H([1, a_1, \ldots, a_e]) \leqslant 2^{e-1} \prod_{j=1}^{e} H(x_j) \qquad \text{from (5.9)}$$

$$= 2^{e-1} H(x)^e \qquad \text{from (5.10)}$$

$$\leqslant (2C)^d \qquad \text{since } H(x) \leqslant C \text{ and } e \leqslant d.$$

Since the a_i's are in $\mathbb{Q}$, it is now clear that for given C and d there are only finitely many possibilities for the polynomial $f_x(T)$. (I.e. We are using the special case of the theorem with $K = \mathbb{Q}$, for which it is easy to prove. See (5.1, 5.3).) Since for a given polynomial there are at most d elements in our set, this proves that the set is finite. ∎

Remark 5.12. Tracing through the proof of (5.11), it is easy enough to give an upper bound, in terms of C and d, for how many points are in the set

$$\{P \in \mathbb{P}^N(\bar{\mathbb{Q}}) : H(P) \leqslant C \quad \text{and} \quad [\mathbb{Q}(P) : \mathbb{Q}] \leqslant d\}.$$

(See exer. 8.7a.) More difficult is to give a precise asymptotic estimate for

$$\#\{P \in \mathbb{P}^N(K) : H_K(P) \leqslant C\}$$

as a function of C for $C \to \infty$. Such an estimate has been given by Schanuel. (See [Scha] or [La 7, Ch. 3, §5].)

§6. Heights on Elliptic Curves

In this section we use the general theory of heights as developed in the previous section to define height functions on elliptic curves. The main theorems (6.2, 6.4) exhibit the interplay between the height of points and the addition law on the elliptic curve. As an immediate corollary, we will deduce the remaining results needed to prove the Mordell–Weil theorem for arbitrary number fields (6.7).

It is convenient to use the "big-O" notation.

Notation. Let f, g be two real-valued functions on a set $\mathscr{S}$. Then we write

$$f = g + O(1)$$

if there are constants C_1 and C_2 so that

$$C_1 \leqslant f(P) - g(P) \leqslant C_2 \qquad \text{for all } P \in \mathscr{S}.$$

If only the lower (respectively upper) inequality is satisfied, then we naturally write $f \geqslant g + O(1)$ (respectively $f \leqslant g + O(1)$).

Let E/K be an elliptic curve. Recall (II.2.2) that any non-constant function $f \in \bar{K}(E)$ determines a surjective morphism (which we also denote by f)

$$f : E \to \mathbb{P}^1$$

$$P \to \begin{cases} [1, 0] & \text{if } P \text{ is a pole of } f \\ [f(P), 1] & \text{otherwise.} \end{cases}$$

It would be reasonable to define a height function on $E(\bar{K})$ by setting $H_f(P) = H(f(P))$. However, the height function H tends to behave multiplicatively (as in (5.6) for example), while for our purposes it will be more convenient to have a height which behaves additively. This prompts the following definitions.

Definition. The (*absolute logarithmic*) *height* on projective space is the function

$$h : \mathbb{P}^N(\bar{\mathbb{Q}}) \to \mathbb{R}$$

$$h(P) = \log H(P).$$

Notice that from (5.4b), $h(P) \geqslant 0$ for all P.

Definition. Let E/K be an elliptic curve and $f \in \bar{K}(E)$ a function. The *height on E* (*relative to* f) is the function

$$h_f : E(\bar{K}) \to \mathbb{R}$$

$$h_f(P) = h(f(P)).$$

We start by transcribing the finiteness result from section 5 into the current setting.

Proposition 6.1. *Let E/K be an elliptic curve and $f \in K(E)$ a non-constant function. Then for any constant C,*

$$\{P \in E(K) : h_f(P) \leq C\}$$

is a finite set.

PROOF. The function f gives a finite-to-one map of the set in question to the set

$$\{Q \in \mathbb{P}^1(K) : H(Q) \leq e^C\}.$$

(Note that since $f \in K(E)$, any $P \in E(K)$ will go to a point $f(P) \in \mathbb{P}^1(K)$.) Now apply (5.11) to this last set. □

The next theorem gives a fundamental relationship between height functions and the addition law on an elliptic curve.

Theorem 6.2. *Let E/K be an elliptic curve and let $f \in K(E)$ be an even function (i.e. $f \circ [-1] = f$). Then for all $P, Q \in E(\bar{K})$,*

$$h_f(P + Q) + h_f(P - Q) = 2h_f(P) + 2h_f(Q) + O(1).$$

(*Here the constants inherent in the $O(1)$ depend on the elliptic curve E and the function f, but are of course independent of P and Q*).

PROOF. Choose a Weierstrass equation for E/K of the form

$$E : y^2 = x^3 + Ax + B.$$

We start by proving the theorem for the particular function $f = x$. The general case will then be an easy corollary.

Since $h_x(O) = 0$ and $h_x(-P) = h_x(P)$, the result clearly holds if $P = O$ or $Q = O$. We now assume that $P, Q \neq O$, and write

$$x(P) = [x_1, 1], \qquad x(Q) = [x_2, 1],$$
$$x(P + Q) = [x_3, 1], \qquad x(P - Q) = [x_4, 1].$$

(Here x_3 or x_4 may equal ∞ if $P = \pm Q$.) Now the addition formula (III.2.3d) and a little bit of algebra yield the relations

$$x_3 + x_4 = \frac{2(x_1 + x_2)(A + x_1 x_2) + 4B}{(x_1 + x_2)^2 - 4x_1 x_2},$$

$$x_3 x_4 = \frac{(x_1 x_2 - A)^2 - 4B(x_1 + x_2)}{(x_1 + x_2)^2 - 4x_1 x_2}.$$

Define a map $g : \mathbb{P}^2 \to \mathbb{P}^2$ by

$$g([t, u, v]) = [u^2 - 4tv, 2u(At + v) + 4Bt^2, (v - At)^2 - 4Btu].$$

Then the formulas for x_3 and x_4 show that there is a commutative diagram

$$
\begin{array}{ccc}
E \times E & \overset{G}{\to} & E \times E \\
\downarrow & & \downarrow \\
\sigma \qquad \mathbb{P}^1 \times \mathbb{P}^1 & & \mathbb{P}^1 \times \mathbb{P}^1 \qquad \sigma, \\
\downarrow & & \downarrow \\
\mathbb{P}^2 & \overset{g}{\to} & \mathbb{P}^2
\end{array}
$$

where

$$G(P, Q) = (P + Q, P - Q),$$

and the vertical map σ is the composition of the two maps

$$E \times E \to \mathbb{P}^1 \times \mathbb{P}^1 \qquad \text{and} \qquad \mathbb{P}^1 \times \mathbb{P}^1 \to \mathbb{P}^2$$

$$(P, Q) \to (x(P), x(Q)) \qquad ([\alpha_1, \beta_1], [\alpha_2, \beta_2]) \to [\beta_1\beta_2, \alpha_1\beta_2 + \alpha_2\beta_1, \alpha_1\alpha_2].$$

(The idea here is to treat t, u, v as $1, x_1 + x_2, x_1 x_2$. Then $g([t, u, v])$ becomes $[1, x_3 + x_4, x_3 x_4]$.)

The next step is to show that g is a morphism, so as to be able to apply (5.6). By definition (cf. I.3.3), this means we must show that except for $t = u = v = 0$, the three homogeneous polynomials defining g have no common zeros. Suppose now that $g([t, u, v]) = [0, 0, 0]$. If $t = 0$, then from

$$u^2 - 4tv = 0 \quad \text{and} \quad (v - At)^2 - 4Btu = 0,$$

we see that $u = v = 0$. Thus we may assume that $t \neq 0$, and so it makes sense to define a new quantity $x = u/2t$. (*Intuition*: If we write t, u, v as $1, x_1 + x_2, x_1 x_2$, then the equation $u^2 - 4tv = 0$ becomes $(x_1 - x_2)^2 = 0$, so $x_1 = x_2 = u/2t$. In other words, we are now dealing with the case that $P = \pm Q$.) Notice that the equation $u^2 - 4tv = 0$ can be written as $x^2 = v/t$. Now dividing the equalities

$$2u(At + v) + 4Bt^2 = 0 \quad \text{and} \quad (v - At)^2 - 4Btu = 0$$

by t^2 and rewriting them in terms of x yields the two equations

$$\psi(x) = 4x(A + x^2) + 4B = 4x^3 + 4Ax + 4B = 0,$$

$$\phi(x) = (x^2 - A)^2 - 8Bx = x^4 - 2Ax^2 - 8Bx + A^2 = 0.$$

[These polynomials should be familiar. Their ratio $\phi(X)/\psi(X)$ is exactly the rational function which appears in the duplication formula (III.2.3d).] To show that $\psi(X)$ and $\phi(X)$ have no common root, one need merely verify the formal identity already used in (4.3),

$$(12X^2 + 16A)\phi(X) - (3X^3 - 5AX - 27B)\psi(X) = 4(4A^3 + 27B^2) \neq 0.$$

(Note how the non-singularity of the Weierstrass equation plays a crucial role here.) This completes the proof that g is a morphism.

We return to our commutative diagram, and compute

$$h(\sigma(P + Q, P - Q)) = h(\sigma \circ G(P, Q))$$
$$= h(g \circ \sigma(P, Q))$$
$$= 2h(\sigma(P, Q)) + O(1) \qquad \text{from (5.6),}$$

since g is a morphism of degree 2. Now to complete the proof of (6.2) for $f = x$, we will show that for all $R_1, R_2 \in E(\bar{K})$ there is a relation

$$h(\sigma(R_1, R_2)) = h_x(R_1) + h_x(R_2) + O(1).$$

Then using this relation twice, once on each side of the equation

$$h(\sigma(P + Q, P - Q)) = 2h(\sigma(P, Q)) + O(1),$$

will give the desired result.

One immediately verifies that if either $R_1 = O$ or $R_2 = O$, then $h(\sigma(R_1, R_2))$ equals $h_x(R_1) + h_x(R_2)$. Otherwise, we may write

$$x(R_1) = [\alpha_1, 1] \quad \text{and} \quad x(R_2) = [\alpha_2, 1],$$

and so

$$h(\sigma(R_1, R_2)) = h([1, \alpha_1 + \alpha_2, \alpha_1\alpha_2]) \quad \text{and} \quad h_x(R_1) + h_x(R_2) = h(\alpha_1) + h(\alpha_2).$$

Then from (5.9) applied to the polynomial $(T + \alpha_1)(T + \alpha_2)$, we obtain the desired estimate

$$h(\alpha_1) + h(\alpha_2) - \log 4 \leqslant h([1, \alpha_1 + \alpha_2, \alpha_1\alpha_2]) \leqslant h(\alpha_1) + h(\alpha_2) + \log 2.$$

Finally, to deal with the case of an arbitrary even function $f \in K(E)$, we prove that

$$h_f = \tfrac{1}{2}(\deg f)h_x + O(1).$$

From this, (6.2) follows immediately by multiplying the known relation for h_x by $\tfrac{1}{2}(\deg f)$. Thus the following lemma will complete the proof of (6.2). □

Lemma 6.3. *Let $f, g \in K(E)$ be even functions. Then*

$$(\deg g)h_f = (\deg f)h_g + O(1).$$

PROOF. Let $x, y \in K(E)$ be Weierstrass coordinates for E/K. The subfield of $K(E)$ consisting of even functions is exactly $K(x)$ (III.2.3.1), so we can find a rational function $\rho(X) \in K(X)$ so that there is a commutative diagram

$$
\begin{array}{ccc}
E & & \\
x \downarrow & \searrow f & \\
\mathbb{P}^1 & \xrightarrow{\rho} & \mathbb{P}^1.
\end{array}
$$

Hence using (5.6) and the fact that ρ is a morphism (II.2.1),

$$h_f = h_x \circ \rho = (\deg \rho)h_x + O(1).$$

But from the diagram,

$$\deg f = \deg x \deg \rho = 2 \deg \rho,$$

so we find

$$2h_f = (\deg f)h_x + O(1).$$

The same reasoning for g yields

$$2h_g = (\deg g)h_x + O(1),$$

and combining these last two equalities gives the desired result.   □

Corollary 6.4. *Let E/K be an elliptic curve and $f \in K(E)$ an even function.*
(a) *Let $Q \in E(\bar{K})$. Then for all $P \in E(\bar{K})$,*

$$h_f(P + Q) \leqslant 2h_f(P) + O(1),$$

where the $O(1)$ depends on E, f, and Q.
(b) *Let $m \in \mathbb{Z}$. Then for all $P \in E(\bar{K})$,*

$$h_f([m]P) = m^2 h_f(P) + O(1),$$

where the $O(1)$ depends on E, f, and m.

PROOF. (a) This follows immediately from (6.3), since $h_f(P - Q) \geqslant 0$.
(b) Since f is even, it suffices to consider $m \geqslant 0$. Further, the result is trivial
for $m = 0, 1$. We finish the proof by induction. Assume it is known for $m - 1$
and m. Replacing P, Q in (6.3) by $[m]P, P$, we find

$$h_f([m + 1]P) = -h_f([m - 1]P) + 2h_f([m]P) + 2h_f(P) + O(1)$$

$$= (-(m - 1)^2 + 2m^2 + 2)h_f(P) + O(1) \qquad \text{by the induction hypothesis}$$

$$= (m + 1)^2 h_f(P) + O(1). \qquad\qquad\qquad\qquad □$$

Remark 6.5. The above results (6.3, 6.4) are clearly also true for an odd
function f, since then f^2 is even, and one easily checks that $h_{f^2} = 2h_f$.
Although we will not prove it, they are true for arbitrary $f \in K(E)$ "to within
ε". To be precise, say for (6.4b), it is true that for every $\varepsilon > 0$ there are
inequalities

$$(1 - \varepsilon)m^2 h_f - O(1) \leqslant h_f \circ [m] \leqslant (1 + \varepsilon)m^2 h_f + O(1),$$

where now the $O(1)$ depends on E, f, m, and ε. (See exer. 9.14c. For a proof in
a much more general setting, see [La 7, Ch. 4, Cor. 3.5].)

Remark 6.6. Theorem 6.2 seems to say that the height function h_f is "more or
less" a quadratic form. In section 9 we will see that there is an actual quadratic form, called the *canonical height*, which differs from h_f by a bounded
amount.

It should be clear that we now have all the tools needed to complete the proof of the Mordell–Weil theorem.

Theorem 6.7 (Mordell–Weil theorem). *Let K be a number field and E/K an elliptic curve. Then the group $E(K)$ is finitely generated.*

PROOF. Choose any even, non-constant function $f \in K(E)$, for example the x-coordinate function on a Weierstrass equation. The Mordell–Weil theorem will now follow immediately from the weak Mordell–Weil theorem (1.1) with $m = 2$ and the descent theorem (3.1), once we show that the height function

$$h_f : E(K) \to \mathbb{R}$$

has the following three properties.

(i) Let $Q \in E(K)$. There is a constant C_1, depending on E, f, and Q, so that for all $P \in E(K)$,

$$h_f(P + Q) \leqslant 2h_f(P) + C_1.$$

(ii) There is a constant C_2, depending on E and f, so that for all $P \in E(K)$,

$$h_f([2]P) \geqslant 4h_f(P) - C_2.$$

(iii) For every constant C_3,

$$\{P \in E(K) : h_f(P) \leqslant C_3\}$$

is a finite set.

But (i) is a restatement of (6.4a), (ii) is immediate from the $m = 2$ case of (6.4b), and (iii) is just (6.1). This completes the proof of the Mordell–Weil theorem. $\qquad\square$

§7. Torsion Points

The Mordell–Weil theorem implies that the group of rational torsion points on an elliptic curve is finite. Of course, this also follows from the corresponding result for local fields.

Since an elliptic curve over a number field K can be treated as an elliptic curve over the completion K_v for each $v \in M_K$, the local integrality conditions for torsion points (VII.3.4) can be pieced together to give the following global statement.

Theorem 7.1. *Let E/K be an elliptic curve with Weierstrass equation*

$$y^2 + a_1 xy + a_3 y = x^3 + a_2 x^2 + a_4 x + a_6$$

such that all of the a_i's are in R. Let $P \in E(K)$ be a point of exact order $m \geqslant 2$.

(a) *If m is not a prime power, then*

$$x(P), \; y(P) \in R.$$

(b) *If $m = p^n$ is a prime power, for each $v \in M_K^0$ let*

$$r_v = \left[\frac{\mathrm{ord}_v(p)}{p^n - p^{n-1}} \right] \qquad ([\;] \text{ is greatest integer}).$$

Then

$$\mathrm{ord}_v(x(P)) \geqslant -2r_v \quad and \quad \mathrm{ord}_v(y(P)) \geqslant -3r_v.$$

In particular, $x(P)$ and $y(P)$ are v-integral if $\mathrm{ord}_v(p) = 0$.

The following corollary was proven independently by Lutz and Nagell, who had discovered divisibility conditions somewhat weaker than (7.1).

Corollary 7.2 ([Lut], [Nag]). *Let $E/\mathbb{Q}$ be an elliptic curve with Weierstrass equation*

$$y^2 = x^3 + Ax + B, \qquad A, \, B \in \mathbb{Z}.$$

Suppose $P \in E(\mathbb{Q})$ is a non-zero torsion point. Then

(a) $$x(P), \; y(P) \in \mathbb{Z}.$$

(b) *Either $[2]P = O$, or else $y(P)^2$ divides $4A^3 + 27B^2$.*

PROOF. (a) Let P have exact order m. If $m = 2$, then $y(P) = 0$, so $x(P) \in \mathbb{Z}$ since it is the root of a monic integral polynomial. If $m > 2$, then the result follows immediately from (7.1), since the quantity r_v in (7.1b) is necessarily 0.
(b) We assume that $[2]P \neq O$, so $y(P) \neq 0$. Then applying (a) to both P and $[2]P$, we have $x(P), \, y(P), \, x([2P]) \in \mathbb{Z}$. Let

$$\phi(X) = X^4 - 2AX^2 - 8BX + A^2$$

and

$$\psi(X) = X^3 + AX + B.$$

Then the duplication formula (III.2.3d) reads

$$x([2P]) = \phi(x(P))/4\psi(x(P)).$$

On the other hand, we have the usual polynomial identity (4.3)

$$f(X)\phi(X) - g(X)\psi(X) = 4A^3 + 27B^2.$$

(I.e. $f(X) = 3X^2 + 4A$ and $g(X) = 3X^3 - 5AX - 27B$.) Now put $X = x(P)$, and use the duplication formula and the fact that $y(P)^2 = \psi(x(P))$ to obtain

$$y(P)^2[4f(x(P))x([2]P) - g(x(P))] = 4A^3 + 27B^2.$$

Since all quantities in this equation are integers, the result follows.   □

Remark 7.3.1. A glance at the proof of (7.2b) will show that we actually proved that any point $P \in E(\mathbb{Q})$ such that $x(P)$ and $x([2]P)$ are both integers has the property that $y(P)^2$ divides $4A^3 + 27B^2$. The same argument works for number fields. Further, even if $x(P)$ or $x([2]P)$ is not integral, any bound for their denominators (such as (7.1b)) will give a corresponding bound for $y(P)$ (see exer. 8.11).

Remark 7.3.2. Recall (VII.3.2) that in practice, one of the quickest methods for bounding the torsion in $E(K)$ is to choose various finite places v for which E has good reduction, and then use the injection (VII.3.1)

$$E(K_v)[m] \to \tilde{E}(k_v)$$

for m relatively prime to $\mathrm{char}(k_v)$.

Example 7.4. The Weierstrass equation

$$E : y^2 = x^3 - 43x + 166$$

has

$$4A^3 + 27B^2 = 425984 = 2^{15} \cdot 13.$$

Hence any torsion point in $E(\mathbb{Q})$ has its y-coordinate in the set

$$\{0, \pm 1, \pm 2, \pm 4, \pm 8, \pm 16, \pm 32, \pm 64, \pm 128\}.$$

A little bit of work with a calculator reveals the points

$$\{(3, \pm 8), (-5, \pm 16), (11, \pm 32)\}.$$

On the other hand, since E has good reduction modulo 3, we know that $E_{\mathrm{tors}}(\mathbb{Q})$ injects into $E(\mathbb{F}_3)$ (cf. VII.3.5); and one checks that $\#E(\mathbb{F}_3) = 7$. This still does not prove anything, since the divisibility condition in (7.2b) is only necessary, not sufficient. But now using the doubling formula for $P = (3, 8)$, one finds

$$x(P) = 3, \qquad x([2]P) = -5, \qquad x([4]P) = 11, \qquad x([8]P) = 3.$$

Hence $[8]P = \pm P$, so P is a torsion point of exact order 7 or 9. (It doesn't have order 3, since $x(P) \neq x([2]P)$.) From above, the only possibility is order 7, so we conclude that $E_{\mathrm{tors}}(\mathbb{Q})$ is a cyclic group of order 7 consisting of the six points listed above together with O.

All of the above discussion has focused on characterizing the torsion subgroup of a given elliptic curve. Another sort of question one might ask is the following. Given a prime p, does there exist an elliptic curve $E/\mathbb{Q}$ such that $E(\mathbb{Q})$ contains a point of order p? The answer in general is no. For example, $E(\mathbb{Q})$ can never contain a point of order 11, a fact which is by no means obvious. Such a statement, which deals uniformly with the set of all elliptic curves, naturally tends to be more difficult to prove than a result such as (7.2), in which the bounds obtained become weaker as the elliptic curve is varied.

The definitive characterization of torsion subgroups over $\mathbb{Q}$ is given by the following theorem, whose proof is unfortunately far beyond the scope of this book.

Theorem 7.5 (Mazur [Maz 1], [Maz 2]). *Let $E/\mathbb{Q}$ be an elliptic curve. Then the torsion subgroup $E_{\text{tors}}(\mathbb{Q})$ is one of the following fifteen groups:*

$$\mathbb{Z}/N\mathbb{Z} \qquad 1 \leqslant N \leqslant 10 \quad or \quad N = 12;$$

$$\mathbb{Z}/2\mathbb{Z} \times \mathbb{Z}/2N\mathbb{Z} \quad 1 \leqslant N \leqslant 4.$$

Further, each of these groups does occur as an $E_{\text{tors}}(\mathbb{Q})$. (For an example of each possible group, see exer. 8.12.)

For arbitrary number fields, there is the following result of Manin.

Theorem 7.6 ([Man 2]). *Let $K/\mathbb{Q}$ be a number field and $p \in \mathbb{Z}$ a prime. There is a constant $N = N(K, p)$ so that for all elliptic curves E/K, the p-primary component of $E(K)$ has order dividing p^N.*

Taken together, (7.5) and (7.6) provide the best evidence to date for the following longstanding conjecture.

Conjecture 7.7. *Let $K/\mathbb{Q}$ be a number field. There is a constant $N = N(K)$ so that for all elliptic curves E/K,*

$$|E_{\text{tors}}(K)| \leqslant N.$$

Remark 7.8. For those torsion subgroups which are allowed in Mazur's theorem (7.5), it is a classical result that the elliptic curves E/K having the specified torsion subgroup all lie in a 1-parameter family. For example, the curves E/K with a point $P \in E(K)$ of order 7 all have Weierstrass equations of the form

$$y^2 + (1 + d - d^2)xy + (d^2 - d^3)y = x^3 + (d^2 - d^3)x^2 \qquad P = (0, 0)$$

with

$$d \in K \text{ and } \Delta = d^7(d - 1)^7(d^3 - 8d^2 + 5d + 1) \neq 0.$$

(See exer. 8.13a, b. A complete list is given in [Ku].) In general, the elliptic curves E/K with a point $P \in E(K)$ of order $m \geqslant 4$ are parametrized by the K-rational points of another curve, called a *modular curve*. (See appendix C §13 and exer. 8.13c.)

§8. The Minimal Discriminant

Let E/K be an elliptic curve. For each non-archimedean absolute value $v \in M_K^0$, we can find a Weierstrass equation for E,

$$y_v^2 + a_{1,v}x_v y_v + a_{3,v}y_v = x_v^3 + a_{2,v}x_v^2 + a_{4,v}x_v + a_{6,v},$$

which is a minimal equation for E at v. Let Δ_v be the discriminant of this equation.

Definition. The *minimal discriminant* of E/K, denoted $\mathscr{D}_{E/K}$, is the (integral) ideal of K given by

$$\mathscr{D}_{E/K} = \prod_{v \in M_K^0} \mathfrak{p}_v^{\mathrm{ord}_v(\Delta_v)}.$$

Here $\mathfrak{p}_v$ is the prime ideal of R associated to v. Thus $\mathscr{D}_{E/K}$ catalogs the valuation of the minimal discriminant of E at every place $v \in M_K^0$. In a certain sense, it is a measure of how arithmetically complicated the elliptic curve E is.

We now ask whether it is possible to find a single Weierstrass equation which is simultaneously minimal for every $v \in M_K^0$. Let

$$y^2 + a_1 xy + a_3 y = x^3 + a_2 x^2 + a_4 x + a_6$$

be any Weierstrass equation for E/K, say with discriminant Δ. Then for each $v \in M_K^0$ we can find a change of coordinates

$$x = u_v^2 x_v + r_v \qquad y = u_v^3 y_v + s_v u_v^2 x_v + t_v$$

which gives the minimal equation listed above. As usual, the two discriminants are related by

$$\Delta = u_v^{12} \Delta_v.$$

Hence if we define an ideal, depending on Δ, by the equation

$$\mathfrak{a}_\Delta = \prod_{v \in M_K^0} \mathfrak{p}_v^{-\mathrm{ord}_v(u_v)},$$

then the minimal discriminant can be written

$$\mathscr{D}_{E/K} = (\Delta)\mathfrak{a}_\Delta^{12}.$$

Lemma 8.1. *With notation as above, the ideal class of $\mathfrak{a}_\Delta$ in the ideal class group of K is independent of Δ.*

PROOF. Take another Weierstrass equation for E/K, say with discriminant Δ'. Then $\Delta = u^{12}\Delta'$ for some $u \in K^*$, so directly from the definitions we see that

$$(\Delta')\mathfrak{a}_{\Delta'}^{12} = \mathscr{D}_{E/K} = (\Delta)\mathfrak{a}_\Delta^{12} = (\Delta')[(u)\mathfrak{a}_\Delta]^{12}.$$

Hence $\mathfrak{a}_{\Delta'} = (u)\mathfrak{a}_\Delta$.   □

Definition. The *Weierstrass class* of E/K, denoted $\bar{\mathfrak{a}}_{E/K}$, is the ideal class of K corresponding to any ideal $\mathfrak{a}_\Delta$ as in (8.1).

Definition. A *global minimal Weierstrass equation* for E/K is a Weierstrass equation

$$y^2 + a_1 xy + a_3 y = x^3 + a_2 x^2 + a_4 x + a_6$$

for E/K such that $a_1, a_2, a_3, a_4, a_6 \in R$ and the discriminant Δ of the equation satisfies $\mathscr{D}_{E/K} = (\Delta)$.

Proposition 8.2. *There exists a global minimal Weierstrass equation for E/K if and only if $\bar{\mathfrak{a}}_{E/K} = (1)$.*

PROOF. Suppose E/K has a global minimal Weierstrass equation, say with discriminant Δ. Then $\mathscr{D}_{E/K} = (\Delta)$, so with notation as above,

$$12 \operatorname{ord}_v(\mathfrak{a}_\Delta) = \operatorname{ord}_v(\mathscr{D}_{E/K}) - \operatorname{ord}_v(\Delta) = 0.$$

Hence $\mathfrak{a}_\Delta = (1)$, so $\bar{\mathfrak{a}}_{E/K} = $ class of $\mathfrak{a}_\Delta = (1)$.

Conversely, suppose $\bar{\mathfrak{a}}_{E/K} = (1)$. Choose any Weierstrass equation for E/K, say with coefficients $a_i \in R$ and discriminant Δ; and as above, for each $v \in M_K^0$ let

$$x = u_v^2 x_v + r_v \qquad y = u_v^3 y_v + s_v u_v^2 x_v + t_v$$

be a change of variables which produces a minimal equation at v, say with coefficients $a_{i,v}$ and discriminant Δ_v. We may clearly assume that $u_v = 1$ and $r_v = s_v = t_v = 0$ for all but finitely many v, say for all v not in some set $S \subset M_K^0$. Note also that all of u_v, r_v, s_v, t_v are v-integral (VII.1.3d).

By definition, the fact that $\bar{\mathfrak{a}}_{E/K} = (1)$ means that the ideal

$$\prod_{v \in M_K^0} \mathfrak{p}_v^{\operatorname{ord}_v(u_v)}$$

is principal, generated by some $u \in K^*$. Then

$$\operatorname{ord}_v(u) = \operatorname{ord}_v(u_v) \qquad \text{for all } v \in M_K^0.$$

Now by the Chinese remainder theorem [La 2, Ch. I, §4], there are elements $r, s, t \in R$ so that for the finitely many $v \in S$, we have

$$\operatorname{ord}_v(r - r_v), \operatorname{ord}_v(s - s_v), \operatorname{ord}_v(t - t_v) > \max_{i=1,2,3,4,6} \{\operatorname{ord}_v(u_v^i a_{i,v})\}.$$

Now consider the new Weierstrass equation for E/K given by the change of coordinates

$$x = u^2 x' + r \qquad y = u^3 y' + s u^2 x' + t,$$

which has coefficients a_i' and discriminant Δ'. Then $\Delta = u^{12} \Delta'$, so from above

$$\operatorname{ord}_v(\Delta') = \operatorname{ord}_v(u^{-12}\Delta) = \operatorname{ord}_v((u_v/u)^{12}\Delta_v) = \operatorname{ord}_v(\Delta_v).$$

Thus the new equation is globally minimal provided that its coefficients are all integral. But this is easily checked using the transformation formulas (III.1.2). If $v \notin S$, then $\operatorname{ord}_v(u) = 0$, so each a_i' is v-integral, since it is a polynomial in $r, s, t, a_1, \ldots, a_6$. For $v \in S$, we illustrate the argument for a_2', the other coefficients being done similarly. Thus

$$\text{ord}_v(u^2 a_2') = \text{ord}_v(a_2 - sa_1 + 3r - s^2)$$
$$= \text{ord}_v[u_v^2 a_{2,v} - (s - s_v)(a_1 + s + s_v) + 3(r - r_v)]$$
$$= \text{ord}_v(u_v^2 a_{2,v}),$$

where the last line follows from the previous one by the choice of r, s and the non-archimedean nature of v. Since

$$\text{ord}_v(u) = \text{ord}_v(u_v) \quad \text{and} \quad \text{ord}_v(a_{2,v}) \geqslant 0,$$

this gives the desired result. $\square$

Corollary 8.3. *If K has class number 1, then every elliptic curve E/K has a global minimal Weierstrass equation. In particular, this is true for $K = \mathbb{Q}$.* (The converse is also true; see exer. 8.14.)

Example 8.4. The equation

$$y^2 = x^3 + 16$$

has discriminant $\Delta = -2^{12} 3^3$. It is not minimal at 2. The substitution

$$x = 4x' \qquad y = 8y' + 4$$

gives the global minimal equation

$$(y')^2 + y' = (x')^3.$$

Example 8.5. Let $K = \mathbb{Q}(\sqrt{-10})$, so K has class number 2, the class group being generated by the prime ideal $\mathfrak{p} = (5, \sqrt{-10})$. Consider the elliptic curve E/K given by the equation

$$E : y^2 = x^3 + 125.$$

This equation has discriminant $\Delta = -2^4 3^3 5^6$, so it is already minimal at every prime of K except possibly for the prime $\mathfrak{p}$, which lies over 5. (See VII.1.1.) For $\mathfrak{p}$, the change of coordinates

$$x = (\sqrt{-10})^2 x' \qquad y = (\sqrt{-10})^3 y'$$

gives an equation

$$(y')^2 = (x')^3 - 2^{-3}$$

which has good reduction at $\mathfrak{p}$. Hence

$$\mathscr{D}_{E/K} = (2^4 3^3)$$

and

$$\bar{\mathfrak{a}}_{E/K} = \text{ideal class of } \mathfrak{p}.$$

In particular, there is no global minimal Weierstrass equation for E/K.

Remark 8.6. If K has class number 1 and E/K is an elliptic curve, then one can find a global minimal Weierstrass equation for E/K by finding local minimal equations (e.g. by using Tate's algorithm [Ta 6]) and then following the proof of (8.2). There is also an algorithm, due to Laska ([Las 1]), which is both fast and easy to implement on a computer.

Even if R has class number greater than 1, it is often useful to know that an elliptic curve E/K has a global Weierstrass equation which is in some sense "almost minimal". The following proposition gives one possibility. (For another, see exer. 8.14c.)

Proposition 8.7. Let $S \subset M_K$ be a finite set of absolute values containing M_K^∞ and all places dividing 2 and 3. Further assume that the ring of S-integers R_S is a principal ideal domain. Then every elliptic curve E/K has a model

$$E : y^2 = x^3 + Ax + B$$

with $A, B \in R_S$ and discriminant $\Delta = -16(4A^3 + 27B^2)$ satisfying

$$\mathscr{D}_{E/K} R_S = \Delta R_S.$$

(*Such a Weierstrass equation might be called* S-minimal.)

PROOF. Choose any Weierstrass equation for E/K of the form

$$E : y^2 = x^3 + Ax + B,$$

and let $\Delta = -16(4A^3 + 27B^2)$. For each $v \in M_K$, $v \notin S$, choose a $u_v \in K^*$ so that the substitution

$$x = u_v^2 x' \qquad y = u_v^3 y'$$

gives a minimal equation at v. Thus

$$v(\mathscr{D}_{E/K}) = v(\Delta) - 12v(u_v) \qquad \text{for all } v \in M_K, v \notin S.$$

Since R_S is a principal ideal domain, there is a $u \in K^*$ such that

$$v(u) = v(u_v) \qquad \text{for all } v \in M_K, v \notin S.$$

Then the equation

$$E : y^2 = x^3 + u^{-4}Ax + u^{-6}B$$

has the desired property. $\qquad\square$

§9. The Canonical Height

Let E/K be an elliptic curve and $f \in K(E)$ an even function. Theorems 6.2 and 6.4 say that the height function h_f is more or less a quadratic form, at least "up to $O(1)$". André Néron asked whether one could find an actual quadratic form which differs from h_f by a bounded amount. He constructed such a function by writing it as a sum of "quasi-quadratic" local functions ([Né 3]). At the same time, Tate came up with a simpler global definition. We will give Tate's construction here. (See appendix C §18 for a discussion of local height functions.)

Proposition 9.1 (Tate). *Let E/K be an elliptic curve, $f \in K(E)$ a non-constant even function, and $P \in E(\bar{K})$. Then the limit*

$$\frac{1}{\deg(f)} \operatorname*{Lim}_{N \to \infty} 4^{-N} h_f([2^N]P)$$

exists, and is independent of f.

PROOF. We show that the sequence is Cauchy. From (6.4b) with $m = 2$, there is a constant C so that for all $Q \in E(\bar{K})$,

$$|h_f([2]Q) - 4h_f(Q)| \leqslant C.$$

Now let $N \geqslant M \geqslant 0$ be integers. Then

$$|4^{-N} h_f([2^N]P) - 4^{-M} h_f([2^M]P)|$$

$$= \left| \sum_{n=M}^{N-1} 4^{-n-1} h_f([2^{n+1}]P) - 4^{-n} h_f([2^n]P) \right|$$

$$\leqslant \sum_{n=M}^{N-1} 4^{-n-1} |h_f([2^{n+1}]P) - 4h_f([2^n]P)|$$

$$\leqslant \sum_{n=M}^{N-1} 4^{-n-1} C \qquad \text{using } Q = [2^n]P \text{ above}$$

$$\leqslant C/4^{M+1}.$$

This shows that the sequence $4^{-N} h_f([2^N]P)$ is Cauchy, so it converges.

Next suppose $g \in K(E)$ is another non-constant even function. Then from (6.3),

$$(\deg g)h_f = (\deg f)h_g + O(1),$$

so

$$(\deg g)4^{-N} h_f([2^N]P) - (\deg f)4^{-N} h_g([2^N]P) = 4^{-N} O(1) \to 0$$

as $N \to \infty$. Hence the limit does not depend on the choice of the function f. $\square$

Definition. The *canonical* (or *Néron–Tate*) height on E/K, denoted $\hat{h}$ or $\hat{h}_E$, is the function

$$\hat{h} : E(\bar{K}) \to \mathbb{R}$$

defined by

$$\hat{h}(P) = \frac{1}{\deg f} \operatorname*{Lim}_{N \to \infty} 4^{-N} h_f([2^N]P).$$

(Here $f \in K(E)$ is any non-constant even function.)

Remark 9.2. From (9.1), the canonical height is well-defined and is independent of the choice of f.

Theorem 9.3 (Néron–Tate). *Let E/K be an elliptic curve and $\hat{h}$ the canonical height on E.*
(a) *For all $P, Q \in E(\bar{K})$,*

$$\hat{h}(P + Q) + \hat{h}(P - Q) = 2\hat{h}(P) + 2\hat{h}(Q) \qquad (parallelogram\ law).$$

(b) *For all $P \in E(\bar{K})$ and $m \in \mathbb{Z}$,*

$$\hat{h}([m]P) = m^2 \hat{h}(P).$$

(c) *$\hat{h}$ is a quadratic form on E. (In other words, $\hat{h}$ is even, and the pairing*

$$\langle\ ,\ \rangle : E(\bar{K}) \times E(\bar{K}) \to \mathbb{R}$$

$$\langle P, Q \rangle = \hat{h}(P + Q) - \hat{h}(P) - \hat{h}(Q)$$

is bilinear.)
(d) *Let $P \in E(\bar{K})$. Then $\hat{h}(P) \geqslant 0$, and*

$$\hat{h}(P) = 0 \qquad if\ and\ only\ if \qquad P\ is\ a\ torsion\ point.$$

(e) *Let $f \in K(E)$ be an even function. Then*

$$(\deg f)\hat{h} = h_f + O(1),$$

where the $O(1)$ depends on E and f.
    Further, if $\hat{h}' : E(\bar{K}) \to \mathbb{R}$ is another function which satisfies (e) for some non-constant function f and (b) for any one integer $m \geqslant 2$, then $\hat{h}' = \hat{h}$.

PROOF. We will start by proving (e), and then return to (a)–(d).
(e) In the course of proving (9.1), we found a constant C (depending on the choice of f) so that for all integers $N \geqslant M \geqslant 0$ and all points $P \in E(\bar{K})$,

$$|4^{-N} h_f([2^N]P) - 4^{-M} h_f([2^M]P)| \leqslant C/4^{M+1}.$$

Taking $M = 0$ and letting $N \to \infty$ gives the desired estimate

$$|(\deg f)\hat{h}(P) - h_f(P)| \leqslant C/4.$$

(a) From (6.2), we have

$$h_f(P + Q) + h_f(P - Q) = 2h_f(P) + 2h_f(Q) + O(1).$$

Replace P, Q by $[2^N]P$, $[2^N]Q$, multiply through by $\dfrac{1}{\deg(f)}4^{-N}$, and let $N \to \infty$. The $O(1)$ term disappears, and we obtain

$$\hat{h}(P + Q) + \hat{h}(P - Q) = 2\hat{h}(P) + 2\hat{h}(Q).$$

(b) From (6.4b),

$$h_f([m]P) = m^2 h_f(P) + O(1).$$

As usual, replace P by $[2^N]P$, multiply by 4^{-N}, and let $N \to \infty$. (Alternative proof: Use (a) and induction on m.)

(c) It is a standard fact from linear algebra that a function satisfying the parallelogram law is quadratic. For completeness, we include a proof.

Putting $P = O$ in the parallelogram law (a) shows that $\hat{h}(-Q) = \hat{h}(Q)$, so $\hat{h}$ is even. By symmetry, it suffices to prove that

$$\langle P + R, Q \rangle = \langle P, Q \rangle + \langle R, Q \rangle,$$

which in terms of $\hat{h}$ becomes

$$\hat{h}(P + R + Q) - \hat{h}(P + R) - \hat{h}(P + Q) - \hat{h}(R + Q) + \hat{h}(P) + \hat{h}(R) + \hat{h}(Q) = 0.$$

Now four applications of the parallelogram law (and the evenness of $\hat{h}$) give

$$\hat{h}(P + R + Q) + \hat{h}(P + R - Q) - 2\hat{h}(P + R) - 2\hat{h}(Q) = 0,$$
$$\hat{h}(P - R + Q) + \hat{h}(P + R - Q) - 2\hat{h}(P) - 2\hat{h}(R - Q) = 0,$$
$$\hat{h}(P - R + Q) + \hat{h}(P + R + Q) - 2\hat{h}(P + Q) - 2\hat{h}(R) = 0,$$
$$2\hat{h}(R + Q) + 2\hat{h}(R - Q) - 4\hat{h}(R) - 4\hat{h}(Q) = 0.$$

The alternating sum of these four equations is the desired result.

(d) The first conclusion is clear, since $h_f(P) \geqslant 0$ for all functions f and all points P. For the second, note that one implication is immediate; since if P is a torsion point, say with $[m]P = O$ for some $m \geqslant 1$, then (b) implies that

$$\hat{h}(P) = m^{-2}\hat{h}([m]P) = m^{-2}\hat{h}(O) = 0.$$

Conversely, let $P \in E(K')$ for some finite extension K'/K, and suppose that $\hat{h}(P) = 0$. Then for every integer m, $\hat{h}([m]P) = m^2 \hat{h}(P) = 0$. Hence from (e) there is a constant C so that for every $m \in \mathbb{Z}$,

$$h_f([m]P) = |(\deg f)\hat{h}([m]P) - h_f([m]P)| \leqslant C.$$

Thus the set $\{P, [2]P, [3]P, \ldots\}$ is contained in

$$\{Q \in E(K') : h_f(Q) \leqslant C\}.$$

But from (6.1), the latter is a finite set, so P must have finite order.

Finally, to prove uniqueness, suppose $\hat{h}'$ satisfies

$$\hat{h}' \circ [m] = m^2 \hat{h}' \quad \text{and} \quad (\deg f)\hat{h}' = h_f + O(1)$$

for some integer $m \geqslant 2$. Repeated application of the first equality yields

$$\hat{h}' \circ [m^N] = m^{2N} \hat{h}' \quad \text{for} \quad N = 1, 2, \dots.$$

Further, since $\hat{h}$ also satisfies (e), we have

$$\hat{h}' - \hat{h} = O(1).$$

Hence

$$\hat{h}' = m^{-2N} \hat{h}' \circ [m^N]$$

$$= m^{-2N} (\hat{h} \circ [m^N] + O(1))$$

$$= \hat{h} + m^{-2N} O(1) \qquad \text{since } \hat{h} \text{ satisfies (b)}.$$

Letting $N \to \infty$ yields $\hat{h}' = \hat{h}$. $\square$

Remark 9.4. Notice that the Mordell–Weil theorem implies that $\mathbb{R} \otimes E(K)$ is a finite dimensional real vector space, while (9.3c,d) implies that $\hat{h}$ is a positive definite quadratic form on the quotient group $E(K)/E_{\text{tors}}(K)$. [Here $E_{\text{tors}}(K)$ is the torsion subgroup of $E(K)$.] Now $E(K)/E_{\text{tors}}(K)$ sits as a lattice in $\mathbb{R} \otimes E(K)$, so it would appear to be clear that the extension of $\hat{h}$ to $\mathbb{R} \otimes E(K)$ is also positive definite. This is true, but as was pointed out by Cassels, one must use more than just (9.3c, d).

Lemma 9.5. *Let V be a finite dimensional real vector space, and let $L \subset V$ be a lattice. Suppose $q : V \to \mathbb{R}$ is a quadratic form which has the following properties:*

(i) *Let $P \in L$. Then $q(P) = 0$ if and only if $P = 0$.*
(ii) *For every constant C,*

$$\{P \in L : q(P) \leqslant C\}$$

   is a finite set.

Then q is positive definite on V.

PROOF. Choose a basis for V so that for $X = (x_1, \dots, x_r) \in V$, q has the form

$$q(X) = \sum_{i=1}^{s} x_i^2 - \sum_{i=1}^{t} x_{s+i}^2,$$

where $s + t \leqslant r = \dim(V)$. (See, e.g. [VdW, §12.7] or [La 8, Ch. XIV, §3, §7].) This basis gives an isomorphism $V \cong \mathbb{R}^n$; let μ be the measure on V corresponding to the usual measure on $\mathbb{R}^n$. We now need the following elementary result, which is due to Minkowski:

   Let $B \subset V$ be a convex set which is symmetric about the origin. If $\mu(B)$ is sufficiently large, then B contains a non-zero lattice point.

For a proof, see for example [H–W, thm. 447] or [La 2, Ch. 5, §3]. Now look at the sets

$$B(\varepsilon, \delta) = \left\{ X = (x_1, \ldots, x_r) \in V : \sum_{i=1}^{s} x_i^2 \leqslant \varepsilon \text{ and } \sum_{i=1}^{t} x_{s+i}^2 \leqslant \delta \right\}.$$

They are convex and symmetric about the origin for any $\varepsilon, \delta > 0$. Let

$$\lambda = \inf\{q(P) : P \in L, P \neq 0\}$$

From (i) and (ii), we have $\lambda > 0$.

Now suppose that q is not positive definite on V, so $s < r$. Then from Minkowski's theorem, the set $B(\frac{1}{2}\lambda, \delta)$ contains a non-zero lattice point P if δ is sufficiently large. (The volume of $B(\frac{1}{2}\lambda, \delta)$ is infinite if $s + t < r$, and grows like $\delta^{t/2}$ as $\delta \to \infty$ if $s + t = r$.) But then

$$q(P) = \sum_{i=1}^{s} x_i^2 - \sum_{i=1}^{t} x_{i+s}^2 \leqslant \frac{1}{2}\lambda,$$

contradicting the definition of λ. Therefore q is positive definite on V.

$\square$

Proposition 9.6. *The canonical height is a positive definite quadratic form on the vector space* $\mathbb{R} \otimes E(K)$.

PROOF. This follows from (9.5) applied to the lattice $E(K)/E_{\text{tors}}(K)$ inside $\mathbb{R} \otimes E(K)$. Condition (i) of (9.5) is exactly (9.3c, d); while condition (ii) of (9.5) follows from (9.3e), which says that bounding $\hat{h}$ is the same as bounding h_f, and then applying (6.1).

$\square$

We now have the following quantities associated to E/K:

$\mathbb{R} \otimes E(K)$	a finite dimensional vector space,
$\hat{h}$	a positive definite quadratic form on $\mathbb{R} \otimes E(K)$,
$E(K)/E_{\text{tors}}(K)$	a lattice in $\mathbb{R} \otimes E(K)$.

Now in such a situation, an extremely important invariant is the volume of a fundamental domain for the lattice, computed with respect to the metric induced by the quadratic form. (For example, the discriminant of a number field K is the volume of its ring of integers with respect to the quadratic form $x \to \text{trace}_{K/\mathbb{Q}}(x^2)$. Similarly, the regulator of K is the volume of its unit group, using the logarithm mapping and the usual metric on Euclidean space.)

Definition. The *Néron–Tate pairing* on E/K is the bilinear form

$$\langle \ , \ \rangle : E(\bar{K}) \times E(\bar{K}) \to \mathbb{R}$$

defined by

$$\langle P, Q \rangle = \hat{h}(P + Q) - \hat{h}(P) - \hat{h}(Q).$$

Definition. The *elliptic regulator of E/K*, denoted $R_{E/K}$, is the volume of a fundamental domain for $E(K)/E_{\text{tors}}(K)$, computed using the quadratic form $\hat{h}$. In other words, choose $P_1, \ldots, P_r \in E(K)$ to generate $E(K)/E_{\text{tors}}(K)$. Then

$$R_{E/K} = \det(\langle P_i, P_j \rangle)_{\substack{1 \leqslant i \leqslant r \\ 1 \leqslant j \leqslant r}}.$$

(If $r = 0$, we set $R_{E/K} = 1$ by convention.)

As an immediate corollary to (9.6), we obtain:

Corollary 9.7. *The elliptic regulator is always positive.*

Remark 9.8. We have defined the elliptic regulator using the absolute height. Sometimes it is defined using the height relative to the given field K. As is immediately clear, this new regulator would differ from the old regulator by a factor of $[K : \mathbb{Q}]^r$.

Since $\hat{h}(P) > 0$ for all non-torsion points $P \in E(K)$, a natural question to ask is how small can $\hat{h}(P)$ be? One would like to say that $\hat{h}(P)$ must be large if the elliptic curve is "complicated" in some sense. The following precise conjecture is a slight generalization of a conjecture of Lang [La 5, p. 92].

Conjecture 9.9. *Let E/K be an elliptic curve with j-invariant j_E and minimal discriminant $\mathscr{D}_{E/K}$. There is a constant $c > 0$, depending only on $[K : \mathbb{Q}]$, so that for all non-torsion points $P \in E(K)$,*

$$\hat{h}(P) > c \max\{h(j_E), \log N_{K/\mathbb{Q}}\mathscr{D}_{E/K}, 1\}.$$

Note that the strength of the conjecture lies in the fact that the constant c is independent of both the elliptic curve E and the point P. Such estimates have applications to counting integral points on elliptic curves (see (IX.3.5) for a discussion). Conjecture 9.9 is known to be true if one restricts attention to elliptic curves whose j-invariant is integral; and more generally such an estimate exists with the constant c depending on $[K : \mathbb{Q}]$ and the number of prime ideals dividing the denominator of j_E. (See [Sil 1] and [Sil 5] for details. A special case is given in exer. 8.17.)

§10. The Rank of an Elliptic Curve

It follows from the Mordell–Weil theorem (6.7) that the *Mordell–Weil group* $E(K)$ of an elliptic curve E/K can be written in the form

$$E(K) \cong E_{\text{tors}}(K) \times \mathbb{Z}^r.$$

As we have seen (§7), the torsion subgroup $E_{\text{tors}}(K)$ is relatively easy to compute, both in theory and in practice. The *rank r* is much more mysterious, and an effective procedure for determining it in all cases is still being sought. There are very few general facts known concerning the rank of elliptic curves, but there are a number of fascinating conjectures. In this section we will briefly discuss some of these conjectures. (See chapter X for a description of some of the methods which have been developed for actually computing the group $E(K)$.)

The rank of a "randomly chosen" elliptic curve over $\mathbb{Q}$ tends to be fairly small, and it is quite difficult to produce such curves of even moderately high rank. None the less, there is the following "folklore" conjecture.

Conjecture 10.1. *There exist elliptic curves $E/\mathbb{Q}$ of arbitrarily large rank.*

The principal evidence for this conjecture comes from work of Shafarevich and Tate ([Sha–T]), who show that the analogous result is true for function fields (i.e. when $\mathbb{Q}$ is replaced by the field of rational functions $\mathbb{F}_p(T)$). Néron has constructed an infinite family of elliptic curves over $\mathbb{Q}$ having rank at least 10 (C.20.1.1), and Mestre ([Mes 2]) has produced examples with higher rank. For example, Mestre shows that the elliptic curve

$$y^2 - 246xy + 36599029y = x^3 - 89199x^2 - 19339780x - 36239244$$

has rank at least 12 over $\mathbb{Q}$; and his ideas can be used to produce curves of even higher rank. (However, they do not seem well-suited to producing infinite families of such curves.)

Attached to an elliptic curve E/K is a certain Dirichlet series $L_{E/K}(s)$, called the *L-series of E/K*. (See (exer. 8.19) and (C §16) for the definition of $L_{E/K}$.) For the moment, it is enough to know that the definition of $L_{E/K}(s)$ involves only the number of points on the reduction $\tilde{E}(k_v)$ for each finite place $v \in M_K$. There is a conjecture, due to Birch and Swinnerton-Dyer, which says that $L_{E/K}(s)$ has a zero at $s = 1$ whose order exactly equals the rank of $E(K)$. Further, the leading coefficient in the Taylor series expansion of $L_{E/K}(s)$ around $s = 1$ should be expressible in terms of various global arithmetic quantities associated to $E(K)$, including the elliptic regulator $R_{E/K}$. Thus in some sense, the conjecture of Birch and Swinnerton-Dyer is a version of the Hasse principle which applies to elliptic curves, since it (hypothetically) shows how information about the v-adic behavior of E for all places $v \in M_K$ determines global information such as the rank of $E(K)$ and the elliptic regulator $R_{E/K}$. (For a more detailed discussion of L-series and the conjecture of Birch and Swinnerton-Dyer, including some of the progress made in proving it, see appendix C §16.)

In addition to having an effective method for computing the rank of an elliptic curve, it would be good to have a theoretical description of just how large a generating set need be. Based partly on analogy with the problem of computing generators for the unit group of a number field and partly on a number of very deep conjectures in analytic number theory, Serge Lang has suggested the following.

Conjecture 10.2 (Lang [La 9]). *Let $E/\mathbb{Q}$ be an elliptic curve of rank r. Then there is a basis $P_1, \ldots, P_r$ for the free part of $E(\mathbb{Q})$ satisfying*

$$\hat{h}(P_i) \leqslant C_\varepsilon |\mathscr{D}_{E/\mathbb{Q}}|^{1/2+\varepsilon} \qquad \text{for all } 1 \leqslant i \leqslant r.$$

Here $\hat{h}$ is the canonical height on E (cf. §9), $\mathscr{D}_{E/\mathbb{Q}}$ is the minimal discriminant of $E/\mathbb{Q}$ (cf. §8), and C_ε is a constant depending only on ε. (Lang's conjecture is actually more precise, see [La 9].)

Since $\hat{h}$ is the *logarithmic* height, (10.2) says that the x-coordinates of the generators might grow exponentially with the discriminant of the curve. (Similarly, the height $H(u)$ of a generator for the unit group in a real quadratic field seems to grow exponentially with the discriminant of the field. Of course, it is easy to choose a sequence of such fields for which $H(u)$ grows polynomially; but on average, one expects the growth to be exponential.) The expected exponential behavior for elliptic curves is illustrated by the following example of Bremner and Cassels [Br–C]. They show that the elliptic curve

$$y^2 = x^3 + 877x$$

has rank 1, and the x-coordinate of a generator P is given by

$$x = (612776083187947368101/7884153586063900210)^2.$$

To compare this example with Lang's conjecture, we compute

$$\log \hat{h}(P)/\log|\mathscr{D}_{E/\mathbb{Q}}| \approx 0.2,$$

which is well within the suggested bound of $\frac{1}{2} + \varepsilon$.

EXERCISES

8.1.   Let E/K be an elliptic curve, $m \geqslant 2$ an integer, $\mathscr{H}_K$ the ideal class group of K, and

$$S = \{v \in M_K^0 : E \text{ has bad reduction at } v\} \cup \{v \in M_K^0 : v(m) \neq 0\} \cup M_K^\infty.$$

Assuming that $E[m] \subset E(K)$, prove the following *quantitative* version of the weak Mordell–Weil theorem:

$$\text{rank}_{\mathbb{Z}/m\mathbb{Z}}(E(K)/mE(K)) \leqslant 2\#S + 2\,\text{rank}_{\mathbb{Z}/m\mathbb{Z}}\mathscr{H}_K[m].$$

8.2.   For each integer $d \geqslant 1$, let $E_d/\mathbb{Q}$ be the elliptic curve

$$E_d : y^2 = x^3 - d^2 x.$$

Prove that

$$E_d(\mathbb{Q}) \cong \text{finite group} \times \mathbb{Z}^r$$

for some integer

$$r \leqslant 2v(2d),$$

where $v(N)$ denotes the number of distinct primes dividing N. [*Hint*: Use exercise 8.1.]

8.3. Let E/K be an elliptic curve and L/K an (infinite) algebraic extension. Suppose that the rank of $E(M)$ is bounded as M ranges over all finite extensions M/K contained in L.

(a) Prove that $E(L) \otimes \mathbb{Q}$ is finite dimensional (as a $\mathbb{Q}$-vector space.)

(b) Assume further that L/K is Galois and $E_{\mathrm{tors}}(L)$ is finite. Prove that $E(L)$ is finitely generated.

8.4. Assume that $\mu_m \subset K$. Prove that the maximal abelian extension of K of exponent m is the field

$$K(a^{1/m} : a \in K).$$

[*Hint*: Use (2.2), which in this case says that every homomorphism $\chi : G_{\bar{K}/K} \to \mu_m$ has the form $\chi(\sigma) = \alpha^\sigma/\alpha$ for some $\alpha \in \bar{K}^*$.]

8.5. Let $\xi \in H^1(G_{\bar{K}/K}, M)$ be unramified at v. Prove that there is a 1-cocycle $c : G_{\bar{K}/K} \to M$ in the cohomology class of ξ such that $c_\sigma = 0$ for all $\sigma \in I_v$. [*Hint*: Use the inflation-restriction sequence (B.2.4) for $I_v \subset G_{\bar{K}/K}$.]

8.6. Prove *Kronecker's theorem*: Let $x \in \bar{\mathbb{Q}}^*$. Then $H(x) = 1$ if and only if x is a root of unity. (This is the multiplicative-group version of (9.3d).)

8.7. (a) Give an explicit upper bound, in terms of N, C, and d, for the number of points in

$$\{P \in \mathbb{P}^N(\bar{\mathbb{Q}}) : H(P) \leqslant C \text{ and } [\mathbb{Q}(P) : \mathbb{Q}] \leqslant d\}.$$

(b) Let

$$v_K(N, C) = \#\{P \in \mathbb{P}^N(K) : H_K(P) \leqslant C\}.$$

Prove that

$$v_\mathbb{Q}(N, C) \sim C^{N+1}/\zeta(N + 1) \qquad \text{as } C \to \infty,$$

where $\zeta(s)$ is the Riemann ζ-function. (For more about $v_K(N, C)$, see (5.12).)

8.8. Prove the following standard facts about height functions.

(a) $H(x_1 x_2 \cdots x_N) \leqslant H(x_1)H(x_2)\cdots H(x_N)$.

(b) $H(x_1 + x_2 + \cdots + x_N) \leqslant NH(x_1)H(x_2)\cdots H(x_N)$.

(c) For $P = [x_0, \ldots, x_N] \in \mathbb{P}^N$ and $Q = [y_0, \ldots, y_M] \in \mathbb{P}^M$, let

$$P*Q = [x_0 y_0, x_0 y_1, \ldots, x_i y_j, \ldots, x_N y_M] \in \mathbb{P}^{MN+M+N}.$$

Then

$$H(P*Q) = H(P)H(Q).$$

(The map $(P, Q) \to P*Q$ is the *Segre embedding* of $\mathbb{P}^N \times \mathbb{P}^M$ in $\mathbb{P}^{MN+M+N}$. See [Har, exer. I.2.14].)

(d) For $P = [x_0, \ldots, x_N] \in \mathbb{P}^N$, let

$$P^{(d)} = [f_0(P), \ldots, f_M(P)] \in \mathbb{P}^M,$$

where $M = \binom{N+d}{N} - 1$, and $f_0(X), \ldots, f_M(X)$ are the M possible monomials of degree d in the $N + 1$ variables $X_0, \ldots, X_N$. Then

$$H(P^{(d)}) = H(P)^d = H([x_0^d, \ldots, x_N^d]).$$

(The map $P \to P^{(d)}$ is the *d-uple embedding* of $\mathbb{P}^N$ in $\mathbb{P}^M$. See [Har, exer. I.2.12].)

(e) If $x \neq 0$, then

$$H(1/x) = H(x).$$

(f) Let K be a number field and let $x_0, \ldots, x_N \in K$ be algebraic *integers*. Then

$$H([x_0, \ldots, x_N]) \leqslant \max_{0 \leqslant i \leqslant N} H(x_i)^{[K:\mathbb{Q}]}.$$

8.9. Let $x_0, \ldots, x_N \in K$, and let $\mathfrak{b}$ be the fractional ideal of K generated by $x_0, \ldots, x_N$. Then

$$H_K([x_0, \ldots, x_N]) = (N_{K/\mathbb{Q}}\mathfrak{b})^{-1} \prod_{v \in M_K^\infty} \max_{0 \leqslant i \leqslant N} \{|x_i|_v\}^{n_v}.$$

8.10. Let F be the rational map (I.3.6) which is a morphism at every point except $[0, 1, 0]$,

$$F : \mathbb{P}^2 \to \mathbb{P}^2$$

$$[x, y, z] \to [x^2, xy, z^2].$$

Prove that for all constants $C, \varepsilon > 0$, there is a point $P \in \mathbb{P}^2(\mathbb{Q})$ so that

$$H(F(P)) < CH(P)^{1+\varepsilon}.$$

In particular, (5.6) becomes false if the map F is merely required to be a rational map.

8.11. Prove the following generalization of (7.2) to arbitrary number fields.
    Let E/K be an elliptic curve given by an equation

$$y^2 = x^3 + Ax + B$$

with $A, B \in R$, and let $\Delta = 4A^3 + 27B^2$. Let $P \in E(K)$ be a point of exact order $m \geqslant 3$, and let $v \in M_K^0$.

(a) If $m = p^n$ is a prime power, then

$$-6r_v \leqslant \text{ord}_v(y(P)^2) \leqslant 6r_v + \text{ord}_v(\Delta),$$

where

$$r_v = \left[\frac{\text{ord}_v(p)}{p^n - p^{n-1}} \right].$$

(b) If $m = 2p^n$ is twice a prime power, then

$$0 \leqslant \text{ord}_v(y(P)^2) \leqslant 2r_v + \text{ord}_v(\Delta),$$

where r_v is as in (a).

(c) If m is not of the form p^n or $2p^n$, then

$$0 \leqslant \text{ord}_v(y(P)^2) \leqslant \text{ord}_v(\Delta).$$

8.12. For each of the following elliptic curves, calculate $E_{\text{tors}}(\mathbb{Q})$.
   (a) $y^2 = x^3 - 2$
   (b) $y^2 = x^3 + 8$
   (c) $y^2 = x^3 + 4$
   (d) $y^2 = x^3 + 4x$
   (e) $y^2 - y = x^3 - x^2$
   (f) $y^2 = x^3 + 1$
   (g) $y^2 = x^3 - 43x + 166$
   (h) $y^2 + 7xy = x^3 + 16x$
   (i) $y^2 + xy + y = x^3 - x^2 - 14x + 29$
   (j) $y^2 + xy = x^3 - 45x + 81$
   (k) $y^2 + 43xy - 210y = x^3 - 210x^2$
   (l) $y^2 = x^3 - 4x$
   (m) $y^2 = x^3 + 2x^2 - 3x$
   (n) $y^2 + 5xy - 6y = x^3 - 3x^2$
   (o) $y^2 + 17xy - 120y = x^3 - 60x^2$

8.13. (a) Let E/K be an elliptic curve and $P \in E(K)$ a point of order at least 4. By an appropriate change of coordinates, show that E has an equation of the form

$$E : y^2 + uxy + vy = x^3 + vx^2$$

with $u, v \in K$ and $P = (0, 0)$.

   (b) Show that there is a one-parameter family of elliptic curves E/K with a K-rational point of order 6. [*Hint:* Set $[3]P = [-3]P$ in (a), and find how u and v must be related.] Same question for points of order 7; order 9; order 12.

   (c) Show that the elliptic curves E/K with a K-rational point of order 11 are parametrized by the K-rational points of a certain curve of genus one.

8.14. (a) Generalize (8.2) as follows. Let E/K be an elliptic curve, and let $\mathfrak{a}$ by any *integral* ideal in the ideal class $\bar{\mathfrak{a}}_{E/K}$. Then there is a Weierstrass equation for E with coefficients $a_i \in R$ and discriminant Δ satisfying

$$(\Delta) = \mathscr{D}_{E/K} \mathfrak{a}^{12}.$$

   (b) Suppose that E/K has everywhere good reduction and the class number of K is relatively prime to 6. Then E/K has a global minimal Weierstrass equation.

   (c) Every elliptic curve E/K has a Weierstrass equation with coefficients $a_i \in R$ and discriminant Δ satisfying

$$|N_{K/\mathbb{Q}} \Delta| \leqslant |\text{Disc } K/\mathbb{Q}|^6 |N_{K/\mathbb{Q}} \mathscr{D}_{E/K}|.$$

(Qualitatively, this says that one can find a Weierstrass equation whose non-minimality is bounded solely in terms of K. Such an equation might be called *quasi-minimal*.)

   (d) Let $\bar{\mathfrak{b}}$ be any ideal class of K. Prove that there is an elliptic curve E/K such that $\bar{\mathfrak{a}}_{E/K} = \bar{\mathfrak{b}}$. In particular, if K does not have class number 1, then there exist elliptic curves over K which do not have global minimal Weierstrass equations. (This gives a converse to (8.3).)

8.15. Prove that there are no elliptic curves $E/\mathbb{Q}$ having everywhere good reduction. [*Hints:* Take a Weierstrass equation with integral coefficients and discriminant $\Delta = \pm 1$. Show a_1 is odd, so $c_4 \equiv 1(8)$. Substitute $c_4 = u \pm 12$ into $c_4^3 - c_6^2 = \pm 1728$. Show $u = 3v$ and $c_6 = 9w$. Then $w = t^2$ or $3t^2$. Rule out the former by finding $w \pmod 8$, and the latter by showing that it leads to v and w being infinitely 3-divisible.]

8.16. Show that the conclusion of (9.5) is false if the quadratic form q is not required to satisfy the finiteness condition (ii).

8.17. Fix non-zero integers A, B with $4A^3 + 27B^2 \neq 0$. For each $d \neq 0$, let $E_d/\mathbb{Q}$ be the elliptic curve

$$E_d : y^2 = x^3 + d^2 A x + d^3 B.$$

Prove that for all square-free integers $d \neq 0$:
(a) j_E is independent of d;
(b) $\log |\mathscr{D}_{E/\mathbb{Q}}| = 6 \log |d| + O(1)$;
(c) Every $P \in E_d(\mathbb{Q})$ satisfies either $[2]P = 0$ or $\hat{h}(P) > \frac{1}{6} \log |d| + O(1)$.
(d) For all but finitely many square-free integers d, the torsion subgroup of $E_d(\mathbb{Q})$ is one of $\{0\}$, $\mathbb{Z}/2\mathbb{Z}$, or $(\mathbb{Z}/2\mathbb{Z})^2$.
(Here the $O(1)$'s may depend on A and B, but they should be independent of d. This exercise provides a proof of conjecture 9.9 for the family of curves E_d.)
[*Hint for* (c): If $P = (r, s) \in E_d(\mathbb{Q})$, then $P' = (r/d, s/d^{3/2}) \in E_1$. Show that $\hat{h}(P) = \hat{h}(P')$, that either $s = 0$ or $h_y(P')$ is greater than $\frac{1}{2} \log |d|$, and that $|\hat{h} - \frac{1}{3} h_y|$ is bounded.]

8.18. Let E/K be an elliptic curve given by a Weierstrass equation

$$E : y^2 = x^3 + Ax + B.$$

(a) Prove that there are *absolute constants* c_1 and c_2 such that for all points $P \in E(\bar{K})$,

$$|h_x([2]P) - 4h_x(P)| \leqslant c_1 h([A, B, 1]) + c_2.$$

Find explicit values for c_1 and c_2. [*Hint:* Combine the proofs of (4.2) and (5.6), keeping track of the dependence on the constants. In particular, notice that the use of the Nullstellensatz in (5.6) can be replaced by the explicit identities given in (4.3).]
(b) Find *absolute constants* c_3 and c_4 so that for all points $P \in E(\bar{K})$,

$$|\tfrac{1}{2} h_x(P) - \hat{h}(P)| \leqslant c_3 h([A, B, 1]) + c_4.$$

[*Hint:* Use (a) and the proof of (9.1).]
(c) Prove that for all integers $m \geqslant 1$ and all points $P, Q \in E(\bar{K})$,

$$|h_x([m]P) - m^2 h_x(P)| \leqslant 2(m^2 + 1)(c_3 h([A, B, 1]) + c_4);$$

and

$$h_x(P + Q) \leqslant 2h_x(P) + 2h_x(Q) + 5(c_3 h([A, B, 1]) + c_4).$$

[*Hint:* Use (b) and (9.3).]

(d) Let $Q_1, \ldots, Q_r$ be a set of generators for $E(K)/2E(K)$. Find *absolute constants* $c_5, c_6,$ and c_7 so that the set of points $P \in E(K)$ satisfying

$$h_x(P) \leqslant c_5 \max \{h_x(Q_i)\} + c_6 h([A, B, 1]) + c_7$$

contains a complete set of generators for $E(K)$. [*Hint*: Follow the proof of (3.1), using (c) to evaluate the constants that appear.]

8.19. *The L-Series Attached to an Elliptic Curve.* Let $E/\mathbb{Q}$ be an elliptic curve, and choose a global minimal Weierstrass equation

$$y^2 + a_1 xy + a_3 y = x^3 + a_2 x^2 + a_4 x + a_6$$

for $E/\mathbb{Q}$ (cf.8.3). For each prime p, let A_p be the number of points on the reduced curve $\tilde{E}$ mod p (remember to include the point at infinity); and let

$$t_p = 1 + p - A_p.$$

The *L-Series associated to $E/\mathbb{Q}$* is defined by the Euler product

$$L_E(s) = \prod_{p | \Delta(E)} (1 - t_p p^{-s})^{-1} \prod_{p \nmid \Delta(E)} (1 - t_p p^{-s} + p^{1-2s})^{-1}.$$

(a) If $L_E(s)$ is expanded as a Dirichlet series $\Sigma c_n n^{-s}$, show that its pth coefficient (for p prime) satisfies $c_p = t_p$.

(b) If E has bad reduction at p (so p divides $\Delta(E)$), prove that $t_p = 1, -1,$ or 0 according as the reduced curve $\tilde{E}$ (mod p) has a node with tangents whose slopes are rational over $\mathbb{F}_p$, a node with tangents quadratic over $\mathbb{F}_p$, or a cusp (cf. exer. 3.5).

(c) Prove that the Euler product for $L_E(s)$ converges for all $s \in \mathbb{C}$ with $\text{Re}(s) > 3/2$. [*Hint*: Use (V.1.1).]

(There are a number of important conjectures concerning the *L*-series attached to elliptic curves. See appendix C §16.)

CHAPTER IX

Integral Points on Elliptic Curves

An elliptic curve may have infinitely many rational points, although the Mordell–Weil theorem at least assures us that the group of rational points is finitely generated. Another natural Diophantine question is that of determining, for a given (affine) Weierstrass equation, which rational points actually have integral coordinates. In this chapter we will prove a theorem of Siegel which says that there are only finitely many such integral points. Siegel gave two proofs of his theorem, which we present in sections 3 and 4. Both proofs make use of techniques from the theory of Diophantine approximation, and so do not provide an effective procedure for actually finding all of the integral points. However, his second method of proof reduces the problem to that of solving the so-called "unit equation", which in turn can be effectively resolved using transcendence theory. We will discuss this method, without giving proofs, in section 5.

Unless otherwise specified, the notations and conventions for this chapter are the same as those for chapter VIII. In addition, we set the following notation:

H, H_K	height functions (see VIII §5)
$n_v = [K_v : \mathbb{Q}_v]$	local degree for $v \in M_K$ (see VIII §5)
$S \subset M_K$	generally a finite set of absolute values containing M_K^∞
R_S	the ring of S-integers of K

$$R_S = \{x \in K : v(x) \geqslant 0 \text{ for all } v \in M_K, v \notin S\}$$

R_S^*	the unit group of R_S.

§1. Diophantine Approximation

The fundamental problem in the subject of Diophantine approximation is the question of how closely an irrational number can be approximated by a rational number.

Example 1.1. For every rational number p/q, we know that the quantity $|(p/q) - \sqrt{2}|$ is positive; and since $\mathbb{Q}$ is dense in $\mathbb{R}$, an appropriate choice of p/q will make it as small as desired. The problem is to make it small without taking p and q too large. The next two elementary results illustrate this idea.

Proposition 1.2 (Dirichlet). *Let $\alpha \in \mathbb{R}$ with $\alpha \notin \mathbb{Q}$. Then there are infinitely many $p/q \in \mathbb{Q}$ such that*

$$\left| \frac{p}{q} - \alpha \right| \leqslant \frac{1}{q^2}.$$

PROOF. Let Q be a large integer, and look at the set

$$\{q\alpha - [q\alpha] : q = 0, 1, \ldots, Q\}.$$

(Here [ ] means greatest integer.) Since α is irrational, this set consists of $Q + 1$ distinct numbers in the interval between 0 and 1; so by the pigeon-hole principle there are integers $0 \leqslant q_1 < q_2 \leqslant Q$ satisfying

$$|(q_1\alpha - [q_1\alpha]) - (q_2\alpha - [q_2\alpha])| \leqslant 1/Q.$$

Hence

$$\left| \frac{[q_2\alpha] - [q_1\alpha]}{q_2 - q_1} - \alpha \right| \leqslant \frac{1}{(q_2 - q_1)Q} \leqslant \frac{1}{(q_2 - q_1)^2}.$$

This provides one rational approximation to α with the desired property, and by increasing Q one can clearly obtain infinitely many.   □

Remark 1.2.1. A result of Hurwitz says that the $1/q^2$ in (1.2) can be replaced by $1/\sqrt{5}q^2$, and that this is best possible. (See, e.g., [H–W, thm. 195].)

Proposition 1.3 (Liouville [Liou]). *Let $\alpha \in \overline{\mathbb{Q}}$ be of degree $d \geqslant 2$ over $\mathbb{Q}$ (i.e. $[\mathbb{Q}(\alpha) : \mathbb{Q}] = d$). There is a constant $C > 0$, depending on α, so that for all rational numbers p/q,*

$$\left| \frac{p}{q} - \alpha \right| \geqslant \frac{C}{q^d}.$$

PROOF. Let

$$f(T) = a_0 T^d + a_1 T^{d-1} + \cdots + a_d \in \mathbb{Z}[T]$$

be the minimal polynomial for α. Let

$$C_1 = \sup\{f'(t) : \alpha - 1 \leqslant t \leqslant \alpha + 1\}.$$

Suppose now that

$$\left|\frac{p}{q} - \alpha\right| \leqslant 1.$$

Then from the mean value theorem,

$$\left|f\left(\frac{p}{q}\right)\right| = \left|f\left(\frac{p}{q}\right) - f(\alpha)\right| \leqslant C_1 \left|\frac{p}{q} - \alpha\right|.$$

On the other hand, $q^d f(p/q) \in \mathbb{Z}$; and $f(p/q) \neq 0$ since f can have no rational roots. Hence

$$\left|q^d f\left(\frac{p}{q}\right)\right| \geqslant 1.$$

Combining the last two inequalities gives

$$\left|\frac{p}{q} - \alpha\right| \geqslant \frac{C}{q^d},$$

which holds for all p/q if we take $C = \min\{1/C_1, 1\}$. $\qquad\square$

Remark 1.3.1. Liouville used his theorem to prove the existence of transcendental numbers. (See exer. 9.2.) Note that it is quite easy to find the constant C in Liouville's theorem explicitly in terms of α. This is in marked contrast to the results which we will consider below.

Proposition (1.2) says that every real number can be approximated by rational numbers to within $1/q^2$, while proposition (1.3) says that an algebraic number of degree d can be approximated no closer than C/q^d. For quadratic irrationalities, there is little more to say; but if $d \geqslant 3$, then one naturally asks what the best exponent is. There is also no particular reason to restrict the approximating values to $\mathbb{Q}$; it is useful to allow them to range over any fixed number field K. Finally, in measuring how close the approximation is, any absolute value should do.

Definition. Let $\tau(d)$ be a positive real-valued function on the natural numbers. A number field K is said to have *approximation exponent* τ if the following condition holds:

Let $\alpha \in \bar{K}$, $d = [K(\alpha) : K]$, and $v \in M_K$ an absolute value on K extended in some fashion to $K(\alpha)$. Then for any constant C, there exist only finitely many $x \in K$ satisfying the inequality

$$|x - \alpha|_v < C H_K(x)^{-\tau(d)}.$$

Thus the elementary estimate of Liouville's theorem (1.3) says that $\mathbb{Q}$ has approximation exponent $\tau(d) = d + \varepsilon$ for any $\varepsilon > 0$. This result has been successively improved by a number of mathematicians. We give a short list.

Liouville	1851	$\tau(d) = d + \varepsilon$
Thue	1909	$\tau(d) = \frac{1}{2}d + 1 + \varepsilon$
Siegel	1921	$\tau(d) = 2\sqrt{d} + \varepsilon$
Gelfond, Dyson	1947	$\tau(d) = \sqrt{2d} + \varepsilon$
Roth	1955	$\tau(d) = 2 + \varepsilon$.

In view of (1.2), Roth's result is essentially best possible, although it is not unlikely that the ε can be replaced by some function $\varepsilon(d)$ such that $\varepsilon(d) \to 0$ as $d \to \infty$. We should also mention that Mahler showed how to handle several absolute values at once, and W. Schmidt ([Schm 2, Ch. VI]) dealt with the more difficult problem of simultaneously approximating several irrationals.

The main ideas which go into the proof of Roth's theorem are quite beautiful; and, at least in theory, relatively elementary. Unfortunately, to develop those ideas fully would take us rather far afield. Hence rather than include the complete proof, we will be content to state here the result that we will be using. Then, in section 8, we will briefly sketch the proof of Roth's theorem without actually giving any of the myriad details.

Theorem 1.4 (Roth's Theorem). *For every $\varepsilon > 0$, every number field K has approximation exponent*

$$\tau(d) = 2 + \varepsilon.$$

PROOF. See §8 for a brief sketch. A nice exposition for $K = \mathbb{Q}$ and the usual (archimedean) absolute value is given in [Schm 2, Ch. V]; the general case is in [La 7, Ch. 7].   □

Example 1.5. How do theorems on Diophantine approximation lead to results concerning Diophantine equations? Consider the simple example of solving the equation

$$x^3 - 2y^3 = a$$

in integers $x, y \in \mathbb{Z}$, where $a \in \mathbb{Z}$ is fixed. Suppose (x, y) is a solution with $y \neq 0$. Let ζ be a primitive cube root of unity, and factor the equation as

$$\left(\frac{x}{y} - \sqrt[3]{2}\right)\left(\frac{x}{y} - \zeta\sqrt[3]{2}\right)\left(\frac{x}{y} - \zeta^2\sqrt[3]{2}\right) = \frac{a}{y^3}.$$

The second and third terms in the product are bounded away from 0, so we obtain an estimate

$$\left|\frac{x}{y} - \sqrt[3]{2}\right| \leq \frac{C}{|y|^3}$$

for some constant C independent of x and y. Now from (1.4), or even Thue's original theorem with $\tau(d) = \frac{1}{2}d + 1 + \varepsilon$, we see that there are only finitely many possibilities for x and y. Hence the equation

$$x^3 - 2y^3 = a$$

has only finitely many solutions in integers. This type of argument will reappear in the proof of (4.1). (See also exer. 9.6.)

Remark 1.6. The statement of (1.4) says that *there exist* only finitely many elements of K with a certain property. This phrasing is especially felicitous, because the proof of (1.4) is not effective. In other words, there is no effective procedure which is guaranteed to produce all of the elements in this finite set. (See (8.1) for a discussion of why this is so.) We note that as a consequence, all of the finiteness results which we will prove in sections 2 and 3 are ineffective, since they rely on (1.4). (Similarly, in (1.5), the proof yields no explicit bound for $|x|$ and $|y|$ in terms of a.) However, there are other methods, based on estimates for linear forms in logarithms, which are effective. We will discuss these, without proof, in section 5.

§2. Distance Functions

A Diophantine inequality such as

$$|x - \alpha|_v < CH_K(x)^{-\tau(d)}$$

consists of two pieces. First, there is the height function $H_K(x)$, which is an *arithmetic* measure of the size of x. We have already studied height functions and their transformation properties in some detail (VIII §5, 6). Second, there is the quantity $|x - \alpha|_v$, which is a *topological* measure of the distance from x to α (i.e. in the v-adic topology). In this section we will define a notion of v-adic distance on curves, deduce some of its basic properties, and reinterpret the Diophantine approximation result from section 1 in terms of this distance.

Definition. Let C/K be a curve and $P, Q \in C(K_v)$. Let $t_Q \in K_v(C)$ be a function with a zero of order $e \geq 1$ at Q. The (v-adic) *distance from P to Q*, denoted $d_v(P, Q)$, is given by

$$d_v(P, Q) = \min\{|t_Q(P)|_v^{1/e}, 1\}.$$

(Of course, if P is a pole of t_Q, then $|t_Q(P)|_v = \infty$, so we naturally set $d_v(P, Q) = 1$.)

Remark 2.1. Clearly the distance function d_v has the right qualitative property; $d_v(P, Q)$ is small if P is v-adically close to Q. On the other hand, it certainly depends on the choice of t_Q, so possibly a better notation would be $d_v(P, t_Q)$. However, since we will only use d_v to measure the rate at which two points approach one another, the following result will show that all of our theorems make sense.

Proposition 2.2. *Let $Q \in C(K_v)$, and let t_Q and t'_Q be functions vanishing at Q. Then with the notation of (2.1),*

$$\underset{\substack{P \in C(K_v) \\ P \to Q}}{\text{Limit}} \frac{\log d_v(P, t'_Q)}{\log d_v(P, t_Q)} = 1.$$

(Here $P \to Q$ means $P \in C(K_v)$ approaches Q in the v-adic topology; i.e., $d_v(P, t_Q) \to 0$.)

PROOF. Let t_Q and t'_Q have zeros of order e and e' respectively at Q. Then the function $\phi = (t'_Q)^e/(t_Q)^{e'}$ has neither a zero nor a pole at Q. Hence $|\phi(P)|_v$ is bounded away from 0 and ∞ as $P \to Q$; so as $P \to Q$,

$$\frac{\log d_v(P, t'_Q)}{\log d_v(P, t_Q)} = 1 + \frac{\log |\phi(P)|_v^{1/ee'}}{\log d_v(P, t_Q)} \to 1. \qquad \square$$

Next we examine the effect of finite maps on the distance between points. The crucial observation is that it depends on the ramification of the map, rather than on its degree (compare (2.3) with (VIII.5.6)).

Proposition 2.3. *Let C_1, C_2/K be curves and $f : C_1 \to C_2$ a finite map defined over K. Let $Q \in C_1(K_v)$, and let $e_f(Q)$ be the ramification index of f at Q (cf. II §2). Then*

$$\underset{\substack{P \in C_1(K_v) \\ P \to Q}}{\text{Limit}} \frac{\log d_v(f(P), f(Q))}{\log d_v(P, Q)} = e_f(Q).$$

PROOF. Let $t_Q \in K_v(C_1)$ and $t_{f(Q)} \in K_v(C_2)$ be uniformizers at the indicated points. By definition of ramification index, we can write

$$t_{f(Q)} \circ f = t_Q^{e_f(Q)} \phi,$$

where $\phi \in K_v(C_1)$ has neither a zero nor a pole at Q. It follows that $|\phi(P)|_v$ is bounded away from 0 and ∞ as $P \to Q$. Therefore

$$\frac{\log d_v(f(P), f(Q))}{\log d_v(P, Q)} = \frac{\log |t_{f(Q)}(f(P))|_v}{\log |t_Q(P)|_v}$$

$$= \frac{e_f(Q) \log |t_Q(P)|_v + \log |\phi(P)|_v}{\log |t_Q(P)|_v}$$

$$\to e_f(Q) \qquad\qquad \text{as} \quad P \to Q. \qquad \square$$

Finally, we reinterpret (1.4) in terms of distance functions.

Corollary 2.4 (of 1.4). *Let C/K be a curve, $f \in K(C)$ a non-constant function, and $Q \in C(\bar{K})$. Then*

$$\underset{\substack{P \in C(K) \\ P \to Q}}{\text{Lim inf}} \frac{\log d_v(P, Q)}{\log H_K(f(P))} \geqslant -2.$$

(Here $P \to Q$ means that P approaches Q in the v-adic topology. We obviously do not allow $P = Q$. If Q is not a (v-adic) accumulation point of $C(K)$, then we define the Lim inf *to be 0.)*

PROOF. Replacing f by $1/f$ if necessary, we may assume that $f(Q) \neq \infty$. (Note that $H_K((1/f)(P)) = H_K(f(P))$.) Then from the definition of d_v, we may take

$$d_v(P, Q) = \min\{|f(P) - f(Q)|_v^{1/e}, 1\},$$

where $e \geqslant 1$ is the order of vanishing of the function $f - f(Q)$ at Q. Hence

$$\underset{P \to Q}{\text{Lim inf}} \frac{\log d_v(P, Q)}{\log H_K(f(P))} = \underset{P \to Q}{\text{Lim inf}} \frac{\log |f(P) - f(Q)|_v}{e \log H_K(f(P))}$$

$$= \frac{1}{e} \underset{P \to Q}{\text{Lim inf}} \left\{ \frac{\log(H_K(f(P))^\tau |f(P) - f(Q)|_v)}{\log H_K(f(P))} - \tau \right\}.$$

Now if we take

$$\tau = 2 + \varepsilon,$$

then (1.4) implies that

$$H_K(f(P))^\tau |f(P) - f(Q)|_v \geqslant 1$$

for all but finitely many $P \in C(K)$. Therefore

$$\underset{P \to Q}{\text{Lim inf}} \frac{\log d_v(P, Q)}{\log H_K(f(P))} \geqslant -\frac{\tau}{e} = -\frac{2 + \varepsilon}{e}.$$

Since $\varepsilon > 0$ is arbitrary, and $e \geqslant 1$, this gives the desired result. $\qquad\square$

§3. Siegel's Theorem

In this section we will prove the following theorem of Siegel, which represents a significant improvement on the Diophantine approximation result (2.4).

Theorem 3.1 (Siegel). *Let E/K be an elliptic curve with $\#E(K) = \infty$, $f \in K(E)$ a non-constant even function, $v \in M_K$, and $Q \in E(\bar{K})$. Then*

$$\underset{\substack{P \in E(K) \\ h_f(P) \to \infty}}{\text{Limit}} \frac{\log d_v(P, Q)}{h_f(P)} = 0.$$

Remark 3.1.1. Although we will only prove (3.1) for even functions, it is in fact true in general. (See exer. 9.14d.)

Before giving the proof of (3.1), let us give some indication of just how strong a theorem it is.

Corollary 3.2.1. *Let E/K be an elliptic curve with Weierstrass coordinate functions x and y, let $S \subset M_K$ be a finite set of places containing M_K^∞, and let R_S be the ring of S-integers of K. Then*

$$\{P \in E(K) : x(P) \in R_S\}$$

is a finite set.

PROOF. We apply (3.1) with the function $f = x$. Thus suppose that $P_1, P_2, \cdots \in E(K)$ is a sequence of distinct points with $x(P_i) \in R_S$. From the definition of the height it follows that

$$h_x(P_i) = \frac{1}{[K : \mathbb{Q}]} \sum_{v \in S} \log \max\{1, |x(P_i)|_v^{n_v}\};$$

since for the terms with $v \notin S$, we have $|x(P_i)|_v \leqslant 1$. Hence by choosing a subsequence of the P_i's, we may assume that

$$h_x(P_i) \leqslant \#S \cdot \log |x(P_i)|_v \qquad \text{for all } i,$$

where $v \in S$ is a fixed absolute value. (Note that $n_v \leqslant [K : \mathbb{Q}]$.) In particular, $|x(P_i)|_v \to \infty$. Since the only pole of x is at O, it follows that $d_v(P_i, O) \to 0$.

Now since x has a pole of order 2 at O, we can take as our distance function

$$d_v(P_i, O) = \min\{|x(P_i)|_v^{-1/2}, 1\}.$$

Then for all sufficiently large i, we have

$$\frac{-\log d_v(P_i, O)}{h_x(P_i)} \geqslant \frac{1}{2\#S}.$$

But this contradicts (3.1), which says that the left-hand side must approach 0 as $i \to \infty$. $\qquad\square$

Clearly the proof of (3.2.1) can be applied to any even function, not just x, since (3.1) is given for all even functions. However, one can actually reduce the case of arbitrary (not necessarily even) functions to the special case given by (3.2.1). This reduction step is also important in its own right, since it is used both in Siegel's second proof of finiteness (4.3.1) and with the effective methods provided by linear forms in logarithms (5.7).

Corollary 3.2.2. *Let C/K be a curve of genus 1, and let $f \in K(C)$ be a nonconstant function. Let S and R_S be as in (3.2.1). Then*

$$\{P \in C(K) : f(P) \in R_S\}$$

is a finite set. Further, (3.2.2) follows formally from (3.2.1).

PROOF. We are clearly proving something stronger if we extend the field K and enlarge the set S. We may thus assume that $C(K)$ contains a pole Q of f. Then (C, Q) is an elliptic curve over K; let x and y be coordinates on a Weierstrass equation for (C, Q), which we may take in the form

$$y^2 = x^3 + Ax + B.$$

Now $f \in K(C) = K(x, y)$ and $[K(x, y) : K(x)] = 2$, so we can write

$$f(x, y) = \frac{\phi(x) + \psi(x)y}{\eta(x)}$$

with polynomials $\phi(x)$, $\psi(x)$, $\eta(x) \in K[x]$. Further, since $\mathrm{ord}_Q(x) = -2$, $\mathrm{ord}_Q(y) = -3$, and $\mathrm{ord}_Q(f) < 0$, it follows that

$$2 \deg \eta < \max\{2 \deg \phi, 2 \deg \psi + 3\}.$$

(I.e. This is the condition for f to have a pole at Q.) Next we compute

$$(f\eta(x) - \phi(x))^2 = (\psi(x)y)^2 = \psi(x)^2(x^3 + Ax + b).$$

Writing this out as a polynomial in x with coefficients in $K[f]$, we see that the highest power of x will come from one of the terms $f^2\eta(x)^2$, $\phi(x)^2$, or $\psi(x)^2 x^3$. From above, the first of these has lower degree (in x) than the latter two, while the leading terms of $\phi(x)^2$ and $\psi(x)^2 x^3$ cannot cancel, since they have different degrees. It follows that x satisfies a *monic* polynomial over $K[f]$. (I.e. x is integral over $K[f]$.) Multiplying this polynomial by an appropriate element of K to "clear denominators", we have shown that x satisfies a relation

$$a_0 x^N + a_1(f)x^{N-1} + \cdots + a_{N-1}(f)x + a_N(f) = 0,$$

where $a_0 \in R_S$ and $a_i(f) \in R_S[f]$ for $1 \leqslant i \leqslant N$. Enlarging the set S, we may further assume that $a_0 \in R_S^*$.

Now suppose that $P \in C(K)$ satisfies $f(P) \in R_S$. Then P is not a pole of x, and the relation

$$a_0 x(P)^N + a_1(f(P))x(P)^{N-1} + \cdots + a_{N-1}(f(P))x(P) + a_N(f(P)) = 0$$

shows that $x(P)$ is integral over R_S. Since also $x(P) \in K$, and R_S is integrally closed in K, it follows that $x(P) \in R_S$. This proves that

$$\{P \in C(K) : f(P) \in R_S\} \subset \{P \in C(K) : x(P) \in R_S\};$$

and so the finiteness assertion of (3.2.2) follows from the finiteness result given in (3.2.1). $\qquad\square$

Example 3.3. Consider the Diophantine equation

$$y^2 = x^3 + Ax + B,$$

where A, $B \in \mathbb{Z}$ and $4A^3 + 27B^2 \neq 0$. The above corollary (3.2.1) says that there are only finitely many solutions with x, $y \in \mathbb{Z}$. What does (3.1) say in this situation, say if we take $Q = O$, $f = x$, and v the archimedean absolute value on $\mathbb{Q}$?

Label the non-zero rational points P_1, P_2, ... in order of non-decreasing height, and write

$$x(P_i) = a_i/b_i \in \mathbb{Q}$$

as a fraction in lowest terms. Then

$$\log d_v(P_i, O) = \tfrac{1}{2} \log \min\{|b_i/a_i|, 1\}$$

and

$$h_x(P_i) = \log \max\{|a_i|, |b_i|\}.$$

(Note that the 1/2 appears because $1/x$ has a zero of order 2 at O.) Now (3.1) implies that

$$\underset{i \to \infty}{\text{Lim}} \frac{\min\{\log|b_i/a_i|, 0\}}{\max\{\log|a_i|, \log|b_i|\}} = 0.$$

Similarly, letting Q be a point with $x(Q) = 0$, we have

$$\log d_v(P_i, Q) = \log \min\{|a_i/b_i|, 1\}$$

(with a factor of $1/2$ if $B = 0$); so again from (3.1) we obtain

$$\underset{i \to \infty}{\text{Lim}} \frac{\min\{\log|a_i/b_i|, 0\}}{\max\{\log|a_i|, \log|b_i|\}} = 0.$$

Now from these two limits, it is an easy matter to deduce that

$$\underset{i \to \infty}{\text{Lim}} \frac{\log|a_i|}{\log|b_i|} = 1.$$

In other words, when looking at the x-coordinates of the rational points on an elliptic curve, the numerators and the denominators tend to have about the same number of digits. This is clearly much stronger than the assertion of (3.2.1), which merely says that there are only finitely many points where the denominator is 1.

Remark 3.4. Although Siegel's theorem (3.2.1) is not effective, which means that it does not yield an explicitly computable upper bound for the height of all integral points, it can be made *quantitative* in the following sense (see, e.g., [Ev–S]):

For a given non-singular Weierstrass equation, there is a constant N, which can be explicitly calculated in terms of the field K and the coefficients of the equation, such that the equation has no more than N integral solutions.

A subtler Diophantine problem, conjectured by Serge Lang, is to give an intrinsic relationship between the number of integral points and the rank of the Mordell–Weil group.

Conjecture 3.5 ([La 5, p. 140]). *Let E/K be an elliptic curve, and choose a quasi-minimal Weierstrass equation for E/K,*

$$E : y^2 = x^3 + Ax + B$$

(cf. exer. 8.14c). Let $S \subset M_K$ be a finite set of places containing M_K^∞, and let R_S be the ring of S-integers in K. There exists a constant C, depending only on K, such that

$$\#\{P \in E(K) : x(P) \in R_S\} \leqslant C^{\#S + \operatorname{rank} E(K)}.$$

This conjecture is known to be true if one restricts attention to elliptic curves with integral j-invariant; and more generally, it holds for a constant C depending on both K and the number of primes of K for which $j(E)$ is not integral. (See [Sil 7].)

We now turn to the proof of (3.1). In broad outline, the argument goes as follows. From the theorem on Diophantine approximation (2.4) we have a bound, in terms of the height of P, on how fast P can approach Q. Suppose now that we write $P = [m]P' + R$ and $Q = [m]Q' + R$. Then the distance from P' to Q' is about the same as the distance from P to Q (using (2.3), since the map $P \to [m]P + R$ is unramified); while the height of P' is much smaller than the height of P. Now applying (2.4) to P' and Q', we will obtain a better estimate; and taking m large enough gives the desired result.

PROOF OF (3.1). Choose a sequence of distinct points $P_i \in E(K)$ so that

$$\operatorname*{Lim}_{i \to \infty} \frac{\log d_v(P_i, Q)}{h_f(P_i)} = L = \operatorname*{Lim\,inf}_{\substack{P \in E(K) \\ h_f(P) \to \infty}} \frac{\log d_v(P, Q)}{h_f(P)}.$$

Since $d_v(P, Q) \leqslant 1$ and $h_f(P) \geqslant 0$ for all points $P \in E(K)$, we have $L \leqslant 0$. It thus suffices to prove that $L \geqslant 0$.

Let m be a large integer. From the (weak) Mordell–Weil theorem (VIII. 1.1), the group $E(K)/mE(K)$ is finite. Hence some coset contains infinitely many points of the sequence P_i. Choosing a subsequence, which we again denote P_i, we can write

$$P_i = [m]P_i' + R,$$

where $P_i', R \in E(K)$ and R does not depend on i. Using the standard properties of height functions, we compute

$$m^2 h_f(P_i') = h_f([m]P_i') + O(1) \qquad \text{(VIII. 6.4b)}$$

$$= h_f(P_i - R) + O(1)$$

$$\leqslant 2h_f(P_i) + O(1) \qquad \text{(VIII. 6.4a)},$$

where the $O(1)$ is independent of i.

Next we do an analogous computation with distance functions. If P_i is bounded away from Q (in the v-adic topology), then $\log d_v(P_i, Q)$ is bounded, so clearly $L = 0$. Otherwise, we can choose a subsequence so that $P_i \to Q$. Then $[m]P_i' \to Q - R$, so the sequence P_i' must have one of the m^2 possible m^{th}-roots of $Q - R$ as an accumulation point. Thus by again taking a subsequence, we can find a $Q' \in E(\bar{K})$ so that

$$P_i' \to Q' \quad \text{and} \quad Q = [m]Q' + R.$$

Note that the map $E \to E$ defined by $P \to [m]P + R$ is everywhere unramified (III. 4.10c). This lets us use (2.3) to compute

$$\operatorname*{Lim}_{i \to \infty} \frac{\log d_v(P_i, Q)}{\log d_v(P_i', Q')} = 1.$$

Combining this with the height inequality from above yields the following. (Note that the $\log d_v$ expressions are negative, which reverses the inequality.)

$$L = \operatorname*{Lim}_{i \to \infty} \frac{\log d_v(P_i, Q)}{h_f(P_i)} \geqslant \operatorname*{Lim}_{i \to \infty} \frac{\log d_v(P_i', Q')}{\frac{1}{2}m^2 h_f(P_i') + O(1)}.$$

Now we apply the theorem on Diophantine approximation (2.4) to the sequence $P_i' \in E(K)$, which v-adically converges to $Q' \in E(\bar{K})$. This yields

$$\operatorname*{Lim\,inf}_{i \to \infty} \frac{\log d_v(P_i', Q')}{[K : \mathbb{Q}]h_f(P_i')} \geqslant -2.$$

(Note that the $[K : \mathbb{Q}]$ factor, which in any case is not important, arises because h_f is the absolute height, while (2.4) is stated using the relative height H_K.) Using this result in the above inequality for L, we obtain

$$L \geqslant -\frac{4[K : \mathbb{Q}]}{m^2}.$$

But K is fixed, while the choice of m was arbitrary. Therefore $L \geqslant 0$, which is the desired conclusion. $\square$

§4. The S-Unit Equation

The proof of Siegel's theorem given in the last section is a special case of Siegel's general result that there are only finitely many S-integral points on any curve of genus at least 1. (See [La 7, ch. 8, thm. 2.4].) Siegel also gave a second proof, which applies only to a more restricted set of curves. However, the set of curves treated does include all elliptic curves. Further, the method is important, because when combined with results on linear forms in logarithms (see section 5), it leads to an effective procedure for finding all S-integral points. For this reason, we will now present Siegel's alternative proof.

The idea of the proof is to reduce the problem of solving for S-integral points on a curve to the problem of solving several equations of the form

$$ax + by = 1$$

in S-units. We start by giving a quick sketch of how the solution of this S-unit equation can be reduced to the Diophantine approximation theorem (1.4). It is this ineffective step which can be replaced by the effective results in section 5.

Theorem 4.1. Let $S \subset M_K$ be a finite set of places, and let $a, b \in K^*$. Then the equation

$$ax + by = 1$$

has only a finite number of solutions in S-units $x, y \in R_S^*$.

INEFFECTIVE PROOF (SKETCH). Let m be a large integer. By Dirichlet's S-unit theorem ([La 2, V §1]), the group $R_S^*/(R_S^*)^m$ is finite; let $c_1, \ldots, c_r \in R_S^*$ be coset representatives. Then any solution (x, y) to the original equation can be written as

$$x = c_i X^m, \qquad y = c_j Y^m$$

for some $X, Y \in R_S^*$ and some choice of c_i, c_j. Thus (X, Y) is a solution to the equation

$$ac_i X^m + bc_j Y^m = 1.$$

Since there are only finitely many choices for c_i, c_j, it certainly suffices to prove that for any $\alpha, \beta \in K^*$, the equation

$$\alpha X^m + \beta Y^m = 1$$

has only finitely many solutions with $X, Y \in R_S$.

Suppose that there were infinitely many such solutions. Then, since

$$H_K(Y) = \prod_{v \in S} \max\{1, |Y|_v^{n_v}\},$$

we can find some $v \in S$ so that for infinitely many of the solutions,

$$|Y|_v \geqslant H_K(Y)^{1/[K \,:\, \mathbb{Q}] \# S}.$$

(Note that $n_v \leqslant [K : \mathbb{Q}]$.) Let

$$\gamma^m = -\beta/\alpha.$$

We will specify below which m^{th} root to take. The idea is that if m is large enough, then X/Y provides too close an approximation to γ.

We can factor our equation as

$$\prod_{\zeta \in \mu_m} \left(\frac{X}{Y} - \zeta\gamma \right) = \frac{1}{\alpha Y^m}.$$

Since there are supposed to be infinitely many solutions, we may assume $H_K(Y)$ is very large; and so $|Y|_v$ will also be large. Then from the equality

$$\prod_{\zeta \in \mu_m} \left| \frac{X}{Y} - \zeta \gamma \right|_v = \frac{1}{|\alpha Y^m|_v},$$

we see that X/Y must be close to one of the $\zeta \gamma$'s; so replacing γ by one of its conjugates, we may assume that $|X/Y - \gamma|_v$ is quite small. But then for $\zeta \neq 1$, $|X/Y - \zeta \gamma|_v$ cannot be too small, since

$$|X/Y - \zeta \gamma|_v \geqslant |\gamma(1 - \zeta)|_v - |X/Y - \gamma|_v.$$

Hence we can find a constant $C_1 > 0$, independent of X/Y, so that

$$|X/Y - \gamma|_v \leqslant C_1 |Y|_v^{-m}.$$

(See exer. 9.5.) Finally, from the expression

$$\alpha(X/Y)^m = (1/Y)^m - \beta,$$

one easily deduces that

$$H_K(X/Y) \leqslant C_2 H_K(Y),$$

where C_2 depends only on α, β, and m. Now combining all of the above estimates, we find

$$|X/Y - \gamma|_v \leqslant C H_K(X/Y)^{-m/[K \,:\, \mathbb{Q}] \# S}.$$

But if we take any $m > 2[K : \mathbb{Q}] \# S$, then Roth's theorem (1.4) says that there are only finitely many possibilities for X/Y. Further, since

$$Y^m = (\alpha(X/Y)^m + \beta)^{-1} \quad \text{and} \quad X = (X/Y)Y,$$

each ratio X/Y corresponds to at most m possible pairs (X, Y). This contradicts our initial assumption that there are infinitely many solutions, and so completes the proof of (4.1). $\qquad\qquad\qquad\qquad\qquad\qquad\qquad\qquad\qquad\qquad\square$

Remark 4.2.1. Notice the great similarity in the method of proof for Siegel's theorem (3.1) and the S-unit equation (4.1). In both cases one starts with a point in a finitely generated group ($P \in E(K)$ for the former, $(x, y) \in R_S^* \times R_S^*$ for the latter). Next one uses the multiplication-by-m map to produce a new point whose height is much smaller, but which is a close approximation to another point defined over some finite extension of K. Finally one invokes a theorem on Diophantine approximation, such as (1.4), to complete the proof.

Remark 4.2.2. The proof of (4.1) given above is ineffective, since it makes use of Roth's theorem (1.4). But just as for Siegel's theorem, it is possible to make (4.1) *quantitative*; that is, to give an upper bound for the number of solutions. A priori, one would expect such a bound to depend on both the field K and the set of primes S. In fact, it is possible to prove the following analogue for the S-unit equation of Lang's conjecture (3.5) for elliptic curves. The proof, which we do not include, is fairly intricate.

Theorem 4.2.3 (Evertse [Ev]). *Let* $S \subset M_K$ *be a finite set of places containing* M_K^∞, *and let* $a, b \in K^*$. *Then the equation*

$$ax + by = 1$$

has at most $3 \times 7^{[K:\mathbb{Q}]+2\#S}$ *solutions in S-units* $x, y \in R_S^*$.

To see most clearly the analogy with (3.5), note that R_S^* is a finitely generated group of rank $\#S - 1$. Thus the bound in conjecture (3.5) takes the form $C^{\mathrm{rank}\,(R_S^*)+\mathrm{rank}\,(E(K))+1}$, while the bound in (4.2.3) can be written as $C^{\mathrm{rank}\,(R_S^*)+1}$.

We now give Siegel's reduction of S-integral points on hyperelliptic curves to solutions of the S-unit equation. Although we will not do so, the reader should note that every step in this reduction process can be made effective.

Theorem 4.3 (Siegel). *Let* $f(x) \in K[x]$ *be a polynomial of degree* $d \geqslant 3$ *with distinct roots* (*in* $\bar{K}$). *Then the equation*

$$y^2 = f(x)$$

has only finitely many solutions in S-integers $x, y \in R_S$.

PROOF. Clearly we are proving something stronger if we take a finite extension of K and enlarge the set S. Thus we may assume that f splits over K, say

$$f(x) = a(x - \alpha_1)\ldots(x - \alpha_d)$$

with $\alpha_i \in K$; and then make S sufficiently large so as to satisfy the following:

(i) $a \in R_S^*$;
(ii) $\alpha_i - \alpha_j \in R_S^*$ for all $i \neq j$;
(iii) R_S is a principal ideal domain.

Now suppose that $x, y \in R_S$ satisfy $y^2 = f(x)$. Let $\mathfrak{p}$ be a prime ideal of R_S. Then $\mathfrak{p}$ can divide at most one $x - \alpha_i$, since if it divides both $x - \alpha_i$ and $x - \alpha_j$, then it divides $\alpha_i - \alpha_j$, contradicting assumption (ii). Further, from (i), $\mathfrak{p}$ does not divide a. It follows from the equation

$$y^2 = a(x - \alpha_1)\ldots(x - \alpha_d)$$

that $\mathrm{ord}_\mathfrak{p}(x - \alpha_i)$ is even, and so the ideal $(x - \alpha_i)R_S$ is the square of an ideal in R_S. But from (iii), R_S is a principal ideal domain. Hence there are elements $z_i \in R_S$ and units $b_i \in R_S^*$ so that

$$x - \alpha_i = b_i z_i^2.$$

Now let L/K be the extension of K obtained by adjoining to K the square root of every element of R_S^*. Note that L/K is a finite extension, since $R_S^*/(R_S^*)^2$ is finite from Dirichlet's S-unit theorem. Further let $T \subset M_L$ be the set of places of L lying over elements of S, and let R_T be the ring of T-integers in L. Now each b_i is a square in R_T, say $b_i = \beta_i^2$, so

$$x - \alpha_i = (\beta_i z_i)^2.$$

Taking the difference of any two of these equations yields

$$\alpha_j - \alpha_i = (\beta_i z_i - \beta_j z_j)(\beta_i z_i + \beta_j z_j).$$

Note that $\alpha_j - \alpha_i \in R_T^*$, while each of the two factors on the right is in R_T. It follows that each of these factors is a unit,

$$\beta_i z_i \pm \beta_j z_j \in R_T^* \qquad \text{for all } i \neq j.$$

Now we use *Siegel's identity*:

$$\frac{\beta_1 z_1 \pm \beta_2 z_2}{\beta_1 z_1 - \beta_3 z_3} \mp \frac{\beta_2 z_2 \pm \beta_3 z_3}{\beta_1 z_1 - \beta_3 z_3} = 1.$$

This is a sum of two elements of R_T^* totaling 1, hence from (4.1) there are only finitely many choices for

$$\frac{\beta_1 z_1 + \beta_2 z_2}{\beta_1 z_1 - \beta_3 z_3} \quad \text{and} \quad \frac{\beta_1 z_1 - \beta_2 z_2}{\beta_1 z_1 - \beta_3 z_3}.$$

Multiplying these two numbers, there are only finitely many possibilities for

$$\frac{\alpha_2 - \alpha_1}{(\beta_1 z_1 - \beta_3 z_3)^2},$$

hence only finitely many for

$$\beta_1 z_1 - \beta_3 z_3,$$

and so only finitely many for

$$\beta_1 z_1 = \frac{1}{2}\left[(\beta_1 z_1 - \beta_3 z_3) + \frac{\alpha_3 - \alpha_1}{\beta_1 z_1 - \beta_3 z_3}\right].$$

But

$$x = \alpha_1 + (\beta_1 z_1)^2,$$

so there are only finitely many possible values of x; and then for each x, at most two y's. $\square$

Corollary 4.3.1. *Let C/K be a curve of genus 1, and let $f \in K(C)$ be a non-constant function. Then there are only finitely many points $P \in C(K)$ such that $f(P) \in R_S$.*

PROOF. Using the reduction procedure given in (3.2.2), it suffices to consider the case that f is the x-coordinate on a Weierstrass equation. But that case is covered by (4.3). $\square$

§5. Effective Methods

In 1949, Gelfond and Schneider independently solved Hilbert's problem concerning the transcendence of $2^{\sqrt{2}}$. They actually proved the following strong transcendence criterion.

Theorem 5.1 (Gelfond, Schneider). *Let* $\alpha, \beta \in \overline{\mathbb{Q}}$ *with* $\alpha \neq 0, 1$ *and* $\beta \notin \mathbb{Q}$. *Then* α^β *is transcendental.*

Gelfond rephrased his result in terms of logarithms. If $\alpha_1, \alpha_2 \in \overline{\mathbb{Q}}^*$ and if $\log \alpha_1$ and $\log \alpha_2$ are linearly independent over $\mathbb{Q}$, then they are linearly independent over $\overline{\mathbb{Q}}$. He further showed that one could give an explicit lower bound for $|\beta_1 \log \alpha_1 + \beta_2 \log \alpha_2|$ whenever this quantity is non-zero, and noted that many Diophantine problems could be solved if one knew an analogous result for sums of arbitrarily many logarithms. Such a theorem was proven by A. Baker in 1966. The proof is quite involved, so we will be content to just quote the following version.

Theorem 5.2 (Baker). *Let* $\alpha_1, \dots, \alpha_n \in K^*$ *and* $\beta_1, \dots, \beta_n \in K$. *For any constant* κ, *define*

$$\tau(\kappa) = \tau(\kappa; \alpha_1, \dots, \alpha_n; \beta_1, \dots, \beta_n) = h([1, \beta_1, \dots, \beta_n]) h([1, \alpha_1, \dots, \alpha_n])^\kappa.$$

(N.B. These are logarithmic height functions.) Fix an embedding $K \subset \mathbb{C}$, *and let* $|\cdot|$ *be the corresponding absolute value. Assume that*

$$\beta_1 \log \alpha_1 + \cdots + \beta_n \log \alpha_n \neq 0.$$

Then there are effectively computable constants $C, \kappa > 0$, *depending only on* n *and* $[K : \mathbb{Q}]$, *such that*

$$|\beta_1 \log \alpha_1 + \cdots + \beta_n \log \alpha_n| > C^{-\tau(\kappa)}.$$

PROOF. See [Ba] or [La 5, VIII, Thm. 1.1]. $\square$

Remark 5.2.1. We have restricted ourselves in (5.2) to the case of an archimedean absolute value. There are analogous results in the non-archimedean case, although minor technical difficulties arise due to the fact that the p-adic logarithm is only defined in a neighborhood of 1. See (5.6) below for a further discussion.

It is not immediately clear how Baker's theorem (5.2) can be applied to give a bound for the solutions to the S-unit equation. We start with the following elementary lemma. (See also exer. 9.8.)

Lemma 5.3. *Let* V *be a finite dimensional vector space over* $\mathbb{R}$. *Given any basis* $\mathbf{e} = \{e_1, \dots, e_n\}$ *for* V, *let* $\|\cdot\|_\mathbf{e}$ *be the sup norm with respect to* $\mathbf{e}$. *(I.e.* $\|x\|_\mathbf{e} = \|\Sigma x_i e_i\| = \max\{|x_i|\}$.) *Suppose that* $\mathbf{f} = \{f_1, \dots, f_n\}$ *is another basis. Then there are constants* $c_1, c_2 > 0$, *depending on* $\mathbf{e}$ *and* $\mathbf{f}$, *so that for all* $x \in V$,

$$c_1 \|x\|_\mathbf{e} \leq \|x\|_\mathbf{f} \leq c_2 \|x\|_\mathbf{e}.$$

PROOF. Let $A = (a_{ij})$ be the change of basis matrix from $\mathbf{e}$ to $\mathbf{f}$, so $e_i = \Sigma a_{ij} f_j$; and let $\|A\| = \max\{|a_{ij}|\}$. Then for any $x = \Sigma x_i e_i \in V$, we have

$x = \Sigma x_i e_i = \Sigma x_i a_{ij} f_j$, so

$$\|x\|_{\mathbf{f}} = \max_j \left\{ \left| \sum_i x_i a_{ij} \right| \right\} \leqslant n \max_{i,j} \{|a_{ij}|\} \max_i \{|x_i|\} = n \|A\| \|x\|_{\mathbf{e}}.$$

This gives one inequality, and the other follows by symmetry.   □

We apply (5.3) to the following situation. Let $S \subset M_K$ be a finite set of places containing M_K^∞, let $s = \#S$, and choose a basis $\alpha_1, \ldots, \alpha_{s-1}$ for the free part of R_S^*. Then every $\alpha \in R_S^*$ can be written uniquely as

$$\alpha = \zeta \alpha_1^{m_1} \ldots \alpha_{s-1}^{m_{s-1}}$$

for integers $m_1, \ldots, m_{s-1}$ and a root of unity ζ. Define *the size of* α *(relative to* $\{\alpha_1, \ldots, \alpha_{s-1}\}$) by

$$m(\alpha) = \max\{|m_i|\}.$$

Lemma 5.4. *With notation as above, there are constants* $c_1, c_2 > 0$, *depending only on* K *and* S, *such that for every* $\alpha \in R_S^*$,

$$c_1 h(\alpha) \leqslant m(\alpha) \leqslant c_2 h(\alpha).$$

PROOF. Let $S = \{v_1, \ldots, v_s\}$, and let $n_i = n_{v_i}$ be the local degree corresponding to v_i. Consider the *S-regulator homomorphism*

$$\rho_S : R_S^* \longrightarrow \mathbb{R}^s$$

$$\alpha \to (n_1 v_1(\alpha), \ldots, n_s v_s(\alpha)).$$

Notice that the image of ρ_S lies in the hyperplane $H = \{x_1 + \cdots + x_s = 0\}$; and by Dirichlet's S-unit theorem, it actually spans H. Let $\|\cdot\|_1$ be the sup norm on $\mathbb{R}^s$ relative to the standard basis, and let $\|\cdot\|_2$ be the sup norm relative to the basis $\{\rho_S(\alpha_1), \ldots, \rho_S(\alpha_{s-1}), (1, 1, \ldots, 1)\}$. (I.e. $\{\rho_S(\alpha_i)\}$ spans H, and we have added one extra vector in order to span all of $\mathbb{R}^s$.) From (5.3), there are constants $c_1, c_2 > 0$ such that

$$c_1 \|x\|_1 \leqslant \|x\|_2 \leqslant c_2 \|x\|_1 \qquad \text{for all } x \in \mathbb{R}^s.$$

Now let $\alpha \in R_S^*$, and write $\rho_S(\alpha) = \Sigma m_i \rho_S(\alpha_i)$. Then directly from the definitions, we have

$$\|\rho_S(\alpha)\|_2 = \max\{|m_i|\} = m(\alpha),$$

$$\|\rho_S(\alpha)\|_1 = \max\{n_i |v_i(\alpha)|\},$$

and

$$h_K(\alpha) = \sum \max\{0, -n_i v_i(\alpha)\}.$$

(Note that the sum for $h_K(\alpha)$ need only run over the absolute values in S, since $v(\alpha) = 0$ for all $v \notin S$.) We must now find a way to compare $\|\rho_S(\alpha)\|_1$ with $h_K(\alpha)$.

More generally, for any $x = (x_1, \ldots, x_s) \in H$, we can compare $\|x\|_1$ with $h(x) = \Sigma \max\{0, -x_i\}$. First, since $\max\{0, -x_i\} \leqslant |x_i|$, we have the obvious estimate

$$h(x) \leqslant s\|x\|_1.$$

On the other hand, if we sum the identity

$$x_i = \max\{0, x_i\} - \max\{0, -x_i\}$$

for $1 \leqslant i \leqslant s$ and use the fact that $x \in H$ (i.e. $\Sigma x_i = 0$), we obtain

$$0 = h(-x) - h(x); \qquad \text{and so} \quad h(x) = h(-x).$$

Therefore

$$
\begin{aligned}
2h(x) &= h(x) + h(-x) \\
&= \sum (\max\{0, -x_i\} + \max\{0, x_i\}) \\
&= \sum |x_i| \\
&\geqslant \max\{|x_i|\} \\
&= \|x\|_1.
\end{aligned}
$$

Thus $\frac{1}{2}\|x\|_1 \leqslant h(x) \leqslant s\|x\|_1$; and combining this with the above results gives an estimate of the desired form,

$$(c_1/s)h_K(\alpha) \leqslant m(\alpha) \leqslant 2c_2 h_K(\alpha). \qquad \square$$

We are now ready to show how the solution of the S-unit equation can be reduced to the problem of bounds for linear forms in logarithms.

Theorem 5.5. *Fix a, $b \in K^*$. There exists an effectively computable constant $C = C(K, S, a, b)$ such that any solution $(\alpha, \beta) \in R_S^* \times R_S^*$ to the S-unit equation*

$$a\alpha + b\beta = 1$$

satisfies $H(\alpha) < C$.

PROOF. Let (α, β) be a solution, and choose the absolute value v in S for which $|\alpha|_v$ is largest. Then, since $|\alpha|_w = 1$ for all $w \notin S$, we have

$$|\alpha|_v^{[K:\mathbb{Q}]s} \geqslant \prod_{w \in S} \max\{1, |\alpha|_w^{n_w}\} = H_K(\alpha);$$

and hence

$$|\alpha|_v \geqslant H(\alpha)^{1/s}.$$

(Here, as usual, $s = \#S$.)

To simplify our discussion, we will now assume that v is archimedean. (This will certainly be true, for example, if $S = M_K^\infty$. For arbitrary S, see the discussion in (5.6) below.) The mean value theorem applied to the function $\log x$ yields

$$\left| \frac{\log x - \log y}{x - y} \right| \leqslant \frac{1}{\min\{|x|, |y|\}}.$$

We use this with $x = a\alpha$, $y = -b\beta$, $x - y = 1$, and obtain

$$|\log a\alpha - \log b\beta| \leqslant \min\{|a\alpha|, |a\alpha - 1|\}^{-1}$$

$$\leqslant 2(|a|H(\alpha)^{1/s})^{-1}.$$

(For the last line, we have assumed that $|\alpha| > 2/|a|$, since otherwise we have the excellent bound $H(\alpha) \leqslant |\alpha|^s \leqslant (2/|a|)^s$.)

Now let $\alpha_1, \ldots, \alpha_{s-1}$ be a basis for R_S^* as above, and write

$$\alpha = \zeta \alpha_1^{m_1} \ldots \alpha_{s-1}^{m_{s-1}} \quad \text{and} \quad \beta = \zeta' \alpha_1^{m_1'} \ldots \alpha_{s-1}^{m_{s-1}'}.$$

Substituting this into the above inequality yields

$$\left| \sum (m_i - m_i') \log \alpha_i + \log(a\zeta/b\zeta') \right| \leqslant c_1 H(\alpha)^{-1/s},$$

where here and in what follows, the constants $c_1, c_2, \ldots$ are effectively computable and depend only on K, S, a, and b.

From the equality $a\alpha + b\beta = 1$, one easily obtains an estimate

$$|h(\alpha) - h(\beta)| \leqslant c_2;$$

and now applying (5.4) yields

$$c_3 m(\alpha) \leqslant m(\beta) \leqslant c_4 m(\alpha).$$

(Since we may clearly assume that $m(\alpha), m(\beta) \geqslant 1$.) In particular,

$$|m_i - m_i'| \leqslant m(\alpha) + m(\beta) \leqslant c_5 h(\alpha).$$

Letting $q_i = m_i - m_i'$ and $\gamma = a\zeta/b\zeta'$, we now have an inequality

$$|q_1 \log \alpha_1 + \cdots + q_{s-1} \log \alpha_{s-1} + \log \gamma| \leqslant c_1 H(\alpha)^{-1/s}$$

with $\alpha_1, \ldots, \alpha_{s-1}, \gamma$ fixed and $q_1, \ldots, q_{s-1}$ integers satisfying $|q_i| \leqslant c_5 h(\alpha)$.

Now use Baker's theorem (5.2). This gives a lower bound of the form

$$|q_1 \log \alpha_1 + \cdots + q_{s-1} \log \alpha_{s-1} + \log \gamma| \geqslant c_6^{-\tau},$$

where

$$\tau = h([1, q_1, \ldots, q_{s-1}]) h([1, \alpha_1, \ldots, \alpha_{s-1}, \gamma])^{\kappa},$$

and κ is a constant depending only on K and s. But from above,

$$h([1, q_1, \ldots, q_{s-1}]) = \log \max\{1, |q_1|, \ldots, |q_{s-1}|\} \leqslant \log(c_5 h(\alpha)).$$

Combining the upper and lower bounds for the linear form in logarithms and using this estimate yields

$$c_7^{-\log(c_5 h(\alpha))} \leqslant c_1 H(\alpha)^{-1/s}.$$

(Note that the basis $\alpha_1, \ldots, \alpha_{s-1}$ depends only on the field K and the set S, so it is alright to absorb the $h([1, \alpha_1, \ldots, \alpha_{s-1}, \gamma])^\kappa$ exponent into the c_7.) Now a little bit of algebra gives

$$H(\alpha) \leqslant c_8 h(\alpha)^{c_9};$$

and since $h(\alpha) = \log H(\alpha)$, this implies the desired bound for $H(\alpha)$. $\square$

Remark 5.6. In order to make the argument given in (5.5) apply to a non-archimedean absolute value, it is necessary to make some minor technical alterations. The main difficulty is that the logarithm function in the p-adic case only converges in a neighborhood of 1. What one does is to take a sub-group of finite index in R_S^* which is generated by S-units which are p-adically close to 1, together with a uniformizer for p. Then, assuming that $|\alpha|_p$ is sufficiently large, one shows that $a\alpha/b\beta$ is p-adically close to 1. Now applying the above argument to some power of $a\alpha/b\beta$ will give a well-defined linear form in p-adic logarithms, and from then on the argument goes just the same. For the final step, of course, one must use a p-adic analogue of Baker's theorem. (For more details of this reduction step, see for example [La 5, VI §1].)

Remark 5.7. In order to obtain an effective bound for those points on an elliptic curve which satisfy $f(P) \in R_S$, where f is an arbitrary non-constant function, it is also necessary to make the reduction step given in (3.2.2) effective. This essentially involves giving an effective version of the Riemann–Roch theorem, which has been done by Coates ([Co]). As the reader might guess from the number of reduction steps involved, the effective bounds which come out of the current proofs are quite large. To indicate their magnitude, we quote the following two results. (See also [Ko–T], (7.2) and (7.4).)

Theorem 5.8. (a) (Baker [Ba, p. 45]) *Let A, B, C, $D \in \mathbb{Z}$ satisfy* $\max\{|A|, |B|, |C|, |D|\} \leqslant H$, *and assume that*

$$E : Y^2 = AX^3 + BX^2 + CX + D$$

is an elliptic curve. Then any point $P = (x, y) \in E(\mathbb{Q})$ with $x, y \in \mathbb{Z}$ satisfies

$$\max\{|x|, |y|\} < \exp((10^6 H)^{10^6}).$$

(b) (Baker, Coates [Ba–C]) *Let $F(X, Y) \in \mathbb{Z}[X, Y]$ be an absolutely irreducible polynomial such that the curve $F(X, Y) = 0$ has genus 1. Assume that F has degree n, and that its coefficients all have absolute value at most H. Then any solution $F(x, y) = 0$ with $x, y \in \mathbb{Z}$ satisfies*

$$\max\{|x|, |y|\} < \exp \exp \exp((2H)^{10^{n^{10}}}).$$

Linear Forms in Elliptic Logarithms

Rather than reducing the problem of integral points on an elliptic curve to the question of solutions to the S-unit equation, and thence as above to bounds for linear forms in logarithms, one can work directly with the analytic parameterization of the elliptic curve. We will now briefly indicate how this is done in the simplest case.

Let $E/\mathbb{Q}$ be an elliptic curve given by a Weierstrass equation

$$E : y^2 = 4x^3 - g_2 x - g_3$$

with $g_2, g_3 \in \mathbb{Z}$. We are interested in bounding the height of points $P \in E(\mathbb{Q})$ which satisfy $x(P) \in \mathbb{Z}$. Let

$$\phi : \mathbb{C}/\Lambda \to E(\mathbb{C})$$

be the analytic parameterization of $E(\mathbb{C})$ given by the Weierstrass $\wp$-function (cf. VI. 5.1.1). We fix a basis $\{\omega_1, \omega_2\}$ for the lattice Λ. Let

$$\psi : E(\mathbb{C}) \to \mathbb{C}$$

be the map inverse to ϕ which takes values in the fundamental parallelogram centered at 0. (Thus ϕ is the *elliptic exponential map*, and choosing a fundamental parallelogram for the *elliptic logarithm* ψ is comparable to choosing a principal value for the ordinary logarithm function.)

Fix a basis $P_1, \ldots, P_r$ for the free part of $E(\mathbb{Q})$. Then given any point $P \in E(\mathbb{Q})$, we can write $P = q_1 P_1 + \cdots + q_r P_r + T$ for certain integers $q_1, \ldots, q_r$ and a torsion point $T \in E_{\text{tors}}(\mathbb{Q})$. It follows that

$$\psi(P) \equiv q_1 \psi(P_1) + \cdots + q_r \psi(P_r) + \psi(T) \,(\text{mod } \Lambda),$$

so there are integers m_1 and m_2 such that

$$\psi(P) = q_1 \psi(P_1) + \cdots + q_r \psi(P_r) + \psi(T) + m_1 \omega_1 + m_2 \omega_2.$$

Now suppose that P is a large integral point; that is, $x(P) \in \mathbb{Z}$ and $|x(P)|$ is large. Then P is close to O (in the archimedean topology), and so $\psi(P)$ is close to 0. More precisely, since $\wp(z) = x(\phi(z))$ behaves like z^{-2} for z close to 0, we see that

$$|\psi(P)|^2 \leqslant c_1 |x(P)|^{-1} = c_1 H(x(P))^{-1}.$$

(Recall that if $x \in \mathbb{Z}$, $x \neq 0$, then $H(x) = |x|$. The constant c_1 will depend on g_2 and g_3.)

On the other hand, since the canonical height is quadratic and positive definite (VIII.9.3 and VIII.9.6), we can estimate

$$\log H(x(P)) = h_x(P) = 2\hat{h}(P) + O(1)$$

$$= 2\hat{h}(\textstyle\sum q_i P_i + T) + O(1)$$

$$\geqslant c_2 \max\{|q_i|\}^2,$$

where c_2 will depend on E and the choice of the basis $P_1, \ldots, P_r$. (See exer. 9.8.) Substituting this above, we obtain an upper bound for our linear form in elliptic logarithms:

$$|q_1 \psi(P_1) + \cdots + q_r \psi(P_r) + \psi(T) + m_1 \omega_1 + m_2 \omega_2| \leqslant c_3^{-\max\{|q_i|\}^2}.$$

Further, since ω_1 and ω_2 are $\mathbb{R}$-linearly independent, it is easy to see that

$$\max\{|m_1|, |m_2|\} \leqslant c_4 \max\{|q_i|\},$$

where c_4 depends on E, $\{P_i\}$, ω_1, and ω_2. Thus we finally obtain

$$|q_1 \psi(P_1) + \cdots + q_r \psi(P_r) + \psi(T) + m_1 \omega_1 + m_2 \omega_2| \leqslant c_5^{-q^2},$$

with $q = \max\{|q_1|, \ldots, |q_r|, |m_1|, |m_2|\}$.

Now any lower bound $C^{-\tau(q)}$ for the left-hand side satisfying $\tau(q)/q^2 \to 0$ as $q \to \infty$ will give the desired finiteness result. The first effective estimate of this sort was proven by Masser ([Mas]) in the case that E has complex multiplication. The general case was dealt with by Wüstholz ([Wü 1], [Wü 2]), who had to overcome great technical difficulties associated with the necessary zero and multiplicity estimates.

It remains to discuss the question of effectivity. The reduction to linear forms in ordinary logarithms via the S-unit equation is fully effective. It is possible to give an explicit upper bound for the height of any S-integral point of $E(K)$ in terms of easily computed quantities associated to K, S, and E. One of these quantities, for example, will be a bound for the heights of generators for the unit group R^*. Now in the analogous reduction to linear forms in elliptic logarithms, one similarly chooses a set of generators for the Mordell–Weil group $E(K)$; and the bound for the integral points then depends on the heights of these generators. Unfortunately, as we have seen (cf. VIII. 3.2 and Ch. X), the proof of the Mordell–Weil theorem is not effective. Thus although the approach to integral points on elliptic curves via elliptic logarithms seems much more natural than the roundabout route through the S-unit equation, it is likely to remain ineffective until an effective proof of the Mordell–Weil theorem is found.

§6. Shafarevich's Theorem

Recall that an elliptic curve E/K has good reduction at a finite place $v \in M_K$ if it has a Weierstrass equation whose coefficients are v-integral and whose discriminant is a v-adic unit (cf. VII §5).

Theorem 6.1 (Shafarevich [Sha]). *Let $S \subset M_K$ be a finite set of places containing M_K^∞. Then up to isomorphism over K, there are only finitely many elliptic curves E/K having good reduction at all primes not in S.*

PROOF. Clearly we are proving something stronger if we enlarge S. We may thus assume that S contains all primes of K lying over 2 and 3. Further, we may enlarge S so that the ring of S-integers R_S has class number 1.

Now under these assumptions, (VIII. 8.7) says that any elliptic curve E/K has a Weierstrass equation of the form

$$E : y^2 = x^3 + Ax + B \qquad A, B \in R_S,$$

with discriminant $\Delta = -16(4A^3 + 27B^2)$ satisfying $\Delta R_S = \mathcal{D}_{E/K} R_S$. (Here $\mathcal{D}_{E/K}$ is the minimal discriminant of E/K. Cf. (VIII §8).) Note that if E has good reduction outside of S, then $\mathrm{ord}_v(\mathcal{D}_{E/K}) = 0$ for all primes v not in S; and so Δ will be in R_S^*.

Assume now that we are given a sequence of elliptic curves $E_1/K, E_2/K, \ldots$, each of which has good reduction outside of S. Associate to each E_i an equation as above with coefficients $A_i, B_i \in R_S$ and discriminant $\Delta_i \in R_S^*$. We break the sequence of E_i's into finitely many subsequences according to the residue class of Δ_i in the finite group $R_S^*/(R_S^*)^{12}$. Restricting attention to one such subsequence, we may assume that $\Delta_i = CD_i^{12}$ for a fixed C and some $D_i \in R_S^*$.

Now the formula $\Delta = -16(4A^3 + 27B^2)$ implies that for each i, the point $(-12A_i/D_i^4, 72B_i/D_i^6)$ is an S-integral point on the elliptic curve

$$Y^2 = X^3 + 27C.$$

Siegel's theorem (3.2.1) says that there are only finitely many such points, and so only finitely many possibilities for A_i/D_i^4 and B_i/D_i^6. But if

$$A_i/D_i^4 = A_j/D_j^4 \quad \text{and} \quad B_i/D_i^6 = B_j/D_j^6,$$

then the change of variables

$$x = (D_i/D_j)^2 x' \qquad y = (D_i/D_j)^3 y'$$

gives an isomorphism from E_i to E_j. Hence the sequence of E_i's contains only finitely many K-isomorphism classes of elliptic curves. ☐

Example 6.1.1. There are no elliptic curves $E/\mathbb{Q}$ having everywhere good reduction (exer. 8.15). For a complete list of the 24 curves $E/\mathbb{Q}$ having good reduction outside of $\{2\}$ and the 784 curves $E/\mathbb{Q}$ having good reduction outside of $\{2, 3\}$, see [B–K, Table 4]. Similar lists have also been compiled for various quadratic number fields; see for example [Las 2] and [Pi].

Shafarevich's theorem (6.1) has a number of important applications. We will content ourselves with the following two corollaries.

Corollary 6.2. *Fix an elliptic curve E/K. Then there are only finitely many elliptic curves E'/K which are K-isogenous to E.*

PROOF. If E and E' are isogenous over K, then (VII.7.2) says that E and E' have the same set of primes of bad reduction. Now apply (6.1). ☐

Corollary 6.3 (Serre). *Let E/K be an elliptic curve with no complex multiplication. Then for all but finitely many primes ℓ, the group of ℓ-torsion points $E[\ell]$ has no non-trivial $G_{\bar{K}/K}$-invariant subgroups. [I.e. The representation of $G_{\bar{K}/K}$ on $E[\ell]$ is irreducible.]*

PROOF. Suppose that $\Phi_\ell \subset E[\ell]$ is a non-trivial $G_{\bar{K}/K}$-invariant subgroup. Since $E[\ell] \cong (\mathbb{Z}/\ell\mathbb{Z})^2$, Φ_ℓ is necessarily cyclic of order ℓ. Further, from (III.4.12), there exists an elliptic curve E_ℓ/K and an isogeny $\phi_\ell : E \to E_\ell$ defined over K with $\ker(\phi_\ell) = \Phi_\ell$.

Since each such E_ℓ is isogenous to E, (6.2) says that the E_ℓ's fall into finitely many K-isomorphism classes. Suppose that $E_\ell \cong E_{\ell'}$ for two primes ℓ and ℓ'. Then the composition

$$E \overset{\phi_\ell}{\to} E_\ell \cong E_{\ell'} \overset{\hat{\phi}_{\ell'}}{\to} E$$

gives an endomorphism of E of degree

$$(\deg \phi_\ell)(\deg \hat{\phi}_{\ell'}) = \ell\ell'.$$

But by assumption, $\text{End}(E) = \mathbb{Z}$, so every endomorphism of E has degree n^2 for some $n \in \mathbb{Z}$. This shows that $\ell = \ell'$, and so the E_ℓ's are pairwise non-isomorphic for distinct primes ℓ. Therefore there are only finitely many primes ℓ for which such a Φ_ℓ and E_ℓ can exist.   □

Example 6.4. For $K = \mathbb{Q}$, results of Mazur ([Maz 2]) and Kenku ([Ke]) give a far more precise statement than (6.2). They show that for a given elliptic curve $E/\mathbb{Q}$, there are at most *eight* $\mathbb{Q}$-isomorphism classes of elliptic curves $E'/\mathbb{Q}$ which are $\mathbb{Q}$-isogenous to E. Further, if $\phi : E \to E'$ is a $\mathbb{Q}$-isogeny for which $\ker(\phi)$ is a cyclic group, then either

$$1 \leqslant \deg \phi \leqslant 19, \quad \text{or} \quad \deg \phi \in \{21, 25, 27, 37, 43, 67, 163\}.$$

It is no coincidence that those d's for which $\mathbb{Q}(\sqrt{-d})$ has class number 1 appear as possibilities for $\deg \phi$. This is because the class number 1 condition allows the elliptic curve E corresponding to the lattice

$$\mathbb{Z} + \mathbb{Z}(\tfrac{1}{2} + \tfrac{1}{2}\sqrt{-d})$$

via (VI. 5.1.1) to be defined over $\mathbb{Q}$. (See C. 11.3.1.) Now one need merely note that multiplication by $\sqrt{-d}$ gives an isogeny from E to itself whose kernel Φ is cyclic of order d and defined over $\mathbb{Q}$. Then $E \to E/\Phi$ is a cyclic isogeny of degree d.

Remark 6.5. An examination of the proof of (6.1) reveals an interesting possibility. If one had some other proof of (6.1) which did not use either Siegel's theorem or Diophantine approximation techniques, then one could deduce that the equation

$$Y^2 = X^3 + D$$

has only finitely many solutions $X, Y \in R_S$. For given such a solution, the equation

$$y^2 = x^3 - Xx - Y$$

would be an elliptic curve with good reduction outside of

$$S \cup \{\text{primes dividing 2 and 3}\}.$$

Hence assuming (6.1), there would be only finitely many such curves, and one could argue back to the finiteness of the number of pairs (X, Y). Building on this idea, Parshin ([Pa]) showed how a generalization of (6.1) to curves of higher genus (which had already been conjectured by Shafarevich [Sha 1]) could be used to prove Mordell's conjecture that curves of genus greater than 1 have only finitely many *rational* points. The subsequent proof of Shafarevich's conjecture by Faltings ([Fa 1]) completed this chain of reasoning. Faltings' proof (together with Parshin's idea) also gives a proof of Siegel's theorem (3.2) which does not involve the use of Diophantine approximation.

§7. The Curve $Y^2 = X^3 + D$

Many of the general results known or conjectured about the arithmetic of elliptic curves were originally noticed and tested on various special sorts of equations, such as the one given in the title of this section. For example, long before the work of Mordell and Siegel led to general finiteness results such as (3.2.1), many special cases had been proven by a variety of methods. (See, e.g., [Mo 4, Ch. 26].) We give two examples where the complete set of solutions can be obtained by relatively elementary means.

Proposition 7.1. (a) *The equation*

$$y^2 = x^3 + 7$$

has no solutions in integers $x, y \in \mathbb{Z}$.
(b) (Fermat) *The only integral solutions to the equation*

$$y^2 = x^3 - 2$$

are $(x, y) = (3, \pm 5)$.

PROOF. (a) Suppose that $x, y \in \mathbb{Z}$ satisfy $y^2 = x^3 + 7$. First, note that x must be odd, since no integer of the form $8k + 7$ is a square. Now rewrite the equation as

$$y^2 + 1 = x^3 + 8 = (x + 2)(x^2 - 2x + 4).$$

Since x is odd,

$$x^2 - 2x + 4 = (x - 1)^2 + 3 \equiv 3 \pmod{4},$$

so we can choose a prime $p \equiv 3 \pmod 4$ which divides $x^2 - 2x + 4$. But then $y^2 + 1 \equiv 0 \pmod p$, which is not possible.

(b) Suppose we have a solution $x, y \in \mathbb{Z}$ to $y^2 = x^3 - 2$. Factor the equation as

$$(y + \sqrt{-2})(y - \sqrt{-2}) = x^3.$$

Since the ring $R = \mathbb{Z}[\sqrt{-2}]$ is a principal ideal domain, and the greatest common divisor of $y + \sqrt{-2}$ and $y - \sqrt{-2}$ (in R) clearly divides $2\sqrt{-2}$, we see that $y + \sqrt{-2}$ can be written as

$$y + \sqrt{-2} \quad = \quad \zeta^3 \quad \text{or} \quad \sqrt{-2}\zeta^3 \quad \text{or} \quad 2\zeta^3$$

for some $\zeta \in R$. Applying complex conjugation gives

$$y - \sqrt{-2} \quad = \quad \bar{\zeta}^3 \quad \text{or} \quad -\sqrt{-2}\bar{\zeta}^3 \quad \text{or} \quad 2\bar{\zeta}^3;$$

and now taking the product yields

$$x^3 \quad = \quad y^2 + 2 \quad = \quad (\zeta\bar{\zeta})^3 \quad \text{or} \quad 2(\zeta\bar{\zeta})^3 \quad \text{or} \quad 4(\zeta\bar{\zeta})^3.$$

Since $\zeta\bar{\zeta} \in \mathbb{Z}$, this shows that only the first case is possible, so

$$y + \sqrt{-2} = \zeta^3 \quad \text{and} \quad y - \sqrt{-2} = \bar{\zeta}^3.$$

Subtracting these equations gives

$$2\sqrt{-2} = \zeta^3 - \bar{\zeta}^3 = (\zeta - \bar{\zeta})(\zeta^2 + \zeta\bar{\zeta} + \bar{\zeta}^2).$$

Now write $\zeta = a + b\sqrt{-2}$ with $a, b \in \mathbb{Z}$. Substituting this above yields

$$2\sqrt{-2} = 2\sqrt{-2}b(3a^2 - 2b^2);$$

so using the fact that a and b are rational integers, we must have

$$b = \pm 1 \quad \text{and} \quad 3a^2 - 2b^2 = \pm 1.$$

Therefore $(a, b) = (\pm 1, \pm 1)$ (with independent $\pm$ signs); and working back, these lead to the values $(x, y) = (3, \pm 5)$. $\qquad\square$

Remark 7.1.1. It is worth remarking that the result in (7.1b) is far more interesting than that of (7.1a). This is because the Mordell–Weil group (over $\mathbb{Q}$) of the elliptic curve $y^2 = x^3 + 7$ turns out to be trivial, so (7.1a) is really a reflection of the fact that the equation has no rational points. On the other hand, the Mordell–Weil group of $y^2 = x^3 - 2$ is infinite cyclic (cf. exer. 10.19), so (7.1b) says that in its infinite set of rational points, there are only two integral points.

Baker applied his methods to obtain an explicit upper bound, in terms of D, for the integral solutions of $y^2 = x^3 + D$. This bound was refined by Stark, who proved the following.

Theorem 7.2 (Stark [Sta]). *For every $\varepsilon > 0$ there is an effectively computable constant C_ε, depending only on ε, so that the following holds. Let $D \in \mathbb{Z}$, $D \neq 0$. Then every solution $x, y \in \mathbb{Z}$ to the equation*

$$y^2 = x^3 + D$$

satisfies

$$\log \max\{|x|, |y|\} \leqslant C_\varepsilon |D|^{1+\varepsilon}.$$

Example 7.3. Stark's estimate (7.2) gives a bound for x and y which is slightly worse than exponential in D. One naturally would like to know whether this is the correct order of magnitude. A number of people have conducted computer searches for large solutions (see, e.g., [La1] or [Ha1]). Among the interesting examples found were

$$378{,}661^2 = 5234^3 + 17$$

$$911{,}054{,}064^2 = 939{,}787^3 - 307$$

$$149{,}651{,}610{,}621^2 = 28{,}187{,}351^3 + 1090.$$

Although these examples show that x and y can be quite large in comparison to D, a close examination of his data led M. Hall to make the following conjecture, which was subsequently partly generalized by Lang.

Conjecture 7.4. (a) (Hall [Ha1]): *For every $\varepsilon > 0$ there is a constant C_ε, depending only on ε, so that the following holds. Let $D \in \mathbb{Z}$, $D \neq 0$. Then every solution $x, y \in \mathbb{Z}$ to the equation*

$$y^2 = x^3 + D$$

satisfies

$$|x| < C_\varepsilon D^{2+\varepsilon}.$$

(b) (Hall–Lang [La 9]) *There are absolute constants C, $\kappa > 0$ such that if $E/\mathbb{Q}$ is an elliptic curve given by a Weierstrass equation*

$$y^2 = x^3 + Ax + B \qquad A, B \in \mathbb{Z},$$

and if $P \in E(\mathbb{Q})$ is an integral point (i.e. $x(P) \in \mathbb{Z}$), then

$$|x(P)| < C \max\{|A|, |B|\}^\kappa.$$

The evidence for these conjectures is fragmentary. They are true for function fields (Davenport [Dav] for (7.4a) and Schmidt [Schm 1] for (7.4b)). Further, Vojta ([Voj]) has shown how (7.4a) is a consequence of his very general Nevanlinna-type conjectures for varieties over number fields; but Vojta's conjectures seem well beyond the reach of current techniques. (Also see exer. 9.10.) Aside from this, very little is known. It is worth pointing out that the effective techniques in section 5 seem intrinsically incapable of leading to estimates like (7.4). Let us explain what the problem is, say for the equation $y^2 = x^3 + D$.

In performing the reduction to the S-unit equation, one deals with a number field K whose discriminant looks like a power of $|D|$. Now the Brauer–Siegel theorem says that $\log(h_K R_K) \sim \frac{1}{2} \log d_K$ as $[K : \mathbb{Q}]/\log d_K \to 0$, where h_K is the class number, R_K is the regulator, and d_K is the absolute discriminant of K. (See, e.g., [La 2, Ch. XVI].) In general there is no reason to expect the class number of K to be large, so the best that one can hope for is to find a bound for the regulator which is a power of $|D|$. Since the regulator is a determinant of the *logarithms* of a basis for the unit group R^*, the resulting bounds for the heights $H(\alpha_i)$ of generators $\alpha_i \in R^*$ will be exponential in $|D|$. This eventually leads to an exponential bound for x and y as in (7.2).

There is a similar problem in trying to prove (7.4) by using linear forms in elliptic logarithms or by following Siegel's method of proof as in (3.1) (even assuming that one could find a strong effective version of Roth's theorem). Of course, neither of these methods is effective, since the Mordell–Weil theorem is not effective. But in any case, it seems likely (cf. VIII. 10.2) that the best possible upper bound for generators of the Mordell–Weil group of $y^2 = x^3 + D$ will have the form $\hat{h}(P) \leqslant C|D|^\kappa$. Here $\hat{h}$ is a logarithmic height, so again this will lead to a bound for the x-coordinate of integral points which is exponential in $|D|$.

The problem in both cases can be explained most clearly by the analogy given in (4.2.1). In solving the S-unit equation and in finding the integral points on an elliptic curve, one is initially given a finitely generated group $(R_S^* \times R_S^*$, resp. $E(K))$ and a certain exceptional subset (solutions to $ax + by = 1$, resp. points with $x(P) \in R_S$). The first step is to choose a basis for the finitely generated group and express the exceptional points in terms of this basis. Now the problem that arises in trying to prove (7.4) (or the analogous estimate for the S-unit equation) is that in general, the best upper bound (conjecturally) obtainable for the heights of the basis elements is exponentially larger than the desired bound for the exceptional points! The moral of this story, assuming the validity of the various conjectures, is that a randomly chosen elliptic curve is unlikely to have any integral points at all.

§8. Roth's Theorem—An Overview

In this section we give a brief sketch of the principal steps which go into the proof of Roth's theorem (1.4). None of these steps are particularly deep, but the details needed to make them rigorous are quite lengthy. (See [Schm 2] or [La 7, Ch. 7].)

We assume given an $\alpha \in \bar{K}$, a $v \in M_K$, and real numbers $C, \varepsilon > 0$. It is desired to prove that there are only finitely many $x \in K$ satisfying

$$|x - \alpha|_v \leqslant C H_K(x)^{-2-\varepsilon}.$$

Step I: An Auxiliary Polynomial

For any given integers $m, d_1, \ldots, d_m$, one uses elementary estimates and the pigeon-hole principle to construct a polynomial

$$P(X_1, \ldots, X_m) \in R[X_1, \ldots, X_m]$$

of degree d_i in X_i which vanishes to fairly high order (in terms of m and the d_i's) at the point $(\alpha, \ldots, \alpha)$. Further, one shows that P can be chosen with coefficients having fairly small heights, the bound for the heights being given explicitly in terms of α, m, and the d_i's.

Step II: An Upper Bound for P

Suppose now that we are given elements $x_1, \ldots, x_m \in K$ satisfying

$$|x_i - \alpha|_v \leqslant CH_K(x_i)^{-2-\varepsilon} \qquad \text{for } 1 \leqslant i \leqslant m.$$

Then using the Taylor series expansion for $P(X_1, \ldots, X_m)$ around $(\alpha, \ldots, \alpha)$ and the fact that P vanishes to high order at $(\alpha, \ldots, \alpha)$, one shows that $|P(x_1, \ldots, x_m)|_v$ is fairly small.

Step III: A Non-Vanishing Result (Roth's Lemma)

Suppose that the degrees $d_1, \ldots, d_m$ are fairly rapidly decreasing (the rate of decrease depending on m), and suppose that $x_1, \ldots, x_m \in K$ have the property that their heights are fairly rapidly increasing (the rate of increase depending on m and $d_1, \ldots, d_m$). Suppose further that $P(X_1, \ldots, X_m) \in R[X_1, \ldots, X_m]$ has degree d_i in X_i and coefficients whose heights are bounded in terms of d_1 and $h(x_1)$. Then one shows that P does not vanish to too high an order at $(x_1, \ldots, x_m)$.

This is the hardest step in Roth's theorem. In Thue's original theorem, he used a polynomial of the form $P(X, Y) = f(X) + g(X)Y$, and obtained an approximation exponent $\tau(d) = \frac{1}{2}d + \varepsilon$. The improvements of Siegel, Gelfond, and Dyson used a general polynomial in 2 variables. It was clear at that time that the way to obtain $\tau(d) = 2 + \varepsilon$ was to use polynomials in more variables; the only stumbling block was the lack of a non-vanishing result such as the one we have just described.

The proof of Roth's lemma is by induction on m, the number of variables in the polynomial P. If P factors as

$$P(X_1, \ldots, X_m) = F(X_1)G(X_2, \ldots, X_m),$$

then the induction proceeds fairly smoothly. Of course, this is unlikely to happen. What one does is construct differential operators $\mathscr{D}_{ij}$ so that the generalized Wronskian determinant $\det(\mathscr{D}_{ij}P)$ is a non-zero polynomial which does factor in the above fashion. It is then a fairly delicate matter to estimate the degrees and heights of the coefficients of the resulting polynomial, and show that they have not grown too large to allow the inductive hypothesis to be applied.

Step IV: The Final Estimate

Suppose that the inequality

$$|x - \alpha|_v \leqslant CH_K(x)^{-2-\varepsilon}$$

has infinitely many solutions $x \in K$. We derive a contradiction as follows.

First choose a value for m depending on ε, C, and $[K(\alpha):K]$. Second choose $x_1, \ldots, x_m \in K$ in succession satisfying

$$|x_i - \alpha|_v \leqslant CH_K(x_i)^{-2-\varepsilon},$$

such that $H_K(x_1)$ is large (depending on m), and $H_K(x_{i+1}) > H_K(x_i)^\kappa$ for some constant κ depending only on m. Third choose a large integer d_1 (depending on m and the $H_K(x_i)$'s), and then choose $d_2, \ldots, d_m$ in terms of d_1 and the $H_K(x_i)$'s. We are now ready to apply the results detailed above.

Using step I, choose a polynomial $P(X_1, \ldots, X_m)$ of degree d_i in X_i which vanishes to high order at $(\alpha, \ldots, \alpha)$. (The order of vanishing will depend on m and the d_i's.) From step III, P does not vanish to too high an order at $(x_1, \ldots, x_m)$, so we choose a low-order non-vanishing partial derivative

$$z = \frac{\partial^{i_1 + \cdots + i_m}}{\partial X_1^{i_1} \ldots \partial X_m^{i_m}} P(x_1, \ldots, x_m) \neq 0.$$

From step II, $|z|_v$ is fairly small. On the other hand, since $z \neq 0$, one can use the product formula to show that $|z|_v$ cannot be too small. Specifically, one shows that $|z|_v \geqslant H_K(z)^{-1}$ (cf. exer. 9.9). Now using elementary (triangle inequality) estimates, one finds a lower bound for $H_K(z)^{-1}$. Combining this with the upper bound provided by step II, some algebra gives a contradiction. It follows that the inequality

$$|x - \alpha|_v \leqslant CH_K(x)^{-2-\varepsilon}$$

has only finitely many solutions.

Remark 8.1. Examining the above proof sketch, especially the sequence of choices in step IV, it is clear why one does not obtain an effective procedure for finding all $x \in K$ satisfying $|x - \alpha|_v \leqslant CH_K(x)^{-2-\varepsilon}$. What the proof shows is that one cannot find a long sequence of such x_i's with heights growing sufficiently rapidly, where the terms "long sequence" and "sufficiently rapidly" can be made completely explicit in terms of K, α, ε, and C. The problem is that the required growth of the height of each x_i is given in terms of the height of its predecessor. What this boils down to is that if one can find a large number of good approximations to α whose heights are sufficiently large, then one can obtain a bound for all of the other possible good approximations to α in terms of the approximations one has. Unfortunately, the bounds which come out of Roth's theorem are so large, it is highly unlikely that there will be even a single good approximation to α of the requisite height.

Using a slight elaboration of the argument given above, it is even possible to give explicit constants C_1 and C_2, depending on K, α, ε, and C, such that the inequality

$$|x - \alpha|_v \leqslant C H_K(x)^{-2-\varepsilon}$$

has at most C_1 solutions $x \in K$ satisfying $H_K(x) > C_2$. (See [Mig], for example.) Further, it is most unlikely that there are any solutions at all with $H_K(x) > C_2$. But the proof of Roth's theorem does not preclude the existence of these large solutions, and it provides no tools with which to find them if they exist!

EXERCISES

9.1. Let $(\phi(n))_{n=1,2,...}$ be a sequence of positive real numbers. We say that a number $\alpha \in \mathbb{R}$ is ϕ-approximable (over $\mathbb{Q}$) if there are infinitely many $p/q \in \mathbb{Q}$ satisfying

$$|\alpha - p/q| < 1/q\phi(q).$$

(E.g. Roth's theorem (1.4) says that no element of $\overline{\mathbb{Q}}$ is $n^{1+\varepsilon}$-approximable.)
(a) Prove that for any $\varepsilon > 0$,

$$\{\alpha \in \mathbb{R} : \alpha \text{ is } n^{1+\varepsilon}\text{-approximable}\}$$

is a set of measure 0.
(b) More generally, prove that if the series $\Sigma 1/\phi(n)$ converges, then

$$\{\alpha \in \mathbb{R} : \alpha \text{ is } \phi\text{-approximable}\}$$

is a set of measure 0.

9.2. (a) Use Liouville's theorem (1.3) to prove that the number $\Sigma 2^{-n!}$ is transcendental.
(b) More generally, let $(e(n))_{n=1,2,...}$ be a sequence of real numbers with the property that for every $d > 0$, there is a constant $C_d > 0$ such that

$$e(n) \geqslant C_d n^d \qquad \text{for all } n = 1, 2, \dots .$$

Prove that for every integer $b \geqslant 2$, the sequence $\Sigma b^{-e(n)}$ defines a transcendental number.

9.3. For each integer $m \neq 0$, let

$$N(m) = \#\{(x, y) \in \mathbb{Z}^2 : y^2 = x^3 + m\}.$$

($N(m)$ is finite from (3.2).)
(a) Prove that $N(m)$ can be arbitrarily large. [Hint: Choose an m_0 so that $y^2 = x^3 + m_0$ has infinitely many rational solutions, and then clear the denominators of a lot of them.]

(b)   More precisely, prove that there is an absolute constant $c > 0$ such that

$$N(m) > c(\log |m|)^{1/3}$$

for infinitely many $m \in \mathbb{Z}$. [*Hint*: Use height functions to estimate the size of the denominators cleared in (a).]

(c)** Prove that $N(m)$ is unbounded as m ranges over sixth-power-free integers (i.e. integers divisible by no non-trivial sixth power).

9.4. Let $E/\mathbb{Q}$ be an elliptic curve, and suppose that $P \in E(\mathbb{Q})$ is a point of infinite order. For each prime $p \in \mathbb{Z}$ for which E has good reduction, let n_p be the order of the reduced point $\tilde{P}$ in the finite group $\tilde{E}(\mathbb{F}_p)$. Prove that there are only finitely many positive integers which do not occur as an n_p for some prime p. [*Hint*: You will need the strong form of Siegel's theorem. Specifically, see (3.3).]

9.5. (a) Let $f(T) = a_0 T^n + \cdots + a_n \in \mathbb{Z}[T]$ be a polynomial with $a_0 a_n \ne 0$ and distinct roots $\xi_1, \ldots, \xi_n \in \mathbb{C}$. Let $A = \max\{|a_0|, \ldots, |a_n|\}$. Prove that for every $t \in \mathbb{Q}$,

$$|f(t)| \geqslant (2n^2 A)^{-n} \min\{|t - \xi_1|, \ldots, |t - \xi_n|\}.$$

(b) Let $f(T) = a_0 T^n + \cdots + a_n \in K[T]$ be a polynomial with distinct roots $\xi_1, \ldots, \xi_n \in \bar{K}$. Let $S \subset M_K$ be a finite set of places of K, each extended in some fashion to $\bar{K}$. Prove that there is a constant $C_f > 0$, depending only on f and S, so that for every $t \in K$,

$$\prod_{v \in S} \min\{1, |f(t)|_v^{n_v}\} \geqslant C_f \prod_{v \in S} \min_{1 \leqslant i \leqslant n} \{1, |t - \xi_i|_v^{n_v}\}.$$

(c) Find an explicit expression for C_f which involves only n and $H_K([a_0, \ldots, a_n])$.

9.6. (a) Let $F(X, Y) \in \mathbb{Z}[X, Y]$ be a homogeneous polynomial of degree $d \geqslant 3$ with non-zero discriminant. Prove that for every non-zero integer b, *Thue's equation*

$$F(X, Y) = b$$

has only finitely many solutions $(x, y) \in \mathbb{Z}^2$. [*Hint*: Let $f(T) = F(T, 1)$, and write $b = F(x, y) = y^n f(x/y)$. Now use (exer. 9.5a) and (1.4).]

(b) More generally, let $F(X, Y) \in K[X, Y]$ be a homogeneous polynomial of degree $d \geqslant 3$ with non-zero discriminant, and let $S \subset M_K$ be a finite set of places containing M_K^∞. Prove that for every $b \in K^*$, the equation

$$F(X, Y) = b$$

has only finitely many solutions $(x, y) \in R_S \times R_S$.

(c) Let $f(X) \in K[X]$ be a polynomial with at least two distinct roots (in $\bar{K}$), let $S \subset M_K$ be as in (b), and let $n \geqslant 3$ be an integer. Prove that the equation

$$Y^n = f(X)$$

has only finitely many solutions $(x, y) \in R_S \times R_S$. [*Hint*: Mimic the proof of (4.3) until you end up with a number of equations of the form $aW^n + bZ^n = c$, and then use (b).]

9.7. Let E/K be an elliptic curve without complex multiplication. Prove that for every prime ℓ, the representation of $G_{\bar{K}/K}$ on the $\mathbb{Q}_\ell$-vector space $T_\ell(E) \otimes \mathbb{Q}_\ell$ is irreducible.

9.8. (a) Let $\| \cdot \|$ be the usual Euclidean norm on $\mathbb{R}^n$, and let $\{v_1, \ldots, v_n\}$ be a basis for $\mathbb{R}^n$. Prove that there is a constant $c > 0$, depending only on n and $\{v_1, \ldots, v_n\}$, such that

$$\left\| \sum a_i v_i \right\| \geqslant c \max\{|a_i|\}.$$

(b) Let $\Lambda \subset \mathbb{R}^n$ be a lattice. Prove that there exists a basis $\{v_1, \ldots, v_n\}$ for Λ and a constant $c_n > 0$ *depending only on* n so that

$$\left\| \sum a_i v_i \right\| \geqslant c_n \sum \|a_i v_i\|.$$

[*Hint*: Ideally, one would like to choose an orthogonal basis for Λ. This may not be possible, but mimic the Gram–Schmidt process to find a basis which is as orthogonal as possible.]

(c) Let $\| \cdot \|_1$ and $\| \cdot \|_2$ be norms on $\mathbb{R}^n$. (I.e. They satisfy $\|v\| \geqslant 0$, $\|v\| = 0$ if and only if $v = 0$, $\|av\| \leqslant |a| \|v\|$, and $\|v + w\| \leqslant \|v\| + \|w\|$.) Prove that there are constants $c_1, c_2 > 0$ such that

$$c_1 \|v\|_1 \leqslant \|v\|_2 \leqslant c_2 \|v\|_1 \qquad \text{for all } v \in \mathbb{R}.$$

Deduce that an estimate as in (a) holds for any norm on $\mathbb{R}^n$.

(d) Let Q be a positive definite quadratic form on $\mathbb{R}^n$. Prove that there is a constant $c > 0$, depending on n and Q, such that for any integral lattice point $(a_1, \ldots, a_n) \in \mathbb{Z}^n \subset \mathbb{R}^n$,

$$Q(a_1, \ldots, a_n) \geqslant c \max\{|a_1|, \ldots, |a_n|\}^2.$$

(e) Let E/K be an elliptic curve and $P_1, \ldots, P_r$ a basis for the free part of $E(K)$. Prove that there is a constant $c > 0$, depending on E and $P_1, \ldots, P_r$, such that for all integers $m_1, \ldots, m_r$,

$$\hat{h}(m_1 P_1 + \cdots + m_r P_r) \geqslant c \max\{|m_1|, \ldots, |m_r|\}^2.$$

9.9. Let $z \in K$, $z \neq 0$.
(a) Prove that for any $v \in M_K$,

$$|z|_v \geqslant H_K(z)^{-1}.$$

(b) More generally, prove that for any (not necessarily finite) set of absolute values $S \subset M_K$,

$$\prod_{v \in S} \min\{1, |z|_v^{n_v}\} \geqslant H_K(z)^{-1}.$$

(This lemma, trivial as it appears, lies at the heart of all known proofs in Diophantine approximation and transcendence theory. In its simplest guise, namely for $K = \mathbb{Q}$, it asserts nothing more than the fact that there are no positive integers less than 1!)

9.10. Prove that there is an (absolute) constant $C > 0$ such that the inequality

$$0 < |y^2 - x^3| < C\sqrt{|x|}$$

has infinitely many solutions $(x, y) \in \mathbb{Z}^2$. [*Hint*: Verify the identity

$$(t^2 - 5)^2((t + 9)^2 + 4) - (t^2 + 6t - 11)^3 = -1728(t - 2).$$

Then take solutions to $u^2 - 2v^2 = -1$, and set $t = 2u - 9$. This leads to a value $C = 432\sqrt{2} + \varepsilon$ for any $\varepsilon > 0$.]

9.11. (a) Let $d \equiv 2 \pmod 4$ and $D = d^3 - 1$. Prove that the equation

$$y^2 = x^3 + D$$

has no solutions $x, y \in \mathbb{Z}$.

(b) For each of the primes p in the set $\{11, 19, 43, 67, 163\}$, find all solutions $x, y \in \mathbb{Z}$ to the equation

$$y^2 = x^3 - p.$$

[*Hint:* Work in the ring $R = \mathbb{Z}[\frac{1}{2}(1 + \sqrt{-p})]$. Note that R is a principal ideal domain, and 2 does not split in R.]

9.12. Let $E/\mathbb{Q}$ be an elliptic curve given by a Weierstrass equation

$$E : y^2 + a_1 xy + a_3 y = x^3 + a_2 x^2 + a_4 x + a_6$$

with $a_1, \ldots, a_6 \in \mathbb{Z}$. Let $P \in E(\mathbb{Q})$ be a point of infinite order; and suppose that for some integer $m \geqslant 1$, $x([m]P) \in \mathbb{Z}$. Prove that $x(P) \in \mathbb{Z}$. (This result is often useful in searching for integral points on elliptic curves of rank 1. See the next exercise for an example.)

9.13. Let $E/\mathbb{Q}$ be the elliptic curve given by the equation

$$E : y^2 + y = x^3 - x.$$

Assume as given that $E(\mathbb{Q})$ has rank 1. (See exer. 10.9 for a proof of this fact.)

(a) Prove that $E_{\text{tors}}(\mathbb{Q}) = \{O\}$, and hence that $E(\mathbb{Q}) \cong \mathbb{Z}$.

(b) Prove that $(0, 0)$ is a generator for $E(\mathbb{Q})$. [*Hint:* Make a sketch of $E(\mathbb{R})$, and show that $(0, 0)$ is not on the identity component. Use (exer. 9.12) to conclude that a generator for $E(\mathbb{Q})$ must be an *integer* point on the non-identity component, and find all such points.]

(c) Find all of the integer points on E. [*Hint:* Let $P = (0, 0)$. Suppose $[m]P$ is integral. Write $m = 2^a n$ with n odd, and use (exer. 9.12) to show that $[n]P$ is integral. Use an argument as in (b) to find all possible values for n, and then do some computations to find the possible a's.]

(d) Solve the following classical number theory problem: Find all positive integers which are simultaneously the product of two consecutive positive integers and the product of three consecutive positive integers.

9.14. Let C/K be a curve, and let $f, g \in K(C)$ be non-constant functions.

(a)* Prove that

$$\mathop{\text{Limit}}_{\substack{P \in C(\bar{K}) \\ h_f(P) \to \infty}} \frac{h_f(P)}{h_g(P)} = \frac{\deg f}{\deg g}.$$

(b) Prove that for every $\varepsilon > 0$ there exists a constant $c = c(f, g, \varepsilon)$ such that

$$|(\deg g)h_f(P) - (\deg f)h_g(P)| < \varepsilon h_f(P) + c \qquad \text{for all } P \in C(\bar{K}).$$

(c) Suppose that C is an elliptic curve. Prove that there is a constant $c = c(f, m, \varepsilon)$ such that

$$|h_f([m]P) - m^2 h_f(P)| < \varepsilon h_f(P) + c \qquad \text{for all } P \in C(\bar{K}).$$

(Note that f need not be even. Compare with (VIII. 6.4b).)

(d) Prove that (3.1) is true for any non-constant function $f \in K(E)$. Use this to prove (3.2.2) directly, without reducing first to (3.2.1).

CHAPTER X

Computing the Mordell–Weil Group

A better title for this chapter might have been "Computing the *Weak Mordell–Weil Group*", since we will be concerned solely with the problem of computing generators for the group $E(K)/mE(K)$. However, given generators for $E(K)/mE(K)$, a finite amount of computation will always yield generators for $E(K)$. (See (VIII.3.2) and (exer. 8.18).) Unfortunately, there is no comparable algorithm currently known which is guaranteed to give generators for $E(K)/mE(K)$ in a finite amount of time!

We start in section 1 by taking the proof of the weak Mordell–Weil theorem given in (VIII §1) and making it quite explicit. In this way the computation of $E(K)/mE(K)$ (in a special case) is reduced to the problem of determining whether each of a certain finite set of auxiliary curves, called *homogeneous spaces*, has a single rational point. Then the question of whether a given homogeneous space has a rational point may often be answered either affirmatively, by finding such a point; or negatively, by showing, for example, that it has no points in some completion K_v of K.

The next two sections develop the general theory of homogeneous spaces (for elliptic curves). Then in section 4 we apply this theory to the problem of computing $E(K)/mE(K)$; or, more generally, $E'(K)/\phi(E(K))$ for any isogeny $\phi : E \to E'$. Again this computation is reduced to the problem of the existence of a single rational point on certain homogeneous spaces. The only impediment to solving this latter problem occurs if some homogeneous space has a K_v-rational point for every completion K_v of K, yet none-the-less has no K-rational points. Unfortunately this precise situation, the failure of the so-called Hasse Principle, can certainly occur. The extent of its failure is quantified by the elements of a certain group, called the *Shafarevich–Tate group*. The question of an effective algorithm for the computation of $E(K)/mE(K)$ is thus finally reduced to the problem of giving a bound for divisibility in the

Shafarevich–Tate group (or even better, proving the conjecture that it is actually a finite group).

In the last section we illustrate our general theory by studying in some detail the family of elliptic curves given by the equations

$$E_D : Y^2 = X^3 + DX \qquad D \in \mathbb{Q}.$$

In particular, we find the torsion subgroup and an upper bound for the rank of $E_D(\mathbb{Q})$, give a large class of examples for which $E_D(\mathbb{Q})$ has rank 0, and show that in certain cases $E_D(\mathbb{Q})$ has an associated homogeneous space which violates the Hasse principle. (I.e. The homogeneous space has points defined over $\mathbb{R}$ and $\mathbb{Q}_p$ for every prime p, but has no $\mathbb{Q}$-rational points.)

Unless explicitly stated to the contrary, the notation for this chapter will be the same as that of chapter VIII. In particular, K will be a number field and M_K a complete set of inequivalent absolute values for K. However, as indicated in the text, this requirement is dropped in sections 2, 3, and 5 of this chapter, where K is allowed to be an arbitrary (perfect) field.

§1. An Example

For this section we let E/K be an elliptic curve, $m \geqslant 2$ an integer, and we assume that $E[m] \subset E(K)$. Recall (VIII §1) that under this assumption there is a pairing

$$\kappa : E(K) \times G_{\bar{K}/K} \to E[m]$$

defined by

$$\kappa(P, \sigma) = Q^\sigma - Q,$$

where $Q \in E$ is chosen so that $[m]Q = P$. Since the kernel of κ on the left is $mE(K)$ (VIII.1.2), we may also think of κ as giving a homomorphism

$$\delta_E : E(K)/mE(K) \to \operatorname{Hom}(G_{\bar{K}/K}, E[m])$$

$$\delta_E(P)(\sigma) = \kappa(P, \sigma).$$

(This is the connecting homomorphism for the long exact sequence in group cohomology; see (VIII §2).)

Next we note that $E[m] \subset E(K)$ implies that $\mu_m \subset K^*$ (III.8.1.1). This follows from the basic properties of the Weil pairing (III §8)

$$e_m : E[m] \times E[m] \to \mu_m,$$

which we will use extensively below.

Finally, since $\mu_m \subset K^*$, Hilbert's theorem 90 (B.2.5c) says that every homomorphism $G_{\bar{K}/K} \to \mu_m$ has the form $\sigma \to \beta^\sigma/\beta$ for some $\beta \in \bar{K}^*$ with $\beta^m \in K^*$. In other words, we have an isomorphism (cf. VIII §2)

$$\delta_K : K^*/K^{*m} \to \operatorname{Hom}(G_{\bar{K}/K}, \mu_m)$$

defined by

$$\delta_K(b)(\sigma) = \beta^\sigma/\beta,$$

where $\beta \in \bar{K}^*$ is chosen so that $\beta^m = b$. (Notice the close resemblence in the definitions of δ_E and δ_K. This is no coincidence. δ_E is the connecting homomorphism for the Kummer sequence associated to the group variety E/K, and δ_K is the connecting homomorphism for the Kummer sequence associated to the group variety $\mathbb{G}_m/K$.)

Using the above maps, we can now make the argument in the proof of the weak Mordell–Weil theorem much more explicit, and in this way derive formulas which will allow us to compute the Mordell–Weil group in certain cases. We start with a theoretical description of this method.

Theorem 1.1. (a) *With notation as above, there is a bilinear pairing*

$$b : E(K)/mE(K) \times E[m] \to K^*/K^{*m}$$

such that

$$e_m(\delta_E(P), T) = \delta_K(b(P, T)).$$

(b) *The pairing in* (a) *is non-degenerate on the left.*
(c) *Let $S \subset M_K$ be the set of infinite places, together with the finite primes at which E has bad reduction and the primes dividing m. Then the image of the pairing in* (a) *lies in the subgroup of K^*/K^{*m} given by*

$$K(S, m) = \{b \in K^*/K^{*m} : \text{ord}_v(b) \equiv 0 \pmod m \text{ for all } v \notin S\}.$$

(d) *The pairing in* (a) *may be computed as follows: For each $T \in E[m]$, choose functions $f_T, g_T \in K(E)$ satisfying the conditions*

$$\text{div}(f_T) = m(T) - m(O), \qquad f_T \circ [m] = g_T^m.$$

(See the definition of the Weil pairing in (III §8).*) Then provided $P \neq T$,*

$$b(P, T) \equiv f_T(P) \pmod{K^{*m}}.$$

[*If $P = T$, one can use linearity. For example, if $[2]T \neq O$, then $b(T, T) = f_T(-T)^{-1}$. More generally, choose any other point $P \in E(K)$ with $P \neq T$, and set $b(T, T) = f_T(T + P)f_T(P)^{-1}$.*]

Remark 1.2. Why do we say that (1.1) provides formulas with which to try to compute the Mordell–Weil group? First, the group $K(S, m)$ in (c) is finite (see the proof of (VIII.1.6)); and in fact it is quite easy to compute explicitly. Second, the functions f_T in (d) are also fairly easy to compute from the equation of the curve. Now, the fact that the pairing in (a) is non-degenerate on the left means that in order to compute $E(K)/mE(K)$, it is "only" necessary to do the following. Fix generators T_1 and T_2 for $E[m]$. Then for each of the finitely many pairs

$$(b_1, b_2) \in K(S, m) \times K(S, m),$$

see if it is possible to solve the equations

$$b_1 z_1^m = f_{T_1}(P) \qquad b_2 z_2^m = f_{T_2}(P)$$

with points $(P, z_1, z_2) \in E(K) \times K^* \times K^*$. To be even more explicit, we can express the function f_T in terms of Weierstrass coordinates x and y; and then we are looking for a solution $(x, y, z_1, z_2) \in K \times K \times K^* \times K^*$ satisfying the simultaneous equations

$$y^2 + a_1 xy + a_3 y = x^3 + a_2 x^2 + a_4 x + a_6$$

$$b_1 z_1^m = f_{T_1}(x, y) \qquad b_2 z_2^m = f_{T_2}(x, y).$$

These equations give a new curve, called a *homogeneous space* for E/K. (See §3 for more details.) What we have done is reduce the problem of calculating $E(K)/mE(K)$ to the problem of the existence or non-existence of a single rational point on each of an explicitly given finite set of curves. Now frequently many of these curves can be immediately eliminated from consideration, because they have no points over some completion K_v of K (which is an easy matter to check). On the other hand, a short search (by hand or computer) will often uncover rational points on some of the others. If in this way one can deal with all of the homogeneous spaces in question, then the determination of $E(K)/mE(K)$ is complete. The problem that arises is that occasionally a homogeneous space will have points defined over every completion K_v, but never-the-less have no K-rational points. It is this situation, the failure of the Hasse principle, which makes the Mordell–Weil theorem ineffective.

Remark 1.3. Notice that the condition $\operatorname{div}(f_T) = m(T) - m(O)$ in (1.1d) is only enough to specify f_T up to multiplication by an arbitrary element of K^*. But the equality $f_T \circ [m] = g_T^m$ with $g_T \in K(E)$ means that in fact f_T is well-determined up to multiplication by an element of K^{*m}. Thus the value $f_T(P)$ in (1.1d) does give a well-defined element of K^*/K^{*m}.

We now proceed to the proof of (1.1), after which we will study the case $m = 2$ in more detail, and use it to compute $E(K)/2E(K)$ for an example.

PROOF OF 1.1. (a) Hilbert's theorem 90 (B.2.5c) shows that the pairing is well-defined. The bilinearity follows from the bilinearity of the Kummer pairing (VIII.1.2b) and the bilinearity of the Weil e_m-pairing (III.8.1a).
(b) To prove non-degeneracy on the left, we suppose that $b(P, T) = 1$ for all $T \in E[m]$. This means that for all $T \in E[m]$ and all $\sigma \in G_{\bar{K}/K}$,

$$e_m(\kappa(P, \sigma), T) = 1.$$

Now the non-degeneracy of the Weil pairing (III.8.1c) implies that $\kappa(P, \sigma) = O$ for all σ; so from (VIII.1.2c), $P \in mE(K)$.
(c) Let $\beta = b(P, T)^{1/m}$. Tracing through the definitions, we see that the field $K(\beta)$ is contained in the field $L = K([m]^{-1}E(K))$ described in (VIII.1.2d).

From (VIII.1.5b), the extension L/K is unramified outside S. But it is easy to see that if $v \in M_K$ is a finite place with $v(m) = 0$, then the extension $K(\beta)/K$ is unramified at v if and only if

$$\mathrm{ord}_v(\beta^m) \equiv 0 \pmod{m}.$$

(Here $\mathrm{ord}_v : K^* \twoheadrightarrow \mathbb{Z}$ is the normalized valuation associated to v.) This says precisely that $b(P, T) \in K(S, m)$.

(d) Choose $Q \in E$ so that $P = [m]Q$, and $\beta \in \bar{K}^*$ so that $b(P, T) = \beta^m$. By definition, we have (for all $\sigma \in G_{\bar{K}/K}$),

$$e_m(\delta_E(P)(\sigma), T) = \delta_K(b(P, T))(\sigma),$$

$$e_m(Q^\sigma - Q, T) = \beta^\sigma/\beta,$$

$$g_T(X + Q^\sigma - Q)/g_T(X) = \beta^\sigma/\beta,$$

$$g_T(Q)^\sigma/g_T(Q) = \beta^\sigma/\beta \qquad \text{putting } X = Q.$$

Since δ_K is an isomorphism, it follows that $g_T(Q)^m \equiv \beta^m \pmod{K^{*m}}$. (Note that $g_T(Q)^m = f_T(P)$ is in K^*.) Therefore

$$f_T(P) = f_T \circ [m](Q) = g_T(Q)^m \equiv \beta^m = b(P, T) \pmod{K^{*m}}. \qquad \square$$

We now consider the special case $m = 2$, which is by far the easiest to work with. Under our assumption that $E[2] \subset E(K)$, we may take a Weierstrass equation in the form

$$y^2 = (x - e_1)(x - e_2)(x - e_3) \qquad \text{with } e_1, e_2, e_3 \in K.$$

Thus $T_1 = (e_1, 0)$, $T_2 = (e_2, 0)$, $T_3 = (e_3, 0)$ are the three non-trivial 2-torsion points. Letting $T = (e, 0)$ represent any one of these points, we claim that the function f_T specified in (1.1d) is $f_T(x, y) = x - e$. This function certainly has the correct divisor,

$$\mathrm{div}(x - e) = 2(T) - 2(O).$$

On the other hand, as one can easily check,

$$x \circ [2] - e = (x^2 - 2ex - 2e^2 + 2(e_1 + e_2 + e_3)e - (e_1 e_2 + e_1 e_3 + e_2 e_3))^2/(2y)^2,$$

so $x - e$ does have both properties needed to be f_T.

Now suppose that we have chosen a pair $(b_1, b_2) \in K(S, 2) \times K(S, 2)$, and wish to determine whether there is a point $P \in E(K)/2E(K)$ satisfying

$$b(P, T_1) = b_1 \quad \text{and} \quad b(P, T_2) = b_2.$$

There will be such a point if and only if there is a solution $(x, y, z_1, z_2) \in K \times K \times K^* \times K^*$ to the system of equations

$$y^2 = (x - e_1)(x - e_2)(x - e_3), \qquad b_1 z_1^2 = x - e_1, \qquad b_2 z_2^2 = x - e_2.$$

We now substitute the latter two equations into the first, and define a new variable z_3 by $y = b_1 b_2 z_1 z_2 z_3$, which is permissible since b_1, b_2, z_1 and z_2

take non-zero values. This yields the three equations

$$b_1 b_2 z_3^2 = x - e_3, \qquad b_1 z_1^2 = x - e_1, \qquad b_2 z_2^2 = x - e_2;$$

and finally eliminating x gives the pair of equations

$$b_1 z_1^2 - b_2 z_2^2 = e_2 - e_1, \qquad b_1 z_1^2 - b_1 b_2 z_3^2 = e_3 - e_1.$$

We now have a finite set of such equations, one for each pair (b_1, b_2), and may use whatever techniques are at our disposal (e.g. v-adic, computer search, etc.) to determine whether they each do or do not have a solution. Notice that if we do find a solution, then the corresponding point in $E(K)/2E(K)$ is immediately recoverable from the equalities

$$x = b_1 z_1^2 + e_1 \qquad y = b_1 b_2 z_1 z_2 z_3.$$

Finally, we must deal with the fact that we can not use the definition $b(P, T) = f_T(P)$ if it should happen that $P = T$. In other words, there are two pairs (b_1, b_2) which do not arise from the above procedure, namely the pairs $(b(T_1, T_1), b(T_1, T_2))$ and $(b(T_2, T_1), b(T_2, T_2))$. They may be computed by linearity as

$$b(T_1, T_1) = b(T_1, T_1 + T_2) b(T_1, T_2)^{-1}$$
$$= b(T_1, T_3) b(T_1, T_2)^{-1}$$
$$= (e_1 - e_3)/(e_1 - e_2);$$

and similarly

$$b(T_2, T_2) = (e_2 - e_3)/(e_2 - e_1).$$

We summarize this entire procedure in the following proposition.

Proposition 1.4 (Complete 2-Descent). *Let E/K be an elliptic curve given by a Weierstrass equation*

$$y^2 = (x - e_1)(x - e_2)(x - e_3) \qquad \text{with } e_1, e_2, e_3 \in K.$$

Let $S \subset M_K$ be a set of places of K including all archimedian places, all places dividing 2, and all places at which E has bad reduction. Further let

$$K(S, 2) = \{b \in K^*/K^{*2} : \operatorname{ord}_v(b) \equiv 0 \pmod{2} \text{ for all } v \notin S\}.$$

There is an injective homomorphism

$$E(K)/2E(K) \to K(S, 2) \times K(S, 2)$$

defined by

$$P = (x, y) \to \begin{cases} (x - e_1, x - e_2) & \text{if } x \neq e_1, e_2 \\ ((e_1 - e_3)/(e_1 - e_2), e_1 - e_2) & \text{if } x = e_1 \\ (e_2 - e_1, (e_2 - e_3)/(e_2 - e_1)) & \text{if } x = e_2 \\ (1, 1) & \text{if } x = \infty \text{ (i.e. if } P = 0). \end{cases}$$

Let $(b_1, b_2) \in K(S, 2) \times K(S, 2)$ be a pair which is not the image of one of the three points O, $(e_1, 0)$, $(e_2, 0)$. Then (b_1, b_2) is the image of a point $P = (x, y) \in E(K)/2E(K)$ if and only if the equations

$$b_1 z_1^2 - b_2 z_2^2 = e_2 - e_1$$

$$b_1 z_1^2 - b_1 b_2 z_3^2 = e_3 - e_1$$

have a solution $(z_1, z_2, z_3) \in K^* \times K^* \times K$; if such a solution exists, then one can take

$$P = (x, y) = (b_1 z_1^2 + e_1, b_1 b_2 z_1 z_2 z_3).$$

PROOF. As explained above, this is a special case of (1.1).   □

Example 1.5. We now use (1.4) to compute $E(\mathbb{Q})/2E(\mathbb{Q})$ for the elliptic curve

$$E : y^2 = x^3 - 12x^2 + 20x = x(x - 2)(x - 10).$$

This equation has discriminant

$$\Delta = 409600 = 2^{14} 5^2,$$

and so has good reduction except at 2 and 5. Reducing the equation modulo 3, one easily checks that $\# \tilde{E}(\mathbb{F}_3) = 4$. Since $E[2] \subset E_{\text{tors}}(\mathbb{Q})$, and $E_{\text{tors}}(\mathbb{Q})$ injects into $\tilde{E}(\mathbb{F}_3)$ (VII.3.5), we see that

$$E_{\text{tors}}(\mathbb{Q}) = E[2].$$

Now let $S = \{2, 5, \infty\} \subset M_{\mathbb{Q}}$; then a complete set of representatives for

$$\mathbb{Q}(S, 2) = \{b \in \mathbb{Q}^*/\mathbb{Q}^{*2} : \text{ord}_p(b) \equiv 0 \ (\text{mod } 2) \text{ for all } p \notin S\}$$

is given by the set

$$\{\pm 1, \pm 2, \pm 5, \pm 10\},$$

which we will identify with $\mathbb{Q}(S, 2)$. Next consider the map

$$E(\mathbb{Q})/2E(\mathbb{Q}) \to \mathbb{Q}(S, 2) \times \mathbb{Q}(S, 2)$$

given in (1.4), say with

$$e_1 = 0, \qquad e_2 = 2, \qquad \text{and } e_3 = 10.$$

There are 64 pairs $(b_1, b_2) \in \mathbb{Q}(S, 2) \times \mathbb{Q}(S, 2)$; and for each pair we must check to see if it comes from an element of $E(\mathbb{Q})/2E(\mathbb{Q})$. For example, using (1.4) we can compute the image of $E[2]$ in $\mathbb{Q}(S, 2) \times \mathbb{Q}(S, 2)$:

$$O \to (1, 1) \qquad (0, 0) \to (5, -2) \qquad (2, 0) \to (2, -1) \qquad (10, 0) \to (10, 2).$$

It remains to determine, for each other pair (b_1, b_2), whether the equations

$$b_1 z_1^2 - b_2 z_2^2 = 2 \qquad b_1 z_1^2 - b_1 b_2 z_3^2 = 10 \qquad (*)$$

have a solution $z_1, z_2, z_3 \in \mathbb{Q}$. (For example, if $b_1 < 0$ and $b_2 > 0$, then $(*)$

Table 10.1

b_2 \ b_1	1	2	5	10	-1	-2	-5	-10
1	0	$(18, -48)^{⑤}$	$\mathbb{Q}_5{}^{⑨}$		$\mathbb{R}^{①}$			
2	$\mathbb{Q}_5{}^{⑧}$	$\mathbb{Q}_5{}^{⑨}$	$(20, 60)^{⑤}$	$(10, 0)^{③}$				
5 / 10	$\mathbb{Q}_5{}^{⑥}$		$\mathbb{Q}_5{}^{⑦}$					
-1	$(1, -3)^{④}$	$(2, 0)^{③}$	$\mathbb{Q}_5{}^{⑨}$		$\mathbb{R}^{②}$			
-2		$\mathbb{Q}_5{}^{⑨}$	$(0, 0)^{③}$	$(\tfrac{10}{9}, -\tfrac{80}{27})^{⑤}$				
-5 / -10	$\mathbb{Q}_5{}^{⑥}$		$\mathbb{Q}_5{}^{⑦}$					

clearly has no rational solutions, since the first equation will not even have a solution in $\mathbb{R}$.)

Proceeding systematically, we list our results in table 10.1. The entry for each pair (b_1, b_2) consists of either a point of $E(\mathbb{Q})$ mapping to (b_1, b_2), or else a (local) field over which the equations (∗) have no solution. (Note that if (z_1, z_2, z_3) is a solution to (∗), then the corresponding point of $E(\mathbb{Q})$ is $(b_1 z_1^2 + e_1, b_1 b_2 z_1 z_2 z_3)$.) The circled numbers in the table refer to the notes which explain each entry. Finally, we note that since the map $E(\mathbb{Q})/2E(\mathbb{Q}) \to \mathbb{Q}(S, 2) \times \mathbb{Q}(S, 2)$ is a *homomorphism*, it is not necessary to check every pair (b_1, b_2). For example, if both (b_1, b_2) and (b_1', b_2') come from $E(\mathbb{Q})$, then so does $(b_1 b_1', b_2 b_2')$. Similarly, if (b_1, b_2) does and (b_1', b_2') does not, then $(b_1 b_1', b_2 b_2')$ does not. This observation can substantially reduce the amount of computation necessary.

(1) If $b_1 < 0$ and $b_2 > 0$, then $b_1 z_1^2 - b_2 z_2^2 = 2$ has no solutions in $\mathbb{R}$.

(2) If $b_1 < 0$ and $b_2 < 0$, then $b_1 z_1^2 - b_1 b_2 z_3^2 = 10$ has no solutions in $\mathbb{R}$.

(3) The 2-torsion points $\{O, (0, 0), (2, 0), (10, 0)\}$ map respectively to $\{(1, 1), (5, -2), (2, -1), (10, 2)\}$.

(4) $(b_1, b_2) = (1, -1)$: By inspection, the equations
$$z_1^2 + z_2^2 = 2 \qquad z_1^2 + z_3^2 = 10$$
have the solution $(1, 1, 3)$. This gives the point $(1, -3) \in E(\mathbb{Q})$.

(5) Adding $(1, -3) \in E(\mathbb{Q})$ to the non-trivial two-torsion points corresponds to multiplying their (b_1, b_2)'s. This gives the pairs $(5, 2)$, $(2, 1)$, and $(10, -2)$ in $\mathbb{Q}(S, 2) \times \mathbb{Q}(S, 2)$, which correspond to $(20, 60)$, $(18, -48)$, and $(10/9, -80/27)$ in $E(\mathbb{Q})$.

(6) $b_1 \not\equiv 0 \pmod 5$ and $b_2 \equiv 0 \pmod 5$: The first equation in (∗) implies that z_1 and z_2 must be 5-adically integral. Then the second equation shows that $z_1 \equiv 0 \pmod 5$, and so from the first equation we obtain $0 \equiv 2 \pmod 5$. Therefore (∗) has no solutions in $\mathbb{Q}_5$.

(7) The eight pairs in (6) are $\mathbb{Q}_5$ non-trivial. (I.e. There are no $\mathbb{Q}_5$ solutions to (∗).) If we multiply them by the $\mathbb{Q}$-trivial pair $(5, 2)$, we obtain eight more $\mathbb{Q}_5$ non-trivial pairs.

(8) $(b_1, b_2) = (1, 2)$: The two equations in $(*)$ are

$$z_1^2 - 2z_2^2 = 2 \quad \text{and} \quad z_1^2 - 2z_3^2 = 10.$$

Since 2 is a quadratic non-residue modulo 5, the second equation implies that $z_1 \equiv z_3 \equiv 0 \pmod 5$. But then the second equation gives $0 \equiv 10 \pmod{25}$. Therefore there are no solutions in $\mathbb{Q}_5$.

(9) Taking the $\mathbb{Q}_5$-non-trivial pair $(1, 2)$ from (8) and multiplying by the seven $\mathbb{Q}$-trivial pairs already in the table gives seven new $\mathbb{Q}_5$-non-trivial pairs which fill the remaining entries.

Conclusion. $E(\mathbb{Q}) \cong \mathbb{Z} \times \mathbb{Z}/2\mathbb{Z} \times \mathbb{Z}/2\mathbb{Z}$.

§2. Twisting—General Theory

For this section (and the next) we drop our requirement that K be a number field, so K will be an arbitrary (perfect) field. As we saw in section 1 while trying to compute the Mordell–Weil group of an elliptic curve E, we were led to the problem of the existence or non-existence of a single rational point on various other curves. These other curves are certain *twists* of E, called homogeneous spaces. In this section we will study the general question of twisting which, since it is no more difficult, we will develop for curves of arbitrary genus. Then in the next section we will look at the homogeneous spaces associated to an elliptic curve.

Definition. Let C/K be a smooth curve (projective, as always.) The *isomorphism group of C*, denoted $\mathrm{Isom}(C)$, is the group of isomorphisms from C to itself (defined over $\bar{K}$). As usual, $\mathrm{Isom}_K(C)$ is the subgroup of $\mathrm{Isom}(C)$ consisting of isomorphisms defined over K. (To ease notation, we will write the composition of maps multiplicatively; thus $\alpha\beta$ instead of $\alpha \circ \beta$.)

Remark 2.1. The group we are denoting $\mathrm{Isom}(C)$ is usually called the *automorphism group of C*, and denoted $\mathrm{Aut}(C)$. However, if E is an elliptic curve, then we have defined $\mathrm{Aut}(E)$ to be the group of isomorphisms from E to E taking O to O. Thus $\mathrm{Aut}(E) \neq \mathrm{Isom}(E)$ since, for example, $\mathrm{Isom}(E)$ contains the translations $\tau_P : E \to E$. We will describe $\mathrm{Isom}(E)$ more fully in section 5.

Definition. A *twist of C/K* is a smooth curve C'/K which is isomorphic to C over $\bar{K}$. We generally identify two twists if they are isomorphic over K. The set of twists of C/K, modulo K-isomorphism, is denoted $Twist(C/K)$.

Now let C'/K be a twist of C/K. This means that there is an isomorphism $\phi : C' \to C$ defined over $\bar{K}$. To measure the failure of ϕ to be defined over K, we might consider the map

$$\xi : G_{\bar{K}/K} \to \mathrm{Isom}(C) \qquad \xi_\sigma = \phi^\sigma \phi^{-1}.$$

It turns out that ξ is a 1-cocycle; and the cohomology class of ξ is uniquely determined by the K-isomorphism class of C'. Further, every cohomology class comes from some twist of C/K. In this way $\text{Twist}(C/K)$ may be identified with a certain cohomology set. We now prove these statements.

Theorem 2.2. *Let C/K be a smooth curve. For each twist C'/K of C/K, choose an isomorphism $\phi : C' \to C$ and define a map $\xi_\sigma = \phi^\sigma \phi^{-1} \in \text{Isom}(C)$ as above.*
(a) *ξ is a 1-cocycle. (I.e. For all $\sigma, \tau \in G_{\bar{K}/K}$,*

$$\xi_{\sigma\tau} = (\xi_\sigma)^\tau \xi_\tau.)$$

We denote the corresponding cohomology class in $H^1(G_{\bar{K}/K}, \text{Isom}(C))$ by $\{\xi\}$.
(b) *The cohomology class $\{\xi\}$ is determined by the K-isomorphism class of C', independent of the choice of ϕ. We thus obtain a natural map*

$$\text{Twist}(C/K) \to H^1(G_{\bar{K}/K}, \text{Isom}(C)).$$

(c) *The map in (b) is a bijection. In other words, the twists of C/K (up to K-isomorphism) are in one-to-one correspondence with the elements of the cohomology set $H^1(G_{\bar{K}/K}, \text{Isom}(C))$.*

Remark 2.3. We emphasize that the group $\text{Isom}(C)$ is often non-abelian (this is always the case for elliptic curves). Hence $H^1(G_{\bar{K}/K}, \text{Isom}(C))$ is in general only a *pointed set*, not a group. (See B §3.) However, if $\text{Isom}(C)$ has a $G_{\bar{K}/K}$-invariant abelian subgroup A, then $H^1(G_{\bar{K}/K}, A)$ is a group, and its image in $H^1(G_{\bar{K}/K}, \text{Isom}(C))$ will give a natural group structure to some subset of $\text{Twist}(C)$. In the next section, we will apply this observation when C is an elliptic curve, taking for A the group of translations.

PROOF. (a) $\xi_{\sigma\tau} = \phi^{\sigma\tau}\phi^{-1} = (\phi^\sigma\phi^{-1})^\tau(\phi^\tau\phi^{-1}) = (\xi_\sigma)^\tau\xi_\tau$.
(b) Let C''/K be another twist of C/K which is K-isomorphic to C'. Choose a $\bar{K}$-isomorphism $\psi : C'' \to C$. We must show that the cocycles $\phi^\sigma\phi^{-1}$ and $\psi^\sigma\psi^{-1}$ are cohomologous. By assumption, there is a K-isomorphism $\theta : C'' \to C'$. Consider the element $\alpha = \phi\theta\psi^{-1} \in \text{Isom}(C)$. We compute

$$\alpha^\sigma(\psi^\sigma\psi^{-1}) = (\phi\theta\psi^{-1})^\sigma(\psi^\sigma\psi^{-1}) = \phi^\sigma\theta^\sigma\psi^{-1}$$

$$= \phi^\sigma\theta\psi^{-1} = (\phi^\sigma\phi^{-1})(\phi\theta\psi^{-1}) = (\phi^\sigma\phi^{-1})\alpha.$$

This proves that $\phi^\sigma\phi^{-1}$ and $\psi^\sigma\psi^{-1}$ are cohomologous.
(c) Suppose that C'/K and C''/K are twists of C/K which give the same cohomology class in $H^1(G_{\bar{K}/K}, \text{Isom}(C))$. This means that if we choose $\bar{K}$-isomorphisms $\phi : C' \to C$ and $\psi : C'' \to C$, then there is a map $\alpha \in \text{Isom}(C)$ such that

$$\alpha^\sigma(\psi^\sigma\psi^{-1}) = (\phi^\sigma\phi^{-1})\alpha \qquad \text{for all } \sigma \in G_{\bar{K}/K}.$$

(I.e. The cocycles for ϕ and ψ are cohomologous.) We now consider the map $\theta : C'' \to C'$ defined by $\theta = \phi^{-1}\alpha\psi$. It is a $\bar{K}$-isomorphism, and we wish to show that it is actually defined over K. For any $\sigma \in G_{\bar{K}/K}$, we compute

$$\theta^\sigma = (\phi^\sigma)^{-1}(\alpha^\sigma \psi^\sigma) = (\phi^\sigma)^{-1}(\phi^\sigma \phi^{-1} \alpha \psi) = \phi^{-1} \alpha \psi = \theta.$$

Therefore C'' and C' are K-isomorphic, and so give the same element of Twist(C/K). This proves that the map Twist$(C/K) \to H^1(G_{\bar{K}/K}, \text{Isom}(C))$ is injective.

To prove surjectivity, we start with a 1-cocycle $\xi : G_{\bar{K}/K} \to \text{Isom}(C)$, and construct a curve C'/K and an isomorphism $\phi : C' \to C$ such that $\xi_\sigma = \phi^\sigma \phi^{-1}$. To do this we consider a field, denoted $\bar{K}(C)_\xi$, which is isomorphic (as a field over $\bar{K}$) to $\bar{K}(C)$, say by an isomorphism $Z : \bar{K}(C) \to \bar{K}(C)_\xi$. The difference between $\bar{K}(C)$ and $\bar{K}(C)_\xi$ lies in the action of Galois; on $\bar{K}(C)_\xi$ it is *twisted by* ξ. In other words, for all $f \in \bar{K}(C)$ and $\sigma \in G_{\bar{K}/K}$,

$$Z(f)^\sigma = Z(f^\sigma \xi_\sigma).$$

(Here we are thinking of f as giving a map $f : C \to \mathbb{P}^1$ (cf. II.2.2), and $f^\sigma \xi_\sigma$ is composition of maps. Equivalently, the map $\xi_\sigma : C \to C$ induces a map $\xi_\sigma^* : \bar{K}(C) \to \bar{K}(C)$, and $f^\sigma \xi_\sigma$ is just another notation for $\xi_\sigma^*(f^\sigma)$.)

Having given the action of $G_{\bar{K}/K}$ on $\bar{K}(C)_\xi$, we may consider the fixed field $\mathscr{F} \subset \bar{K}(C)_\xi$ consisting of all elements of $\bar{K}(C)_\xi$ fixed by $G_{\bar{K}/K}$. We now show in several steps that this field $\mathscr{F}$ is the function field of the desired twist of C.

(i) $\mathscr{F} \cap \bar{K} = K$

Suppose that $Z(f) \in \mathscr{F} \cap \bar{K}$. In particular, since Z induces the identity on $\bar{K}$, $f \in \bar{K}$. Now the fact that $Z(f) \in \mathscr{F}$ combined with the fact that f is a constant function (and so unaffected by isomorphisms of C) implies

$$Z(f) = Z(f)^\sigma = Z(f^\sigma \xi_\sigma) = Z(f^\sigma).$$

Since this holds for all $\sigma \in G_{\bar{K}/K}$, it follows that $f \in K$.

(ii) $\bar{K}\mathscr{F} = \bar{K}(C)_\xi$

This is an immediate consequence of (II.5.8.1) applied to the $\bar{K}$-vector space $\bar{K}(C)_\xi$.

It follows from (ii) that $\mathscr{F}$ has transcendence degree 1 over K; so using (i) and (II.2.5), there exists a smooth curve C'/K such that $\mathscr{F} \cong K(C')$. Further, (ii) implies that $\bar{K}(C') = \bar{K}\mathscr{F} = \bar{K}(C)_\xi \cong \bar{K}(C)$, so C' and C are isomorphic over $\bar{K}$ (II.2.4.1). In other words, C' is a twist of C, and the final step in proving surjectivity is to show that it gives the cohomology class $\{\xi\}$.

(iii) *Let* $\phi : C' \to C$ *be a* $\bar{K}$-*isomorphism corresponding* (II.2.4b) *to the isomorphism*

$$Z : \bar{K}(C) \to \bar{K}(C)_\xi = \bar{K}\mathscr{F} = \bar{K}(C').$$

(*I.e.* $\phi^* = Z$.) *Then for all* $\sigma \in G_{\bar{K}/K}$, $\xi_\sigma = \phi^\sigma \phi^{-1}$.

Having identified ϕ^* with Z, the relation $Z(f)^\sigma = Z(f^\sigma \xi_\sigma)$ used in defining the map Z can be rewritten as $(f\phi)^\sigma = f^\sigma \xi_\sigma \phi$. In other words, for all $f \in \bar{K}(C)$,

$$f^\sigma \phi^\sigma = (f\phi)^\sigma = f^\sigma \xi_\sigma \phi.$$

But this implies that $\phi^\sigma = \xi_\sigma \phi$, which is exactly the desired result. $\qquad\square$

Example 2.4. Let E/K be an elliptic curve, $K(\sqrt{d})$ a quadratic extension of K, and $\chi: G_{\bar{K}/K} \to \{\pm 1\}$ the quadratic character associated with $K(\sqrt{d})/K$. (I.e. $\chi(\sigma) = \sqrt{d^\sigma}/\sqrt{d}$. Note that char$(K) \neq 2$.) Then we can define a cocycle

$$\xi: G_{\bar{K}/K} \to \text{Isom}(E) \qquad \xi_\sigma = [\chi(\sigma)].$$

Let C/K be the corresponding twist of E/K. We now find an equation for C/K.

Choosing a Weierstrass equation for E/K of the form $y^2 = f(x)$, we write $\bar{K}(E) = \bar{K}(x, y)$ and $\bar{K}(C) = \bar{K}(x, y)_\xi$. Since $[-1](x, y) = (x, -y)$, the action of $\sigma \in G_{\bar{K}/K}$ on $\bar{K}(x, y)_\xi$ may be summarized by

$$\sqrt{d^\sigma} = \chi(\sigma)\sqrt{d}, \qquad x^\sigma = x, \qquad y^\sigma = \chi(\sigma)y.$$

Thus the functions $x' = x$ and $y' = y/\sqrt{d}$ in $\bar{K}(x, y)_\xi$ are fixed by $G_{\bar{K}/K}$, hence are in $K(C)$. They satisfy the equation

$$dy'^2 = f(x'),$$

which is the equation of an elliptic curve defined over K. Further, the identification $(x', y') \to (x', y'\sqrt{d})$ shows that this curve is isomorphic to E over $K(\sqrt{d})$. It is now an easy matter to check that the associated cocycle is ξ, and so verify that we have found an equation for C/K. C is called the *twist of E by the quadratic character* χ. We will return to this example in more detail in section 5.

§3. Homogeneous Spaces

Recall from (VIII §2) that associated to an elliptic curve E/K, we have a Kummer sequence

$$0 \to E(K)/mE(K) \to H^1(G_{\bar{K}/K}, E[m]) \to H^1(G_{\bar{K}/K}, E)[m] \to 0.$$

The proof of the weak Mordell–Weil theorem hinged on the essential fact that the image of the first term inside the second consists of elements which are unramified outside a certain finite set of primes. In this section we analyze the third term in this sequence by associating to each element of $H^1(G_{\bar{K}/K}, E)$ a certain twist of E called a *homogeneous space*. Rather than starting with the cohomology, we will begin by defining homogeneous spaces and describing their basic properties. After this will come the cohomological interpretation, which says that homogeneous spaces are those twists which correspond to cocycles with values in the group of translations.

Definition. Let E/K be an elliptic curve. A (*principal*) *homogeneous space for* E/K is a smooth curve C/K together with a simply transitive algebraic group action of E on C defined over K. [I.e. A homogeneous space for E/K really consists of a pair (C, μ), where C/K is a smooth curve and

$$\mu : C \times E \to C$$

is a morphism defined over K with the following three properties:

(i) $\mu(p, O) = p$ for all $p \in C$.
(ii) $\mu(\mu(p, P), Q) = \mu(p, P + Q)$ for all $p \in C$ and $P, Q \in E$.
(iii) For all $p, q \in C$ there is a unique $P \in E$ satisfying $\mu(p, P) = q$.]

We will often denote $\mu(p, P)$ with the more intuitive notation $p + P$. Then property (ii) is just the associative law, $(p + P) + Q = p + (P + Q)$. Of course, one has to determine from context whether $+$ means addition on E or the action of E on C.

In view of the simple transitivity of the action, we may also define a *subtraction map* on C by the rule

$$v : C \times C \to E$$

$$v(q, p) = \text{(the unique } P \in E \text{ such that } \mu(p, P) = q).$$

As we will see below, v is also a morphism and defined over K. (This also follows from elementary intersection theory on $C \times C$. Note that it is not even clear a priori that v is a rational map.) As with μ, we will often write $v(q, p)$ as $q - p$.

One immediately verifies that addition and subtraction on a homogeneous space have the right properties.

Lemma 3.1. *Let C/K be a homogeneous space for E/K. Then for all $p, q \in C$ and $P, Q \in E$:*

(a) $\mu(p, O) = p$ *and* $v(p, p) = O$.

(b) $\mu(p, v(q, p)) = q$ *and* $v(\mu(p, P), p) = P$.

(c) $v(\mu(q, Q), \mu(p, P)) = v(q, p) + Q - P$.

[I.e. Using the alternative notation,

(a) $p + O = p$ *and* $p - p = O$.

(b) $p + (q - p) = q$ *and* $(p + P) - p = P$.

(c) $(q + Q) - (p + P) = (q - p) + Q - P$.

In other words, using the $+$ and $-$ signs provides the right intuition.]

PROOF. (a) The equality $\mu(p, O) = p$ is part of the definition of homogeneous space. Now using this and the definition of v,

$$\mu(p, O) = p = \mu(p, v(p, p));$$

so the simple transitivity implies that $v(p, p) = O$.

(b) The relation $\mu(p, v(q, p)) = q$ is the definition of v. Then, from

$$\mu(p, v(\mu(p, P), p)) = \mu(p, P),$$

we conclude that $v(\mu(p, P), p) = P$.

(c) We start with

$$q = \mu(p, v(q, p)).$$

Adding on Q gives

$$\mu(q, Q) = \mu(p, v(q, p) + Q)$$
$$= \mu(p, P + v(q, p) + Q - P)$$
$$= \mu(\mu(p, P), v(q, p) + Q - P).$$

From the definition of v, this is equivalent to

$$v(\mu(q, Q), \mu(p, P)) = v(q, p) + Q - P. \qquad \square$$

Next we show that a homogeneous space C/K for E/K is always a twist of E/K, so we may apply the results of the previous section. We also characterize the addition and subtraction on C in terms of a given $\bar{K}$-isomorphism $E \to C$; this will enable us to prove that the subtraction map is a K-morphism.

Proposition 3.2. *Let E/K be an elliptic curve, and let C/K be a homogeneous space for E/K. Fix a point $p_0 \in C$, and define a map*

$$\theta : E \to C \qquad \theta(P) = p_0 + P.$$

(a) *θ is an isomorphism defined over $K(p_0)$. In particular, C/K is a twist of E/K.*

(b) *For all $p \in C$ and $P \in E$,*

$$p + P = \theta(\theta^{-1}(p) + P).$$

(Note that the first $+$ is the action of E on C, while the second $+$ is addition on E.)

(c) *For all $p, q \in C$,*

$$q - p = \theta^{-1}(q) - \theta^{-1}(p).$$

(d) *The subtraction map*

$$v : C \times C \to E \qquad v(q, p) = q - p$$

is a morphism defined over K.

PROOF. (a) The action of E on C is defined over K. Hence for any $\sigma \in G_{\bar{K}/K}$ satisfying $p_0^\sigma = p_0$, we have

$$\theta(P)^\sigma = (p_0 + P)^\sigma = p_0^\sigma + P^\sigma = p_0 + P^\sigma = \theta(P^\sigma).$$

This shows that θ is defined over $K(p_0)$. Further, the simple transitivity of the action implies that θ has degree 1; hence by (II.2.4.1), θ is an isomorphism.

(b) $$\theta(\theta^{-1}(p) + P) = p_0 + \theta^{-1}(p) + P = p + P.$$

(We are using the fact that $\theta^{-1}(p)$ is the unique point of E which gives p when added to p_0.)

(c) $$\theta^{-1}(q) - \theta^{-1}(p) = (p_0 + \theta^{-1}(q)) - (p_0 + \theta^{-1}(p)) = q - p.$$

(d) The fact that v is a morphism follows from (c). (Note that subtraction on E is a morphism (III.3.6).) To check that v is defined over K, we let $\sigma \in G_{\bar{K}/K}$ and use (c) to compute

$$\begin{aligned}
(q - p)^\sigma &= (\theta^{-1}(q) - \theta^{-1}(p))^\sigma \\
&= \theta^{-1}(q)^\sigma - \theta^{-1}(p)^\sigma \\
&= [p_0 + \theta^{-1}(q)]^\sigma - [p_0 + \theta^{-1}(p)]^\sigma \\
&= q^\sigma - p^\sigma.
\end{aligned}$$

(The second and third equalities follow from the facts that subtraction on E is defined over K and the action of E on C is defined over K.) This completes the proof that v is defined over K.   □

Definition. Two homogeneous spaces C/K and C'/K for E/K are *equivalent* if there is an isomorphism $\theta : C \to C'$ defined over K which is compatible with the action of E on C and C'. [In other words, for all $p \in C$ and $P \in E$,

$$\theta(p + P) = \theta(p) + P.]$$

The equivalence class containing E, acting on itself by translation, is called the *trivial class*. The collection of equivalence classes of homogeneous spaces for E/K is called the *Weil–Châtelet group for E/K*, and is denoted $WC(E/K)$. (We will see below why it is a group.)

We now characterize the trivial homogeneous spaces.

Proposition 3.3. *Let C/K be a homogeneous space for E/K. Then C/K is in the trivial class if and only if $C(K)$ is not empty.*

PROOF. If C/K is in the trivial class, then there is a K-isomorphism $\theta : E \to C$, and so $\theta(O) \in C(K)$.

Conversely, suppose that $p_0 \in C(K)$. Then from (3.2a), the map

$$\theta : E \to C \qquad \theta(P) = p_0 + P$$

is an isomorphism defined over $K(p_0) = K$. The necessary compatibility condition on θ is

$$p_0 + (P + Q) = (p_0 + P) + Q,$$

which is part of the definition of homogeneous space.   □

Remark 3.4. Notice that (3.3) says that the problem of checking the triviality of a homogeneous space is exactly equivalent to answering the fundamental Diophantine question of whether a given curve has any rational points. Thus our next step, namely the identification of $WC(E/K)$ with a certain cohomology group, may be regarded as the development of a tool to help us study this difficult Diophantine problem.

Lemma 3.5. *Let* $\theta : C/K \to C'/K$ *be an equivalence of homogeneous spaces for* E/K. *Then*

$$\theta(q) - \theta(p) = q - p \qquad \text{for all } p, q \in C.$$

PROOF. This is just a matter of grouping points so that the additions and subtractions are defined.

$$\theta(q) - \theta(p) = ([\theta(q) + (p - q)] - \theta(p)) + (q - p)$$
$$= (\theta[q + (p - q)] - \theta(p)) + (q - p) = q - p. \qquad \square$$

Theorem 3.6. *Let* E/K *be an elliptic curve. There is a natural bijection*

$$WC(E/K) \to H^1(G_{\bar{K}/K}, E)$$

defined as follows:
    Let C/K *be a homogeneous space, and choose any point* $p_0 \in C$. *Then*

$$\{C/K\} \to \{\sigma \to p_0^\sigma - p_0\}.$$

(Here the brackets indicate an equivalence class.)

Remark 3.6.1. Since $H^1(G_{\bar{K}/K}, E)$ is a group, (3.6) defines a group structure on the set $WC(E/K)$. One can also give the group law on $WC(E/K)$ geometrically, without using cohomology (exer. 10.2), which is in fact the way it was originally defined ([We 5]).

PROOF. First we check that the map is well-defined. It is immediate that $\sigma \to p_0^\sigma - p_0$ is a cocycle:

$$p_0^{\sigma\tau} - p_0 = (p_0^{\sigma\tau} - p_0^\tau) + (p_0^\tau - p_0) = (p_0^\sigma - p_0)^\tau + (p_0^\tau - p_0).$$

Now suppose that C'/K is another homogeneous space which is equivalent to C/K. Let $\theta : C \to C'$ be a K-isomorphism giving the equivalence, and let $p_0' \in C'$. Then using (3.5), we compute

$$p_0^\sigma - p_0 = \theta(p_0^\sigma) - \theta(p_0)$$
$$= (p_0'^\sigma - p_0') + [(\theta(p_0) - p_0')^\sigma - (\theta(p_0) - p_0')].$$

Hence the cocycles $p_0^\sigma - p_0$ and $p_0'^\sigma - p_0'$ differ by the coboundary generated by $\theta(p_0) - p_0' \in E$, so they give the same cohomology class in $H^1(G_{\bar{K}/K}, E)$.
    Next we check injectivity. Suppose that the cocycles $p_0^\sigma - p_0$ and $p_0'^\sigma - p_0'$

corresponding to C/K and C'/K are cohomologous. Thus there is a point $P_0 \in E$ such that

$$p_0^\sigma - p_0 = (p_0'^\sigma - p_0') + (P_0^\sigma - P_0) \qquad \text{for all } \sigma \in G_{\bar{K}/K}.$$

Consider the map

$$\theta : C \to C' \qquad \theta(p) = p_0' + (p - p_0) + P_0.$$

It is clear that θ is an isomorphism (over $\bar{K}$), and that it is compatible with the action of E. To see that θ is defined over K, we check

$$\theta(p)^\sigma = p_0'^\sigma + (p^\sigma - p_0^\sigma) + P_0^\sigma$$

$$= p_0' + (p^\sigma - p_0) + P_0$$

$$+ [(p_0'^\sigma - p_0') + P_0^\sigma - P_0 - (p_0^\sigma - p_0)]$$

$$= \theta(p^\sigma).$$

This proves that C and C' are equivalent.

Finally we prove surjectivity. Thus let $\xi : G_{\bar{K}/K} \to E$ be a 1-cocycle representing an element in $H^1(G_{\bar{K}/K}, E)$. If we embed E in $\mathrm{Isom}(E)$ by sending $P \in E$ to the translation $\tau_P \in \mathrm{Isom}(E)$, then we may look at the image of ξ in $H^1(G_{\bar{K}/K}, \mathrm{Isom}(E))$. From (2.2), there is a curve C/K and a $\bar{K}$-isomorphism $\phi : C \to E$ such that for all $\sigma \in G_{\bar{K}/K}$,

$$\phi^\sigma \circ \phi^{-1} = \text{translation by } -\xi_\sigma.$$

(The reason we use $-\xi$ instead of ξ will become apparent below.)

Define a map

$$\mu : C \times E \to C \qquad \mu(p, P) = \phi^{-1}(\phi(p) + P).$$

We now show that this gives C/K the structure of a homogeneous space over E/K, and that the cohomology class associated to C/K is $\{\xi\}$.

First, to see that μ is simply transitive, let $p, q \in C$. Then by definition, $\mu(p, P) = q$ if and only if $\phi^{-1}(\phi(p) + P) = q$; and so the only choice for P is $P = \phi(q) - \phi(p)$. Second, to check that μ is defined over K, we let $\sigma \in G_{\bar{K}/K}$ and compute

$$\mu(p, P)^\sigma = (\phi^{-1})^\sigma(\phi^\sigma(p^\sigma) + P^\sigma)$$

$$= \phi^{-1}([\phi(p^\sigma) - \xi_\sigma) + P^\sigma] + \xi_\sigma)$$

$$= \mu(p^\sigma, P^\sigma).$$

Third, to compute the cohomology class associated to C/K, we may choose *any* $p_0 \in C$ and look at the cocycle $\sigma \to p_0^\sigma - p_0$. In particular, if we take $p_0 = \phi^{-1}(O)$, then

$$p_0^\sigma - p_0 = (\phi^\sigma)^{-1}(O) - \phi^{-1}(O)$$

$$= \phi^{-1}(O + \xi_\sigma) - \phi^{-1}(O)$$

$$= \xi_\sigma.$$

This completes the proof of (3.6).   □

Example 3.7. Let E/K be an elliptic curve and $K(\sqrt{d})/K$ a quadratic extension (so $\text{char}(K) \neq 2$). Suppose that $T \in E(K)$ is a non-trivial point of order 2. Then the homomorphism

$$\xi : G_{\bar{K}/K} \to E$$

$$\sigma \to \begin{cases} O & \text{if } \sqrt{d}^{\sigma} = \sqrt{d} \\ T & \text{if } \sqrt{d}^{\sigma} = -\sqrt{d} \end{cases}$$

is a 1-cocycle. We will now construct the homogeneous space corresponding to the element $\{\xi\} \in H^1(G_{\bar{K}/K}, E)$.

Since $T \in E(K)$, we may choose a Weierstrass equation for E/K of the form

$$E : y^2 = x^3 + ax^2 + bx \qquad \text{with } T = (0, 0).$$

Then the translation-by-T map has the simple form

$$\tau_T(P) = (x, y) + (0, 0) = (b/x, -by/x^2).$$

Thus if $\sigma \in G_{\bar{K}/K}$ represents the non-trivial automorphism of $K(\sqrt{d})/K$, then the action of σ on the twisted field $\bar{K}(E)_{\xi}$ may be summarized by

$$\sqrt{d}^{\sigma} = -\sqrt{d}, \qquad x^{\sigma} = b/x, \qquad y^{\sigma} = -by/x^2.$$

We must find the subfield of $K(\sqrt{d})(x, y)_{\xi}$ fixed by σ.

The functions

$$\sqrt{d}x/y \quad \text{and} \quad \sqrt{d}(x - b/x)$$

are easily seen to be invariant. Anticipating the form of our final answer, we will consider instead the functions

$$z = \sqrt{d}x/y \quad \text{and} \quad w = \sqrt{d}(x - b/x)(x/y)^2.$$

To find the relation that they satisfy, we compute

$$d(w/z^2)^2 = (x - b/x)^2 = (x + b/x)^2 - 4b$$

$$= ((y/x)^2 - a)^2 - 4b = (d/z^2 - a)^2 - 4b.$$

Thus (z, w) are affine coordinates for the hyperelliptic curve

$$C : dw^2 = d^2 - 2adz^2 + (a^2 - 4b)z^4.$$

(For general facts about hyperelliptic curves, see (II.2.5.1) and (exer. 2.14).) We claim that C/K is the twist of E/K corresponding to the cocycle ξ.

First, recall from (II.2.5.1) that C will be a smooth *affine* curve provided that the polynomial $d^2 - 2adz^2 + (a^2 - 4b)z^4$ has four distinct roots (in $\bar{K}$). Further, (II.2.5.2) says that if this quartic polynomial has distinct roots, then there is a smooth curve in $\mathbb{P}^3$ which has an affine piece isomorphic to C; and further, this smooth curve will consist of C together with the two points $[0, 0, \pm\sqrt{(a^2 - 4b)/d}, 1]$ out at infinity. (N.B. The projective closure of C in $\mathbb{P}^2$ is always singular.) Now it is easy to check that the quartic has distinct roots if and only if $b(a^2 - 4b) \neq 0$. On the other hand, since E is non-singular, we know that $\Delta(E) = 16b^2(a^2 - 4b) \neq 0$. Therefore C is an affine piece of a smooth curve in $\mathbb{P}^3$; and to ease notation, we will also use C to denote this smooth curve $C \subset \mathbb{P}^3$.

Next, we have the map (defined over $K(\sqrt{d})$)

$$\phi : E \to C$$

$$(x, y) \to (z, w) = (\sqrt{d}x/y, \sqrt{d}(x - b/x)(x/y)^2).$$

Note that since

$$x/y = xy/y^2 = y/(x^2 + ax + b),$$

ϕ may also be written as

$$\phi(x, y) = \left(\frac{\sqrt{d}\,y}{x^2 + ax + b}, \frac{\sqrt{d}(x^2 - b)}{x^2 + ax + b} \right).$$

This allows us to evaluate

$$\phi(0, 0) = (0, -\sqrt{d}) \quad \text{and} \quad \phi(O) = (0, \sqrt{d}).$$

To show that ϕ is an isomorphism, we compute its inverse:

$$\sqrt{d}w/z^2 = x - b/x = 2x - (x + b/x)$$

$$= 2x - ((y/x)^2 - a) = 2x - (d/z^2 - a).$$

This gives x in terms of z and w, and then $y = \sqrt{d}x/z$. Thus

$$\phi^{-1} : C \to E$$

$$(z, w) \to \left(\frac{\sqrt{d}w - az^2 + d}{2z^2}, \frac{dw - a\sqrt{d}z^2 + d\sqrt{d}}{2z^3} \right).$$

Since C and E are smooth, ϕ is an isomorphism (II.2.4.1).

Finally, to compute the element of $H^1(G_{\bar{K}/K}, E)$ corresponding to C/K, we may choose *any* point $p \in C$ and compute the cocycle

$$\sigma \to p^\sigma - p = \phi^{-1}(p^\sigma) - \phi^{-1}(p).$$

For instance, we may take $p = (0, \sqrt{d}) \in C$. Clearly $p^\sigma - p = O$ if $\sqrt{d}^\sigma = \sqrt{d}$. On the other hand, if $\sqrt{d}^\sigma = -\sqrt{d}$, then from above

$$p^\sigma - p = \phi^{-1}(0, -\sqrt{d}) - \phi^{-1}(0, \sqrt{d}) = (0, 0).$$

Therefore $p^\sigma - p = \xi_\sigma$ for all $\sigma \in G_{\bar{K}/K}$, so $\{C/K\} \in WC(E/K)$ maps to $\{\xi\} \in H^1(G_{\bar{K}/K}, E)$. [Of course, it was just "luck" that we obtained an equality $p^\sigma - p = \xi_\sigma$. In general, the difference of these two cocycles would be some coboundary.]

We conclude this section by showing that if C/K is a homogeneous space for E/K, then $\text{Pic}^0(C)$ may be canonically identified with E. This means that E is the *Jacobian of C/K*. Since every curve C/K of genus 1 is a homogeneous space for some elliptic curve E/K (exer. 10.3), this shows that the abstract group $\text{Pic}^0(C)$ can always be represented as the group of points on an elliptic curve. The analogous result for curves of higher genus, in which $\text{Pic}^0(C)$ is represented by an abelian variety of dimension equal to the genus of C, is considerably harder to prove.

Theorem 3.8. *Let C/K be a homogeneous space for an elliptic curve E/K. Choose a point $p_0 \in C$, and consider the summation map*

$$\text{sum} : \text{Div}^0(C) \to E$$

$$\Sigma n_i(p_i) \to \Sigma [n_i](p_i - p_0).$$

(a) *There is an exact sequence*

$$1 \to \bar{K}^* \to \bar{K}(C)^* \overset{\text{div}}{\to} \text{Div}^0(C) \overset{\text{sum}}{\to} E \to 0.$$

(b) *The summation map is independent of the choice of the point p_0.*
(c) *The summation map commutes with the natural action of $G_{\bar{K}/K}$ on $\text{Div}^0(C)$ and E. Hence it gives an isomorphism of $G_{\bar{K}/K}$-modules (also denoted sum)*

$$\text{sum} : \text{Pic}^0(C) \overset{\sim}{\to} E.$$

In particular.

$$\text{Pic}^0_K(C) \cong E(K).$$

PROOF. (a) Using (II.3.4), we see that we must check that *sum* is a surjective homomorphism and has as kernel the set of principal divisors. It is clear that *sum* is a homomorphism. Let $P \in E$ and $D = (p_0 + P) - (p_0) \in \text{Div}^0(C)$. Then

$$\text{sum}(D) = ((p_0 + P) - p_0) - (p_0 - p_0) = P,$$

so *sum* is surjective.

Next let $D = \Sigma n_i(p_i) \in \text{Div}^0(C)$ satisfy $\text{sum}(D) = 0$. Then the *divisor* $\Sigma n_i(p_i - p_0) \in \text{Div}^0(E)$ sums to 0, so (III.3.5) it is principal, say

$$\sum n_i(p_i - p_0) = \text{div}(f) \qquad \text{for some } f \in \bar{K}(E)^*.$$

We have an isomorphism

$$\phi : C \to E \qquad \phi(p) = p - p_0;$$

and so by (II.3.6b),

$$\text{div}(\phi^* f) = \phi^* \text{div}(f) = \sum n_i \phi^*((p_i - p_0)) = \sum n_i(p_i) = D.$$

Therefore D is principal.

Finally, if $D = \text{div}(g)$ is principal, then

$$\sum n_i(p_i - p_0) = (\phi^{-1})^* \text{div}(g) = \text{div}((\phi^{-1})^* g);$$

and so $\text{sum}(D) = 0$. This shows that the kernel of *sum* is the set of principal divisors.

(b) Let $sum' : \mathrm{Div}^0(C) \to E$ be the summation map defined using the base-point $p_0' \in C$. Then

$$sum(D) - sum'(D) = \sum [n_i]((p_i - p_0) - (p_i - p_0'))$$
$$= \sum [n_i](p_0' - p_0)$$
$$= 0,$$

since $\Sigma n_i = \deg(D) = 0$.

(c) Let $\sigma \in G_{\overline{K}/K}$. Then

$$sum(D)^\sigma = \sum [n_i](p_i^\sigma - p_0^\sigma) = sum(D^\sigma),$$

since from (b) we know that the sum is the same if we use p_0^σ as our basepoint instead of p_0. Now from (a) and the definition of $\mathrm{Pic}^0(C)$, we have a group isomorphism $sum : \mathrm{Pic}^0(C) \to E$, and the fact that sum commutes with $G_{\overline{K}/K}$ says precisely that it is an isomorphism of $G_{\overline{K}/K}$-modules. Finally, the last statement in (3.8c) follows by taking $G_{\overline{K}/K}$-invariants.   □

§4. The Selmer and Shafarevich–Tate Groups

We return now to the problem of calculating the Mordell–Weil group of an elliptic curve E/K defined over a number field K. As we have seen (VIII.3.2 and exer. 8.18), it is enough to find generators for the finite group $E(K)/mE(K)$ for any one integer $m \geqslant 2$.

Suppose that we are given another elliptic curve E'/K and a non-zero isogeny $\phi : E \to E'$ defined over K. (For example, we could always take $E' = E$ and $\phi = [m]$.) Then there is an exact sequence of $G_{\overline{K}/K}$-modules,

$$0 \to E[\phi] \to E \xrightarrow{\phi} E' \to 0,$$

where $E[\phi]$ denotes the kernel of ϕ. Taking Galois cohomology yields the long exact sequence

$$0 \to \quad E(K)[\phi] \quad \to \quad E(K) \quad \xrightarrow{\phi} \quad E'(K)$$
$$\xrightarrow{\delta} H^1(G_{\overline{K}/K}, E[\phi]) \to H^1(G_{\overline{K}/K}, E) \to H^1(G_{\overline{K}/K}, E') \to ;$$

and from this we form the fundamental short exact sequence

$$0 \to E'(K)/\phi(E(K)) \xrightarrow{\delta} H^1(G_{\overline{K}/K}, E[\phi]) \to H^1(G_{\overline{K}/K}, E)[\phi] \to 0. \qquad (*)$$

Note that (3.6) says that the last term in $(*)$ may be identified with the ϕ-torsion in the Weil–Châtelet group $WC(E/K)$.

The next step is to replace the second and third terms of $(*)$ with certain finite groups. This is accomplished by local considerations. For each $v \in M_K$, we fix an extension of v to $\overline{K}$, which serves to fix an embedding $\overline{K} \subset \overline{K}_v$ and a decomposition group $G_v \subset G_{\overline{K}/K}$. Now G_v acts on $E(\overline{K}_v)$ and $E'(\overline{K}_v)$; and so repeating the above argument yields exact sequences

$$0 \to E'(K_v)/\phi(E(K_v)) \xrightarrow{\delta} H^1(G_v, E[\phi]) \to H^1(G_v, E)[\phi] \to 0. \qquad (*_v)$$

Now the natural inclusions $G_v \subset G_{\bar{K}/K}$ and $E(\bar{K}) \subset E(\bar{K}_v)$ give restriction maps on cohomology, and so we end up with the following commutative diagram (where we have replaced each $H^1(G, E)$ by the corresponding Weil–Châtelet group):

$$
\begin{array}{ccccccccc}
0 \to & E'(K)/\phi(E(K)) & \xrightarrow{\delta} & H^1(G_{\bar{K}/K}, E[\phi]) & \to & WC(E/K)[\phi] & \to 0 \\
& \downarrow & & \downarrow & & \downarrow & & (**) \\
0 \to & \prod_{v \in M_K} E'(K_v)/\phi(E(K_v)) & \xrightarrow{\delta} & \prod_{v \in M_K} H^1(G_v, E[\phi]) & \to & \prod_{v \in M_K} WC(E/K_v)[\phi] & \to 0.
\end{array}
$$

Our ultimate goal is to compute the image of $E'(K)/\phi(E(K))$ in $H^1(G_{\bar{K}/K}, E[\phi])$; or equivalently, the kernel of the map

$$H^1(G_{\bar{K}/K}, E[\phi]) \to WC(E/K)[\phi].$$

Now using (3.3), this last problem is the same as determining whether certain homogeneous spaces possess a K-rational point, which may be a very difficult question. On the other hand, by the same reasoning, the determination of each local kernel

$$\ker\{H^1(G_v, E[\phi]) \to WC(E/K_v)[\phi]\}$$

is straightforward; since the question of whether a curve has a point over a complete local field K_v reduces (by Hensel's lemma) to checking whether it has a point in some finite ring $R_v/\mathcal{M}_v^e$ (for some easily computable integer e), and so requires only a finite amount of computation. This prompts the following definitions.

Definition. Let $\phi : E/K \to E'/K$ be an isogeny as above. The ϕ-*Selmer group of* E/K is the subgroup of $H^1(G_{\bar{K}/K}, E[\phi])$ defined by

$$S^{(\phi)}(E/K) = \ker\left\{H^1(G_{\bar{K}/K}, E[\phi]) \to \prod_{v \in M_K} WC(E/K_v)\right\}.$$

The *Shafarevich–Tate group of* E/K is the subgroup of $WC(E/K)$ defined by

$$\text{Ш}(E/K) = \ker\left\{WC(E/K) \to \prod_{v \in M_K} WC(E/K_v)\right\}.$$

Remark 4.1.1. Since the exact sequences $(*_v)$ given above depend on choosing an extension of each $v \in M_K$ to $\bar{K}$, it may appear that the groups $S^{(\phi)}(E/K)$ and $\text{Ш}(E/K)$ will depend on that choice. However, in order to determine whether an element of $WC(E/K)$ becomes trivial in $WC(E/K_v)$, one must check whether the associated homogeneous space (which is a curve defined over K) has any points defined over K_v. This last question is clearly independent of any choice of extension of v to $\bar{K}$, since v itself determines the embedding of K in K_v. Therefore $S^{(\phi)}(E/K)$ and $\text{Ш}(E/K)$ depend only on E and K. (Alterna-

tively, one can check directly on cocycles that the cohomological definition of
$S^{(\phi)}$ and Ш does not depend on the extension of the v's to $\bar{K}$. We will leave this
for the reader. See also (exer. B.6).)

Remark 4.1.2. A good way to think of Ш(E/K) is as the group of homogeneous spaces for E/K which possess a K_v-rational point for every $v \in M_K$.
I.e., Ш(E/K) is the group of homogeneous spaces which are everywhere locally
trivial, modulo equivalence.

Theorem 4.2. *Let $\phi: E/K \to E'/K$ be an isogeny of elliptic curves defined over
K.*
(a) *There is an exact sequence*

$$0 \to E'(K)/\phi(E(K)) \to S^{(\phi)}(E/K) \to \text{Ш}(E/K)[\phi] \to 0.$$

(b) *The Selmer group $S^{(\phi)}(E/K)$ is finite.*

PROOF. (a) This is immediate from the diagram $(**)$ and the definition of the
Selmer and Shafarevich–Tate groups.
(b) Notice that if $E = E'$ and $\phi = [m]$, then the finiteness of $S^{(m)}(E/K)$ implies
the weak Mordell–Weil theorem. On the other hand, in order to prove that
$S^{(\phi)}(E/K)$ is finite for a general map ϕ, we must essentially reprove the weak
Mordell–Weil theorem. The argument goes as follows.

Let $\xi \in S^{(\phi)}(E/K)$, and let $v \in M_K$ be a finite place of K not dividing
$m = \deg(\phi)$ such that E'/K has good reduction at v. We claim that ξ is
unramified at v. (See (VIII §2) for the definition of an unramified cocycle.)

To check this, let $I_v \subset G_v$ be the inertia group for v. Since $\xi \in S^{(\phi)}(E/K)$, we
know that ξ is trivial in $WC(E/K_v)$. Hence from the sequence $(*_v)$ given
above, there is a point $P \in E(\bar{K}_v)$ such that

$$\xi_\sigma = \{P^\sigma - P\} \qquad \text{for all } \sigma \in G_v.$$

(Note that $P^\sigma - P \in E[\phi]$.) In particular, this holds for all σ in the inertia
group. But if $\sigma \in I_v$, then looking at the "reduction modulo v" map $E \to \tilde{E}_v$
yields

$$\widetilde{P^\sigma - P} = \tilde{P}^\sigma - \tilde{P} = \tilde{O},$$

since by definition inertia acts trivially on $\tilde{E}_v$. Thus $P^\sigma - P$ is in the kernel of
reduction modulo v. But $P^\sigma - P$ is also in $E[\phi]$, which is contained in $E[m]$;
and from (VIII.1.4), $E(K)[m]$ injects in $\tilde{E}_v$. Therefore $P^\sigma = P$, and so

$$\xi_\sigma = \{P^\sigma - P\} = 0 \qquad \text{for all } \sigma \in I_v.$$

This proves that every element in $S^{(\phi)}(E/K)$ is unramified at all but a fixed,
finite set of places $v \in M_K$. The finiteness of $S^{(\phi)}(E/K)$ now follows from the
next lemma.   □

Lemma 4.3. *Let M be a finite (abelian) $G_{\bar{K}/K}$-module, $S \subset M_K$ a finite set of
places, and define*

$$H^1(G_{\bar{K}/K}, M; S) = \{\xi \in H^1(G_{\bar{K}/K}, M) : \xi \text{ is unramified outside } S\}.$$

Then $H^1(G_{\bar{K}/K}, M; S)$ is finite.

PROOF. Since M is finite and $G_{\bar{K}/K}$ acts continuously on M, there is a subgroup of finite index in $G_{\bar{K}/K}$ which fixes every element of M. Using the inflation-restriction sequence (B.2.4), we see that it is alright to replace K by a finite extension, so we may assume that the action of $G_{\bar{K}/K}$ on M is trivial. Then

$$H^1(G_{\bar{K}/K}, M; S) = \text{Hom}(G_{\bar{K}/K}, M; S).$$

Now let m be the exponent of M (i.e. $mx = 0$ for all $x \in M$); and let L/K be the maximal abelian extension of K having exponent m which is unramified outside of S. Since M has exponent m, the natural map

$$\text{Hom}(G_{L/K}, M) \to \text{Hom}(G_{\bar{K}/K}, M; S)$$

is clearly an isomorphism. But from (VIII.1.6), L/K is a *finite* extension. Therefore $\text{Hom}(G_{\bar{K}/K}, M; S)$ is finite.   □

We record as a corollary the main fact about the Selmer group derived during the course of proving (4.2). (Note that by (VII.7.2), isogenous elliptic curves have the same set of primes of bad reduction.)

Corollary 4.4. Let $\phi : E/K \to E'/K$ be as in (4.2), and let $S \subset M_K$ be a finite set of places containing

$$M_K^\infty \cup \{v : E \text{ has bad reduction at } v\} \cup \{v : v \text{ divides } \deg(\phi)\}.$$

Then

$$S^{(\phi)}(E/K) \subset H^1(G_{\bar{K}/K}, E[\phi]; S).$$

Remark 4.5. At least in theory, and often in practice, the Selmer group is effectively computable. The point is that the finite group $H^1(G_{\bar{K}/K}, E[\phi]; S)$ may be effectively computed. Then to determine whether a given element $\xi \in H^1(G_{\bar{K}/K}, E[\phi]; S)$ is in $S^{(\phi)}(E/K)$, one takes the corresponding homogeneous space $\{C/K\} \in WC(E/K)$ and checks whether $C(K_v) \neq \varnothing$ for each of the finitely many $v \in S$. This last problem may be reduced, by Hensel's lemma, to a finite amount of computation.

Example 4.5.1. We reformulate the example of section 1 in these terms (leaving some details to the reader). Thus let E/K be an elliptic curve with $E[m] \subset E(K)$, let $S \subset M_K$ be the usual set of places, and let $K(S, m)$ be as in (1.1c). Choosing a basis for $E[m]$, we may identify $E[m]$ with $\mu_m \times \mu_m$ (as $G_{\bar{K}/K}$-modules); and then

$$H^1(G_{\bar{K}/K}, E[m]; S) \cong K(S, m) \times K(S, m).$$

(I.e. Use the isomorphism $K^*/K^{*m} \to H^1(G_{\bar{K}/K}, \mu_m)$.) Restricting attention now to the case $m = 2$, the homogeneous space associated to a pair $(b_1, b_2) \in K(S, m) \times K(S, m)$ is the curve in $\mathbb{P}^3$ given by the equations (cf. (1.4))

$$C : b_1 z_1^2 - b_2 z_2^2 = (e_2 - e_1)z_0^2 \qquad b_1 z_1^2 - b_1 b_2 z_3^2 = (e_3 - e_1)z_0^2.$$

For any given pair (b_1, b_2), it is now an easy matter to check if $C(K_v) \neq \varnothing$ for each $v \in S$, and so to calculate $S^{(2)}(E/K)$. For example, the conclusion of (1.5) may be summarized by stating that for the curve

$$E : y^2 = x^3 - 12x^2 + 20x,$$

$$S^{(2)}(E/\mathbb{Q}) \cong (\mathbb{Z}/2\mathbb{Z})^3 \quad \text{and} \quad \text{III}(E/\mathbb{Q})[2] = 0.$$

(The conclusion about III follows from the exact sequence (4.2a), since in (1.5) we actually showed that every element of $S^{(2)}(E/\mathbb{Q})$ is the image of a point of $E(\mathbb{Q})$.)

Suppose now that we have computed the Selmer group $S^{(\phi)}(E/K)$ for some isogeny ϕ. Thus each $\xi \in S^{(\phi)}(E/K)$ corresponds to a homogeneous space C_ξ/K which has a point defined over every local field K_v. Suppose further that we are lucky and can show that $\text{III}(E/K)[\phi] = 0$. This means that on each of the curves C_ξ we are able to find a K-rational point. It follows from (4.2a) that $E'(K)/\phi(E(K)) \cong S^{(\phi)}(E/K)$, and all that remains is to explain how to find generators for $E'(K)/\phi(E(K))$ in terms of the points we found on each $C_\xi(K)$. This is accomplished by the following proposition.

Proposition 4.6. *Let $\phi : E/K \to E'/K$ be an isogeny, let ξ be a cocycle representing an element of $H^1(G_{\bar{K}/K}, E[\phi])$, and let C/K be a homogeneous space representing the image of ξ in $WC(E/K)$. Choose an isomorphism $\theta : C \to E$ (defined over $\bar{K}$) satisfying*

$$\theta^\sigma \circ \theta^{-1} = (\text{translation by } \xi_\sigma) \qquad \text{for all } \sigma \in G_{\bar{K}/K}.$$

(a) *The map $\phi \circ \theta : C \to E'$ is defined over K.*
(b) *If $P \in C(K)$, and so $\{C/K\}$ is trivial in $WC(E/K)$, then the point $\phi \circ \theta(P) \in E'(K)$ maps to ξ under the connecting homomorphism $\delta : E'(K) \to H^1(G_{\bar{K}/K}, E[\phi])$.*

PROOF. (a) Let $\sigma \in G_{\bar{K}/K}$ and $P \in C$. Then, since ϕ is defined over K and $\xi_\sigma \in E[\phi]$, we have

$$(\phi \circ \theta(P))^\sigma = (\phi \circ \theta^\sigma)(P^\sigma) = \phi(\theta(P^\sigma) + \xi_\sigma) = \phi \circ \theta(P^\sigma).$$

Therefore $\phi \circ \theta$ is defined over K.
(b) This is just a matter of unwinding definitions. Thus

$$\delta(\phi \circ \theta(P))_\sigma = \theta(P)^\sigma - \theta(P) = \theta(P^\sigma) + \xi_\sigma - \theta(P) = \xi_\sigma. \qquad \square$$

Remark 4.7. We have been working with arbitrary isogenies $\phi : E \to E'$. But in order to compute the Mordell–Weil group of E', we need generators for

$E'(K)/mE'(K)$ for some integer m; just knowing $E'(K)/\phi(E(K))$ is not enough. The solution to this dilemma is to consider also the dual isogeny $\hat{\phi}: E' \to E$. Using the procedure outlined above, one computes both Selmer groups $S^{(\phi)}(E/K)$ and $S^{(\hat{\phi})}(E'/K)$; and then, with a little bit of luck, one finds generators for the two groups $E'(K)/\phi(E(K))$ and $E(K)/\hat{\phi}(E'(K))$. Having done this, it is then a simple matter to obtain generators for $E(K)/mE(K)$ (where $m = \deg \phi$) by using the following elementary exact sequence (note $\hat{\phi} \circ \phi = [m]$):

$$0 \to \frac{E'(K)[\hat{\phi}]}{\phi(E(K)[m])} \to \frac{E'(K)}{\phi(E(K))} \xrightarrow{\hat{\phi}} \frac{E(K)}{mE(K)} \to \frac{E(K)}{\hat{\phi}(E'(K))} \to 0.$$

Example 4.8. *Two-isogenies.* We are going to illustrate the above theory by completely analyzing the case of isogenies of degree 2. Let $\phi: E \to E'$ have degree 2. Then the kernel $E[\phi] = \{O, T\}$ is defined over K, so $T \in E(K)$. Thus E has a K-rational 2-torsion point, so by moving that point to $(0, 0)$, we can find a Weierstrass equation for E/K of the form

$$E: y^2 = x^3 + ax^2 + bx.$$

Now let $S \subset M_K$ be the usual set of places. Identifying $E[\phi]$ with μ_2 (as $G_{\bar{K}/K}$-modules), we see that $K^*/K^{*2} \cong H^1(G_{\bar{K}/K}, E[\phi])$; and so

$$H^1(G_{\bar{K}/K}, E[\phi]; S) \cong K(S, 2)$$

(using the notation of (1.1c) and (4.3).) More precisely, if $d \in K(S, 2)$, then tracing through the above identifications shows that the corresponding cocycle is

$$\sigma \to \begin{cases} O & \text{if } \sqrt{d}^\sigma = \sqrt{d} \\ T & \text{if } \sqrt{d}^\sigma = -\sqrt{d}. \end{cases}$$

The homogeneous space C_d/K associated to this cocycle was computed in (3.7); it is given by the equation

$$C_d: dw^2 = d^2 - 2adz^2 + (a^2 - 4b)z^4.$$

Now in order to compute the Selmer group $S^{(\phi)}(E/K)$, we need merely check if $C_d(K_v) \neq \emptyset$ for each of the finitely many $d \in K(S, 2)$ and $v \in S$.

Next, E'/K has a Weierstrass equation of the form

$$E': Y^2 = X^3 - 2aX^2 + (a^2 - 4b)X,$$

where the isogeny $\phi: E \to E'$ is given by the formula (III.4.5)

$$\phi(x, y) = (y^2/x^2, y(b - x^2)/x^2).$$

In (3.7) we gave an isomorphism $\theta: C_d \to E$ (defined over $K(\sqrt{d})$). Computing the composition $\phi \circ \theta$ yields the map

$$\phi \circ \theta: C_d \to E' \qquad \phi \circ \theta(z, w) = (d/z^2, -dw/z^3)$$

described in (4.6). Finally, just as was done in (1.4) (see also exer. 10.1), one can compute the connecting homorphism

$$\delta : E'(K) \to H^1(G_{\bar{K}/K}, E[\phi]) \cong K^*/K^{*2}$$

$$\delta(O) = 1, \qquad \delta(0, 0) = a^2 - 4b, \text{ and } \quad \delta(X, Y) = X \quad \text{if } X \neq 0, \infty.$$

We summarize the preceding discussion in the following proposition.

Proposition 4.9. (Descent via Two-Isogeny.) *Let E/K and E'/K be elliptic curves given by equations*

$$E : y^2 = x^3 + ax^2 + bx \quad and \quad E' : Y^2 = X^3 - 2aX^2 + (a^2 - 4b)X;$$

and let

$$\phi : E \to E' \qquad \phi(x, y) = (y^2/x^2, y(b - x^2)/x^2)$$

be the isogeny of degree 2 with kernel $E[\phi] = \{O, (0, 0)\}$. Let

$$S = M_K^\infty \cup \{primes\ dividing\ 2b(a^2 - 4b)\}.$$

There is an exact sequence

$$0 \to E'(K)/\phi(E(K)) \xrightarrow{\delta} K(S, 2) \to WC(E/K)[\phi]$$

$$O \to 1$$

$$(0, 0) \to a^2 - 4b \qquad d \to \{C_d/K\},$$

$$(X, Y) \to X$$

where C_d/K is the homogeneous space for E/K given by the equation

$$C_d : dw^2 = d^2 - 2adz^2 + (a^2 - 4b)z^4.$$

The ϕ-Selmer group is then

$$S^{(\phi)}(E/K) \cong \{d \in K(S, 2) : C_d(K_v) \neq \varnothing \text{ for all } v \in S\}.$$

Finally, the map

$$\psi : C_d \to E' \qquad \psi(z, w) = (d/z^2, -dw/z^3)$$

has the property that if $P \in C_d(K)$, then

$$\delta(\psi(P)) \equiv d \ (mod\ K^{*2}).$$

Remark 4.9.1. Note that since the isogenous curve E' in (4.9) has the same form as E, everything in (4.9) applies also to the dual isogeny $\hat{\phi} : E' \to E$. Then, using the exact sequence in (4.7), we may be able to compute $E(K)/2E(K)$.

Remark 4.9.2. If E/K is an elliptic curve which has a K-rational 2-torsion point, then E also has an isogeny of degree 2 defined over K (III.4.5). Thus the procedure described in (4.8) may be applied to any elliptic curve with

$E(K)[2] \neq 0$. In particular, (4.9) in some sense subsumes (1.4), where we had assumed that $E[2] \subset E(K)$.

Example 4.10. We now use (4.9) to compute $E(\mathbb{Q})/2E(\mathbb{Q})$ for the elliptic curve

$$E : y^2 = x^3 - 6x^2 + 17x.$$

This equation has discriminant $\Delta = -147968 = -2^9 17^2$, so our set S is $\{\infty, 2, 17\}$, and we may identify $\mathbb{Q}(S, 2)$ with $\{\pm 1, \pm 2, \pm 17, \pm 34\}$. The curve which is 2-isogenous to E has the equation

$$E' : Y^2 = X^3 + 12X^2 - 32X;$$

and for $d \in \mathbb{Q}(S, 2)$, the corresponding homogeneous space is

$$C_d : dw^2 = d^2 + 12dz^2 - 32z^4.$$

From (4.9), the point $(0, 0) \in E'(\mathbb{Q})$ maps to $\delta(0, 0) = -32 \equiv -2 \pmod{\mathbb{Q}^{*2}}$, so $-2 \in S^{(\phi)}(E/\mathbb{Q})$. We now check the other possible values for d.

$\boxed{d = 2}$ $\qquad\qquad$ $C_2 : 2w^2 = 4 + 24z^2 - 32z^4$

Dividing by 2 and letting $z = Z/2$ gives the equation

$$C_2 : w^2 = 2 + 3Z^2 - Z^4,$$

which by inspection has the rational point $(Z, w) = (1, 2)$. Then using (4.9), the point $(z, w) = (\frac{1}{2}, 2) \in C_2(\mathbb{Q})$ maps to $\psi(\frac{1}{2}, 2) = (8, -32) \in E'(\mathbb{Q})$; and as predicted by the theory, $\delta(8, -32) = 8 \equiv 2 \pmod{\mathbb{Q}^{*2}}$.

$\boxed{d = 17}$ $\qquad\qquad$ $C_{17} : 17w^2 = 17^2 + 12 \cdot 17z^2 - 32z^4$.

Suppose that $C_{17}(\mathbb{Q}_{17}) \neq \varnothing$. Since $\mathrm{ord}_{17}(17w^2)$ is odd and $\mathrm{ord}_{17}(32z^4)$ is even, we see that necessarily $z, w \in \mathbb{Z}_{17}$. But then the equation for C_{17} implies first that $z \equiv 0 \pmod{17}$, then that $w \equiv 0 \pmod{17}$, and finally that $17^2 \equiv 0 \pmod{17^3}$. This contradiction shows that $C_{17}(\mathbb{Q}_{17}) = \varnothing$, so $17 \notin S^{(\phi)}(E/\mathbb{Q})$.

We now know that

$$1, -2, 2 \in S^{(\phi)}(E/\mathbb{Q}) \quad \text{and} \quad 17 \notin S^{(\phi)}(E/\mathbb{Q}).$$

Since $S^{(\phi)}(E/\mathbb{Q})$ is a subgroup of $\mathbb{Q}(S, 2)$, it follows that $S^{(\phi)}(E/\mathbb{Q}) = \{\pm 1, \pm 2\}$. We have also shown that $E'(\mathbb{Q})$ surjects onto $S^{(\phi)}(E/\mathbb{Q})$, and so from (4.2a), $\mathrm{III}(E/\mathbb{Q})[\phi] = 0$.

We now repeat the above computation with the roles of E and E' reversed. Thus for $d \in \mathbb{Q}(S, 2)$, we look at the homogeneous space

$$C'_d : dw^2 = d^2 - 24dz^2 + 272z^4.$$

As above, the point $(0, 0) \in E(\mathbb{Q})$ maps to $\delta(0, 0) = 272 \equiv 17 \pmod{\mathbb{Q}^{*2}}$. Next, if $d < 0$, then clearly $C'_d(\mathbb{R}) = \varnothing$, so $d \notin S^{(\hat{\phi})}(E'/\mathbb{Q})$. Finally, for $d = 2$, if we let $z = Z/2$, then C'_2 has the equation

$$C'_2 : 2w^2 = 4 - 12Z^2 + 17Z^4.$$

But if $C'_2(\mathbb{Q}_2) \neq \varnothing$, then necessarily $Z, w \in \mathbb{Z}_2$; and then from the equation for C'_2 we deduce successively $Z \equiv 0 \pmod{2}$, $w \equiv 0 \pmod{2}$, $4 \equiv 0 \pmod{2^3}$. Therefore $C'_2(\mathbb{Q}_2) = \varnothing$, and so $2 \notin S^{(\hat\phi)}(E'/\mathbb{Q})$. Hence $S^{(\hat\phi)}(E'/\mathbb{Q}) = \{1, 17\}$ and $Ш(E'/\mathbb{Q})[\hat\phi] = 0$.

To recapitulate, we now know that

$$E'(\mathbb{Q})/\phi(E(\mathbb{Q})) \cong (\mathbb{Z}/2\mathbb{Z})^2 \quad \text{and} \quad E(\mathbb{Q})/\hat\phi(E'(\mathbb{Q})) \cong \mathbb{Z}/2\mathbb{Z},$$

the former being generated by $\{(0, 0), (8, -32)\}$ and the latter by $\{(0, 0)\}$. The exact sequence (4.7) then yields

$$E(\mathbb{Q})/2E(\mathbb{Q}) \cong (\mathbb{Z}/2\mathbb{Z})^2 \cong E'(\mathbb{Q})/2E'(\mathbb{Q});$$

and so

$$E(\mathbb{Q}) \cong E'(\mathbb{Q}) \cong \mathbb{Z} \times \mathbb{Z}/2\mathbb{Z}.$$

Remark 4.11. In all of the examples up to this point, we have been lucky in the sense that for every locally trivial homogeneous space that has appeared, we have been able to find (by inspection) a global rational point. Another way to say this is that we have yet to see a non-trivial element in the Shafarevich–Tate group. The first examples of such spaces are due to Lind [Lin] and (independently, but shortly later) Reichardt [Rei], who proved that the curve

$$2w^2 = 1 - 17z^4$$

has no $\mathbb{Q}$-rational point. (One easily checks that it has a point defined over every $\mathbb{Q}_p$.) We will prove a more general result below (6.5). Shortly thereafter, Selmer [Sel 1] made an extensive study of the curves $ax^3 + by^3 + cz^3 = 0$, which are homogeneous spaces for the elliptic curves $x^3 + y^3 + dz^3 = 0$. He gave many examples of locally trivial, globally non-trivial homogeneous spaces, of which the simplest is

$$3x^3 + 4y^3 + 5z^3 = 0.$$

It is a difficult problem, in general, to divide the Selmer group into the piece coming from rational points on the elliptic curve and the piece giving non-trivial elements in the Shafarevich–Tate group. At present, there is no algorithm known which is guaranteed to solve this problem. The procedure which we now describe will often work in practice, although it tends to lead to fairly elaborate computations in algebraic number fields.

Recall that for each integer $m \geqslant 2$ there is an exact sequence (4.2a)

$$E(K) \xrightarrow{\delta} S^{(m)}(E/K) \to Ш(E/K)[m] \to 0;$$

and the finite group $S^{(m)}(E/K)$ is effectively computable, at least in theory (4.5). If we knew some way of computing $Ш(E/K)[m]$, then we would be able to find generators for $E(K)/mE(K)$, and thence for $E(K)$. Unfortunately, a general procedure for computing $Ш(E/K)[m]$ is still being sought. However, for each integer $n \geq 1$ we can fit together the above exact sequences to form a commutative diagram

$$E(K) \to S^{(m^n)}(E/K) \to Ш(E/K)[m^n] \to 0$$

$$\downarrow id \qquad \downarrow \qquad \qquad \downarrow \text{mult. by } m^{n-1}$$

$$E(K) \to S^{(m)}(E/K) \to Ш(E/K)[m] \to 0.$$

Now at least in principle, the middle column of this diagram is effectively computable. This allows us to make the following refinement to the exact sequence in (4.2a).

Proposition 4.12. Let E/K be an elliptic curve. For integers $m \geq 2$ and $n \geq 1$, let $S^{(m,n)}(E/K)$ be the image of $S^{(m^n)}(E/K)$ in $S^{(m)}(E/K)$. Then there is an exact sequence.

$$0 \to E(K)/mE(K) \to S^{(m,n)}(E/K) \to m^{n-1}Ш(E/K)[m^n] \to 0.$$

PROOF. Immediate from the commutative diagram given above. □

Now to find generators for $E(K)$, one can apply the following procedure. Compute successively the *relative Selmer groups*

$$S^{(m)}(E/K) = S^{(m,1)}(E/K) \supset S^{(m,2)}(E/K) \supset S^{(m,3)}(E/K) \supset \cdots$$

and the *rational-point groups*

$$T_{(m,1)}(E/K) \subset T_{(m,2)}(E/K) \subset T_{(m,3)}(E/K) \subset \cdots,$$

where $T_{(m,r)}(E/K)$ is the subgroup of $S^{(m)}(E/K)$ generated by all points $P \in E(K)$ with height $h_x(P) \leq r$. Eventually, with sufficient perserverence, one hopes to arrive at an equality

$$S^{(m,n)}(E/K) = T_{(m,r)}(E/K).$$

Once this occurs, then one knows that $m^{n-1}Ш(E/K)[m^n] = 0$, and that the points with height $h_x(P) \leq r$ generate $E(K)/mE(K)$. The problem is that, as far as is known, there is nothing to prevent $Ш(E/K)$ from containing an element which is infinitely m-divisible; that is, a $\xi \in Ш(E/K)$, $\xi \neq 0$, such that for every $n \geq 1$ there is a $\xi_n \in Ш(E/K)$ such that $\xi = m^n\xi_n$. If such an element were to exist, then the above procedure would never terminate! However, opposed to such a gloomy scenario is the following optimistic conjecture.

Conjecture 4.13. Let E/K be an elliptic curve. Then $Ш(E/K)$ is finite.

This conjecture has been proven for certain elliptic curves by Kolyvagin and Rubin. Note that the successful carrying out of the procedure described above will only show that the m-primary component of Ш is finite; this has of course been done in many cases. (For example, we showed that for the elliptic curve in (4.10), $Ш(E/\mathbb{Q})$ has trivial 2-primary component.)

We close by quoting the following beautiful result of Cassels, which says something quite interesting about the order of this group which is not yet known to be finite.

Theorem 4.14 ([Ca 3], [Ta 2]). *Let E/K be an elliptic curve. There exists an alternating, bilinear pairing*

$$\Gamma : Ш(E/K) \times Ш(E/K) \to \mathbb{Q}/\mathbb{Z}$$

whose kernel is precisely the group of divisible elements of Ш. (I.e. If $\Gamma(\alpha, \beta) = 0$ for all $\beta \in Ш$, then there exist arbitrarily large integers N and elements $\alpha_N \in Ш$ such that $\alpha = N\alpha_N$.)

In particular, if $Ш(E/K)$ is finite (or, more generally, if any p-primary component of $Ш(E/K)$ is finite), then its order is a perfect square. (See exer. 10.20.)

§5. Twisting—Elliptic Curves

Again we let K be an arbitrary (perfect) field, and let E/K be an elliptic curve. As we have seen (2.2), if we consider E merely as a curve and ignore the basepoint O, then the twists of E/K correspond to the elements of the (pointed) cohomology set $H^1(G_{\bar{K}/K}, \text{Isom}(E))$. Now $\text{Isom}(E)$ has two obvious subgroups, namely $\text{Aut}(E)$ and E, where we identify E with the set of translations $\{\tau_P\}$ in $\text{Isom}(E)$. Notice also that $\text{Aut}(E)$ acts naturally on E. The next proposition describes $\text{Isom}(E)$.

Proposition 5.1. *The map*

$$E \times \text{Aut}(E) \to \text{Isom}(E)$$

$$(P, \alpha) \to \tau_P \circ \alpha$$

is a bijection of sets. It identifies $\text{Isom}(E)$ with the product of E and $\text{Aut}(E)$ twisted by the natural action of $\text{Aut}(E)$ on E. [I.e. $\text{Isom}(E)$ is the set $E \times \text{Aut}(E)$ with the group law

$$(P, \alpha)\cdot(Q, \beta) = (P + \alpha Q, \alpha \circ \beta).]$$

PROOF. Let $\phi \in \text{Isom}(E)$. Then $\tau_{-\phi(O)} \circ \phi \in \text{Aut}(E)$, so writing

$$\phi = \tau_{\phi(O)} \circ (\tau_{-\phi(O)} \circ \phi)$$

shows that the map is surjective. On the other hand, if $\tau_P \circ \alpha = \tau_Q \circ \beta$, than evaluating at O gives $P = Q$, and then also $\alpha = \beta$. This proves injectivity.

Finally, the twisted nature of the group law follows from the calculation

$$\tau_P \circ \alpha \circ \tau_Q \circ \beta = \tau_P \circ \tau_{\alpha Q} \circ \alpha \circ \beta. \qquad \square$$

We have already made an extensive study of those twists of E/K arising from translations, namely the group $H^1(G_{\bar{K}/K}, E) \cong WC(E/K)$ studied in sections 3 and 4. We now look at the twists of E/K coming from isomorphisms of E *as an elliptic curve*; that is, isomorphisms of the pair (E, O). In other words, we consider the twists of E/K corresponding to elements of $H^1(G_{\bar{K}/K}, \mathrm{Aut}(E))$. We start with a general proposition, and then (for $\mathrm{char}(K) \neq 2, 3$) derive explicit equations.

Remark 5.2. In the literature, the phrase "let C be a twist of E" often means that C corresponds to an element of $H^1(G_{\bar{K}/K}, \mathrm{Aut}(E))$. More properly, such a C should be called a twist of the pair (E, O), since the group of isomorphisms of (E, O) with itself is precisely $\mathrm{Aut}(E)$. However, one can generally resolve any ambiguity from context without too much trouble.

Proposition 5.3. *Let E/K be an elliptic curve.*
(a) *The natural inclusion $\mathrm{Aut}(E) \subset \mathrm{Isom}(E)$ induces an inclusion*

$$H^1(G_{\bar{K}/K}, \mathrm{Aut}(E)) \subset H^1(G_{\bar{K}/K}, \mathrm{Isom}(E)).$$

Identifying the latter set with $\mathrm{Twist}(E/K)$ by (2.2), we will denote the former by $\mathrm{Twist}((E, O)/K)$.
(b) *Let $C/K \in \mathrm{Twist}((E, O)/K)$. Then $C(K) \neq \varnothing$, and so C/K can be given the structure of an elliptic curve over K. [N.B. C is not generally K-isomorphic to E. Contrast with (3.3).]*
(c) *Conversely, if E'/K is an elliptic curve which is isomorphic to E over $\bar{K}$, then E'/K represents an element of $\mathrm{Twist}((E, O)/K)$.*

PROOF. (a) Let $i : \mathrm{Aut}(E) \to \mathrm{Isom}(E)$ be the natural inclusion. From (5.1), there is a homomorphism $j : \mathrm{Isom}(E) \to \mathrm{Aut}(E)$ such that $j \circ i = 1$. It follows that the induced map

$$H^1(G_{\bar{K}/K}, \mathrm{Aut}(E)) \overset{i}{\to} H^1(G_{\bar{K}/K}, \mathrm{Isom}(E))$$

is one-to-one.
(b) Let $\phi : C \to E$ be an isomorphism defined over $\bar{K}$ such that the cocycle

$$\sigma \to \phi^\sigma \circ \phi^{-1}$$

represents the element of $H^1(G_{\bar{K}/K}, \mathrm{Aut}(E))$ corresponding to C/K. Then $\phi^\sigma \circ \phi^{-1}(O) = O$, so

$$\phi^{-1}(O) = \phi^{-1}(O)^\sigma \qquad \text{for all } \sigma \in G_{\bar{K}/K}.$$

Hence $\phi^{-1}(O) \in C(K)$, so $(C, \phi^{-1}(O))$ is an elliptic curve defined over K.
(c) Let $\phi : E' \to E$ be an isomorphism of elliptic curves defined over $\bar{K}$. In

particular, $\phi(O') = O$, where $O \in E(K)$ and $O' \in E'(K)$ are the respective zero points of E and E'. Then for any $\sigma \in G_{\bar{K}/K}$,

$$\phi^\sigma \circ \phi^{-1}(O) = \phi^\sigma(O') = \phi(O')^\sigma = O^\sigma = O.$$

Thus $\phi^\sigma \circ \phi^{-1} \in \mathrm{Aut}(E)$, so the cocycle corresponding to E'/K lies in $H^1(G_{\bar{K}/K}, \mathrm{Aut}(E))$ as desired. $\square$

If the characteristic of K is not 2 or 3, then the elements of $\mathrm{Twist}((E, O)/K)$ can be described quite explicitly.

Proposition 5.4. *Assume that* $\mathrm{char}(K) \neq 2, 3$, *and let*

$$n = \begin{cases} 2 & \text{if } j(E) \neq 0, 1728 \\ 4 & \text{if } j(E) = 1728 \\ 6 & \text{if } j(E) = 0. \end{cases}$$

Then $\mathrm{Twist}((E, O)/K)$ *is canonically isomorphic to* K^*/K^{*n}.
   More precisely, choose a Weierstrass equation

$$E : y^2 = x^3 + Ax + B$$

for E/K, *and let* $D \in K^*$. *Then the elliptic curve* $E_D \in \mathrm{Twist}((E, O)/K)$ *corresponding to* D (mod K^{*n}) *has the Weierstrass equation*

(i) $E_D : y^2 = x^3 + D^2 Ax + D^3 B$ *if* $j(E) \neq 0, 1728$;
(ii) $E_D : y^2 = x^3 + DAx$ *if* $j(E) = 1728$ (*so* $B = 0$);
(iii) $E_D : y^2 = x^3 + DB$ *if* $j(E) = 0$ (*so* $A = 0$).

Corollary 5.4.1. *Define an equivalence* $\sim$ *on the set* $K \times K^*$ *by*

$$(j, D) \sim (j', D') \qquad \text{if } j = j' \text{ and } D/D' \in (K^*)^{n(j)},$$

where $n(j) = 2$ (*resp.* 4, *resp.* 6) *if* $j \neq 0, 1728$ (*resp.* $j = 1728$, *resp.* $j = 0$). *Then the* K-*isomorphism classes of elliptic curves* E/K *are in one-to-one correspondence with elements of the quotient*

$$K \times K^*/\sim.$$

PROOF. From (III.10.2), we have an isomorphism

$$\mathrm{Aut}(E) \cong \boldsymbol{\mu}_n$$

of $G_{\bar{K}/K}$-modules. It follows from (B.2.5c) that

$$\mathrm{Twist}((E, O)/K) = H^1(G_{\bar{K}/K}, \mathrm{Aut}(E)) \cong H^1(G_{\bar{K}/K}, \boldsymbol{\mu}_n) \cong K^*/K^{*n}.$$

The calculation of an equation for the twist E_D is straightforward. The case $j(E) \neq 0, 1728$ was already done in (2.4). We will do $j(E) = 1728$ here, and leave $j(E) = 0$ for the reader.
   Thus let $D \in K^*$, choose a fourth root $\delta \in \bar{K}$ of D, and define a cocycle

$$\xi : G_{\bar{K}/K} \to \boldsymbol{\mu}_4 \qquad \xi_\sigma = \delta^\sigma/\delta.$$

We also fix an isomorphism

$$[\;\;] : \mu_4 \to \text{Aut}(E) \qquad [\zeta](x, y) = (\zeta^2 x, \zeta y).$$

Then E_D corresponds to the cocycle $\sigma \to [\xi_\sigma]$ in $H^1(G_{\bar{K}/K}, \text{Aut}(E))$.
    Now the action of $G_{\bar{K}/K}$ on the twisted field $\bar{K}(E)_\xi$ is given by

$$\delta^\sigma = \xi_\sigma \delta \qquad x^\sigma = \xi_\sigma^2 x \qquad y^\sigma = \xi_\sigma y.$$

The subfield fixed by $G_{\bar{K}/K}$ thus contains the functions

$$X = \delta^{-2} x \quad \text{and} \quad Y = \delta^{-1} y,$$

and these functions satisfy the equation

$$Y^2 = DX^3 + AX.$$

This gives the desired equation for the twist E_D/K, and the substitution
$(X, Y) = (D^{-1} X', D^{-1} Y')$ puts it into the required form.
    The corollary follows by combining the proposition and (5.3c) with
(III.1.4bc), which says that up to $\bar{K}$-isomorphism, the elliptic curves E/K are
in one-to-one correspondence with their j-invariants $j(E) \in K$.   □

§6. The Curve $Y^2 = X^3 + DX$

Many of the deepest theorems and conjectures in the arithmetic theory of
elliptic curves have had as their testing grounds one of the families of curves
given in (5.4i, ii, iii). To illustrate the theory that we have developed, let us see
what we can say about the family of elliptic curves $E/\mathbb{Q}$ with j-invariant
$j(E) = 1728$.
    One such curve is given by the equation

$$y^2 = x^3 + x;$$

and then from (5.3) and (5.4) we see that every such curve has an equation

$$E : y^2 = x^3 + Dx,$$

where D ranges over representatives for the cosets in $\mathbb{Q}^*/\mathbb{Q}^{*4}$. Thus if we
specify that D be a fourth-power-free integer, then it is uniquely determined
by E. Notice that the equation for E has discriminant $\Delta(E) = -64D^3$, so E
has good reduction at all primes not dividing $2D$.
    Let p be a prime not dividing $2D$, and consider the reduced curve $\tilde{E}$ over
the finite field $\mathbb{F}_p$. From (V.4.1), $\tilde{E}$ is supersingular if and only if the coefficient
of x^{p-1} in $(x^3 + Dx)^{(p-1)/2}$ is zero. In particular, if $p \equiv 3 \pmod 4$, then $\tilde{E}/\mathbb{F}_p$ is
supersingular; and so from (exer. 5.10) we conclude that

$$\#\tilde{E}(\mathbb{F}_p) = p + 1 \qquad \text{for all } p \equiv 3 \pmod 4.$$

(See exer. 10.17 for an elementary derivation of this result.)

Next we recall (VII.3.5) that if $p \neq 2$ and E has good reduction at p, then $E_{\mathrm{tors}}(\mathbb{Q})$ injects into the reduction $\tilde{E}(\mathbb{F}_p)$. It follows from above that $\# E_{\mathrm{tors}}(\mathbb{Q})$ divides $p + 1$ for all but finitely many primes $p \equiv 3 \pmod 4$; hence $\# E_{\mathrm{tors}}(\mathbb{Q})$ divides 4. Since $(0, 0) \in E(\mathbb{Q})[2]$, the only possibilities for $E_{\mathrm{tors}}(\mathbb{Q})$ are $\mathbb{Z}/2\mathbb{Z}$, $(\mathbb{Z}/2\mathbb{Z})^2$, and $\mathbb{Z}/4\mathbb{Z}$.

Now $E[2] \subset E(\mathbb{Q})$ if and only if the polynomial $x^3 + Dx$ factors completely over $\mathbb{Q}$, so if and only if $-D$ is a perfect square. Similarly, $E(\mathbb{Q})$ will have a point of order 4 if and only if $(0, 0) \in 2E(\mathbb{Q})$. The duplication formula for E reads

$$x(2P) = (x^2 - D)^2/(4x^3 + 4Dx),$$

so we see that

$$(0, 0) = [2](D^{1/2}, (4D^3)^{1/4}).$$

Hence assuming that D is a fourth-power-free integer, we conclude that $(0, 0) \in 2E(\mathbb{Q})$ if and only if $D = 4$; in which case $(0, 0) = [2](2, \pm 4)$.

Next, since $E(\mathbb{Q})$ contains a 2-torsion point, we may use (4.9) to try to calculate $E(\mathbb{Q})/2E(\mathbb{Q})$. E is isogenous to the curve

$$E' : Y^2 = X^3 - 4DX$$

via the isogeny

$$\phi : E \to E' \qquad \phi(x, y) = (y^2/x^2, y(D - x^2)/x^2).$$

The set $S \subset M_{\mathbb{Q}}$ consists of ∞ and the primes dividing $2D$; and for each $d \in \mathbb{Q}(S, 2)$, the corresponding homogeneous space $C_d/\mathbb{Q} \in WC(E/\mathbb{Q})$ is given by the equation

$$C_d : dw^2 = d^2 - 4Dz^4.$$

Similarly, working with the dual isogeny $\hat{\phi} : E' \to E$ leads to the homogeneous spaces $C_d'/\mathbb{Q} \in WC(E'/\mathbb{Q})$ with equations

$$C_d' : dW^2 = d^2 + DZ^4.$$

(Actually (4.9) leads to the equation $dW^2 = d^2 + 16DZ^4$, but we are free to replace Z by $Z/2$.)

Let $v(2D)$ be the number of distinct primes dividing $2D$. Since $\mathbb{Q}(S, 2)$ is generated by -1 and the primes dividing $2D$, we have the estimate

$$\dim_2 E(\mathbb{Q})/2E(\mathbb{Q}) \leqslant 2 + 2v(2D) - \dim_2 E'(\mathbb{Q})[\hat{\phi}] + \dim_2 \phi(E(\mathbb{Q})[2]).$$

Here $\dim_2$ denotes the dimension of an $\mathbb{F}_2$-vector space. Now clearly $E'(\mathbb{Q})[\hat{\phi}] \cong \mathbb{Z}/2\mathbb{Z}$. Next, to deal with the other two terms, we consider two cases.

(1) $E(\mathbb{Q})[2] \cong \mathbb{Z}/2\mathbb{Z}$.
    Then $\phi(E(\mathbb{Q})[2]) \cong 0$ and $\dim_2 E(\mathbb{Q})/2E(\mathbb{Q}) = \mathrm{rank}\ E(\mathbb{Q}) + 1$.
(2) $E(\mathbb{Q})[2] \cong \mathbb{Z}/2\mathbb{Z} \times \mathbb{Z}/2\mathbb{Z}$.
    Then $\phi(E(\mathbb{Q})[2]) \cong \mathbb{Z}/2\mathbb{Z}$ and $\dim_2 E(\mathbb{Q})/2E(\mathbb{Q}) = \mathrm{rank}\ E(\mathbb{Q}) + 2$.

Substituting these values into the above inequality yields in both cases the estimate

$$\text{rank } E(\mathbb{Q}) \leqslant 2v(2D).$$

Notice that we have not yet checked any of the homogeneous spaces C_d or C_d' for local triviality. But by inspection, if $d < 0$, then either $C_d(\mathbb{R}) = \varnothing$ or $C_d'(\mathbb{R}) = \varnothing$. Thus our estimate may be cut by 1, giving the slight improvement

$$\text{rank } E(\mathbb{Q}) \leqslant 2v(2D) - 1.$$

We summarize the preceding discussion in the following proposition.

Proposition 6.1. *For each fourth-power-free integer D, let $E_D/\mathbb{Q}$ be the elliptic curve*

$$E_D : y^2 = x^3 + Dx.$$

(a)
$$E_{D,\text{tors}}(\mathbb{Q}) \cong \begin{cases} \mathbb{Z}/4\mathbb{Z} & \text{if } D = 4 \\ \mathbb{Z}/2\mathbb{Z} \times \mathbb{Z}/2\mathbb{Z} & \text{if } -D \text{ is a perfect square} \\ \mathbb{Z}/2\mathbb{Z} & \text{otherwise.} \end{cases}$$

(b)
$$\text{rank } E_D(\mathbb{Q}) \leqslant 2v(2D) - 1.$$

Let us now restrict attention to the special case that $D = p$ is an odd prime. Then the following proposition gives a complete description of the resulting Selmer groups and deduces corresponding upper bounds for the rank of $E(\mathbb{Q})$ and the dimension of $Ш(E/\mathbb{Q})[2]$.

Proposition 6.2. *For each odd prime p, let $E_p/\mathbb{Q}$ be the elliptic curve*

$$E_p : y^2 = x^3 + px,$$

and let $\phi : E_p \to E_p'$ be the isogeny of degree 2 with kernel $E_p[\phi] = \{O, (0,0)\}$.

(a) $E_{p,\text{tors}}(\mathbb{Q}) \cong \mathbb{Z}/2\mathbb{Z}.$

(b) $S^{(\hat{\phi})}(E_p'/\mathbb{Q}) \cong \mathbb{Z}/2\mathbb{Z}.$

$$S^{(\phi)}(E_p/\mathbb{Q}) \cong \begin{cases} \mathbb{Z}/2\mathbb{Z} & \text{if } p \equiv 7, 11 \pmod{16} \\ (\mathbb{Z}/2\mathbb{Z})^2 & \text{if } p \equiv 3, 5, 13, 15 \pmod{16} \\ (\mathbb{Z}/2\mathbb{Z})^3 & \text{if } p \equiv 1, 9 \pmod{16}. \end{cases}$$

(c) $\text{rank } E_p(\mathbb{Q}) + \dim_2 Ш(E_p/\mathbb{Q})[2] = \begin{cases} 0 & \text{if } p \equiv 7, 11 \pmod{16} \\ 1 & \text{if } p \equiv 3, 5, 13, 15 \pmod{16} \\ 2 & \text{if } p \equiv 1, 9 \pmod{16}. \end{cases}$

PROOF. To ease notation, we let $E = E_p$ and $E' = E_p'$.
(a) This was proven above (6.1a).

(b) As usual, we take representatives $\{\pm 1, \pm 2, \pm p, \pm 2p\}$ for the cosets in $\mathbb{Q}(S, 2)$. From (4.9), the images of the 2-torsion points in the Selmer groups are given by

$$-p \in S^{(\phi)}(E/\mathbb{Q}) \quad \text{and} \quad p \in S^{(\hat{\phi})}(E'/\mathbb{Q}).$$

Further, if $d < 0$, then by inspection $C'_d(\mathbb{R}) = \varnothing$, so $d \notin S^{(\hat{\phi})}(E'/\mathbb{Q})$.

Next we consider the homogeneous space

$$C'_2 : 2W^2 = 4 + pZ^4.$$

If $(Z, W) \in C'_2(\mathbb{Q}_2)$, then necessarily $Z, W \in \mathbb{Z}_2$; and then we conclude that $Z \equiv 0 \pmod{2}$, so $W \equiv 0 \pmod{2}$, so $0 \equiv 4 \pmod{8}$. Therefore $C'_2(\mathbb{Q}_2) = \varnothing$, and hence $2 \notin S^{(\hat{\phi})}(E'/\mathbb{Q})$. We now know that

$$p \in S^{(\hat{\phi})}(E'/\mathbb{Q}) \qquad -1, \pm 2, -p, -2p \notin S^{(\hat{\phi})}(E'/\mathbb{Q}),$$

from which it follows that $S^{(\hat{\phi})}(E'/\mathbb{Q}) = \{1, p\} \cong \mathbb{Z}/2\mathbb{Z}$.

It remains to calculate $S^{(\phi)}(E/\mathbb{Q})$; and from the form the answer takes, it is clear that there will be many cases to consider. The best approach is to consider the various $d \in \mathbb{Q}(S, 2)$, and check for which primes the homogeneous space C_d is locally trivial. Note that from (4.9), d will be in $S^{(\phi)}(E/\mathbb{Q})$ if and only if $C_d(\mathbb{Q}_p) \neq \varnothing$ and $C_d(\mathbb{Q}_2) \neq \varnothing$. (I.e. It suffices to check whether C_d is locally trivial at the primes p and 2.) We will make frequent use of Hensel's lemma (exer. 10.12), which gives a criterion for when a solution of an equation modulo q^n lifts to a solution in $\mathbb{Q}_q$.

$\boxed{d = -1}$ $\qquad\qquad\qquad C_{-1} : w^2 + 1 = 4pz^4$

(i) If $(z, w) \in C_{-1}(\mathbb{Q}_p)$, then necessarily $z, w \in \mathbb{Z}_p$, and so $w^2 \equiv -1 \pmod{p}$. Conversely, by (exer. 10.12), any solution to $w^2 \equiv -1 \pmod{p}$ will lift to a point in $C_{-1}(\mathbb{Q}_p)$. Therefore

$$C_{-1}(\mathbb{Q}_p) \neq \varnothing \quad \Leftrightarrow \quad p \equiv 1 \pmod{4}.$$

(ii) From (i), we may assume that $p \equiv 1 \pmod{4}$. If $p \equiv 1 \pmod{8}$, we let $(z, w) = (Z/4, W/8)$. Then our equation becomes $W^2 + 64 = pZ^4$, and the solution $(Z, W) = (1, 1)$ to the congruence

$$W^2 + 64 \equiv pZ^4 \pmod{8}$$

lifts to a point in $C_{-1}(\mathbb{Q}_2)$. Similarly, if $p \equiv 5 \pmod{8}$, then we let $(z, w) = (Z/2, W/2)$; and again we have a solution $(Z, W) = (1, 1)$ to a congrunce

$$W^2 + 4 \equiv pZ^4 \pmod{8}$$

which lifts to a point in $C_{-1}(\mathbb{Q}_2)$. This proves that if $p \equiv 1 \pmod{4}$, then $C_{-1}(\mathbb{Q}) \neq \varnothing$.

Combining the results of (i) and (ii) yields

$$-1 \in S^{(\phi)}(E/\mathbb{Q}) \quad \Leftrightarrow \quad p \equiv 1 \pmod{4}.$$

$\boxed{d = -2}$ $\qquad\qquad\qquad C_{-2} : w^2 + 2 = 2pz^4$

(i) If $(z, w) \in C_{-2}(\mathbb{Q}_p)$, then $z, w \in \mathbb{Z}_p$ and $w^2 \equiv -2 \pmod{p}$. Conversely, a solution to $w^2 \equiv -2 \pmod{p}$ lifts to a point of $C_{-2}(\mathbb{Q}_p)$. Therefore

$$C_{-2}(\mathbb{Q}_p) \neq \emptyset \quad \Leftrightarrow \quad p \equiv 1, 3 \pmod{8}.$$

(ii) If $(z, w) \in C_{-2}(\mathbb{Q}_2)$, then $z, w \in \mathbb{Z}_2$ and $w \equiv 0 \pmod{2}$. Letting $(z, w) = (Z, 2W)$, we must check if there are any solution $Z, W \in \mathbb{Z}_2$ to the equation

$$2W^2 + 1 = pZ^4.$$

From (i), it suffices to consider those primes $p \equiv 1, 3 \pmod{8}$. Now the congruence $2W^2 + 1 \equiv pZ^4 \pmod{16}$ has no solutions if $p \equiv 11 \pmod{16}$, so

$$p \equiv 11 \pmod{16} \quad \Rightarrow \quad C_{-2}(\mathbb{Q}_2) = \emptyset.$$

On the other hand, in order to use (exer. 10.12), we must find solutions modulo $2^5 = 32$ if we want to lift them to points in $C_{-2}(\mathbb{Q}_2)$. The following table gives solutions (Z, W) to the congruence

$$2W^2 + 1 = pZ^4 \pmod{32}$$

for each of the remaining values of $p \pmod{32}$.

p (mod 32)	1	3	9	17	19	25
(Z, W)	$(1, 0)$	$(3, 11)$	$(1, 2)$	$(3, 0)$	$(1, 3)$	$(3, 2)$

Combining (i) and (ii), we have proven that

$$-2 \in S^{(\phi)}(E/\mathbb{Q}) \quad \Leftrightarrow \quad p \equiv 1, 3, 9 \pmod{16}.$$

$\boxed{d = 2}$ $\qquad\qquad C_2 : w^2 = 2 - 2pz^4$

This is entirely similar to the case $d = -2$ just completed. A point $(z, w) \in C_2(\mathbb{Q}_p)$ will have $z, w \in \mathbb{Z}_p$ and $w^2 \equiv 2 \pmod{p}$, and any such solution will lift, so

$$C_2(\mathbb{Q}_p) \neq \emptyset \quad \Leftrightarrow \quad p \equiv 1, 7 \pmod{8}.$$

Now if $p \equiv 1 \pmod{8}$, then from above $-1, -2 \in S^{(\phi)}(E/\mathbb{Q})$, so certainly $2 \in S^{(\phi)}(E/\mathbb{Q})$. It remains to consider the case $p \equiv 7 \pmod{8}$.

A point $(z, w) \in C_2(\mathbb{Q}_2)$ will have $(z, w) = (Z, 2W)$ with $Z, W \in \mathbb{Z}_2$ and

$$2W^2 = 1 - pZ^4.$$

There are no solutions modulo 16 if $p \equiv 7 \pmod{16}$. On the other hand, if $p \equiv 15 \pmod{16}$, then the solutions

$$2 \cdot 3^2 \equiv 1 - p \cdot 1^4 \pmod{32} \quad \text{if } p \equiv 15 \pmod{32},$$

$$2 \cdot 1^2 \equiv 1 - p \cdot 1^4 \pmod{32} \quad \text{if } p \equiv 31 \pmod{32},$$

lift to points in $C_2(\mathbb{Q}_2)$. Putting all of this together, we have shown that

$$2 \in S^{(\phi)}(E/\mathbb{Q}) \quad \Leftrightarrow \quad p \equiv 1, 9, 15 \ (\mathrm{mod}\ 16).$$

We have now determined exactly which of $-1, \pm 2$ are in $S^{(\phi)}(E/\mathbb{Q})$ in terms of the residue of p modulo 16. Since also $-p \in S^{(\phi)}(E/\mathbb{Q})$, it is now a simple matter to reconstruct the table for $S^{(\phi)}(E/\mathbb{Q})$ given in (b). [In fact, one obtains even more information, namely a precise list of which elements of $\mathbb{Q}(S, 2)$ are in $S^{(\phi)}(E/\mathbb{Q})$.]

(c) We use (4.7) and (4.2a) to compute

$$\dim_2 E'(\mathbb{Q})[\hat{\phi}]/\phi(E(\mathbb{Q})[2]) + \dim_2 E(\mathbb{Q})/2E(\mathbb{Q})$$

$$= \dim_2 E'(\mathbb{Q})/\phi(E(\mathbb{Q})) + \dim_2 E(\mathbb{Q})/\hat{\phi}(E'(\mathbb{Q}))$$

$$= \dim_2 S^{(\phi)}(E/\mathbb{Q}) - \dim_2 Ш(E/\mathbb{Q})[\phi]$$

$$+ \dim_2 S^{(\hat{\phi})}(E'/\mathbb{Q}) - \dim_2 Ш(E'/\mathbb{Q})[\hat{\phi}].$$

From (a), we see that

$$E'(\mathbb{Q})[\hat{\phi}]/\phi(E(\mathbb{Q})[2]) \cong \mathbb{Z}/2\mathbb{Z} \quad \text{and} \quad E(\mathbb{Q})/2E(\mathbb{Q}) \cong (\mathbb{Z}/2\mathbb{Z})^{1+\mathrm{rank}\ E(\mathbb{Q})}.$$

Further, since $E(\mathbb{Q})/\hat{\phi}E'(\mathbb{Q}) \cong S^{(\hat{\phi})}(E'/\mathbb{Q}) \cong \mathbb{Z}/2\mathbb{Z}$ from (b), the exact sequence (4.2a) implies that $Ш(E'/\mathbb{Q})[\hat{\phi}] = 0$. Hence the exact sequence

$$0 \to Ш(E/\mathbb{Q})[\phi] \to Ш(E/\mathbb{Q})[2] \xrightarrow{\phi} Ш(E'/\mathbb{Q})[\hat{\phi}] = 0$$

gives

$$\dim_2 Ш(E/\mathbb{Q})[2] = \dim_2 Ш(E/\mathbb{Q})[\phi];$$

and combining this with the above results yields

$$1 + (1 + \mathrm{rank}\ E(\mathbb{Q})) = \dim_2 S^{(\phi)}(E/\mathbb{Q}) + \dim_2 S^{(\hat{\phi})}(E'/\mathbb{Q}) - \dim_2 Ш(E/\mathbb{Q})[2].$$

Now (c) is immediate from the calculation of $S^{(\phi)}(E/\mathbb{Q})$ and $S^{(\hat{\phi})}(E'/\mathbb{Q})$ given in (b). $\qquad\qquad\Box$

Corollary 6.2.1. *There are infinitely many elliptic curves $E/\mathbb{Q}$ with*

$$\mathrm{rank}\ E(\mathbb{Q}) = 0 \quad \text{and} \quad Ш(E/\mathbb{Q})[2] = 0.$$

PROOF. From (6.2), the elliptic curves $y^2 = x^3 + px$ with $p \equiv 7, 11 \ (\mathrm{mod}\ 16)$ have this property. $\qquad\qquad\Box$

Remark 6.3. One of the consequences of (6.2) is that if p is a prime with $p \equiv 5 \ (\mathrm{mod}\ 8)$, then the elliptic curve

$$E_p : y^2 = x^3 + px$$

has rank at most 1. Further, examining the proof of (6.2), it will have rank 1 if and only if the homogeneous space

$$C_{-1} : w^2 + 1 = 4pz^4$$

has a $\mathbb{Q}$-rational point; and if there is such a point, then we can find a point of infinite order in $E(\mathbb{Q})$ by using the map (cf. 4.9)

$$\hat{\phi} \circ \psi : C_{-1} \to E \qquad \hat{\phi} \circ \psi(z, w) = (w^2/4z^2, w(w^2 + 2)/8z^3).$$

Taking the first few values for p, one does indeed find a point in $C_{-1}(\mathbb{Q})$, and these give the points of infinite order in $E(\mathbb{Q})$ listed in the following table.

p	5	13	29	37
(x, y)	$(1/4, 9/8)$	$(9/4, 51/8)$	$(25/4, 165/8)$	$(22801/900, 3540799/27000)$

Suppose that we knew, a priori, that the Shafarevich–Tate group $Ш(E_p/\mathbb{Q})$ were finite; or even that its 2-primary component were finite. Then the existence of the Cassels' pairing (4.14) would imply in particular that $\dim_2 Ш(E_p/\mathbb{Q})[2]$ is even, and so that $E_p(\mathbb{Q})$ has rank 1 for *all* primes $p \equiv 5 \pmod 8$. (This would also follow from a conjecture of Selmer ([Sel 2]) concerning the difference in the number of "first and second descents". It is also a consequence of the conjectures of Birch and Swinnerton-Dyer (C.16.5). The fact that rank $E_p(\mathbb{Q}) = 1$ has been verified numerically for all such primes less than 1000 ([Br–C]). To give the reader an idea of the magnitude of the solutions which can occur, we mention that for $p = 877$, the Mordell–Weil group of the elliptic curve

$$y^2 = x^3 + 877x$$

has as generators the points $(0, 0)$ and (x_0, y_0), where $x_0 = r^2/s^2$ with

$$r = 612, 776, 083, 187, 947, 368, 101$$

and

$$s = 7, 884, 153, 586, 063, 900, 210.$$

Similarly, if $p \equiv 3, 15 \pmod{16}$ and the 2-primary component of $Ш(E_p/\mathbb{Q})$ is finite, then (6.2) and (4.14) again imply that $E_p(\mathbb{Q})$ has rank 1. The fact that the rank is 1 in these cases may be verified numerically by searching for points in $C_{-2}(\mathbb{Q})$ and $C_2(\mathbb{Q})$ respectively. (See, for example, the tables in [B–Sw 1].)

Remark 6.4. If $p \equiv 7, 11 \pmod{16}$, then (6.2c) says that $E_p(\mathbb{Q})$ has order 2; while if $p \equiv 3, 5, 13, 15 \pmod{16}$, then (6.2c) combined with the reasonable conjecture that $Ш(E_p/\mathbb{Q})[2^\infty]$ is finite tells us that $E_p(\mathbb{Q}) \cong \mathbb{Z}/2\mathbb{Z} \times \mathbb{Z}$. In the remaining case, namely $p \equiv 1 \pmod 8$, there appear to be two possibilities. First, $E_p(\mathbb{Q})$ might have rank 2. This can certainly occur. For example, the curves

$$y^2 = x^3 + 73x \quad \text{and} \quad y^2 = x^3 + 89x$$

both have rank 2, independent points being given by

$$(9/16, 411/64), (36, 222) \in E_{73}(\mathbb{Q})$$

and

$$(25/16, 765/64), (4/9, 170/27) \in E_{89}(\mathbb{Q}).$$

Second, $E_p(\mathbb{Q})$ might have rank 0, which would mean that $\text{Ш}(E_p/\mathbb{Q})[2] \cong (\mathbb{Z}/2\mathbb{Z})^2$. (Note that rank $E_p(\mathbb{Q}) = 1$ is precluded if we assume that Ш is finite.) The following proposition gives a fairly general condition under which the second possibility holds. It also provides our first examples of homogeneous spaces which are everywhere locally trivial, but have no global rational points.

Proposition 6.5. *Let* $p \equiv 1 \pmod 8$ *be a prime for which 2 is not a quartic residue.*
(a) *The curves*

$$w^2 + 1 = 4pz^4 \qquad w^2 + 2 = 2pz^4 \qquad w^2 + 2pz^4 = 2$$

have points defined over every completion of $\mathbb{Q}$*, but have no* $\mathbb{Q}$*-rational points.*
(b) *The elliptic curve*

$$E_p : y^2 = x^3 + px$$

satisfies

$$\text{rank } E_p(\mathbb{Q}) = 0 \quad and \quad \text{Ш}(E_p/\mathbb{Q})[2] \cong (\mathbb{Z}/2\mathbb{Z})^2.$$

Remark 6.5.1. Any prime $p \equiv 1 \pmod 8$ can be written as $p = A^2 + B^2$ with $A, B \in \mathbb{Z}$ satisfying $AB \equiv 0 \pmod 4$. A theorem of Gauss, which we will prove below (6.6), says that 2 is then a quartic residue modulo p if and only if $AB \equiv 0 \pmod 8$. Thus for example, 2 is a quartic non-residue for the primes

$$17 = 1^2 + 4^2 \qquad 41 = 5^2 + 4^2 \qquad 97 = 9^2 + 4^2 \qquad 193 = 7^2 + 12^2;$$

and so these primes satisfy the conclusions of (6.5).

PROOF. During the course of proving (6.2b), we showed that the Selmer group $S^{(\phi)}(E_p/\mathbb{Q}) \subset \mathbb{Q}^*/\mathbb{Q}^{*2}$ is given by $\{\pm 1, \pm 2, \pm p, \pm 2p\}$. Further, $-p$ is the image of the 2-torsion point $(0, 0) \in E_p'(\mathbb{Q})$. Thus in order to show that $\text{Ш}(E_p/\mathbb{Q})[\phi]$ has order 4, it suffices to prove that the homogeneous spaces C_{-1}, C_2, and C_{-2} have no $\mathbb{Q}$-rational points. These are the three curves listed in (a); and so once we prove that they have no $\mathbb{Q}$-rational points, all of (6.5) will follow from (6.2). The following proof is based on ideas of Lind and Mordell ([Lin], [Ca 7]. See also [Rei], [Mo 3], and [B–Sw 1].)

Case I.

$$C_{\pm 2} : \pm w^2 = 2 - 2pz^4$$

Suppose that $(z, w) \in C_{+2}(\mathbb{Q})$. Writing z and w in lowest terms, we see that they necessarily have the form $(z, w) = (r/t, 2s/t^2)$, where $r, s, t \in \mathbb{Z}$ satisfy

$$\pm 2s^2 = t^4 - pr^4 \quad \text{and} \quad \gcd(r, s, t) = 1.$$

Let q be an odd prime dividing s. Then $(p|q) = 1$, so $(q|p) = 1$ by quadratic reciprocity. (Here $(a|b)$ is the Legendre symbol.) Since also $(2|p) = 1$, we see that $(s|p) = 1$, so $(s^2|p)_4 = 1$. (I.e. s^2 is a quartic residue modulo p.) Now the above equation implies that $(\pm 2|p)_4 = 1$. But -1 is always a quartic residue for $p \equiv 1 \pmod 8$, while by assumption 2 is a quartic non-residue modulo p. This contradiction proves that $C_{+2}(\mathbb{Q}) = \emptyset$.

Case II.

$$C_{-1} : -w^2 = 1 - 4pz^4$$

Writing $(z, w) \in C_{-1}(\mathbb{Q})$ in (almost) lowest terms as $(z, w) = (r/2t, s/2t^2)$, we have

$$s^2 + 4t^4 = pr^4 \qquad \gcd(r, t) = 1.$$

(We do not preclude the possibility that r is even.) Since $p \equiv 1 \pmod 4$, there are integers $A \equiv 1 \pmod 2$ and $B \equiv 0 \pmod 2$ such that

$$p = A^2 + B^2.$$

It is then a simple matter to verify the identity

$$(pr^2 + 2Bt^2)^2 = p(Br^2 + 2t^2)^2 + A^2s^2,$$

from which we obtain the factorization

$$(pr^2 + 2Bt^2 + As)(pr^2 + 2Bt^2 - As) = p(Br^2 + 2t^2)^2.$$

Now it is not difficult to check that $\gcd(pr^2 + 2Bt^2 + As, pr^2 + 2Bt^2 - As)$ is either a square or twice a square. (Up to a multiple of 2, it equals $\gcd(A, s)^2$.) Hence the above factorization implies that there are integers u and v satisfying

$$
\begin{aligned}
pr^2 + 2Bt^2 \pm As &= pu^2 && 2pu^2 \\
pr^2 + 2Bt^2 \mp As &= v^2 \quad \text{or} \quad && 2v^2 \\
Br^2 + 2t^2 &= uv && 2uv.
\end{aligned}
$$

Eliminating s from these equations, we obtain the two systems

$$2pr^2 + 4Bt^2 = pu^2 + v^2$$
$$Br^2 + 2t^2 = uv;$$

and

$$pr^2 + 2Bt^2 = pu^2 + v^2$$
$$Br^2 + 2t^2 = 2uv.$$

Now the fact that $p \equiv 1 \pmod 8$ and 2 is a quartic non-residue modulo p means that $B \equiv 4 \pmod 8$. (This will be proven below (6.6).) Reducing our two systems of equations modulo 8, it is now a simple matter to verify that in both cases, any solution must satisfy $r \equiv t \equiv 0 \pmod 2$. This contradicts our original assumption that $\gcd(r, t) = 1$, and so completes the proof that $C_{-1}(\mathbb{Q}) = \varnothing$. $\square$

We now prove the theorem of Gauss giving the quartic character of 2 which was used above. The proof that we give is due to Dirichlet, see also [Mo 3].

Proposition 6.6. *Let p be a prime, $p \equiv 1 \pmod 8$. Write $p = A^2 + B^2$ as a sum of two squares. Then*

$$(2|p)_4 = (-1)^{AB/4}.$$

(I.e. 2 is a quartic residue modulo p if and only if $AB \equiv 0 \pmod 8$.)

PROOF. Using the fact that $A^2 + B^2 \equiv 0 \pmod p$, we compute

$$(A + B)^{(p-1)/2} \equiv (2AB)^{(p-1)/4} \qquad\qquad (\text{mod } p)$$

$$\equiv 2^{(p-1)/4}(-1)^{(p-1)/8}A^{(p-1)/2} \pmod p.$$

In terms of residue symbols, this becomes

$$(A + B|p) = (-1)^{(p-1)/8}(2|p)_4(A|p).$$

By symmetry, we may assume that A is odd; and then the fact that $p \equiv 1 \pmod 4$ implies that

$$(A|p) = (p|A) = (B^2|A) = 1.$$

Hence

$$(A + B|p) = (-1)^{(p-1)/8}(2|p)_4.$$

Finally, we observe that

$$(A + B|p) = (p|A + B) = (2|A + B)(2p|A + B)$$

$$= (2|A + B) = (-1)^{((A+B)^2-1)/8},$$

since the identity

$$2p = (A + B)^2 + (A - B)^2 \qquad \text{implies that } (2p|A + B) = 1.$$

Substituting this above yields

$$(2|p)_4 = (-1)^e,$$

where

$$e = ((A + B)^2 - 1)/8 - (p - 1)/8 = AB/4. \qquad\qquad \square$$

EXERCISES

10.1. Let $\phi : E/K \to E'/K$ be an isogeny of degree m of elliptic curves over an arbitrary (perfect) field. Assume that $E'[\hat{\phi}] \subset E'(K)$. Generalize (1.1) as follows.

(a) Prove that there is a bilinear pairing

$$b : E'(K)/\phi(E(K)) \times E'[\hat{\phi}] \to K(S, m)$$

defined by

$$e_\phi(\delta_\phi(P), T) = \delta_K(b(P, T)).$$

(Here e_ϕ is the generalized Weil pairing (exer. 3.15), and $\delta_\phi : E'(K) \to H^1(G_{\bar{K}/K}, E[\phi])$ is the usual connecting homomorphism.)

(b) Prove that this pairing is non-degenerate on the left.

(c) For $T \in E'[\hat{\phi}]$, let $f_T \in K(E')$ and $g_T \in K(E)$ be functions satisfying

$$\mathrm{div}(f_T) = m(T) - m(O) \qquad f_T \circ \phi = g_T^m.$$

Prove that

$$b(P, T) = f_T(P) \,(\mathrm{mod}\ K^{*m}) \qquad \text{provided } P \neq O, T.$$

(d) In particular, if $\deg(\phi) = 2$, so $E'[\hat{\phi}] = \{O, T\}$, then

$$b(P, T) = x(P) - x(T) \,(\mathrm{mod}\ K^{*2}).$$

(We thus recover part of (4.9).)

10.2. Let K be an arbitrary (perfect) field, let E/K be an elliptic curve, and let C_1/K and C_2/K be homogeneous spaces for E/K.

(a) Prove that there exists a homogeneous space C_3/K for E/K and a morphism

$$\phi : C_1 \times C_2 \to C_3$$

defined over K such that for all $p_1 \in C_1$, $p_2 \in C_2$, and $P_1, P_2 \in E$,

$$\phi(p_1 + P_1, p_2 + P_2) = \phi(p_1, p_2) + P_1 + P_2.$$

(b) Prove that C_3 is unique up to equivalence of homogeneous spaces.

(c) Prove that

$$\{C_1\} + \{C_2\} = \{C_3\},$$

the sum taking place in $WC(E/K)$.

10.3. Let C/K be a curve of genus 1 defined over an arbitrary (perfect) field.

(a) Prove that there exists an elliptic curve E/K such that C/K is a homogeneous space for E/K. [*Hint*: Use exercise 3.22 to show that $C/K \in$ Twist(E/K). Then find an element $\{\xi\} \in H^1(G_{\bar{K}/K}, \mathrm{Aut}(E))$ so that C/K is a homogeneous space for the twist of E by ξ.]

(b) Prove that E is unique up to K-isomorphism.

10.4. Let K be an arbitrary (perfect) field and E/K an elliptic curve.

(a) Prove that there is a natural action of $\mathrm{Aut}_K(E)$ on $WC(E/K)$ defined as follows:

Let $\{C/K, \mu\} \in WC(E/K)$ and $\alpha \in \mathrm{Aut}_K(E)$. Then

$$\{C/K, \mu\}^{\alpha} = \{C/K, \mu \circ (1 \times \alpha)\}.$$

[I.e. Take the same curve C, but define a new action of E on C by the rule

$$\mu^{\alpha}(p, P) = \mu(p, \alpha P).]$$

(b) Conversely, if $\{C/K, \mu\}$ and $\{C/K, \mu'\}$ are elements of $WC(E/K)$, prove that there exists an $\alpha \in \operatorname{Aut}_K(E)$ such that $\mu' = \mu \circ (1 \times \alpha)$.

(c) Conclude that for a given curve C/K of genus 1, there are only finitely many non-equivalent ways of making C/K into a homogeneous space. In particular, if $j(C) \neq 0, 1728$, then there are at most two. (See also exer. B.5.)

10.5. Let $\phi : E/K \to E'/K$ be a separable isogeny of elliptic curves defined over an arbitrary (perfect) field K, and let C/K be a homogeneous space for E/K. Then the finite group $E[\phi]$ acts on C; let $C' = C/E[\phi]$ be the quotient curve (exer. 3.13).

(a) Prove that C' is a curve of genus 1 defined over K.

(b) Prove that C'/K is a homogeneous space for E'/K; and that under the natural map $\phi : WC(E/K) \to WC(E'/K)$, we have $\phi\{C/K\} = \{C'/K\}$.

(c) In particular, if $\{C/K\} \in WC(E/K)[\phi]$, then C' is isomorphic to E' over K. Prove that this isomorphism can be chosen so that the natural projection $C \to C/E[\phi] \cong E'$ is the map $\phi \circ \theta$ defined in (4.6a).

10.6. *WC Over Finite Fields.* Let $\mathbb{F}_q$ be a finite field with q elements, let $C/\mathbb{F}_q$ be a curve of genus 1, and pick any point of $C(\overline{\mathbb{F}}_q)$ as origin to make C into an elliptic curve. Let $\phi : C \to C$ be the qth power Frobenius map on C.

(a) Prove that there is an endomorphism $f \in \operatorname{End}(C)$ and a point $P_0 \in C(\overline{\mathbb{F}}_q)$ such that $\phi(P) = f(P) + P_0$.

(b) Prove that f is inseparable and conclude that there exists a point $P_1 \in C(\overline{\mathbb{F}}_q)$ satisfying $(1 - f)(P_1) = P_0$.

(c) Prove that $\phi(P_1) = P_1$, and hence that $P_1 \in C(\overline{\mathbb{F}}_q)$.

(d) Let $E/\mathbb{F}_q$ be an elliptic curve. Prove that $WC(E/\mathbb{F}_q) = 0$.

10.7. *WC Over $\mathbb{R}$.* Let $E/\mathbb{R}$ be an elliptic curve.

(a) Prove that $WC(E/\mathbb{R}) \cong \mathbb{Z}/2\mathbb{Z}$ if $\Delta(E/\mathbb{R}) > 0$, and $WC(E/\mathbb{R}) = 0$ if $\Delta(E/\mathbb{R}) < 0$.

(b) Find an equation for a homogeneous space representing the non-trivial element of $WC(E/\mathbb{R})$ in terms of a given Weierstrass equation for E.

10.8. Let E/K be an elliptic curve, $m \geq 2$ an integer, and assume that $E[m] \subset E(K)$. Let $v \in M_K$ be a prime not dividing m. Prove that the restriction map

$$WC(E/K)[m] \to WC(E/K_v)[m]$$

is surjective. [*Hint*: Show that the map on the $H^1(*, E[m])$'s is surjective.]

10.9. Let E/K be an elliptic curve, let $T \in E[m]$, and suppose that the field $L = K(T)$ has maximal degree, namely $[L : K] = m^2 - 1$. Consider the chain of maps

$$\alpha : E(K) \xrightarrow{\delta} H^1(G_{\overline{K}/K}, E[m]) \xrightarrow{\text{res}} H^1(G_{\overline{K}/L}, E[m]) \to H^1(G_{\overline{K}/L}, \mu_m) \cong L^*/L^{*m}.$$
$$\xi_{\sigma} \to e_m(\xi_{\sigma}, T)$$

(Here e_m is the Weil pairing.)

(a) Let $f_T \in L(E)$ be as in (1.1d). (I.e. $\text{div}(f_T) = m(T) - m(O)$ and $f_T \circ [m] \in L(E)^{*m}$.) Prove that

$$\alpha(P) = f_T(P) \pmod{L^{*m}}.$$

(b) Prove that for all $P \in E(K)$,

$$N_{L/K}(\alpha(P)) \in K^{*m}.$$

(c) Let $S \subset M_L$ be a set of places of L containing all archimedean places, all places dividing m, and all places at which E/L has bad reduction. Show that if $P \in E(K)$ and $v \in M_L$ with $v \notin S$, then

$$\text{ord}_v(\alpha(P)) \equiv 0 \pmod{m}.$$

(d) For $m = 2$, prove that the kernel of α is exactly $2E(K)$. Hence in this case there is an *injective* homomorphism from $E(K)/2E(K)$ into the group

$$\{a \in L^*/L^{*2} : N_{L/K}(a) \in K^{*2} \text{ and } \text{ord}_v(a) \equiv 0 \pmod{2} \text{ for all } v \notin S\}$$

given by the map

$$P \to x(P) - x(T).$$

This map may often be used to compute $E(K)/2E(K)$.
[*Hint*: Write out $x(P) - x(T) = (r + sx(T) + tx(T)^2)$, and use the resulting relations on $r, s, t \in K$ to show that P is in $2E(K)$.]

(e) Use (d) to compute $E(\mathbb{Q})/2E(\mathbb{Q})$ for the curve

$$E : y^2 + y = x^3 - x.$$

[*Hint*: Let $K/\mathbb{Q}$ be the totally real cubic extension generated by a root of $4x^3 - 4x + 1 = 0$. Start by showing that K has class number 1, and that every totally positive unit in K is a square.]

10.10. Let C/K be a curve of genus 1, and suppose that $C(K_v) \neq \varnothing$ for every $v \in M_K$. Prove that the map

$$\text{Div}_K(C) \to \text{Pic}_K(C)$$

is surjective. [*Hint*: Take Galois cohomology of the exact sequence

$$1 \to \bar{K}^* \to \bar{K}(C)^* \to \text{Div}(C) \to \text{Pic}(C) \to 0.$$

Use Noether's generalization of Hilbert's theorem 90,

$$H^1(G_{\bar{K}/K}, \bar{K}(C)^*) = 0;$$

and the (cohomological version) of the Brauer–Hasse–Noether theorem ([Ta3 §9.6]), which says that an element of $H^2(G_{\bar{K}/K}, \bar{K}^*)$ is trivial if and only if it is trivial in $H^2(G_{\bar{K}_v/K_v}, \bar{K}_v^*)$ for every $v \in M_K$.]

10.11. *Index and Period in WC.* Let K be an arbitrary (perfect) field, E/K an elliptic curve, and C/K a homogeneous space for E/K. Define the *period of C/K* to be the exact order of $\{C/K\}$ in $WC(E/K)$; and the *index of C/K* to be the degree of the smallest extension L/K for which $C(L) \neq \varnothing$. (E.g. (3.3) says precisely that the period equals 1 if and only if the index equals 1.)

(a) Prove that the period may also be characterized as the smallest integer

$m \geqslant 1$ for which there exists a point $p \in C$ such that $p^\sigma - p \in E[m]$ for every $\sigma \in G_{\bar{K}/K}$.

(b) Prove that the index may also be characterized as the smallest degree among the *positive* divisors in $\mathrm{Div}_K(C)$.

(c) Prove that the period divides the index.

(d) Prove that the period and the index are divisible by the same set of primes.

(e)* Give an example with $K = \mathbb{Q}$ showing that the period may be strictly smaller than the index.

(f) Prove that if K is a number field, and if C/K represents an element of $\text{III}(E/K)$, then the period and the index are equal. [*Hint*: Use (a), (b), (c), and exer. 10.10.]

10.12. *Hensel's Lemma*. The following version of Hensel's lemma is often useful for proving that a homogeneous space is locally trivial. Let R be a ring which is complete with respect to a discrete valuation v.

(a) Let $f(T) \in R[T]$ and $a_0 \in R$ satisfy
$$v(f(a_0)) > 2v(f'(a_0)).$$
Define a sequence $a_n \in R$ by
$$a_{n+1} = a_n - f(a_n)/f'(a_n).$$
Prove that $\{a_n\}$ converges to an element $a \in R$ satisfying
$$f(a) = 0 \quad \text{and} \quad v(a - a_0) \geqslant v(f(a_0)/f'(a_0)^2) > 0.$$

(b) Now let $F(X_1, \ldots, X_N) \in R[X_1, \ldots, X_N]$, and suppose that the point $(a_1, \ldots, a_N) \in R^N$ satisfies
$$v(F(a_1, \ldots, a_N)) > 2v((\partial F/\partial X_i)(a_1, \ldots, a_N))$$
for some $1 \leqslant i \leqslant N$. Then F has a root in R^N.

(c) Show that the curve
$$3X^3 + 4Y^3 + 5Z^3 = 0$$
in $\mathbb{P}^2$ has a point defined over $\mathbb{Q}_p$ for every prime p.

10.13. Use (1.4) to compute $E(\mathbb{Q})/2E(\mathbb{Q})$ for each of the following elliptic curves.
(a) $E : y^2 = x(x - 1)(x + 3)$.
(b) $E : y^2 = x(x - 12)(x - 36)$.

10.14. Use (4.9) to compute $E(\mathbb{Q})/2E(\mathbb{Q})$ for each of the following elliptic curves.
(a) $E : y^2 = x^3 + 6x^2 + x$.
(b) $E : y^2 = x^3 + 14x^2 + x$.
(c) $E : y^2 = x^3 + 9x^2 - x$.

10.15. Let E/K be an elliptic curve, $\xi \in H^1(G_{\bar{K}/K}, \mathrm{Aut}(E))$, and E_ξ the twist of E corresponding to ξ. Let $v \in M_K$ be a finite place for which E has good reduction. Prove that E_ξ has good reduction at v if and only if ξ is unramified at v. (See VIII §2 for the definition of unramified.) [*Hint*: If the residue characteristic is not 2 or 3, then one can easily use explicit Weierstrass equations. In general, use the criterion of Néron–Ogg–Shafarevich (VII.7.1).]

10.16. Let E/K be an elliptic curve, let $D \in K^*$ be such that $L = K(\sqrt{D})$ is a quadratic extension of K, and let E_D/K be the twist of E/K given by (5.4(i)). Prove

$$\text{rank } E(L) = \text{rank } E(K) + \text{rank } E_D(K).$$

10.17. Let $p \equiv 3 \pmod 4$ be a prime, and let $D \in \mathbb{F}_p^*$.
   (a) Show directly that the equation

$$C : v^2 = u^4 - 4D$$

   has $p - 1$ solutions $(u, v) \in \mathbb{F}_p \times \mathbb{F}_p$. [*Hint*: Since $p \equiv 3 \pmod 4$, the map $u^2 \to u^4$ is an automorphism of $\mathbb{F}_p^{*2}$.]
   (b) Let $E/\mathbb{F}_p$ be the elliptic curve

$$E : y^2 = x^3 + Dx.$$

   Use the map

$$\phi : C \to E \qquad \phi(u, v) = (\tfrac{1}{2}(u^2 + v), \tfrac{1}{2}u(u^2 + v))$$

   to prove that

$$\# E(\mathbb{F}_p) = p + 1.$$

10.18. Do a computation analogous to that of (6.2) to determine the Selmer groups and a bound for the ranks of the following families of elliptic curves $E/\mathbb{Q}$. (Here p is an odd prime.)
   (a) $E : y^2 = x^3 - 2px$. (The curve with $p = 41$ has rank 3.)
   (b) $E : y^2 = x^3 + p^2x$.

10.19. Let $E/\mathbb{Q}$ be an elliptic curve with $j(E) = 0$.
   (a) Prove that there is a unique sixth-power-free integer D such that E is given by the Weierstrass equation

$$E : y^2 = x^3 + D.$$

   (b) Let $p \equiv 2 \pmod 3$ be a prime not dividing $6D$. Prove that

$$\# \tilde{E}(\mathbb{F}_p) = p + 1.$$

   (c) Prove that $\# E_{\text{tors}}(\mathbb{Q})$ divides 6.
   (d) More precisely, show that $E_{\text{tors}}(\mathbb{Q})$ is given by the following list.

$$E_{\text{tors}}(\mathbb{Q}) \cong \begin{cases} \mathbb{Z}/6\mathbb{Z} & \text{if } D = 1 \\ \mathbb{Z}/3\mathbb{Z} & \text{if } D \neq 1 \text{ is a cube, or } D = -432 \\ \mathbb{Z}/2\mathbb{Z} & \text{if } D \neq 1 \text{ is a square} \\ 1 & \text{otherwise.} \end{cases}$$

10.20. Let A be a finite abelian group, and suppose that there exists a bilinear, alternating, non-degenerate pairing

$$\Gamma : A \times A \to \mathbb{Q}/\mathbb{Z}.$$

Prove that $\# A$ is a perfect square.

APPENDIX A

Elliptic Curves in Characteristics 2 and 3

In this appendix we prove some of the results for elliptic curves in characteristics 2 and 3 which were omitted in the main body of the text. To simplify the computations, we start by giving normal forms for the Weierstrass equations of such curves.

Proposition 1.1. *Let E/K be a curve given by a Weierstrass equation. Then under the boxed assumptions, there is a substitution*

$$x = u^2 x' + r \qquad y = u^3 y' + u^2 sx' + t \qquad \text{with } u \in K^* \text{ and } r, s, t \in K$$

such that E/K has a Weierstrass equation of the indicated form.

(a) $\boxed{\text{char } K \neq 2, 3}$

$$y^2 = x^3 + a_4 x + a_6 \qquad \Delta = -16(4a_4^3 + 27a_6^2) \qquad j = 1728 \frac{4a_4^3}{4a_4^3 + 27a_6^2}$$

(b) $\boxed{\text{char } K = 3 \quad \text{and} \quad j(E) \neq 0}$

$$y^2 = x^3 + a_2 x^2 + a_6 \quad \Delta = -a_2^3 a_6 \quad j = -a_2^3/a_6$$

$\boxed{\text{char } K = 3 \quad \text{and} \quad j(E) = 0}$

$$y^2 = x^3 + a_4 x + a_6 \quad \Delta = -a_4^3 \quad j = 0$$

(c) $\boxed{\text{char } K = 2 \quad \text{and} \quad j(E) \neq 0}$

$$y^2 + xy = x^3 + a_2 x^2 + a_6 \quad \Delta = a_6 \quad j = 1/a_6$$

$$\boxed{\text{char } K = 2 \quad \text{and} \quad j(E) = 0}$$

$$y^2 + a_3 y = x^3 + a_4 x + a_6 \qquad \Delta = a_3^4 \qquad j = 0.$$

PROOF. (a) See (III §1).

(b) Take a general Weierstrass equation and complete the square on the left. This gives an equation of the form

$$y^2 = x^3 + a_2 x^2 + a_4 x + a_6$$

with invariants

$$\Delta = a_2^2 a_4^2 - a_2^3 a_6 - a_4^3 \qquad j = a_2^3/\Delta.$$

(Remember that char $K = 3$.) If $j = 0$, then $a_2 = 0$, so the equation already has the right shape. On the other hand, if $j \neq 0$, then $a_2 \neq 0$; and so the substitution $x = x' + a_4/a_2$ will eliminate the linear term.

(c) Again starting with a general Weierstrass equation

$$y^2 + a_1 xy + a_3 y = x^3 + a_2 x^2 + a_4 x + a_6,$$

one easily computes (in characteristic 2)

$$j = a_1^{12}/\Delta.$$

If $j \neq 0$, so $a_1 \neq 0$, then the substitution

$$x = a_1^2 x' + a_3/a_1 \qquad y = a_1^3 y' + (a_1^2 a_4 + a_3^2)/a_1^3$$

gives an equation in the desired form. Similarly, if $j = a_1 = 0$, then the substitution

$$x = x' + a_2 \qquad y = y'$$

will have the desired effect.

(Note that there is no deep theory involved in finding these substitutions. One merely looks at the transformation formulas (III.1.2), sets various coefficients equal to 0 or 1, and chooses appropriate u, r, s, t.) $\qquad \square$

It is now a simple matter to complete the proofs of (III.1.4) and (III.10.1), parts of which we restate here.

Proposition 1.2. (a) *A curve given by a Weierstrass equation is non-singular if and only if the discriminant of the equation is non-zero.*

(b) *Two elliptic curves E/K and E'/K are isomorphic over $\bar{K}$ if and only if they have the same j-invariant.*

(c) *Let E/K be an elliptic curve. Then $\mathrm{Aut}(E)$ is a finite group of order*

2	*if $j(E) \neq 0, 1728$*
4	*if $j(E) = 1728$ and char $K \neq 2, 3$*
6	*if $j(E) = 0$ and char $K \neq 2, 3$*

| 12 | if $j(E) = 0 = 1728$ and char $K = 3$ |
| 24 | if $j(E) = 0 = 1728$ and char $K = 2$. |

(*See also exercise A.1.*)

PROOF. (a) From the proof of (III.1.4a), all that remains is to show that if char$(K) = 2$ and $\Delta = 0$, then the curve is singular. But this is immediate from the normal forms given in (1.1c).

(b), (c) Again referring to the proofs of (III.1.4b) and (III.10.1), we need only deal with the cases of char$(K) = 2$ or 3. We use the normal forms given in (1.1b,c) and consider 4 cases.

Case I. char $K = 3$ *and* $j(E) \neq 0$. E and E' have Weierstrass equations of the form

$$y^2 = x^3 + a_2 x^2 + a_6.$$

The only substitutions preserving this sort of equation are

$$x = u^2 x' \qquad y = u^3 y'.$$

Since $j(E) = j(E')$, we have $a_2^3 a_6' = a_2'^3 a_6 \neq 0$, so taking $u^2 = a_2/a_2'$ will give an isomorphism from E to E'. Further, if $E = E'$, then we must have $u^2 = 1$, so Aut$(E) \cong \{\pm 1\}$.

Case II. char $K = 3$ *and* $j(E) = 0$. E and E' are given by equations of the form

$$y^2 = x^3 + a_4 x + a_6.$$

The substitutions preserving this form look like

$$x = u^2 x' + r \qquad y = u^3 y.$$

Note we have $a_4, a_4' \neq 0$. Then an isomorphism from E to E' is given by choosing u and r to satisfy

$$u^4 = a_4'/a_4 \qquad r^3 + a_4 r + a_6 - u^6 a_6' = 0.$$

Further, if $E = E'$, then an automorphism of E has $u^4 = 1$ and $r^3 + a_4 r + (1 - u^2)a_6 = 0$. Since $a_4 \neq 0$, there are exactly 12 such pairs (u, r) making up Aut(E).

Case III. char $K = 2$ *and* $j(E) \neq 0$. In this case E and E' are given by equations of the form

$$y^2 + xy = x^3 + a_2 x^2 + a_6.$$

The substitutions preserving this form look like

$$x = x' \qquad y = y' + sx'.$$

Since $j(E) = j(E')$, we have $a_6 = a_6' \neq 0$, so an isomorphism from E to E' is given by taking s to be a root of the equation

$$s^2 + s + a_2 + a_2' = 0.$$

Similarly, the automorphisms of E are obtained by taking $s \in \{0, 1\}$.

Case IV. char $K = 2$ *and* $j(E) = 0$. E and E' have equations of the form

$$y^2 + a_3 y = x^3 + a_4 x + a_6,$$

and allowable substitutions look like

$$x = u^2 x' + s^2 \qquad y = u^3 y' + u^2 s x' + t.$$

By assumption, $a_3, a_3' \neq 0$, so to map E to E', we choose u, s, t to satisfy the equations

$$u^3 = a_3/a_3' \qquad s^4 + a_3 s + a_4 - u^4 a_4' = 0$$
$$t^2 + a_3 t + s^6 + a_4 s^2 + a_6 - u^6 a_6' = 0.$$

Finally, the automorphism group of E is given by the set of triples (u, s, t) satisfying the equations

$$u^3 = 1 \qquad s^4 + a_3 s + (1 - u)a_4 = 0 \qquad t^2 + a_3 t + s^6 + a_4 s^2 = 0.$$

Since $a_3 \neq 0$, we see that $\text{Aut}(E)$ has order 24. $\qquad \square$

The next proposition gives a normal form for Weierstrass equations which is similar to Legendre form, but is valid in characteristic 2. Having done this, we can then easily complete the proofs of (VII.5.4c) and (VII.5.5).

Proposition 1.3 (Deuring Normal Form). *Let E/K be an elliptic curve over a field with* char $K \neq 3$. *Then E has a Weierstrass equation over $\bar{K}$ of the form*

$$E_\alpha : y^2 + \alpha x y + y = x^3 \qquad \alpha \in \bar{K}, \alpha^3 \neq 27.$$

This equation has discriminant and j-invariant

$$\Delta = \alpha^3 - 27 \qquad j = \alpha^3(\alpha^3 - 24)^3/(\alpha^3 - 27).$$

PROOF. The computation of Δ and j for E_α is an exercise. In order to show that E has an equation of the form E_α, one can find appropriate substitutions. However, using (1.2b), we have a quicker route available. Thus let $\alpha \in \bar{K}$ be a solution to the equation

$$\alpha^3(\alpha^3 - 24)^3 - (\alpha^3 - 27)j(E) = 0.$$

Since char$(K) \neq 3$, we see that $\alpha^3 \neq 27$, so E_α will be an elliptic curve with the same j-invariant as E. If follows from (1.2b) that E and E_α are isomorphic (over $\bar{K}$). $\qquad \square$

Corollary 1.4. *Let E/K be an elliptic curve defined over a local field. (I.e. K is given with a discrete valuation.)*

(a) *There exists a finite extension K'/K such that E has either good or split multiplicative reduction over K'.*
(b) *E has potential good reduction if and only if its j-invariant is integral.*

PROOF. Let R be the ring of integers of K, $\mathcal{M}$ its maximal ideal, and $k = R/\mathcal{M}$ its residue field. From the proofs of (VII.5.4c) and (VII.5.5), we are left to deal with char$(k) = 2$. In any case, we may assume that char$(k) \neq 3$. Replacing K by a finite extension, we choose an equation for E in Deuring normal form

$$E : y^2 + \alpha xy + y = x^3 \qquad \alpha^3 \neq 27.$$

This equation has

$$c_4 = \alpha(\alpha^3 - 24) \quad \text{and} \quad \Delta = \alpha^3 - 27.$$

(a) We consider three cases.

Case I. $\alpha \in R$, $\alpha^3 \not\equiv 27 \pmod{\mathcal{M}}$. Then $\Delta \not\equiv 0 \pmod{\mathcal{M}}$, so the given equation has good reduction.

Case II. $\alpha \in R$, $\alpha^3 \equiv 27 \pmod{\mathcal{M}}$. Then $\Delta \equiv 0 \pmod{\mathcal{M}}$ and $c_4 \equiv 81 \not\equiv 0 \pmod{\mathcal{M}}$, so by (VII.5.1b), the given equation for E has multiplicative reduction. To obtain split multiplicative reduction then requires, at worst, taking a quadratic extension of K.

Case III. $\alpha \notin R$. Let π be a uniformizer for R, and choose an integer $r \geqslant 1$ so that $\pi^r \alpha \in R^*$. Then the substitution $x = \pi^{-2r} x'$, $y = \pi^{-3r} y'$ gives an equation

$$y'^2 + \beta x' y' + \pi^{3r} y' = x'^3,$$

where $\beta = \pi^r \alpha \in R^*$. This equation has

$$c_4' = \beta(\beta^3 - 24\pi^{3r}) \equiv \beta^4 \not\equiv 0 \pmod{\mathcal{M}}$$

and

$$\Delta = \pi^{9r}(\beta^3 - 27\pi^{3r}) \equiv 0 \pmod{\mathcal{M}},$$

so again from (VII.5.1b), it has multiplicative reduction. Further, the reduced curve is given by $y(y + \beta x) \equiv x^3 \pmod{\mathcal{M}}$, so the reduction is split multiplicative.
(b) By assumption, $j(E)$ and α are related by

$$\alpha^3(\alpha^3 - 24)^3 - (\alpha^3 - 27)j(E) = 0.$$

From this equation and the integrality of $j(E)$, we see that α is integral. Further, since the characteristic of k is different from 3, we have $\alpha^3 \not\equiv 27 \pmod{\mathcal{M}}$. Thus the Deuring normal equation has integral coefficients and good reduction.   □

Exercises

A.1. Let E/K be an elliptic curve with $j(E) = 0$. Strengthen (1.2) by showing that the automorphism group of E may be described as follows:
   (a) If char$(K) = 3$, then Aut(E) is the twisted product of C_4 (a cyclic group of order 4) and C_3. C_3 is a normal subgroup, and C_4 acts on C_3 in the unique non-trivial way.
   (b) If char$(K) = 2$, then Aut(E) is the twisted product of C_3 and a quaternion group. The quaternion group is a normal subgroup; and if we write the quaternion group as $\{\pm 1, \pm i, \pm j, \pm k\}$, then a generator for C_3 acts by permuting i, j, and k.

A.2. Let K be a field of characteristic 2, and let E/K be an elliptic curve with $j(E) \ne 0$ given by a Weierstrass equation

$$y^2 + xy = x^3 + a_2 x^2 + a_6.$$

Let $\xi \in H^1(G_{\bar{K}/K}, \mathrm{Aut}(E)) = \mathrm{Hom}(G_{\bar{K}/K}, \mathbb{Z}/2\mathbb{Z})$, and let L/K be the corresponding quadratic extension. Show that the twist of E by ξ (cf. (X §5)) is given by an equation

$$y^2 + xy = x^3 + (a_2 + D)x^2 + a_6,$$

where $D \in K$ and L/K is the Artin–Schreier extension generated by a root of

$$t^2 - t - D = 0.$$

A.3. Let E/K be an elliptic curve with Weierstrass coordinate functions x and y. Show that the differential dx is holomorphic if and only if char$(K) = 2$ and $j(E) = 0$.

A.4. Let E/K and E'/K be elliptic curves over a *not necessarily perfect* field K. Suppose that $j(E) = j(E')$. Prove that E and E' are isomorphic over a *separable* extension L of K of degree dividing 24. If $j(E) \ne 0$, 1728, then L can be chosen to have degree 2.

Group Cohomology (H^0 and H^1)

In this appendix we give the basic facts about group cohomology which are used in chapter VIII §2 and chapter X. Since only H^0 and H^1 are needed in this book, we have restricted our attention to these two groups. The reader desiring more information about group cohomology might look at [A–W], [Gru], [Se 8], or [Se 9].

§1. Cohomology of Finite Groups

Let G be a finite group, and let M be an abelian group on which G acts. We denote the action of $\sigma \in G$ on $m \in M$ by $m \to m^\sigma$. Then M is a *(right) G-module* if the action of G on M satisfies

$$m^1 = m \qquad (m + m')^\sigma = m^\sigma + m'^\sigma \qquad (m^\sigma)^\tau = m^{\sigma\tau}.$$

If M and N are G-modules, a *G-homomorphism* is a homomorphism $\phi : M \to N$ of abelian groups commuting with the action of G; that is

$$\phi(m^\sigma) = \phi(m)^\sigma \qquad \text{for all } m \in M \text{ and } \sigma \in G.$$

For a given G-module, one is often interested in calculating the largest sub-module on which G acts trivially.

Definition. The 0th *cohomology group of the G-module* M, denoted M^G or $H^0(G, M)$, is defined by

$$H^0(G, M) = \{m \in M : m^\sigma = m \text{ for all } \sigma \in G\}.$$

It is the submodule of M consisting of all *G-invariant elements*.

Let

$$0 \to P \overset{\phi}{\to} M \overset{\psi}{\to} N \to 0$$

be an *exact sequence of G-modules*. (I.e. ϕ and ψ are G-module homomorphisms with ϕ injective, ψ surjective, and Image(ϕ) = Kernel(ψ).) Then one easily checks that taking G-invariants gives another exact sequence

$$0 \to P^G \to M^G \to N^G;$$

but the map on the right need no longer be surjective. In order to measure this lack of surjectivity, we make the following definitions.

Definition. Let M be a G-module. The *group of 1-cochains (from G to M)* is defined by

$$C^1(G, M) = \{\text{maps } \xi : G \to M\}.$$

The *group of 1-cocycles (from G to M)* is given by

$$Z^1(G, M) = \{\xi \in C^1(G, M) : \xi_{\sigma\tau} = \xi_\sigma^\tau + \xi_\tau \text{ for all } \sigma, \tau \in G\}.$$

The *group of 1-coboundaries (from G to M)* is defined by

$B^1(G, M) = \{\xi \in C^1(G, M) :$ there exists an $m \in M$ such that

$$\xi_\sigma = m^\sigma - m \text{ for all } \sigma \in G\}.$$

One easily checks that $B^1(G, M) \subset Z^1(G, M)$. Then the 1st *cohomology group of the G-module M* is the quotient group

$$H^1(G, M) = Z^1(G, M)/B^1(G, M).$$

In other words, $H^1(G, M)$ is the group of 1-cocycles $\xi : G \to M$, modulo the equivalence relation that two cocycles are identified if their difference is of the form $\sigma \to m^\sigma - m$ for some $m \in M$.

Remark 1.1. Notice that if the action of G on M is trivial, then

$$H^0(G, M) = M \quad \text{and} \quad H^1(G, M) = \text{Hom}(G, M).$$

These both follow immediately from the definitions; for the latter, the 1-cocycles are homomorphisms, and all of the 1-coboundaries are 0.

Let $\phi : M \to N$ be a G-module homomorphism. Then composition with ϕ clearly takes $Z^1(G, M)$ to $Z^1(G, N)$ and $B^1(G, M)$ to $B^1(G, N)$. Thus ϕ induces a map on cohomology $\phi : H^1(G, M) \to H^1(G, N)$.

Proposition 1.2. *Let*

$$0 \to P \overset{\phi}{\to} M \overset{\psi}{\to} N \to 0$$

be an exact sequence of G-modules. Then there is a long exact sequence

$$0 \to H^0(G, P) \to H^0(G, M) \to H^0(G, N) \xrightarrow{\delta} H^1(G, P) \to H^1(G, M) \to H^1(G, N),$$

where the connecting homomorphism δ *is defined as follows.*

Let $n \in H^0(G, N) = N^G$. *Choose an* $m \in M$ *such that* $\psi(m) = n$, *and define a cochain* $\xi \in C^1(G, M)$ *by*

$$\xi_\sigma = m^\sigma - m.$$

Then in fact $\xi \in Z^1(G, P)$, *and* $\delta(n)$ *is the cohomology class in* $H^1(G, P)$ *of the* 1-*cocycle* ξ.

PROOF. A straightforward (but tedious) diagram chase, which we leave to the reader (exer. B.1). (Or see any of the references listed above.)   □

Suppose now that H is a subgroup of G. Then any G-module M is automatically an H-module. Further, if $\xi : G \to M$ is a 1-cochain, then by restricting the domain of ξ to H, we obtain an H-to-M cochain. It is clear that this process takes cocycles to cocycles and coboundaries to coboundaries, and so we obtain a *restriction homomorphism*

$$\text{Res} : H^1(G, M) \to H^1(H, M).$$

Suppose further that H is a normal subgroup of G. Then the submodule M^H of M consisting of the elements fixed by H has a natural structure of G/H-module. Now let $\xi : G/H \to M^H$ be a 1-cochain from G/H to M^H. Then composing with the projection $G \to G/H$ and with the inclusion $M^H \subset M$ gives a G-to-M 1-cochain

$$G \to G/H \xrightarrow{\xi} M^H \subset M.$$

Again it is easy to see that if ξ is a cocycle or coboundary, then the new G-to-M cochain has the same property. Hence we obtain an *inflation homomorphism*

$$\text{Inf} : H^1(G/H, M^H) \to H^1(G, M).$$

Proposition 1.3. *Let M be a G-module and let H be a normal subgroup of G. Then the following sequence is exact.*

$$0 \to H^1(G/H, M^H) \xrightarrow{\text{Inf}} H^1(G, M) \xrightarrow{\text{Res}} H^1(H, M).$$

PROOF. From the definitions, it is clear that $\text{Res} \circ \text{Inf} = 0$.

Next let $\xi : G/H \to M^H$ be a 1-cocycle with $\text{Inf}\{\xi\} = 0$. (We use braces $\{\cdot\}$ to indicate the cohomology class of a cocycle.) Thus there is an $m \in M$ such that $\xi_\sigma = m^\sigma - m$ for all $\sigma \in G$. But ξ depends only on $\sigma \pmod H$, so $m^\sigma - m = m^{\tau\sigma} - m$ for all $\tau \in H$. Thus $m^\tau - m = 0$ for all $\tau \in H$, so $m \in M^H$, and hence ξ is a G/H-to-M^H coboundary.

Finally, suppose that $\xi : G \to M$ is a 1-cocycle with $\text{Res}\{\xi\} = 0$. Thus there is an $m \in M$ such that $\xi_\tau = m^\tau - m$ for all $\tau \in H$. Subtracting the G-to-M

coboundary $\sigma \to m^\sigma - m$ from ξ, we may assume that $\xi_\tau = 0$ for all $\tau \in H$. Then the cocycle condition applied to $\sigma \in G$ and $\tau \in H$ yields

$$\xi_{\tau\sigma} = \xi_\tau^\sigma + \xi_\sigma = \xi_\sigma.$$

Thus ξ_σ depends only on the class of σ in G/H. Next, since H is normal, there is a $\tau' \in H$ such that $\sigma\tau = \tau'\sigma$. Then using the cocycle condition again together with the fact that ξ is a map on G/H gives

$$\xi_\sigma = \xi_{\tau'\sigma} = \xi_{\sigma\tau} = \xi_\sigma^\tau + \xi_\tau = \xi_\sigma^\tau.$$

This proves that ξ gives a map from G/H to M^H, and so $\{\xi\} \in H^1(G/H, M^H)$.

□

§2. Galois Cohomology

Let K be a perfect field (as usual), let $\bar{K}$ be an algebraic closure of K, and let $G_{\bar{K}/K}$ be the Galois group of $\bar{K}$ over K. Recall that $G_{\bar{K}/K}$ is equal to the inverse limit of $G_{L/K}$ as L varies over all finite Galois extensions of K. Thus $G_{\bar{K}/K}$ is a profinite group (inverse limit of finite groups), and as such it comes equipped with a topology in which a basis of open sets around the identity consists of the collection of normal subgroups having finite index in $G_{\bar{K}/K}$. (I.e. The subgroups which are kernels of maps $G_{\bar{K}/K} \to G_{L/K}$ for finite Galois extensions L/K.)

Definition. A (*discrete*) $G_{\bar{K}/K}$-*module* is an abelian group M on which $G_{\bar{K}/K}$ acts such that the action is continuous for the profinite topology on $G_{\bar{K}/K}$ and the discrete topology on M. (Equivalently, the action of $G_{\bar{K}/K}$ on M has the property that for all $m \in M$, the stabilizer of m,

$$\{\sigma \in G : m^\sigma = m\},$$

is a subgroup of finite index in $G_{\bar{K}/K}$.) Since all of our $G_{\bar{K}/K}$-modules will be discrete, we will normally just refer to them as $G_{\bar{K}/K}$-modules.

Example 2.1.1. $\bar{K}$ and $\bar{K}^*$ with the natural action of $G_{\bar{K}/K}$ are $G_{\bar{K}/K}$-modules. This is because for any $x \in \bar{K}$, $K(x)/K$ is a finite extension, so the stabilizer of x will have finite index.

Example 2.1.2. More generally, let $\mathcal{D}/K$ be any (abelian) algebraic group. Then $\mathcal{D} = \mathcal{D}(\bar{K})$ is a $G_{\bar{K}/K}$-module, since again the coordinates of any point of $\mathcal{D}$ will generate a finite extension of K.

The 0th-cohomomogy of a $G_{\bar{K}/K}$-module is defined just as in the case of finite groups.

Definition. The 0th-*cohomology of the $G_{\overline{K}/K}$-module* M is the group of $G_{\overline{K}/K}$-invariant elements of M,

$$M^{G_{\overline{K}/K}} = H^0(G_{\overline{K}/K}, M) = \{m \in M : m^\sigma = m \text{ for all } \sigma \in G_{\overline{K}/K}\}.$$

We could also define H^1 exactly as in the case of finite groups, but instead we use the fact that our group is profinite and our module discrete in order to put some restriction on the allowable cocycles.

Definition. Let M be a $G_{\overline{K}/K}$-module. A map $\xi : G_{\overline{K}/K} \to M$ is *continuous* if it is continuous for the profinite topology on $G_{\overline{K}/K}$ and the discrete topology on M. (I.e. If for each $m \in M$, $\xi^{-1}(m)$ contains a subgroup of finite index of $G_{\overline{K}/K}$.) We define the *group of continuous 1-cocycles from $G_{\overline{K}/K}$ to M*, denoted $Z^1_{\text{cont}}(G_{\overline{K}/K}, M)$, to be the group of continuous maps $\xi : G_{\overline{K}/K} \to M$ satisfying the cocycle condition

$$\xi_{\sigma\tau} = \xi_\sigma^\tau + \xi_\tau.$$

(This is a subgroup of the full group of 1-cocycles $Z^1(G_{\overline{K}/K}, M)$.) Notice that since M is discrete, any coboundary $\sigma \to m^\sigma - m$ will automatically be continuous. The 1st-*cohomology of the $G_{\overline{K}/K}$-module* M is defined by

$$H^1(G_{\overline{K}/K}, M) = Z^1_{\text{cont}}(G_{\overline{K}/K}, M)/B^1(G_{\overline{K}/K}, M).$$

Remark 2.2. Just as in the case of finite groups, if $G_{\overline{K}/K}$ acts trivially on M, then we have

$$H^0(G_{\overline{K}/K}, M) = M \quad \text{and} \quad H^1(G_{\overline{K}/K}, M) = \text{Hom}_{\text{cont}}(G_{\overline{K}/K}, M).$$

(Here Hom_{cont} means the group of continuous homomorphisms.)

The fundamental exact sequences (1.2) and (1.3) in the cohomology of finite groups carry over word-for-word to the profinite case.

Proposition 2.3. *Let*

$$0 \to P \xrightarrow{\phi} M \xrightarrow{\psi} N \to 0$$

be an exact sequence of $G_{\overline{K}/K}$-modules. Then there is a long exact sequence

$$0 \to H^0(G_{\overline{K}/K}, P) \to H^0(G_{\overline{K}/K}, M) \to H^0(G_{\overline{K}/K}, N)$$
$$\xrightarrow{\delta} H^1(G_{\overline{K}/K}, P) \to H^1(G_{\overline{K}/K}, M) \to H^1(G_{\overline{K}/K}, N),$$

where the connecting homomorphism δ is defined as in (1.2).

Now let M be a $G_{\overline{K}/K}$-module, and let L/K be a finite Galois extension. Then $G_{\overline{K}/L}$ is a subgroup of $G_{\overline{K}/K}$ of finite index, and so M is naturally a $G_{\overline{K}/L}$-module. This leads to a *restriction map* on cohomology,

$$\text{Res} : H^1(G_{\overline{K}/K}, M) \to H^1(G_{\overline{K}/L}, M).$$

Further, $G_{\bar{K}/L}$ is a normal subgroup of $G_{\bar{K}/K}$, the quotient being the finite group $G_{L/K}$. The invariant submodule $M^{G_{\bar{K}/L}}$ has a natural structure of $G_{L/K}$-module. Then any 1-cocycle $\xi : G_{L/K} \to M^{G_{\bar{K}/L}}$ becomes a (continuous) $G_{\bar{K}/K}$ 1-cocycle via the composition

$$G_{\bar{K}/K} \to G_{L/K} \xrightarrow{\xi} M^{G_{\bar{K}/L}} \subset M.$$

This gives an *inflation map*

$$\mathrm{Inf} : H^1(G_{L/K}, M^{G_{\bar{K}/L}}) \to H^1(G_{\bar{K}/K}, M).$$

Proposition 2.4. *With notation as above, there is an exact sequence*

$$0 \to H^1(G_{L/K}, M^{G_{\bar{K}/L}}) \xrightarrow{\mathrm{Inf}} H^1(G_{\bar{K}/K}, M) \xrightarrow{\mathrm{Res}} H^1(G_{\bar{K}/L}, M).$$

PROOF. Virtually identical to the proof of (1.3). ☐

The next proposition gives fundamental facts about the cohomology of the additive and multiplicative groups of a field.

Proposition 2.5. *Let K be a field.*

(a) $H^1(G_{\bar{K}/K}, \bar{K}^+) = 0.$

(b) (*Hilbert Theorem 90*)

$$H^1(G_{\bar{K}/K}, \bar{K}^*) = 0.$$

(c) *Assume that* $\mathrm{char}(K)$ *does not divide* m (*or* $\mathrm{char}(K) = 0$). *Then*

$$H^1(G_{\bar{K}/K}, \mu_m) \cong K^*/K^{*m}.$$

PROOF. (a) [Se 9, Ch. X, Prop. 1].
(b) [Se 9, Ch. X, Prop. 2].
(c) Consider the exact sequence

$$1 \to \mu_m \to \bar{K}^* \xrightarrow{m} \bar{K}^* \to 1$$

of $G_{\bar{K}/K}$-modules. Applying (2.3) yields the long exact sequence

$$\to K^* \xrightarrow{m} K^* \xrightarrow{\delta} H^1(G_{\bar{K}/K}, \mu_m) \to H^1(G_{\bar{K}/K}, \bar{K}^*) \to.$$

From (b), $H^1(G_{\bar{K}/K}, \bar{K}^*) = 0$, which gives the desired result. ☐

§3. Non-Abelian Cohomology

Again we start with a finite group G and a group M on which G acts, but now we no longer require that M be abelian. (To emphasize this fact, we will write M multiplicatively.) As above, the 0th-*cohomology group of M* is defined to be the subgroup of G-invariant elements:

$$H^0(G, M) = M^G = \{m \in M : m^\sigma = m \text{ for all } \sigma \in G\}.$$

Further, we define the *set of* 1-*cocycles of G into M* to be the set of maps

$$\xi : G \to M \qquad \text{satisfying } \xi_{\sigma\tau} = (\xi_\sigma)^\tau \xi_\tau \qquad \text{for all } \sigma, \tau \in G.$$

[N.B. The 1-cocycles do *not* in general form a group. The non-commutativity of M may prevent the product of two cocycles from being a cocycle.] We say that two 1-cocycles ξ and ζ are *cohomologous* if there is an $m \in M$ such that

$$m^\sigma \xi_\sigma = \zeta_\sigma m \qquad \text{for all } \sigma \in G.$$

One easily checks that this gives an equivalence relation on the set of 1-cocycles. The 1st-*cohomology set of M*, denoted $H^1(G, M)$, is the set of 1-cocycles modulo this relation. We note that $H^1(G, M)$ has a distinguished element, namely the equivalence class of the identity cocycle. It is thus a *pointed set*; that is, a set with a distinguished element.

Continuing as in section 2, we say that the Galois group $G_{\bar{K}/K}$ acts *discretely* on a (possibly non-abelian) group M if the stabilizer of any element of M is a subgroup of finite index in $G_{\bar{K}/K}$. We can again define a *continuous* 1-*cocycle from $G_{\bar{K}/K}$ to M* to be a map $\xi : G_{\bar{K}/K} \to M$ which satisfies the cocycle condition and is continuous for the profinite topology on $G_{\bar{K}/K}$ and the discrete topology on M. Two cocycles ξ and ζ are again deemed *cohomologous* if $m^\sigma \xi_\sigma = \zeta_\sigma m$ for some $m \in M$, and the 0th-*cohomology group and* 1st-*cohomology set of M* are defined as above by

$$H^0(G_{\bar{K}/K}, M) = M^{G_{\bar{K}/K}} = \{m \in M : m^\sigma = m \text{ for all } \sigma \in G\},$$

and

$$H^1(G_{\bar{K}/K}, M) = \frac{\text{set of continuous 1-cocycles from } G_{\bar{K}/K} \text{ to } M}{\text{equivalence of cohomologous 1-cocycles}}.$$

Example 3.1. If $\mathcal{D}/K$ is any algebraic group, then there is a natural action of $G_{\bar{K}/K}$ on $\mathcal{D} = \mathcal{D}(\bar{K})$; and as explained above (2.1.2), this action will be discrete. Clearly

$$H^0(G_{\bar{K}/K}, \mathcal{D}) = \mathcal{D}(K)$$

is the subgroup of K-rational points. The structure of the set $H^1(G_{\bar{K}/K}, \mathcal{D})$ is harder to describe, but for the special case of the general linear group there is the following generalization of Hilbert's Theorem 90.

Proposition 3.2. *For all integers $n \geqslant 1$,*

$$H^1(G_{\bar{K}/K}, GL_n(\bar{K})) = \{1\}.$$

PROOF. [Se 9, Ch. X, Prop. 3].

EXERCISES

B.1. Prove that the sequence in (1.2) is exact.

B.2. Let G be a finite group and M a G-module.
   (a) If G has order n, prove that every element of $H^1(G, M)$ is killed by n.
   (b) If M is finitely generated as a G-module, prove that $H^1(G, M)$ is finite.

B.3. Let G be a finite group, M a G-module, and H a normal subgroup of G.
   (a) Show that there is a natural action of G/H on $H^1(H, M)$.
   (b) Prove that the image of the restriction map $\text{Res} : H^1(G, M) \to H^1(H, M)$ lies in the subgroup of $H^1(H, M)$ fixed by G/H. This allows (1.3) to be refined to

$$0 \to H^1(G/H, M^H) \overset{\text{Inf}}{\to} H^1(G, M) \overset{\text{Res}}{\to} H^1(H, M)^{G/H}.$$

B.4. Let M be a (discrete) $G_{\bar{K}/K}$-module. If $F/L/K$ is a tower of fields, then there are inflation maps

$$H^1(G_{L/K}, M^{G_{\bar{K}/L}}) \to H^1(G_{F/K}, M^{G_{\bar{K}/F}}).$$

Prove that these form a direct system, and that there is an isomorphism

$$H^1(G_{\bar{K}/K}, M) \cong \text{Lim } H^1(G_{L/K}, M^{G_{\bar{K}/L}}),$$

where the direct limit is taken over all finite Galois extensions L/K. (This provides an alternative definition for the cohomology of $G_{\bar{K}/K}$-modules.)

B.5. Let G be a finite group, and let E and A be groups on which G acts. Assume that E is abelian, and that A acts on E in a manner compatible with the action of G. (I.e. $(\alpha x)^\sigma = \alpha^\sigma x^\sigma$ for all $\alpha \in A$, $x \in E$, and $\sigma \in G$.) The *twisted product of E and A,* denoted $E \ltimes A$, is the group whose underlying set is $E \times A$, and whose group law is given by

$$(x, \alpha) * (y, \beta) = (x(\alpha y), \alpha\beta).$$

Notice that G acts on $E \ltimes A$ via $(x, \alpha)^\sigma = (x^\sigma, \alpha^\sigma)$.
   (a) Prove that there are exact sequences

$$1 \to E \to E \ltimes A \to A \to 1$$

   and

$$1 \to E^G \to (E \ltimes A)^G \to A^G \to 1.$$

   (b) Any $\alpha \in A^G$ gives a G-isomorphism $\alpha : E \to E$, and so induces an automorphism of $H^1(G, E)$. Show that two elements $\xi_1, \xi_2 \in H^1(G, E)$ have the same image under the natural map $H^1(G, E) \to H^1(G, E \ltimes A)$ if and only if there is an $\alpha \in A^G$ such that $\alpha \xi_1 = \xi_2$.

B.6. Let G be a finite group, M a G-module, and H_1 and H_2 subgroups of G. Suppose further that H_1 and H_2 are conjugate. (I.e. $H_1 = \sigma H_2 \sigma^{-1}$ for some $\sigma \in G$.) Prove that the restriction maps

$$\text{Res} : H^1(G, M) \to H^1(H_1, M) \quad \text{and} \quad \text{Res} : H^1(G, M) \to H^1(H_2, M)$$

have the same kernel.

Further Topics: An Overview

In this volume we have tried to give an essentially self-contained introduction to the basic theory of the arithmetic of elliptic curves. Unfortunately, due to limitations of time and space, many important topics have had to be omitted. This appendix contains a *very brief* introduction to some of the material which could not be included in the main body of the text. Further details may be found in the references listed at the end of each section.

Since the ten topics covered in this appendix were originally supposed to form chapters XI through XX of this book, they have been numbered as sections 11 through 20. The contents of appendix C are as follows:

§11. Complex Multiplication

The Kronecker–Weber theorem says that the maximal abelian extension $\mathbb{Q}^{ab}$ of $\mathbb{Q}$ is generated by roots of unity; and so the class field theory of $\mathbb{Q}$ is given explicitly by an isomorphism

$$G_{\mathbb{Q}^{ab}/\mathbb{Q}} \cong \prod_p \mathbb{Z}_p^*.$$

The theory of complex multiplication provides a similar description for the abelian extensions of quadratic imaginary fields.

Let $\mathscr{K}/\mathbb{Q}$ be a quadratic imaginary field, $\mathscr{R} \subset \mathscr{K}$ the ring of integers of $\mathscr{K}$, and $\mathscr{C}\ell(\mathscr{R})$ the ideal class group of $\mathscr{R}$. If we fix an embedding $\mathscr{K} \subset \mathbb{C}$, then each ideal Λ of $\mathscr{R}$ is a lattice $\Lambda \subset \mathbb{C}$, and we can consider the elliptic curve $\mathbb{C}/\Lambda$. From (VI.4.1),

$$\mathrm{End}(\mathbb{C}/\Lambda) \cong \{\alpha \in \mathbb{C} : \alpha\Lambda \subset \Lambda\} = \mathscr{R}.$$

Further, (VI.4.1.1) says that up to isomorphism, $\mathbb{C}/\Lambda$ only depends on the ideal class $\{\Lambda\} \in \mathscr{C}\ell(\mathscr{R})$.

Conversely, suppose that $E/\mathbb{C}$ satisfies $\mathrm{End}(E) \cong \mathscr{R}$. Then (VI.5.1.1) implies that $E(\mathbb{C}) \cong \mathbb{C}/\Lambda$ for a unique ideal class $\{\Lambda\} \in \mathscr{C}\ell(\mathscr{R})$. We have proven the following.

Proposition 11.1. *With notation as above, there is a one-to-one correspondence between ideal classes in $\mathscr{C}\ell(\mathscr{R})$ and isomorphism classes of elliptic curves $E/\mathbb{C}$ with $\mathrm{End}(E) \cong \mathscr{R}$.*

Corollary 11.1.1. (a) *There are only finitely many isomorphism classes of elliptic curves $E/\mathbb{C}$ with $\mathrm{End}(E) \cong \mathscr{R}$.*
(b) *Let $E/\mathbb{C}$ be an elliptic curve with $\mathrm{End}(E) \cong \mathscr{R}$. Then $j(E)$ is algebraic over $\mathbb{Q}$.*

PROOF. (a) Clear from (11.1), since $\mathscr{C}\ell(\mathscr{R})$ is finite.
(b) Let $\sigma \in \mathrm{Aut}(\mathbb{C}/\mathbb{Q})$. Then $\mathrm{End}(E^\sigma) \cong \mathrm{End}(E) \cong \mathscr{R}$. It follows from (a) that $\{E^\sigma : \sigma \in \mathrm{Aut}(\mathbb{C}/\mathbb{Q})\}$ contains only finitely many isomorphism classes of elliptic curves. Since $j(E^\sigma) = j(E)^\sigma$, we see that the set $\{j(E)^\sigma : \sigma \in \mathrm{Aut}(\mathbb{C}/\mathbb{Q})\}$ is finite. It follows that $j(E)$ is algebraic over $\mathbb{Q}$. $\square$

Actually, we can say quite a bit more about the j-invariant of an elliptic curve with complex multiplication. For any $\{\Lambda\} \in \mathscr{C}\ell(\mathscr{R})$, let us denote the j-invariant of $\mathbb{C}/\Lambda$ by $j(\Lambda)$.

Theorem 11.2 (Weber, Fueter). *Let $\{\Lambda\} \in \mathscr{C}\ell(\mathscr{R})$.*
(a) *$j(\Lambda)$ is an algebraic integer.*
(b) *$[\mathscr{K}(j(\Lambda)) : \mathscr{K}] = [\mathbb{Q}(j(\Lambda)) : \mathbb{Q}]$.*
(c) *The field $\mathscr{H} = \mathscr{K}(j(\Lambda))$ is the maximal unramified abelian extension of $\mathscr{K}$. (I.e. $\mathscr{H}$ is the Hilbert class field of $\mathscr{K}$.)*
(d) *Let $\{\Lambda_1\}, \ldots, \{\Lambda_h\}$ be a complete set of representatives for $\mathscr{C}\ell(\mathscr{R})$. Then $j(\Lambda_1), \ldots, j(\Lambda_h)$ form a complete set of $G_{\mathscr{H}/\mathscr{K}}$ conjugates for $j(\Lambda)$.*

PROOF. (a) The original proof of the integrality of $j(\Lambda)$ uses the theory of modular functions. (See, for example, [Shi 1, §4.6] or [La 3, ch. 5, thm. 4].) An algebraic proof (which generalizes to higher dimensions) can be given using the criterion of Néron–Ogg–Shafarevich ([Se–T, thm. 6]. See also (exer. 7.10).)

(b), (c), (d) [La 3, ch. 10, thm. 1], [Se 4], or [Shi 1, thm. 5.7].   □

Example 11.3.1. Suppose that $E/\mathbb{Q}$ is an elliptic curve with complex multiplication, and suppose that $\text{End}(E)$ is the full ring of integers $\mathscr{R}$ in the field $\mathscr{K} = \text{End}(E) \otimes \mathbb{Q}$. (Note that $\mathscr{K}$ is necessarily quadratic imaginary (VI.5.5).) Since $j(E) \in \mathbb{Q}$, it follows from (11.2c) that

$$\mathscr{H} = \mathscr{K}(j(E)) = \mathscr{K};$$

and so $\mathscr{K}$ has class number 1.

Conversely, if $\mathscr{K}/\mathbb{Q}$ is a quadratic imaginary field with class number 1, then (11.2bc) implies that for any $\{\Lambda\} \in \mathscr{C}\ell(\mathscr{R})$, we have $j(\Lambda) \in \mathbb{Q}$. (E.g. We could take $\Lambda = \mathscr{R}$.) Hence $\mathbb{C}/\Lambda$ is (analytically) isomorphic to an elliptic curve $E/\mathbb{Q}$ with $j(E) = j(\Lambda)$ and $\text{End}(E) \cong \mathscr{R}$.

Now Baker, Heegner, and Stark have shown that there are exactly 9 quadratic imaginary fields whose ring of integers has class number 1, namely $\mathbb{Q}(\sqrt{-d})$ for $d \in \{1, 2, 3, 7, 11, 19, 43, 67, 163\}$. Hence there are only 9 possible j-invariants for elliptic curves E defined over $\mathbb{Q}$ for which $\text{End}(E)$ is the full ring of integers in $\text{End}(E) \otimes \mathbb{Q}$.

Remark 11.3.2. If we relax the requirement that $\text{End}(E)$ be the full ring of integers of $\mathscr{K}$, and allow it to be an arbitrary order of $\mathscr{K}$, then $\text{End}(E)$ will have the form $\text{End}(E) \cong \mathbb{Z} + f\mathscr{R}$ for some $f \in \mathbb{Z}$ (exer. 3.20). One can show in this case that

$$[\mathscr{K}(j(E)) : \mathscr{K}] = \#\mathscr{C}\ell(\mathbb{Z} + f\mathscr{R}),$$

where $\mathscr{C}\ell(\mathbb{Z} + f\mathscr{R})$ is the group of projective $(\mathbb{Z} + f\mathscr{R})$-modules of rank 1. In particular, if $j(E) \in \mathbb{Q}$, then $\mathscr{C}\ell(\mathbb{Z} + f\mathscr{R}) = (1)$; and one can then check that there are only four possibilities with $f \geqslant 2$, namely

$$\mathbb{Q}(\sqrt{-1}), \mathbb{Q}(\sqrt{-3}), \mathbb{Q}(\sqrt{-7}) \qquad \text{with } f = 2,$$

and

$$\mathbb{Q}(\sqrt{-3}) \qquad\qquad\qquad \text{with } f = 3.$$

Combining this with (11.3.1), we see that up to isomorphism over $\overline{\mathbb{Q}}$, there are exactly 13 elliptic curves $E/\mathbb{Q}$ having complex multiplication. Of course, each $\overline{\mathbb{Q}}$-isomorphism class contains infinitely many $\mathbb{Q}$-isomorphism classes (X.5.4). (For example, the family of curves $E/\mathbb{Q}$ with $\text{End}(E) \cong \mathbb{Z}[\sqrt{-1}]$ is studied in (X §6).)

Returning now to the situation in (11.2), let $\{\Lambda\} \in \mathscr{C}\ell(\mathscr{R})$. Then from (11.2), the Galois group $G_{\mathscr{H}/\mathscr{K}}$ acts on $\mathscr{K}(j(\Lambda))$. This action can be described quite precisely in terms of the Artin map.

Theorem 11.4 (Hasse). *Let* $\{\Lambda_j$ $\mathscr{l}(\mathscr{R})$ *and* $\mathscr{H} = \mathscr{K}(j(\Lambda))$ *be as in* (11.2). *For each prime ideal* $\mathfrak{p}$ *of* $\mathscr{K}$, *let* F $\mathfrak{v}) \in G_{\mathscr{H}/\mathscr{K}}$ *be the Frobenius element corresponding to* $\mathfrak{p}$. *Suppose that tr.* *an elliptic curve (defined over* $\mathscr{H}$) *with j-invariant* $j(\Lambda)$ *which has good reduction at all primes of* $\mathscr{H}$ *lying over* $\mathfrak{p}$. *Then*

$$j(\Lambda)^{\mathrm{Frob}(\mathfrak{p})} = j(\Lambda \cdot \mathfrak{p}^{-1}).$$

(*Here* $\Lambda \cdot \mathfrak{p}^{-1}$ *is the usual product of fractional ideals of* $\mathscr{K}$.)

PROOF. [La 3, ch. 10, thm. 1], [Se 4], or [Shi 1, thm. 5.7]. $\qquad\square$

Suppose now that E/K is an elliptic curve with complex multiplication over K. (I.e. $\mathrm{End}_K(E) \neq \mathbb{Z}$.) Then the fact that $G_{\bar{K}/K}$ and $\mathrm{End}_K(E)$ commute with one another in their action on the Tate module $T_\ell(E)$ will imply that the action of $G_{\bar{K}/K}$ is abelian. (This is essentially Schur's lemma. See exer. 3.24.) Thus the field $K(E_{\mathrm{tors}})$ obtained by adjoining to K the coordinates of all of the torsion points of E will be an abelian extension of K.

Let us return now to the case that $\{\Lambda\} \in \mathscr{Cl}(\mathscr{R})$, $\mathscr{H} = \mathscr{K}(j(\Lambda))$, and $E/\mathscr{H}$ is an elliptic curve with j-invariant $j(\Lambda)$. Then $\mathscr{H}(E_{\mathrm{tors}})$ is an abelian extension of $\mathscr{H}$, but it will not in general be an abelian extension of $\mathscr{K}$. However, it turns out that $\mathscr{H}(E_{\mathrm{tors}})$ contains $\mathscr{K}^{ab}$, and $\mathscr{H}(E_{\mathrm{tors}})/\mathscr{K}^{ab}$ is an abelian extension whose Galois group is (generally) a product of groups of order 2. In order to produce $\mathscr{K}^{ab}$ itself, we instead adjoin (essentially) just the x-coordinates of the torsion points.

To make this precise, for any elliptic curve E/K, let us define a *Weber function* on E/K to be a morphism defined over K of the form

$$\phi_E : E \to E/\mathrm{Aut}(E) \cong \mathbb{P}^1.$$

(For the definition of the quotient curve $E/\mathrm{Aut}(E)$, see (exer. 3.13).) Classically, if E is given by a Weierstrass equation

$$E : y^2 = 4x^3 - g_2 x - g_3 \qquad g_2, g_3 \in \mathbb{C}$$

with discriminant $\Delta = g_2^3 - 27g_3^2$, then one defines *the Weber function* quite explicitly by the formula

$$\phi_E(P) = \begin{cases} (g_2 g_3/\Delta)x(P) & \text{if } j(E) \neq 0, 1728 \\ (g_2^2/\Delta)x(P)^2 & \text{if } j(E) = 1728 \\ (g_3/\Delta)x(P)^3 & \text{if } j(E) = 0. \end{cases}$$

Notice that although g_2 and g_3 are allowed to be in $\mathbb{C}$, the map $\phi_E : E \to \mathbb{P}^1$ is independent of the choice of Weierstrass equation for E, and will thus be defined over any field of definition for E.

Theorem 11.5. *Let* $\mathscr{K}$ *be a quadratic imaginary field*, $\mathscr{R} \subset \mathscr{K}$ *its ring of integers, and let* $E/\mathbb{C}$ *be an elliptic curve with* $\mathrm{End}(E) \cong \mathscr{R}$.
(a) *The maximal unramified abelian extension of* $\mathscr{K}$ *is* $\mathscr{K}(j(E))$.
(b) *The maximal abelian extension* $\mathscr{K}^{ab}$ *of* $\mathscr{K}$ *is given by*

$$\mathcal{K}^{ab} = \mathcal{K}(j(E); \phi_E(T), T \in E_{\text{tors}}).$$

[*I.e. $\mathcal{K}^{ab}$ is the field obtained by adjoining to $\mathcal{K}$ the j-invariant of E and the value of a Weber function at all of the torsion points of E.*]

PROOF. (a) This is a restatement of (11.2c).
(b) [La 3, ch. 10, thm. 2], [Se 4], or [Shi 1, cor. 5.6].   □

Remark 11.6. Let $\{\Lambda\}$ be any ideal class of $\mathcal{R}$, for example $\Lambda = \mathcal{R}$. Then in (11.5), we could take E to be the elliptic curve with $E(\mathbb{C}) \cong \mathbb{C}/\Lambda$ given by the Weierstrass equation

$$E : y^2 = 4x^3 - g_2(\Lambda)x - g_3(\Lambda).$$

(For the definition of $g_2(\Lambda)$ and $g_3(\Lambda)$ in terms of infinite series, see (VI §3).) Then the Weber function

$$\phi_\Lambda : \mathbb{C}/\Lambda \to \mathbb{C}$$

is given analytically by

$$\phi_\Lambda(z) = \begin{cases} (g_2(\Lambda)g_3(\Lambda)/\Delta(\Lambda))\wp(z, \Lambda) & \text{if } j(\Lambda) \neq 0, 1728 \\ (g_2(\Lambda)^2/\Delta(\Lambda))\wp(z, \Lambda)^2 & \text{if } j(\Lambda) = 1728 \\ (g_3(\Lambda)/\Delta(\Lambda))\wp(z, \Lambda)^3 & \text{if } j(\Lambda) = 0. \end{cases}$$

Now (11.5) says that $\mathcal{K}^{ab}$ is generated by $j(\Lambda)$ and $\phi_\Lambda(t)$ for $t \in \mathbb{Q}\Lambda \subset \mathbb{C}$. Thus $\mathcal{K}^{ab}$ is given explicitly by the values of an analytic function evaluated at points of finite order on the complex torus $\mathbb{C}/\Lambda$. Notice the similarity with the situation over $\mathbb{Q}$, where $\mathbb{Q}^{ab}$ is generated by the values of the analytic function $\phi(z) = e^{2\pi i z}$ at the points of finite order on the cylinder $\mathbb{C}/\mathbb{Z}$.

Remark 11.7. Just as in (11.4), one can use the Artin map to describe the action of $G_{\mathcal{K}^{ab}/\mathcal{K}}$ on the elements $\phi_E(T)$ which generate $\mathcal{K}^{ab}/\mathcal{K}$. See, for example, [Shi 1, thm. 5.4] or [La 3, ch. 10, lemma 1 and thm. 3].

References. [La 3], [Se 4], [Shi 1]. For generalizations to abelian varieties, see [Shi–T], [Se–T], [La 10].

§12. Modular Functions

As we have seen (VI.5.1.1), every elliptic curve $E/\mathbb{C}$ is analytically isomorphic to a complex torus $\mathbb{C}/\Lambda$, where $\Lambda \subset \mathbb{C}$ is a lattice which is determined up to homothety by E. Associated to the lattice Λ are the Eisenstein series $G_{2k}(\Lambda)$, discriminant $\Delta(\Lambda)$, and j-invariant $j(\Lambda)$. One easily verifies the homogeneity properties (exer. 6.6)

$$G_{2k}(\alpha\Lambda) = \alpha^{-2k}G_{2k}(\Lambda) \qquad \Delta(\alpha\Lambda) = \alpha^{-12}\Delta(\Lambda) \qquad j(\alpha\Lambda) = j(\Lambda).$$

These functions have as their domain the space of lattices. Using homogeneity, it is enough to study them in the space of lattices modulo homothety. In order to do this, we set the following notation:

$$\mathbb{H} = \{\tau \in \mathbb{C} : \operatorname{Im}(\tau) > 0\}$$

$$\Lambda_\tau = \mathbb{Z} + \mathbb{Z}\tau \qquad \text{for } \tau \in \mathbb{H}$$

$$G_{2k}(\tau) = G_{2k}(\Lambda_\tau) \qquad \Delta(\tau) = \Delta(\Lambda_\tau) \qquad j(\tau) = j(\Lambda_\tau).$$

Clearly every lattice Λ is homothetic to Λ_τ for some $\tau \in \mathbb{H}$. In order to describe when two τ's give the same lattice, we note that the group

$$SL_2(\mathbb{Z}) = \{(\begin{smallmatrix} a & b \\ c & d \end{smallmatrix}) : a, b, c, d \in \mathbb{Z}, ad - bc = 1\}$$

acts on $\mathbb{H}$ by linear fractional transformation

$$\gamma = (\begin{smallmatrix} a & b \\ c & d \end{smallmatrix}) : \mathbb{H} \to \mathbb{H} \qquad \gamma(\tau) = (a\tau + b)/(c\tau + d).$$

This action is described by the following proposition.

Proposition 12.1. (a) *The group $SL_2(\mathbb{Z})$ acts properly discontinuously on $\mathbb{H}$.*
(b) *The region*

$$\mathscr{F} = \{\tau \in \mathbb{H} : |\operatorname{Re}(\tau)| \leq \tfrac{1}{2} \text{ and } |\tau| \geq 1\}$$

is a fundamental domain for $\mathbb{H}/SL_2(\mathbb{Z})$. (I.e. The natural map $\mathscr{F} \to \mathbb{H}/SL_2(\mathbb{Z})$ is surjective, and its restriction to the interior of $\mathscr{F}$ is injective.)
(c) *Let*

$$S = (\begin{smallmatrix} 0 & -1 \\ 1 & 0 \end{smallmatrix}) \quad and \quad T = (\begin{smallmatrix} 1 & 1 \\ 0 & 1 \end{smallmatrix}).$$

Then $S^2 = -1, (ST)^3 = -1$. The modular group $PSL_2(\mathbb{Z}) = SL_2(\mathbb{Z})/\pm 1$ is the free product of the cyclic groups of order 2 and 3 generated by S and ST. In particular, S and T generate $PSL_2(\mathbb{Z})$.

PROOF. [Ap, thm. 2.1, 2.3], [Se 7, VII §1].   □

Corollary 12.1.1. *Every lattice $\Lambda \subset \mathbb{C}$ is homothetic to a lattice Λ_τ for some $\tau \in \mathscr{F}$.*

Figure 12.1 illustrates the fundamental domain $\mathscr{F}$ and its translates under various elements of $SL_2(\mathbb{Z})$.

Remark 12.2. Any two bases $\{\omega_1, \omega_2\}$ and $\{\omega_1', \omega_2'\}$ for a lattice Λ are related by a change of basis formula

$$\omega_1' = a\omega_1 + b\omega_2 \qquad \omega_2' = c\omega_1 + d\omega_2$$

with $a, b, c, d \in \mathbb{Z}$ and $ad - bc = \pm 1$. If we use homotheties to replace these bases by ones of the form $\{1, \tau\}$ and $\{1, \tau'\}$ with $\tau, \tau' \in \mathbb{H}$, then the above change of basis action on the ω's becomes exactly the linear fractional action of $SL_2(\mathbb{Z})$ on the τ's described above.

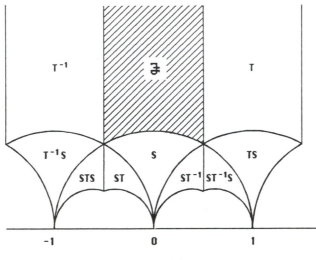

Figure 12.1

The function $G_{2k}(\Lambda)$ depends only on the lattice Λ, and not on any particular choice of basis. However, if $\Lambda_\tau = \Lambda_{\tau'}$ for some $\tau, \tau' \in \mathbb{H}$, then $G_{2k}(\tau)$ and $G_{2k}(\tau')$ may not be equal. Tracing through the definitions, one checks that

$$G_{2k}(\gamma\tau) = (c\tau + d)^{2k} G_{2k}(\tau) \qquad \text{for } \gamma = \left(\begin{smallmatrix} a & b \\ c & d \end{smallmatrix}\right) \in SL_2(\mathbb{Z}).$$

Notice that if $c = 0$, then $G_{2k}(\gamma\tau) = G_{2k}(\tau)$. In other words,

$$G_{2k}(T^n\tau) = G_{2k}(\tau + n) = G_{2k}(\tau) \qquad \text{for all } n \in \mathbb{Z}.$$

This means that G_{2k} has a Fourier expansion

$$G_{2k}(\tau) = \sum_{n=-\infty}^{\infty} c(n)q^n,$$

where we write $q = e^{2\pi i \tau}$.

Definition. A meromorphic function on $\mathbb{H}$ is called a *modular function of weight k* (*for* $SL_2(\mathbb{Z})$) if it satisfies:

(i) $f(\tau) = (c\tau + d)^{-k} f(\gamma\tau)$ for all $\gamma = \left(\begin{smallmatrix} a & b \\ c & d \end{smallmatrix}\right) \in SL_2(\mathbb{Z})$;
(ii) The Fourier expansion of f in the variable $q = e^{2\pi i \tau}$ has the form

$$f(\tau) = \sum_{n=n_0}^{\infty} c(n)q^n$$

for some (finite) integer $n_0 = n_0(f)$.

We say that f is a *modular form of weight k* if f is holomorphic on $\mathbb{H}$ and $n_0(f) = 0$, in which case we also say that f is *holomorphic at* ∞ and set $f(\infty) = c(0)$. (Notice that $q \to 0$ as $\tau \to i\infty$.) If further $f(\infty) = 0$, then we call f a *cusp form*.

Remark 12.3. Note that if $\alpha = \begin{pmatrix} -1 & 0 \\ 0 & -1 \end{pmatrix}$, then $\alpha\tau = \tau$ for every $\tau \in \mathbb{H}$. Hence if f is a modular function of weight k with k an odd integer, then $f(\tau) = (-1)^{-k} f(\alpha\tau) = -f(\tau)$, and so f is identically 0. Thus a non-trivial modular function for $SL_2(\mathbb{Z})$ is necessarily of even weight. The next proposition provides some examples of modular functions.

Proposition 12.4. (a) *The j-function $j(\tau)$ is a modular function of weight 0 which is holomorphic on $\mathbb{H}$. It Fourier series has the form*

$$j(\tau) = \frac{1}{q} + 744 + \sum_{n=1}^{\infty} c(n)q^n \qquad \text{with } c(n) \in \mathbb{Z}.$$

(b) *The Eisenstein series $G_{2k}(\tau)$ is a modular form of weight $2k$. Its Fourier series is given by*

$$G_{2k}(\tau) = 2\zeta(2k) + 2\frac{(2\pi i)^{2k}}{(2k-1)!} \sum_{n=1}^{\infty} \sigma_{2k-1}(n)q^n.$$

(*Here $\zeta(s) = \Sigma n^{-s}$ is the Riemann zeta function; and $\sigma_\alpha(n)$ is the divisor function, $\sigma_\alpha(n) = \Sigma_{d|n} d^\alpha$.*)
(c) *The discriminant function $\Delta(\tau)$ is a cusp form of weight 12. Its Fourier series has the form*

$$\Delta(\tau) = (2\pi)^{12} \sum_{n=1}^{\infty} \tau(n)q^n \qquad \text{with } \tau(1) = 1 \text{ and } \tau(n) \in \mathbb{Z}.$$

(*The integer-valued function $n \to \tau(n)$ is called the* Ramanujan τ-function.)

PROOF. [Ap, thm. 1.18, 1.19, 1.20], [Se 7, VII prop. 4, 5, 8].   □

Remark 12.4.1. The Fourier coefficients of $j(\tau)$ and $\Delta(\tau)$ have many interesting congruence properties. For example,

$$\tau(n) \equiv \sigma_{11}(n) \,(\text{mod } 691)$$

for all $n = 1, 2, \ldots$, a result due to Ramanujan. We will not pursue this topic, but see, for example, [Ap, ch. 4] or [Se 7, VII, §3.3, §4.5].

Remark 12.4.2. The Fourier series for the Eisenstein series $G_{2k}(\tau)$ is often rewritten using the identity

$$\sum_{n=1}^{\infty} \sigma_\alpha(n)q^n = \sum_{n=1}^{\infty} n^\alpha q^n/(1 - q^n).$$

It commonly appears in the literature in both forms.

The discriminant function $\Delta(\tau)$ also has the following beautiful product expansion.

Theorem 12.5. (a) (Jacobi)

$$\Delta(\tau) = (2\pi)^{12} q \prod_{n=1}^{\infty} (1 - q^n)^{24}.$$

(b) *Define the* Dedekind η-function *by*

$$\eta(\tau) = q^{1/24} \prod_{n=1}^{\infty} 1 - q^n.$$

(Here $q^{1/24}$ is defined to be $e^{\pi i \tau/12}$. Notice that $\Delta(\tau) = (2\pi)^{12}\eta(\tau)^{24}$.) Then

$$\eta(\tau + 1) = e^{\pi i/12}\eta(\tau) \quad and \quad \eta(-1/\tau) = (-i\tau)^{1/2}\eta(\tau).$$

(We take the branch of the square-root function which is positive on the positive real axis.)

PROOF. [Ap, thm. 3.1, 3.3], [Se 7, VII, thm. 6].   □

Remark 12.5.1. Since the maps $\tau \to \tau + 1$ and $\tau \to -1/\tau$ generate the action of $SL_2(\mathbb{Z})$ on $\mathbb{H}$ (12.1.c), one can check that

$$\eta(\gamma\tau) = \varepsilon\{-i(c\tau + d)\}^{1/2}\eta(\tau) \qquad \text{for all } \gamma = \left(\begin{smallmatrix} a & b \\ c & d \end{smallmatrix}\right) \in SL_2(\mathbb{Z}),$$

where $\varepsilon = \varepsilon(a, b, c, d)$ satisfies $\varepsilon^{24} = 1$. There is a formula for $\varepsilon(a, b, c, d)$ which involves Dedekind sums. See [Ap, ch. 3, especially thm. 3.4] for details.

Elliptic functions, such at the Weierstrass $\wp$-function, can be treated as functions of two variables, the second variable being the lattice. We define

$$\wp(z; \tau) = \wp(z; \Lambda_\tau) \qquad \wp'(z; \tau) = \wp'(z; \Lambda_\tau) \qquad \sigma(z; \tau) = \sigma(z; \Lambda_\tau).$$

These functions have the following q-expansions.

Proposition 12.6. Let $q = e^{2\pi i \tau}$ and $u = e^{2\pi i z}$. Then

$$(2\pi i)^{-2}\wp(z; \tau) = \sum_{n=-\infty}^{\infty} q^n u/(1 - q^n u)^2 + 1/12 - 2\sum_{n=1}^{\infty} q^n/(1 - q^n)^2;$$

$$(2\pi i)^{-3}\wp'(z; \tau) = \sum_{n=-\infty}^{\infty} q^n(1 + q^n u^2)/(1 - q^n u)^3;$$

$$(2\pi i)\sigma(z; \tau) = e^{\eta z^2/2}(u^{1/2} - u^{-1/2}) \prod_{n=1}^{\infty} (1 - q^n u)(1 - q^n u^{-1})/(1 - q^n)^2.$$

(In this last formula, $\eta = \eta(1)$ is one of the quasi-periods associated to the lattice $\mathbb{Z} + \mathbb{Z}\tau$. See (exer. 6.4b).)

PROOF. [La 3, ch. 18, §2], [Rob, II §5].   □

Remark 12.7. An elliptic curve $E/\mathbb{C}$ is analytically isomorphic to a torus $\mathbb{C}/\Lambda$, and we can choose the lattice Λ to be of the form $\Lambda = \Lambda_\tau = \mathbb{Z} + \mathbb{Z}\tau$. Consider the exponential map $\exp(2\pi i \cdot): \mathbb{C} \to \mathbb{C}^*$. The image of Λ_τ under this map is the cyclic subgroup $q^{\mathbb{Z}} = \{q^n : n \in \mathbb{Z}\}$ of $\mathbb{C}^*$. Thus composing with the exponential map, we obtain an analytic isomorphism

$$\mathbb{C}^*/q^{\mathbb{Z}} \xrightarrow{\sim} E(\mathbb{C}).$$

If we let u be the parameter on $\mathbb{C}^*$ (i.e. $u = e^{2\pi i z}$), then this map is given by $[\wp, \wp', 1]$, where $\wp$ and $\wp'$ are given in terms of u by (12.6).

We now describe the field of modular functions and the algebra of modular forms.

Definition. Let

$$M_k = \{\text{modular forms of weight } 2k\},$$

$$M_{k,0} = \{\text{cusp forms of weight } 2k\}.$$

Notice that M_k and $M_{k,0}$ are $\mathbb{C}$-vector spaces. Further, if $f \in M_k$ and $g \in M_{k'}$, then $fg \in M_{k+k'}$. Thus the ring

$$M = \sum_{k=0}^{\infty} M_k$$

has a natural structure as a graded $\mathbb{C}$-algebra.

Theorem 12.8. (a) $j(\tau)$ *is a modular function of weight* 0. *Every modular function of weight* 0 *is a rational function of* $j(\tau)$.
(b) *The map*

$$\mathbb{C}[X, Y] \to M \qquad P(X, Y) \to P(G_4, G_6)$$

is an isomorphism of graded $\mathbb{C}$-*algebras, where we assign weights* $wt(X) = 4$ *and* $wt(Y) = 6$. *In particular, every modular form is a polynomial in* G_4 *and* G_6.

(c) $$\dim_{\mathbb{C}} M_k = \begin{cases} 0 & \text{if } k < 0 \\ [k/6] & \text{if } k \equiv 1 \ (\text{mod } 6), \ k \geqslant 0 \\ [k/6] + 1 & \text{if } k \not\equiv 1 \ (\text{mod } 6), \ k \geqslant 0. \end{cases}$$

(d) *Multiplication by* $\Delta(\tau)$ *defines an isomorphism of* M_{k-6} *onto* $M_{k,0}$.

PROOF. [Ap, thm. 2.8, §6.4, §6.5], [Se 7, VII, §3.2, §3.3]. □

The study of the spaces M_k and $M_{k,0}$ is facilitated by the existence of certain linear operators. For each integer $n \geqslant 1$, we define the *Hecke operator* $T(n)$ on the space M_k of modular forms of weight $2k$ by the formula

$$(T(n)f)(\tau) = n^{2k-1} \sum_{d|n} d^{-2k} \sum_{b=0}^{d-1} f((n\tau + bd)/d^2).$$

(For a more intrinsic definition, see [Ap, §6.8], [Se 7, VII §5.1], or [Shi 1, Ch. 3].)

Proposition 12.9. (a) *If* f *is a modular form (respectively cusp form) of weight* $2k$, *then* $T(n)f$ *is also. In other words,* $T(n)$ *induces linear maps*

$$T(n): M_k \to M_k \quad \text{and} \quad T(n): M_{k,0} \to M_{k,0}.$$

(b) *For all integers m and n,*

$$T(m)T(n) = T(n)T(m).$$

(c) *If m and n are relatively prime, then*

$$T(mn) = T(m)T(n).$$

(d) *For all primes p and integers $r \geqslant 1$,*

$$T(p^{r+1}) = T(p^r)T(p) - p^{2k-1}T(p^{r-1}).$$

PROOF. [Ap, thm. 6.11, 6.13], [Se 7, VII §5.1, §5.3].   □

Application 12.10. Of particular interest are those modular forms which are *simultaneous eigenfunctions* for every Hecke operator $T(n)$. In other words,

$$T(n)f = \lambda(n)f \quad \text{for all } n = 1, 2, \ldots,$$

where $\lambda(1), \lambda(2), \ldots$ are certain constants. If this occurs, then it is not hard to show that the Fourier expansion of f, $f = \Sigma c(n)q^n$, satisfies

$$c(n) = c(1)\lambda(n) \quad \text{for all } n = 1, 2, \ldots.$$

(See [Ap, thm. 6.14, 6.15], [Se 7, VII §5.4].) In particular, if f is not constant, then $c(1) \neq 0$, and f is uniquely determined by $c(1)$ and the eigenvalues $\{\lambda(n)\}$.

Example 12.10.1. Consider the space $M_{6,0}$ of cusp forms of weight 12. From (12.8c) and (12.4c), it has dimension 1, and is generated by the discriminant function

$$\Delta = (2\pi)^{12} \prod_{n=1}^{\infty} (1 - q^n)^{24} = (2\pi)^{12} \sum_{n=1}^{\infty} \tau(n)q^n.$$

Since $T(n)\Delta$ is also in $M_{6,0}$, it follows that $T(n)\Delta$ is a multiple of Δ. From (12.10) we conclude that

$$T(n)\Delta = \tau(n)\Delta \quad \text{for all } n = 1, 2, \ldots.$$

(Note that $\tau(1) = 1$.) Now the identities (12.9cd) satisfied by the Hecke operators $T(n)$ lead to analogous formulas for the Ramanujan function:

$$\tau(mn) = \tau(m)\tau(n) \qquad \qquad \text{if } \gcd(m, n) = 1;$$

$$\tau(p^{r+1}) = \tau(p^r)\tau(p) - p^{11}\tau(p^{r-1}) \quad \text{for } p \text{ prime, } r \geqslant 1.$$

These beautiful identities, conjectured by Ramanujan, were first proven by Mordell. There is also the estimate, demonstrated by Deligne as a consequence of his proof of the Weil conjectures, which states that

$$|\tau(p)| \leqslant 2p^{11/2} \quad \text{for } p \text{ prime.}$$

Since $j(\tau)$ is a modular function of weight 0 (12.4a), it defines a function on the quotient space $\mathbb{H}/SL_2(\mathbb{Z})$. Now $\mathbb{H}/SL_2(\mathbb{Z})$ has a natural structure as a Riemann surface, and one can show that $j(\tau)$ defines a holomorphic function on that surface. (See [Shi 1, §1.3, §1.4, §1.5].)

Proposition 12.11. *The map*

$$j: \mathbb{H}/SL_2(\mathbb{Z}) \to \mathbb{C}$$

is a complex analytic isomorphism of (open) Riemann surfaces.

PROOF. [Se 7, VII prop. 5]. □

Corollary 12.11.1 (Uniformization Theorem). *Let $E/\mathbb{C}$ be an elliptic curve. Then there exists a lattice $\Lambda \subset \mathbb{C}$ and a complex analytic isomorphism $\mathbb{C}/\Lambda \to E(\mathbb{C})$.*

PROOF. Let J be the j-invariant of E. From (12.11), there is a $\tau \in \mathbb{H}$ such that $j(\tau) = J$. Then the elliptic curve

$$E_\tau: y^2 = 4x^3 - g_2(\tau)x - g_3(\tau)$$

has j-invariant J, so $E_\tau \cong E$ from (III.1.4b). On the other hand, (VI.3.6b) describes a complex analytic isomorphism $\mathbb{C}/(\mathbb{Z} + \mathbb{Z}\tau) \to E_\tau(\mathbb{C})$, which gives the desired result. □

From (12.11), we see that the Riemann surface $\mathbb{H}/SL_2(\mathbb{Z})$ is not compact. Its natural compactification is $\mathbb{P}^1(\mathbb{C})$, obtained by adding a single extra point at infinity. However, with a view toward eventual generalizations, we will take the following approach. Define

$$\mathbb{H}^* = \mathbb{H} \cup \mathbb{P}^1(\mathbb{Q}).$$

Here one should think of the points $[x, 1] \in \mathbb{P}^1(\mathbb{Q})$ as forming the usual copy of $\mathbb{Q}$ in $\mathbb{C}$; and the point $[1, 0] \in \mathbb{P}^1(\mathbb{Q})$ as a point at infinity. Notice that $SL_2(\mathbb{Z})$ acts on $\mathbb{P}^1(\mathbb{Q})$ in the usual manner,

$$\gamma: [x, y] \to [ax + by, cx + dy].$$

The quotient space $\mathbb{H}^*/SL_2(\mathbb{Z})$ can be given the structure of a Riemann surface, and one can show that the j-function then defines a complex analytic isomorphism

$$j: \mathbb{H}^*/SL_2(\mathbb{Z}) \to \mathbb{P}^1(\mathbb{C}).$$

(See [Shi 1, §1.3, §1.4, §1.5] for details.) Notice that since $SL_2(\mathbb{Z})$ acts transitively on $\mathbb{P}^1(\mathbb{Q})$, the net effect has been to add a single point, called a *cusp*, to $\mathbb{H}/SL_2(\mathbb{Z})$.

Congruence Subgroups

In studying modular functions for $SL_2(\mathbb{Z})$, one soon discovers the need to deal with functions which are modular only for certain subgroups of $SL_2(\mathbb{Z})$.

Definition. For each integer $N \geqslant 1$, we define subgroups of $SL_2(\mathbb{Z})$ as follows:

$$\Gamma_0(N) = \{ (\begin{smallmatrix} a & b \\ c & d \end{smallmatrix}) \in SL_2(\mathbb{Z}) : c \equiv 0 \ (\mathrm{mod}\ N) \};$$

$$\Gamma_1(N) = \{ (\begin{smallmatrix} a & b \\ c & d \end{smallmatrix}) \in SL_2(\mathbb{Z}) : c \equiv 0 \ (\mathrm{mod}\ N), a \equiv d \equiv 1 \ (\mathrm{mod}\ N) \};$$

$$\Gamma(N) = \{ (\begin{smallmatrix} a & b \\ c & d \end{smallmatrix}) \in SL_2(\mathbb{Z}) : b \equiv c \equiv 0 \ (\mathrm{mod}\ N), a \equiv d \equiv 1 \ (\mathrm{mod}\ N) \}.$$

More generally, a *congruence subgroup of* $SL_2(\mathbb{Z})$ is defined to be a subgroup Γ of $SL_2(\mathbb{Z})$ which contains $\Gamma(N)$ for some integer $N \geqslant 1$.

If Γ is a congruence subgroup of $SL_2(\mathbb{Z})$, then Γ acts on $\mathbb{H}^*$, and we can form the quotient space $\mathbb{H}^*/\Gamma$. $\mathbb{H}^*/\Gamma$ has a natural structure as a Riemann surface ([Shi 1, §1.3, §1.5]). The action of Γ on $\mathbb{P}^1(\mathbb{Q}) \subset \mathbb{H}^*$ gives finitely many orbits; the images of these orbits in $\mathbb{H}^*/\Gamma$ are called the *cusps of* Γ.

Example 12.12. If p is prime, then $\mathbb{H}^*/\Gamma_0(p)$ contains two cusps, represented by the points $[1, 0]$ and $[0, 1]$ in $\mathbb{P}^1(\mathbb{Q})$.

Definition. Let Γ be a congruence subgroup of $SL_2(\mathbb{Z})$. A meromorphic function f on $\mathbb{H}$ is called a *modular function of weight k for* Γ if

(i) $f(\tau) = (c\tau + d)^{-k} f(\gamma\tau)$ for all $\gamma = (\begin{smallmatrix} a & b \\ c & d \end{smallmatrix}) \in \Gamma$; and
(ii) f is meromorphic at each of the cusps of $\mathbb{H}^*/\Gamma$. (See [Shi 1, §2.1] for the precise definition.)

A modular function is called a *modular form* if it is holomorphic on $\mathbb{H}$ and at each of the cusps of $\mathbb{H}^*/\Gamma$; and it is a *cusp form* if it is a modular form which vanishes at every cusp.

Example 12.13. The function

$$f(\tau) = \eta(\tau)^2 \eta(11\tau)^2$$

is a cusp form of weight 2 for the group $\Gamma_0(11)$. (Here $\eta(\tau)$ is the Dedekind η-function (12.5b).)

If $f(\tau)$ is a modular form of weight 2 for Γ, then one easily checks that the differential form $f(\tau)\, d\tau$ on $\mathbb{H}$ is invariant under the action of Γ. (This follows from the identity $d((a\tau + b)/(c\tau + d)) = (c\tau + d)^{-2}\, d\tau$.) If, further, f is a cusp form, then one can show that $f(\tau)\, d\tau$ is holomorphic at each of the cusps of Γ, and so defines a holomorphic 1-form on the quotient space $\mathbb{H}^*/\Gamma$.

Proposition 12.14. *Let* Γ *be a congruence subgroup of* $SL_2(\mathbb{Z})$. *There is a natural isomorphism between the space of weight 2 cusp forms for* Γ *and the space of holomorphic 1-forms on the Riemann surface* $\mathbb{H}^*/\Gamma$.

PROOF. [Shi 1, §2.4].   □

Remark 12.15. It is not difficult to calculate the genus of $\mathbb{H}^*/\Gamma$; and thereby, using (12.14), to find the dimension of the space of weight 2 cusp forms for Γ. For example, if p is prime with $p \equiv 11 \pmod{12}$, then $\mathbb{H}^*/\Gamma_0(p)$ has genus equal to $(p + 1)/12$. For general formulas, see [Shi 1, prop. 1.40, 1.43].

The Hecke operators defined above also act on the space of modular forms relative to congruence subgroups.

Proposition 12.16. *Let Γ be a congruence subgroup of $SL_2(\mathbb{Z})$, say $\Gamma \supset \Gamma(N)$; and let $f(\tau)$ be a modular form of weight $2k$ for Γ. Then for each integer $n \geqslant 1$ relatively prime to N, the function $T(n)f$ defined by the formula given above is again a modular form of weight $2k$ for Γ. Further, if f is a cusp form, then so is $T(n)f$.*

PROOF. [Shi 1, prop. 3.37].   □

Remark 12.17. Just as in the case of the full modular group $SL_2(\mathbb{Z})$, one studies those modular forms relative to Γ which are simultaneous eigenfunctions for all of the Hecke operators. For example, the Riemann surface $\mathbb{H}^*/\Gamma_0(11)$ has genus 1 (12.15), so the space of cusp forms of weight 2 for $\Gamma_0(11)$ is of dimension 1 (12.14). It follows that the function

$$f(\tau) = \eta(\tau)^2 \eta(11\tau)^2$$

given in (12.13) is an eigenfunction of $T(n)$ for every integer n satisfying $\gcd(n, 11) = 1$.

References. [Ah1, ch. 7], [Ap, ch. 2,3,6], [B–Sw 2], [Ko], [La 3], [Og 4], [Rob, ch. I, §3,4], [Se 7, ch. VII], [Shi 1, ch. 1,2,3].

§13. Modular Curves

Let Γ be a congruence subgroup of $SL_2(\mathbb{Z})$. If $\Gamma = SL_2(\mathbb{Z})$, then we have seen (§12) that the points of the Riemann surface $\mathbb{H}/\Gamma$ are in one-to-one correspondence with the isomorphism classes of elliptic curves defined over $\mathbb{C}$. This correspondence associates to the point $\tau \pmod{\Gamma}$ of $\mathbb{H}/\Gamma$ the elliptic curve $E_\tau \cong \mathbb{C}/(\mathbb{Z} + \mathbb{Z}\tau)$. We will now describe a similar interpretation for the points of $\mathbb{H}/\Gamma$ in the case that Γ is a more general congruence subgroup.

For example, consider the subgroup $\Gamma_1(N)$, which we recall consists of all matrices $\gamma = \left(\begin{smallmatrix} a & b \\ c & d \end{smallmatrix}\right)$ such that $c \equiv 0 \pmod{N}$ and $a \equiv d \equiv 1 \pmod{N}$. Since $\Gamma_1(N) \subset SL_2(\mathbb{Z})$, we can again associate to each $\tau \in \mathbb{H}/\Gamma_1(N)$ the elliptic curve E_τ. This is nothing more than the natural map $\mathbb{H}/\Gamma_1(N) \to \mathbb{H}/SL_2(\mathbb{Z})$. But a point of $\mathbb{H}/\Gamma_1(N)$ contains additional information. Consider the point $T_\tau \in E_\tau$ corresponding to $1/N \in \mathbb{C}/(\mathbb{Z} + \mathbb{Z}\tau)$. (Thus $T_\tau \in E_\tau[N]$.) Then for any $\gamma \in SL_2(\mathbb{Z})$, the isomorphism

$$f : \mathbb{C}/(\mathbb{Z} + \mathbb{Z}\tau) \to \mathbb{C}/(\mathbb{Z} + \mathbb{Z}\gamma(\tau))$$

$$z \to z/(c\tau + d)$$

maps $1/N$ to $1/N(c\tau + d)$. (Note that $\gamma(\tau) = (a\tau + b)/(c\tau + d)$.) If we further assume that $\gamma \in \Gamma_1(N)$, then

$$\frac{1}{N} - \frac{1}{N(c\tau + d)} = \frac{(c/N)\tau + (d-1)/N}{c\tau + d} \in f(\mathbb{Z} + \mathbb{Z}\tau) = \mathbb{Z} + \mathbb{Z}\gamma(\tau).$$

Thus the point $1/N \in \mathbb{C}/(\mathbb{Z} + \mathbb{Z}\tau)$ remains fixed when the basis for the lattice is changed by an element of $\Gamma_1(N)$. Hence a point of $\mathbb{H}/\Gamma_1(N)$ gives an elliptic curve $E_\tau/\mathbb{C}$ together with a specified point $T_\tau \in E_\tau$ of exact order N. Further, given any elliptic curve $E/\mathbb{C}$ and any point $T \in E$ of exact order N, there is a point $\tau \in \mathbb{H}/\Gamma_1(N)$ and an isomorphism $E_\tau \to E$ such that $T_\tau \to T$. Using fancier terminology, we say that the Riemann surface $\mathbb{H}/\Gamma_1(N)$ is a *moduli space* for the *moduli problem* of determining equivalence classes of pairs (E, T), where E is an elliptic curve defined over $\mathbb{C}$, and $T \in E$ is a point of exact order N. (Two pairs (E, T) and (E', T') are deemed equivalent if there is an isomorphism $E \cong E'$ which takes T to T'.)

Similarly, if $\gamma \in \Gamma_0(N)$ and $\tau \in \mathbb{H}/\Gamma_0(N)$, then one easily checks that the subgroup

$$\left\{ \frac{1}{N}, \frac{2}{N}, \ldots, \frac{N-1}{N} \right\} \subset \frac{\mathbb{C}}{\mathbb{Z} + \mathbb{Z}\tau}$$

remains invariant under the action of γ. As above, $\mathbb{H}/\Gamma_0(N)$ is a moduli space for the problem of determining (equivalence classes of) pairs (E, C), where E is an elliptic curve and $C \subset E$ is a cyclic subgroup of exact order N. Note that from (III.4.12), there is a one-to-one correspondence between finite subgroups $\Phi \subset E$ and isogenies $\phi : E \to E_1$, given by the association $\Phi \leftrightarrow \ker \phi$. Thus the points of $\mathbb{H}/\Gamma_0(N)$ can also be viewed as classifying triples (E, E', ϕ), where $\phi : E \to E'$ is an isogeny whose kernel is cyclic of order N.

Finally, we consider the moduli problem associated to the congruence subgroup $\Gamma(N)$. If $\gamma \in \Gamma(N)$ and $\tau \in \mathbb{H}/\Gamma(N)$, then as above one checks that the points $1/N$ and τ/N in $\mathbb{C}/(\mathbb{Z} + \mathbb{Z}\tau)$ remain invariant under the action of γ. Thus associated to a point of $\mathbb{H}/\Gamma(N)$ is an elliptic curve $\mathbb{C}/(\mathbb{Z} + \mathbb{Z}\tau)$, together with a basis $\{1/N, \tau/N\}$ for the group of N-torsion points. However, a point of $\mathbb{H}/\Gamma(N)$ contains one further piece of information. Recall (III §8) that there is a pairing e_N on the group of N-torsion points of an elliptic curve. Then one can check that

$$e_N(1/N, \tau/N) = e^{2\pi i/N}.$$

Thus not only do we get a basis for the N-torsion, but the two points making up that basis pair, via the Weil pairing, to a specific primitive N^{th} root of unity.

Now for arithmetic applications, it is important to understand when an elliptic curve $E/\mathbb{C}$ or a point $T \in E(\mathbb{C})$ is defined over some subfield of $\mathbb{C}$, such as a number field. For example, although the Riemann surface $\mathbb{H}/SL_2(\mathbb{Z})$ only classifies elliptic curves over $\mathbb{C}$, we have a complex analytic isomorphism (C.12.11)

$$j : \mathbb{H}/SL_2(\mathbb{Z}) \to \mathbb{A}^1,$$

where $\mathbb{A}^1$ is a variety which is defined over $\mathbb{Q}$. Further, the elliptic curve E_τ corresponding to $\tau \in \mathbb{H}/SL_2(\mathbb{Z})$ is isomorphic, over $\mathbb{C}$, to an elliptic curve *defined over* $\mathbb{Q}(j(\tau))$. There is a general theory which deals with fields of definition for the spaces $\mathbb{H}/\Gamma$ and their associated moduli problems, but we will be content with the following description for the three special sorts of congruence subgroups considered above.

Theorem 13.1. *Let $N \geqslant 1$ be an integer.*
(a) *There exists a smooth projective curve $X_0(N)/\mathbb{Q}$ and a complex analytic isomorphism*

$$j_{N,0} : \mathbb{H}^*/\Gamma_0(N) \to X_0(N)(\mathbb{C})$$

such that the following holds:

Let $\tau \in \mathbb{H}/\Gamma_0(N)$, and let $K = \mathbb{Q}(j_{N,0}(\tau))$. From above, τ corresponds to an equivalence class of pairs (E, C), where E is an elliptic curve and $C \subset E$ is a cyclic subgroup of order N. Then this equivalence class contains a pair such that both E and C are defined over K. (I.e. E is an elliptic curve defined over K, and $C \subset E(\bar{K})$ is mapped to itself by $G_{\bar{K}/K}$.)
(b) *There exists a smooth projective curve $X_1(N)/\mathbb{Q}$ and a complex analytic isomorphism*

$$j_{N,1} : \mathbb{H}^*/\Gamma_1(N) \to X_1(N)(\mathbb{C})$$

such that the following holds:

Let $\tau \in \mathbb{H}/\Gamma_1(N)$, and let $K = \mathbb{Q}(j_{N,1}(\tau))$. From above, τ corresponds to an equivalence class of pairs (E, T), where E is an elliptic curve and $T \in E$ is a point of exact order N. Then this equivalence class contains a pair such that E is defined over K and $T \in E(K)$.
(c) *Fix a primitive N^{th} root of unity $\zeta \in \mathbb{C}$. There is a smooth projective curve $X(N)/\mathbb{Q}(\zeta)$ and a complex analytic isomorphism*

$$j_N : \mathbb{H}^*/\Gamma(N) \to X(N)(\mathbb{C})$$

such that the following holds:

Let $\tau \in \mathbb{H}/\Gamma(N)$, and let $K = \mathbb{Q}(\zeta, j_N(\tau))$. As explained above, τ corresponds to an equivalence class of triples (E, T_1, T_2), where E is an elliptic curve, and $\{T_1, T_2\}$ are generators for $E[N]$ satisfying $e_N(T_1, T_2) = \zeta$. (Here e_N is the Weil pairing. See (III §8).) Then this equivalence class contains a triple such that E is defined over K and $T_1, T_2 \in E(K)$.

PROOF. [Shi 1, §6.7].   □

Remark 13.2. If Γ is any congruence subgroup of $SL_2(\mathbb{Z})$, then one can find in a similar manner a smooth projective curve $X(\Gamma)$ defined over some number field $K(\Gamma)$ and a complex analytic isomorphism $j_\Gamma : \mathbb{H}^*/\Gamma \to X(\Gamma)(\mathbb{C})$. See [Shi 1, §6.7] for details. Recall that the cusps of Γ are defined to be the image of $\mathbb{P}^1(\mathbb{Q})$ in $\mathbb{H}^*/\Gamma$.

Definition. With notation as in (13.2), the curve $X(\Gamma)$ is called a *modular curve*. The *set of cusps of* $X(\Gamma)$ consists of the finite set of points $j_\Gamma(\mathbb{P}^1(\mathbb{Q})/\Gamma)$. (I.e. The cusps of $X(\Gamma)$ are the image under j_Γ of the cusps of Γ.) We denote the complement of the set of cusps of $X(\Gamma)$ by $Y(\Gamma)$. $Y(\Gamma)$ is a smooth affine curve.

Notation. For the congruence subgroups $\Gamma_0(N)$, $\Gamma_1(N)$, and $\Gamma(N)$ considered in (13.1), the curve $Y(\Gamma)$ is usually denoted by $Y_0(N)$, $Y_1(N)$, and $Y(N)$ respectively.

Example 13.3. Let N be an odd prime. Then $X_0(N)$ has two cusps, both of which are rational over $\mathbb{Q}$ (i.e. in $X_0(N)(\mathbb{Q})$). Similarly, $X_1(N)$ will have $N - 1$ cusps; but now only half of the cusps will be in $X_1(N)(\mathbb{Q})$. The other $\frac{1}{2}(N - 1)$ cusps of $X_1(N)$ are defined over the maximal real subfield of $\mathbb{Q}(\zeta_N)$.

Example 13.4. The curve $X_1(7)$ is isomorphic to $\mathbb{P}^1$. To make this precise, we can associate to each point $[t, 1] \in \mathbb{P}^1$ the pair (E_t, P_t), where E_t is the curve (defined over $\mathbb{Q}(t)$) given by the equation

$$E_t : y^2 + (1 + t - t^2)xy + (t^2 - t^3)y = x^3 + (t^2 - t^3)x^2,$$

and $P_t \in E_t$ is the point $P_t = (0, 0)$. The curve E_t will be an elliptic curve provided that the discriminant

$$\Delta(t) = t^7(t - 1)^7(t^3 - 8t^2 + 5t + 1)$$

does not vanish. Further, if $\Delta(t) \neq 0$, then one easily checks (using the addition law) that $[7]P_t = 0$. The curve $X_1(7)$ has six cusps, corresponding to the values $t = 0$, $t = 1$, $t = \infty$, and the three roots of $t^3 - 8t^2 + 5t + 1 = 0$. The reader may verify that each of these latter three roots generates the maximal real subfield of $\mathbb{Q}(\zeta_7)$, thereby verifying (13.3) in this case.

Remark 13.5. To illustrate one application of modular curves, we use them to rephrase conjecture (VIII.7.7). That conjecture says that for every number field K, there is an integer $N(K)$ such that for every elliptic curve E/K, $E(K)$ has no torsion points of order greater than $N(K)$. Notice that if $E(K)$ has a point P of order N, then the pair (E, P) corresponds to a non-cuspidal point of $X_1(N)(K)$. (i.e. a K-rational point of the modular curve $X_1(N)$). Thus (VIII.7.7) is equivalent to the statement that for any number field K, the set of rational points $X_1(N)(K)$ consists entirely of cusps for all sufficiently large N. The question of rational torsion points on elliptic curves is thus transformed into the question of rational points on modular curves. It is this idea which provides the starting point for the results of Mazur (VIII.7.5) and Manin (VIII.7.6).

Some modular curves are actually elliptic curves themselves. For example, the curve $X_0(11)$ has genus 1; and since it has two cusps defined over $\mathbb{Q}$ (13.3), we can use one of the cusps to make $X_0(11)$ into an elliptic curve. Now an elliptic curve such as $X_0(11)$ has a lot of additional structure, due to the fact that it is a modular curve; and it is possible to use that extra information to study the arithmetic of $X_0(11)$. Unfortunately, the genus of $X_0(N)$ grows (slightly irregularly) with N, so there are only finitely many curves $X_0(N)$ of any given genus. However, it sometimes happens that there is a map $\phi : X_0(N) \to E$, defined over $\mathbb{Q}$, from $X_0(N)$ onto an elliptic curve $E/\mathbb{Q}$. In this case, we say that E is a *Weil curve*, or that E is *parametrized by modular functions*. Such elliptic curves have a very rich structure, which can be used to study their arithmetic properties. We will discuss these curves in more detail later (16.4), and will just state here the following (weak) version of the conjecture of Taniyama and Weil.

Conjecture 13.6 (Taniyama–Weil). *Every elliptic curve defined over $\mathbb{Q}$ is a Weil curve. (I.e. If $E/\mathbb{Q}$ is an elliptic curve, then there exists an integer N and a surjective morphism $\phi : X_0(N) \to E$ defined over $\mathbb{Q}$.)*

References. [B–Sw 2], [Ka–M], [Maz 1], [Maz 2], [Shi 1].

§14. Tate Curves

For this section and the next, we let K be a local field, complete with respect to a discrete valuation v. Recall that for elliptic curves over $\mathbb{C}$, the existence of a lattice $\Lambda \subset \mathbb{C}$ and a uniformization $E(\mathbb{C}) \cong \mathbb{C}/\Lambda$ provides a powerful tool for the study of $E(\mathbb{C})$. If one attempts to mimic this construction for K, one is immediately stymied, since such a field can have no non-trivial discrete subgroups. However, if Λ is normalized as $\mathbb{Z} + \mathbb{Z}\tau$, then applying the exponential map $\exp(2\pi i \cdot)$ to $\mathbb{C}/\Lambda$ gives a new isomorphism $E(\mathbb{C}) \cong \mathbb{C}^*/q^{\mathbb{Z}}$. Here $q = e^{2\pi i \tau}$, and $q^{\mathbb{Z}}$ is the subgroup of $\mathbb{C}^*$ generated by q (cf. 12.7). Now the analogous situation for K looks more promising, since the multiplicative group K^* has lots of discrete subgroups, namely those of the form $q^{\mathbb{Z}}$ with $|q|_v \neq 1$. Further, all of the classical q-expansions for the various elliptic and modular functions (cf. §12) will converge in the v-adic case provided that q is chosen to satisfy $|q|_v < 1$.

For example, consider the elliptic curve (called the *Tate curve*)

$$E_q : y^2 + xy = x^3 + a_4 x + a_6$$

whose coefficients are given by the power series (considered, for the moment, as formal power series in $\mathbb{Z}[\![q]\!]$)

$$a_4 = -5 \sum_{n \geqslant 1} n^3 q^n / (1 - q^n) \qquad a_6 = -\frac{1}{12} \sum_{n \geqslant 1} (7n^5 + 5n^3) q^n / (1 - q^n).$$

The associated discriminant and j-invariant are given by the familiar formulas from the complex case (see §12)

$$\Delta = q \prod_{n \geqslant 1} (1 - q^n)^{24} \qquad j = \frac{1}{q} + 744 + 196884q + \cdots.$$

(Note that except for the leading term, $j \in \mathbb{Z}[\![q]\!]$.) Further, the elliptic curve E_q has the point (in the power series ring $\mathbb{Z}[\![q, u]\!]$) defined by

$$x = x(u, q) = \sum_{n \in \mathbb{Z}} q^n u / (1 - q^n u)^2 - 2 \sum_{n \geqslant 1} n q^n / (1 - q^n)$$

$$y = y(u, q) = \sum_{n \in \mathbb{Z}} q^{2n} u^2 / (1 - q^n u)^3 + \sum_{n \geqslant 1} n q^n / (1 - q^n).$$

Now one need merely observe that all of the above formulas make sense if q and u are taken to be elements of K^*, provided that $|q|_v < 1$. In other words, the various power series will converge in the v-adic metric. We thus obtain a v-adic analytic uniformization

$$\phi : K^*/q^{\mathbb{Z}} \to E_q(K)$$

$$u \to (x(u, q), y(u, q)).$$

(Of course, we set $\phi(1) = O$.) More generally, the power series $x(u, q)$ and $y(u, q)$ will converge for any $u \in \bar{K}$, and so will induce a map

$$\phi : \bar{K}^*/q^{\mathbb{Z}} \to E_q(\bar{K}).$$

(Note that although $\bar{K}$ will not be v-adically complete, the convergence of the power series is taking place in the finite extension $K(u)$. As an alternative, one can work in the v-adic completion of $\bar{K}$, which turns out to be algebraically closed.)

Another important point to notice is that since the action of $G_{\bar{K}/K}$ on $\bar{K}$ is v-adically continuous, this action will commute with the convergence of power series. In other words, ϕ is an isomorphism of $G_{\bar{K}/K}$-modules, so it can be used to make arithmetic deductions. (In this respect, at least, the non-archimedean uniformization is more useful than the corresponding situation over $\mathbb{C}$.)

The uniformization theorem (VI.5.1.1) (combined with the exponential map) says that every elliptic curve over $\mathbb{C}$ is (analytically) isomorphic to $\mathbb{C}^*/q^{\mathbb{Z}}$ for some $q \in \mathbb{C}^*$ with $|q| < 1$. It is clear that this cannot be true over K. For examining the power series for $j = j(q)$, we see that $|q|_v < 1$ implies that $|j(q)|_v > 1$. Thus every curve E_q has non-integral j-invariant. More precisely, the reduction $\tilde{E}_q$ of E_q modulo v has the Weierstrass equation

$$\tilde{E}_q : y^2 + xy = x^3;$$

so E_q has split multiplicative reduction at v.

Theorem 14.1 (Tate). *Let K be a field complete with respect to a discrete valuation v.*
(a) *For every $q \in K^*$ with $|q|_v < 1$, the map*

$$\phi : \bar{K}^*/q^{\mathbb{Z}} \to E_q(\bar{K})$$

described above is an isomorphism of $G_{\bar{K}/K}$-modules.
(b) *For every $j_0 \in K^*$ with $|j_0|_v > 1$, there is a $q \in K^*$ with $|q|_v < 1$ such that the elliptic curve E_q/K has j-invariant j_0. E_q is characterized by $j(E_q) = j_0$ and the fact that it has split multiplicative reduction at v.*
(c) *Let R be the ring of integers of K. Then under the isomorphism $E_q(K) \cong K^*/q^{\mathbb{Z}}$, we have the identifications*

$$(E_q)_0(K) \cong R^* \quad and \quad (E_q)_1(K) \cong \{u \in R^* : u \equiv 1 \,(\mathrm{mod}\ v)\}.$$

(d) *Let E/K be an elliptic curve with non-integral j-invariant which does not have split multiplicative reduction. From (b), there is a $q \in K^*$ such that $j(E) = j(E_q)$. Then there is a unique quadratic extension L/K such that E is isomorphic to E_q over L. Further,*

$$E(K) \cong \{u \in L^* : \mathrm{Norm}_{L/K}(u) \in q^{\mathbb{Z}}\}/q^{\mathbb{Z}}.$$

The extension L/K is unramified if and only if E has (non-split) multiplicative reduction, in which case the residue field extension of L/K is generated by the tangents to the node of the reduction $\tilde{E}$ of E at v.

PROOF. This was originally discovered by Tate, but never officially published by him. Accounts can be found in [Rob, II §5] and [Roq]. $\qquad\square$

References. [Rob], [Roq], [La 3, ch. 15].

§15. Néron Models and Tate's Algorithm

As in the last section, we let K be complete with respect to a discrete valuation v, and let R be the ring of integers and k the residue field of K. Let E/K be an elliptic curve, and choose a minimal Weierstrass equation for E at v. Suppose now that we consider this equation as defining a scheme E over $\mathrm{Spec}(R)$. The resulting scheme may not be regular (i.e. smooth), for if E has bad reduction at v, then the singular point on the special fiber $\tilde{E}$ of E may be a singular point of the scheme. By resolving the singularity, one obtains a scheme $\mathscr{C}/\mathrm{Spec}(R)$ whose generic fiber is E/K and whose special fiber is a union of curves (with multiplicities) over k.

Theorem 15.1 (Kodaira, Néron). *Let E/K be as above.*

(a) *There is a regular projective two-dimensional scheme $\mathscr{C}/\mathrm{Spec}(R)$ whose generic fiber $\mathscr{C} \times_{\mathrm{Spec}(R)} \mathrm{Spec}(K)$ is isomorphic (over K) to E/K. Suppose further that $\mathscr{C}$ is minimal (i.e., the map $\mathscr{C} \to \mathrm{Spec}(R)$ cannot be factored as $\mathscr{C} \to \mathscr{C}' \to \mathrm{Spec}(R)$ in such a way that $\mathscr{C} \times_{\mathrm{Spec}(R)} \mathrm{Spec}(K) \to \mathscr{C}' \times_{\mathrm{Spec}(R)} \mathrm{Spec}(K)$ is an isomorphism.) Then $\mathscr{C}$ is unique.*

(b) *Let $\mathscr{E} \subset \mathscr{C}$ be the subscheme of $\mathscr{C}$ obtained by discarding all of the singular points of the special fiber $\tilde{\mathscr{C}} = \mathscr{C} \times_{\mathrm{Spec}(R)} \mathrm{Spec}(k)$. (I.e. We discard all multiple fibral components and all intersections of fibral components. Note that these are not singular points of $\mathscr{C}$ itself, which is regular.) Then $\mathscr{E}$ is a group scheme over $\mathrm{Spec}(R)$ whose generic fiber $\mathscr{E} \times_{\mathrm{Spec}(R)} \mathrm{Spec}(K)$ is isomorphic, as a group variety, to E/K. $\mathscr{E}$ is called the Néron minimal model of E/K.*

(c) *The natural map $\mathscr{E}(R) \to E(K)$ is an isomorphism. (I.e. Every section $\mathrm{Spec}(K) \to E$ on the generic fiber extends to a section $\mathrm{Spec}(R) \to \mathscr{E}$.)*

(d) *Let $\tilde{\mathscr{E}} = \mathscr{E} \times_{\mathrm{Spec}(R)} \mathrm{Spec}(k)$ be the special fiber of $\mathscr{E}$. Then $\tilde{\mathscr{E}}$ is an algebraic group over k, and we let $\tilde{\mathscr{E}}^0/k$ be its identity component (so $\tilde{\mathscr{E}}$ is an extension of $\tilde{\mathscr{E}}^0$ by a finite group.) Note that there is a reduction map $\mathscr{E}(R) \to \tilde{\mathscr{E}}(k)$. Then with the identification $\mathscr{E}(R) \cong E(K)$ from (c),*

(i) $\qquad\qquad\qquad \tilde{\mathscr{E}}^0(k) \cong \tilde{E}_{ns}(k) \cong E_0(K)/E_1(K).$

(ii) $\qquad\qquad\qquad \tilde{\mathscr{E}}(k)/\tilde{\mathscr{E}}^0(k) \cong E(K)/E_0(K).$

PROOF. [Né 2].   □

Remark 15.1.1. In some sources, $\mathscr{C}$ is called the Néron minimal model of E/K. However, for abelian varieties of higher dimension, the Néron minimal model always refers to a group scheme analogous to our $\mathscr{E}$, and there is no natural analogue of $\mathscr{C}$. The ambiguity for elliptic curves results because the minimal model of E, considered as a *curve*, is $\mathscr{C}$; while the minimal model of E, treated as a *group variety*, is $\mathscr{E}$.

Notice that (15.1d(ii)) gives a description of $E(K)/E_0(K)$ in terms of the group of components of a certain algebraic group $\tilde{\mathscr{E}}/k$. It turns out that there are only a handful of possibilities for $\tilde{\mathscr{E}}$. More precisely, one can write down all of the possibilities for the special fiber $\tilde{\mathscr{C}} = \mathscr{C} \times_{\mathrm{Spec}(R)} \mathrm{Spec}(k)$; and then one obtains $\tilde{\mathscr{E}}$ by discarding all of the components of multiplicity greater than 1 and all of the points where components intersect. The results are as follows.

Theorem 15.2 (Kodaira, Néron). *With notation as in (15.1), all of the possibilities for the special fiber $\tilde{\mathscr{C}}$ and the group of components $\tilde{\mathscr{E}}(k)/\tilde{\mathscr{E}}^0(k)$ are given in table 15.1.*

Except for the case I_0, each of the pictured components is a rational curve (i.e. a copy of $\mathbb{P}^1$). Further, Δ_v is the minimal discriminant of E at v, and f_v is the exponent of the conductor (which is defined below in §16).

Table 15.1

Kodaira symbol	I_0	I_v $(v>0)$	II	III	IV	I_0^*	$I_v^*(v>0)$	IV^*	III^*	II^*
Special fiber $\tilde{\mathcal{C}}$ (the numbers indicate multiplicities)	*(diagram)*	*(diagram)*	*(diagram)*	*(diagram)*	*(diagram)*	*(diagram)*	*(diagram)*	*(diagram)*	*(diagram)*	*(diagram)*
m = number of irreducible components	1	v	1	2	3	5	$5+v$	7	8	9
$E(K)/E_0(K)$ $\cong \tilde{\mathcal{E}}(k)/\tilde{\mathcal{E}}^0(k)$	(0)	$\frac{\mathbb{Z}}{v\mathbb{Z}}$	(0)	$\frac{\mathbb{Z}}{2\mathbb{Z}}$	$\frac{\mathbb{Z}}{3\mathbb{Z}}$	$\frac{\mathbb{Z}}{2\mathbb{Z}}\times\frac{\mathbb{Z}}{2\mathbb{Z}}$	$\frac{\mathbb{Z}}{2\mathbb{Z}}\times\frac{\mathbb{Z}}{2\mathbb{Z}}$ v even ‒‒‒‒‒‒ $\frac{\mathbb{Z}}{4\mathbb{Z}}$ v odd	$\frac{\mathbb{Z}}{3\mathbb{Z}}$	$\frac{\mathbb{Z}}{2\mathbb{Z}}$	(0)
$\tilde{\mathcal{E}}^0(k)$	$\tilde{E}(k)$	k^*	k^+	k^+	k^+	k^+	k^+	k^+	k^+	k^+
Entries below this line are valid only for char(k) ≠ 2,3										
$\mathrm{ord}_v(\Delta_v)$	0	v	2	3	4	6	$6+v$	8	9	10
f_v = exponent of conductor = $\mathrm{ord}_v(\Delta_v)+1-m$	0	1	2	2	2	2	2	2	2	2
behavior of j	$v(j)\geq0$	$\mathrm{ord}_v(j)=-v$	$\tilde{j}=0$	$\tilde{j}=1728$	$\tilde{j}=0$	$v(j)\geq0$	$\mathrm{ord}_v(j)=-v$	$\tilde{j}=0$	$\tilde{j}=1728$	$\tilde{j}=0$

PROOF. [Né 2].   □

Corollary 15.2.1. *The group $E(K)/E_0(K)$ is finite. If E has split multiplicative reduction, then it is cyclic of order $-\mathrm{ord}_v(j(E))$. In all other cases, it has order at most 4.*

PROOF. The first statement follows from (15.1d). For the second, if E has split multiplicative reduction, then $E(K) \cong K^*/q^{\mathbb{Z}}$ and $E(K)/E_0(K) \cong \mathbb{Z}/\mathrm{ord}_v(q)\mathbb{Z}$ (14.1ac). Further, $\mathrm{ord}_v(q) = -\mathrm{ord}_v(j(E))$ (cf. §14). Next, if E has non-split multiplicative reduction, then one easily checks using (14.1d) that $E(K)/E_0(K)$ has order 1 or 2. Finally, if E has additive reduction, then the result follows by inspection of table 15.1.   □

Remark 15.3. Note that except when k has characteristic 2 or 3, everything about E (i.e. reduction type, exponent of conductor, the group $E(K)/E_0(K)$) can be read off from table 15.1 once one has a minimal Weierstrass equation for E. Further, a given Weierstrass equation will be minimal if and only if either $\text{ord}_v(\Delta) < 12$ or $\text{ord}_v(c_4) < 4$ (exer. 7.1a), so in this case it is easy to check for minimality. In general, for k of arbitrary characteristic, one can use a straightforward (but somewhat lengthy) algorithm of Tate ([Ta 6]) to compute the special fiber $\tilde{\mathscr{C}}$, and then read the desired results from the corresponding column in table (15.1).

References. [Né 2], [Ta 6].

§16. *L*-Series

The *L*-series of an elliptic curve is a generating function which records information about the reduction of the curve modulo every prime. Known results are fragmentary, but conjecturally such *L*-series contain a large amount of information concerning the set of global points on the curve, (which may be somewhat surprising, in view of the failure of the Hasse principle for curves of genus 1.) Further, there are intimate relations, both known and conjectural, between these *L*-series and the theory of modular forms. In this section we will explain these conjectures and give some of the evidence which support their validity.

Let E/K be an elliptic curve, and let $v \in M_K$ be a finite place at which E has good reduction. We denote the residue field of K at v by k_v, the reduction of E at v by $\tilde{E}_v$, and let $q_v = \#k_v$ be the norm of the prime ideal corresponding to v. Recall (V §2) that the *zeta function* of $\tilde{E}_v/k_v$ is the power series

$$Z(\tilde{E}_v/k_v; T) = \exp\left(\sum_{n=1}^{\infty} \#\tilde{E}_v(k_{v,n}) T^n/n\right).$$

(Here $k_{v,n}$ is the unique extension of k_v of degree n.) Further, we proved (V.2.4) that $Z(\tilde{E}_v/k_v; T)$ is actually a rational function

$$Z(\tilde{E}_v/k_v; T) = L_v(T)/(1 - T)(1 - q_v T),$$

where

$$L_v(T) = 1 - a_v T + q_v T^2 \in \mathbb{Z}[T] \quad \text{and} \quad a_v = q_v + 1 - \#\tilde{E}_v(k_v).$$

We extend the definition of $L_v(T)$ to the case that E has bad reduction by setting

$$L_v(T) = \begin{cases} 1 - T & \text{if } E \text{ has split multiplicative reduction at } v \\ 1 + T & \text{if } E \text{ has non-split multiplicative reduction at } v \\ 1 & \text{if } E \text{ has additive reduction at } v. \end{cases}$$

Then in all cases we have the relation

$$L_v(1/q_v) = \#\tilde{E}_{ns}(k_v)/q_v.$$

Definition. The *L-series of E/K* is defined by the Euler product

$$L_{E/K}(s) = \prod_{v \in M_K^0} L_v(q_v^{-s})^{-1}.$$

This product converges and gives an analytic function for all $\mathrm{Re}(s) > \frac{3}{2}$. (Use the fact (V.2.4) that $|a_v| \leqslant 2\sqrt{q_v}$.) It is conjectured that far more is true.

Conjecture 16.1. *The L-series $L_{E/K}(s)$ has an analytic continuation to the entire complex plane, and it satisfies a functional equation relating the values at s and $2 - s$.*

This conjecture is known to be true for those elliptic curves having complex multiplication (Deuring [De 3], Weil [We 4]), in which case $L_{E/K}(s)$ is shown to equal a Hecke *L*-series with Grössencharacter. It is also known for those elliptic curves over $\mathbb{Q}$ which are parametrized by modular functions (Eichler, Shimura, see below), where $L_{E/\mathbb{Q}}(s)$ turns out to be the Mellin transform of a modular form.

Next we define the *conductor of E/K*. It is a certain integral ideal of K which is the same for isogenous elliptic curves. If $v \in M_K$, the *exponent of the conductor of E at v* is defined by

$$f_v = \begin{cases} 0 & \text{if } E \text{ has good reduction at } v \\ 1 & \text{if } E \text{ has multiplicative reduction at } v \\ 2 + \delta_v & \text{if } E \text{ has additive reduction at } v, \end{cases}$$

where δ_v is a measure of the "wild ramification" in the action of the inertia group on $T_\ell(E)$ (cf. [Se–T], [Og 3]). In particular, $\delta_v = 0$ provided $\mathrm{char}(k_v) \neq 2, 3$. Further, f_v may be computed by using *Ogg's formula*.

Proposition 16.2 (Ogg [Og 3]). *Let m_v be the number of irreducible components (ignoring multiplicities) on the special fiber of the minimal (complete) Néron model of E at v (cf. §15). Then*

$$f_v = \mathrm{ord}_v(\mathscr{D}_{E/K}) + 1 - m_v.$$

(Here $\mathscr{D}_{E/K}$ is the minimal discriminant of E/K.)

Definition. The *conductor of E/K* is the integral ideal of K defined by

$$N_{E/K} = \prod_{v \in M_K^0} \mathfrak{p}_v^{f_v}.$$

In order to simplify the exposition in the rest of this section, we will now restrict attention to the case $K = \mathbb{Q}$. Notice that we can then take the conductor $N_E = N_{E/\mathbb{Q}}$ to be a positive integer. Define a new function

$$\xi_E(s) = N_E^{s/2}(2\pi)^{-s}\Gamma(s)L_E(s).$$

Then (16.1) has the following more precise formulation.

Conjecture 16.3. *The function $\xi_E(s)$ has an analytic continuation to the entire complex plane, and it satisfies the functional equation*

$$\xi_E(s) = w\xi_E(2 - s) \qquad with\ w = \pm 1.$$

As noted above, (16.3) is known to be true if E is parametrized by modular functions. Weil and Taniyama have conjectured that every elliptic curve over $\mathbb{Q}$ has this property. More precisely, we have the following.

Conjecture 16.4 (Taniyama–Weil). *Let $E/\mathbb{Q}$ be an elliptic curve of conductor N, let $L_E(s) = \Sigma c_n n^{-s}$ be its L-series, and let $f(\tau) = \Sigma c_n e^{2\pi i n\tau}$ be the inverse Mellin transform of L_E.*
(a) $f(\tau)$ is a weight 2 cusp form for the congruence subgroup $\Gamma_0(N)$ of $SL_2(\mathbb{Z})$.
(b) For each prime $p \nmid N$, let $T(p)$ be the corresponding Hecke operator; and let W be the operator $(Wf)(\tau) = f(-1/N\tau)$. Then

$$T(p)f = c_p f \quad and \quad Wf = wf,$$

where $w = \pm 1$ is the sign of the functional equation (16.3).
(c) Let ω be an invariant differential on $E/\mathbb{Q}$. There exists a morphism $\phi : X_0(N) \to E$, defined over $\mathbb{Q}$, such that $\phi^*(\omega)$ is a multiple of the differential form on $X_0(N)$ represented by $f(\tau)\,d\tau$.

Weil has shown [We 6] that if $L_E(s)$ and sufficiently many of its twists satisfy a functional equation as in (16.3), then $f(\tau)$ is a cusp form for $\Gamma_0(N)$. Shimura ([Shi 2], [Shi 3]) has verified that this holds for elliptic curves with complex multiplication. Conjecture (16.4) has also been verified numerically for curves of low conductor (cf. [Og 1], [Og 2], and the tables in [B–K]).

Another important conjecture about the L-series of elliptic curves concerns their special value at $s = 1$. Before stating it, we set the following notation:

$E/\mathbb{Q}$	an elliptic curve
ω	the invariant differential $dx/(2y + a_1 x + a_3)$ on a global minimal Weierstrass equation for $E/\mathbb{Q}$. (Cf. VIII §8.)
Ω	$\int_{E(\mathbb{R})} \lvert\omega\rvert$ [Either the real period, or twice the real period, depending on whether or not $E(\mathbb{R})$ is connected.]
$\text{III}(E/\mathbb{Q})$	the Shafarevich–Tate group of $E/\mathbb{Q}$. (Cf. X §4.)
$R(E/\mathbb{Q})$	the elliptic regulator of $E(\mathbb{Q})/E_{\text{tors}}(\mathbb{Q})$, computed using the canonical height pairing. (Cf. VIII §9.)
c_p	$\#E(\mathbb{Q}_p)/E_0(\mathbb{Q}_p)$ [Thus $c_p = 1$ when E has good reduction at p. See (§15) and (VII §6) for a geometric description of c_p.]

Conjecture 16.5 (Birch and Swinnerton-Dyer). *(a) $L_E(s)$ has a zero at $s = 1$ of order equal to the rank of $E(\mathbb{Q})$.*
(b) Let $r = \text{rank } E(\mathbb{Q})$. Then with notation as above,

$$\lim_{s \to 1} (s - 1)^{-r} L_E(s) = \Omega \# \text{III}(E/\mathbb{Q}) 2^r R(E/\mathbb{Q}) (\# E_{\text{tors}}(\mathbb{Q}))^{-2} \prod_p c_p.$$

As described by Tate in 1974, "this remarkable conjecture relates the behavior of a function L at a point where it is not at present known to be defined to the order of a group Ш which is not known to be finite!" There is a great deal of evidence for this conjecture, of which we will mention the following.

Evidence 16.5.1. Conjecture (16.5) has been checked numerically in a large number of cases. Since Ш is not known to be finite, what this means for (16.5b) is that the conjecture is used to compute a hypothetical value for Ш. This always turns out to be the square of an integer (as it should, due to the existence of the Cassel's pairing (X.4.14)); and it agrees with the calculated value of the 2 and/or 3 primary component of Ш. (See [B–Sw 1], which contains the original numerical evidence, and [Ste].)

Evidence 16.5.2. Isogenous elliptic curves have the same number of points modulo p for all primes p (exer. 5.4), and so they have the same L-series. (One must also check the primes of bad reduction.) Consequently, if (16.5b) is true, then the quantity

$$\Omega \# \text{Ш}(E/\mathbb{Q})R(E/\mathbb{Q})(\# E_{\text{tors}}(\mathbb{Q}))^{-2}\prod c_p$$

must be an isogeny invariant. This has been verified by Cassels ([Ca 6]), and extended to abelian varieties by Tate, in both cases under the assumption that Ш is finite. It is worth noting that *none* of the individual terms in the product need be the same for isogenous curves.

Evidence 16.5.3. Coates and Wiles ([Co–W], see also [Arth], [Ru]) have shown that if $E/\mathbb{Q}$ has complex multiplication and $E(\mathbb{Q})$ is infinite, then $L_E(1) = 0$.

Evidence 16.5.4. Greenberg ([Gre]) has shown that if $E/\mathbb{Q}$ has complex multiplication and $\text{ord}_{s=1} L_E(s)$ is odd, then either rank $E(\mathbb{Q}) \geqslant 1$, or else $\text{Ш}(E/\mathbb{Q})[p^\infty]$ is infinite for a set of primes p of density $\frac{1}{2}$. (This last possibility seems most unlikely, to say the least. See also [Roh].)

Evidence 16.5.5. For certain elliptic curves $E/\mathbb{Q}$ which are parametrized by modular functions and satisfy $L_E(1) = 0$, Gross and Zagier ([Gr–Z 2]) have given a limit formula relating $L'_E(1)$ to the canonical height of a certain point $P \in E(\mathbb{Q})$ (called a Heegner point). In particular, they show that if $L'_E(1) \neq 0$, then $E(\mathbb{Q})$ has rank at least 1; and if it has rank exactly 1, then the equality in (16.5b) is true up to multiplication by a rational number. (I.e. They show that $L'_E(1)/\Omega R(E/\mathbb{Q})$ is a rational number. Notice that in this case there is no need to assume that Ш is finite.)

References. [Arth], [B–Sw 1,2], [Ca 6], [Co–W], [De 3], [Gre], [Gr–Z 2,3], [Og 1,2,3,4], [Roh], [Ru], [Se–T], [Shi 1,2,3], [Ste], [We 4,6].

§17. Duality Theory

In (X §4) we discussed the bilinear pairing on the Shafarevich–Tate group. There is a complementary duality theorem in the local case which goes as follows.

Theorem 17.1 (Tate [Ta 1], [Ta 2]). *Let $v \in M_K$ be a non-archimedean absolute value, and let E/K_v be an elliptic curve. Then there exists a bilinear, non-degenerate pairing*

$$\langle \ , \ \rangle : E(K_v) \times WC(E/K_v) \to \mathbb{Q}/\mathbb{Z}.$$

More precisely, if $E(K_v)$ is given the v-adic topology and $WC(E/K_v)$ the discrete topology, then $\langle \ , \ \rangle$ induces a duality of locally compact groups. (I.e. The pairing $\langle \ , \ \rangle$ is continuous, the continuous homomorphisms $E(K_v) \to \mathbb{Q}/\mathbb{Z}$ all have the form $\langle \cdot, \xi \rangle$ for some $\xi \in WC(E/K_v)$; and similarly the continuous homomorphisms $WC(E/K_v) \to \mathbb{Q}/\mathbb{Z}$ all have the form $\langle P, \cdot \rangle$ for some $P \in E(K_v)$.)

The global duality theory is not quite as satisfactory, due to the fact that it is not known in general whether the Shafarevich–Tate group can have divisible elements.

Theorem 17.2 (Cassels [Ca 3], Tate [Ta 2]). *Let E/K be an elliptic curve. There exists an alternating, bilinear pairing*

$$\text{Ш}(E/K) \times \text{Ш}(E/K) \to \mathbb{Q}/\mathbb{Z}$$

whose kernel on either side is precisely the group of divisible elements of Ш.

Corollary 17.2.1. *If* Ш(E/K) *is finite, or more generally if any p-primary component* Ш$(E/K)[p^\infty]$ *is finite, then its order is a perfect square.*

References. [Ca 3], [Se 8], [Ta 1], [Ta 2].

§18. Local Height Functions

In his original construction of the canonical height, Néron ([Né 3]) proceeded by constructing a local height pairing for each absolute value $v \in M_K$; and then he formed the (global) canonical height by taking the sum of the local heights. A nice exposition of this theory for elliptic curves was given by Tate ([Ta 4]) and published in [La 5]. The theory of local height functions is important in the study of the more delicate properties of the canonical height. (See, for example, [Gr–Z 2] or [Sil 1].) It is also useful for numerical computation of the canonical height of points on elliptic curves.

Theorem 18.1. *Let $v \in M_K$, and let E/K_v be an elliptic curve given by a Weierstrass equation*

$$E : y^2 + a_1 xy + a_3 y = x^3 + a_2 x^2 + a_4 x + a_6.$$

There is a unique function

$$\lambda_v : E(K_v) - \{O\} \to \mathbb{R},$$

called the local height function *for E at v, with the following properties:*

(i) λ_v *is continuous for the v-adic topology on $E(K_v)$ and the usual topology on* $\mathbb{R}$.

(ii) $\underset{P \to O}{\text{Limit}} (\lambda_v(P) + \frac{1}{2} v(x(P)))$ *exists, where $P \to O$ in the v-adic topology.*

(iii) *For all $P \in E(K_v)$ with $[2]P \neq O$,*

$$\lambda_v([2]P) = 4\lambda_v(P) + v(2y(P) + a_1 x(P) + a_3) - \tfrac{1}{4} v(\Delta).$$

(Here Δ is the discriminant of the given Weierstrass equation.)
Further, property (iii) *may be replaced by the "quasi-parallelogram law"*
(iii') *For all $P, Q \in E(K_v)$ with $P, Q, P \pm Q \neq O$,*

$$\lambda_v(P + Q) + \lambda_v(P - Q) = 2\lambda_v(P) + 2\lambda_v(Q) + v(x(P) - x(Q)) - \tfrac{1}{6} v(\Delta).$$

Remark 18.1.1. Note that the function λ_v in (18.1) does not depend on the choice of Weierstrass equation for E, because the conditions (i)–(iii) are invariant under change of coordinates.

Theorem 18.2. *Let E/K be an elliptic curve. Then for all points $P \in E(K) - \{O\}$, the canonical height $\hat{h}(P)$ is given by*

$$\hat{h}(P) = \frac{1}{[K : \mathbb{Q}]} \sum_{v \in M_K} n_v \lambda_v(P).$$

There are explicit formulas for the local height function in all cases, but we will be content with the following statement.

Theorem 18.3. *Let E/K be an elliptic curve and $v \in M_K$.*

(a) *Case I. v archimedean*
Choose a lattice $\Lambda \subset \mathbb{C}$ and an isomorphism $E(\bar{K}_v) \cong \mathbb{C}/\Lambda$. Let $\sigma(z, \Lambda)$ be the Weierstrass σ-function, and let $\Delta(\Lambda)$ and $\eta : \mathbb{C} \to \mathbb{R}$ be as in (exers. 6.6 and 6.4). If $P \in E(K_v)$ corresponds to $z \in \mathbb{C}/\Lambda$, then

$$\lambda_v(P) = -\log|\Delta(\Lambda)^{1/12} e^{-z\eta(z)/2} \sigma(z, \Lambda)|_v.$$

(b) *Case II. v non-archimedean and $P \in E_0(K_v)$*
Let x and y be coordinate functions on a minimal Weierstrass equation for E at v. Then

$$\lambda_v(P) = \max\{-\tfrac{1}{2} v(x(P)), 0\} + \tfrac{1}{12} v(\Delta).$$

(c) *Case III. v non-archimedean, E has split multiplicative reduction at v, and*
$P \notin E_0(K_v)$

Fix an isomorphism $E(K_v)/E_0(E_v) \cong \mathbb{Z}/N\mathbb{Z}$, *where* $N = -\mathrm{ord}_v(j(E))$ *(cf.
VII.6.1). Suppose that* $P \in E(K_v)$ *corresponds to* $n \in \mathbb{Z}/N\mathbb{Z}$ *for some*
$1 \leqslant n \leqslant N - 1$. *Then*

$$\lambda_v(P) = -\tfrac{1}{2}B_2(n/N)v(j(E)),$$

where $B_2(T) = T^2 - T + 1/6$ *is the second Bernoulli polynomial.*

Remark 18.3.1. There are also formulas for $\lambda_v(P)$ in the other cases of bad
reduction; but note that in any case, one can always apply (18.3) after replacing K by some finite extension (VII.5.4c).

References. [La 5], [Né 3]. For a reformulation (and generalization) in terms
of arithmetic intersection theory, see for example [Chi] or [Fa 2].

§19. The Image of Galois

Let E/K be an elliptic curve defined over a number field, and let ℓ be a prime.
Many of the arithmetic properties of E are determined by the ℓ-adic representation

$$\rho_\ell : G_{\bar{K}/K} \to \mathrm{Aut}(T_\ell(E)).$$

Two of the most important results concerning ρ_ℓ are the following.

Theorem 19.1 (Serre [Se 5], [Se 6]). *Assume that* E *does not have complex
multiplication.*

(a) *The image of* ρ_ℓ *is of finite index in* $\mathrm{Aut}(T_\ell(E))$ *for all primes* ℓ.
(b) *The image of* ρ_ℓ *equals* $\mathrm{Aut}(T_\ell(E))$ *for all but finitely many primes* ℓ.

Theorem 19.2 (Faltings [Fa 1]). *Let* E/K *and* E'/K *be elliptic curves. Then the
natural map*

$$\mathrm{Hom}_K(E, E') \otimes \mathbb{Z}_\ell \to \mathrm{Hom}_K(T_\ell(E), T_\ell(E'))$$

is an isomorphism. (Here the right-hand side is the group of $\mathbb{Z}_\ell$-*linear homomorphisms from* $T_\ell(E)$ *to* $T_\ell(E')$ *which commute with the action of* $G_{\bar{K}/K}$. *Note that
we proved injectivity in (III.7.4); the real difficulty lies in showing that the map
is surjective.)*

References. [Se 5], [Se 6], [Fa 1].

§20. Function Fields and Specialization Theorems

Let V/K be a variety defined over a number field. Then we can consider elliptic curves defined over the function field $K(V)$. These will be curves given by a Weierstrass equation

$$E : y^2 + a_1 xy + a_3 y = x^3 + a_2 x^2 + a_4 x + a_6,$$

where $a_1, \ldots, a_6 \in K(V)$. Now for almost all points $t \in V$ (i.e. outside of some proper subvariety of V), all of the functions $a_1, \ldots, a_6$ will be defined at t. We will then be able to define a *specialization of E* by

$$E_t : y^2 + a_1(t)xy + a_3(t)y = x^3 + a_2(t)x^2 + a_4(t)x + a_6(t).$$

Similarly, if $P = (x, y) \in E(K(V))$, then the functions $x, y \in K(V)$ will be defined for almost all $t \in V$, and so we can *specialize P* to a point

$$P_t = (x(t), y(t)) \in E_t.$$

Now it is a fact (although we did not prove it) that the group $E(K(V))$ is finitely generated. (I.e. The Mordell–Weil theorem holds in this case.) Thus by choosing a finite set of generators for $E(K(V))$, we can define (for almost all choices of $t \in V$) a *specialization homomorphism*

$$\sigma_t : E(K(V)) \to E_t.$$

Note further that if $t \in V(K)$, then the image of σ_t lies in $E_t(K)$. By using a generalization of Hilbert's irreducibility theorem, Néron proved that the specialization homomorphism is frequently injective.

Theorem 20.1 (Néron [Né 1], [La 7, ch.9]). *Let E be an elliptic curve defined over the field $K(\mathbb{P}^n)$. Then there are infinitely many points $t \in \mathbb{P}^n(K)$ for which the specialization homomorphism*

$$\sigma_t : E(K(\mathbb{P}^n)) \to E_t(K)$$

is injective.

Corollary 20.1.1. *There exist infinitely many elliptic curves $E/\mathbb{Q}$ such that $E(\mathbb{Q})$ has rank at least* 10.

Remark 20.2. In order to use (20.1) to produce curves $E/\mathbb{Q}$ with large rank, one must find elliptic curves over $\mathbb{Q}(T_1, \ldots, T_n)$ with large rank. Taking $n = 18$ and letting $C/\mathbb{Q}(T_1, \ldots, T_{18})$ be the cubic curve passing through the nine points $(T_1, T_2), \ldots, (T_{17}, T_{18})$, it is not hard to show that the Jacobian of C has rank (at least) 9. To obtain rank 10 (as in [Né 1]), one must do some additional work. There are other methods which have now been used to find specific elliptic curves over $\mathbb{Q}$ with even larger rank ([Mes 1], [Mes 2]), but these methods do not give infinite families of such curves.

In the case that the variety V is a curve, Néron's theorem (20.1) can be strengthened as follows.

Theorem 20.3 (Silverman [Sil 3], [La 7, ch.12]). *Let C/K be a curve, and let E be an elliptic curve defined over the function field $K(C)$ such that $j(E) \notin K$. Then the specialization map*

$$\sigma_t : E(K(C)) \rightarrow E_t$$

is (well-defined) and injective for all but finitely many points $t \in C(K)$. (More generally, it is injective for all but finitely many points of $\cup \, C(L)$, where the union is over all fields L/K whose degree is bounded by a fixed number.)

References. [La 7], [Né 1], [Sil 3], [Sil 5], [Ta 8].

Notes on Exercises

Many of the exercises in this book are standard results which were not included in the text due to lack of space; while others are special cases of results which appear in the literature. The following list thus serves two purposes: it is an attempt by the author to give credit for the theorems which appear in the exercises, and an aid for the reader who wishes to delve more deeply into some aspect of the theory. However, since any attempt to assign credit is bound to be incomplete in some respects, the author herewith tenders his apologies to anyone who feels that they have been slighted.

Except for an occasional computational problem, we have not included solutions (nor even hints). Indeed, since it is hoped that this book will lead the student on into the realm of active mathematics, the benefits of working without aid clearly outweigh any advantage that might be gained by having solutions readily available.

CHAPTER I

(1.1) (a) $B(A^3 - 27B) = 0$ (b) $4A^3 + 27B^2 = 0$
(1.2) (a) $(0, 0)$ (b) $(0, 0)$ (c) $(0, 0)$ (d) $(0, 0, 1)$
(1.3) [Har, I.5.1]
(1.5) (b) $P_3 = (-8/9, 109/27)$
(1.7) (b) $\psi = [Y, X]$ (c) No

CHAPTER II

(2.1) [A−M, prop. 9.2]
(2.4) (b) [La 6, lemma, page 7]
(2.5) [Har, II.6.10.1] and [Har, IV.1.3.5]

(2.6)   This volume (III §3).

(2.9)   This example is due to Hurwitz [Hur]. See also [Ca 7, §22].

(2.11) This proof of Weil reciprocity is due to E. Kani.

CHAPTER III

(3.4) $P_2 = -[2]P_1 + P_3, \ P_4 = P_1 - P_3, \ P_5 = -[2]P_1, \ P_6 = -P_1 + [2]P_3,$
$P_7 = [3]P_1 - P_3, \ P_8 = -[4]P_1 + [3]P_3$

(3.6)   [Har, IV.3.2(b)]

(3.7)   [Ca 1], [Ca 7, lemma 7.2], [La 5, II thm. 2.1]

(3.8)   (b) This volume (VI §6).

(3.9)   [Rob, II.1.24], [Rob, II.2.9]

(3.13) [Mum, II §7 page 66]

(3.18) (d) [De 1]

(3.20) [Shi 1, prop. 4.11]

(3.21) [Har, IV §4]

(3.23) This volume (A.1.3).

CHAPTER IV

(4.1)   (a) [Frö 2, I §3 prop. 1]

(4.2)   (b) [Haz, thm. I.6.1]

(4.4)   [Frö 2, IV §2, thm. 2]

CHAPTER V

(5.3)   [Har, C.4.1]

(5.4)   (a) Due to F. K. Schmidt. See [Ca 7, lemma 15.1].   (b) [Ta 7]

(5.8)   [Mum, thm. page 217]

(5.11) This proof of a weak version of [Se 11, §4.3] was suggested by Serre.

CHAPTER VI

(6.3)   (d) [Wh–Wa, ch. XX, misc. ex. 33]

(6.4)   (a)–(e) [Ahl, ch. 7 §3.2], [Wh–Wa, ch. XX], [La 5, ch. I §6]
(f) [La 5, ch. I §7]. (Log $|G(z)|$ is a Green's function.)

(6.8)   For more information about complex multiplication and class field theory, see (C §11) and the references listed there.

(6.11)–(6.13) The literature on elliptic integrals is vast. A nice summary may be found in [Wh–Wa, ch. XXII].

(6.14) [Cox]

CHAPTER VII

(7.2)   [Ta 6, §3]

(7.4)   [Ta 6, §4]

(7.9)   [Se–T, thm. 2 and corollaries]
(7.10) For more on complex multiplication, see (C §11) and the references listed there.
(7.11) This volume (A.1.4).

CHAPTER VIII

(8.2)   This volume (X.6.1b).
(8.3)   This problem was suggested by D. Rohrlich.
(8.6)   [La 7, ch. 3, prop. 1.1]
(8.7)   [Scha]
(8.9)   [La 7, page 54]
(8.11) [Ca 7, thm. 17.2]
(8.12) One example of each group allowed by (VIII.7.5).
(8.14) (a), (b) [Set]   (d) [Sil 6]
(8.15) Due to Tate. See [Og 1].
(8.17) (c) [Sil 1]   (d) [O1s]
(8.18) Due to Dem'janenko and Zimmer. See [La 9] and [Zi].

CHAPTER IX

(9.3)   (b) [Mah], [Sil 4]
(9.5)   [Le–M] Lewis and Mahler, Acta Arith. 6 (1961), 333–363
(9.6)   Due to A. Thue.
(9.7)   [Se 5, IV §2]
(9.8)   [La 7, ch. 5 §7]
(9.10) [Dan]
(9.11) [Mo 4, ch. 26]
(9.13) This argument appears in an unpublished letter from Tate to Serre.
        One can also do (c) and (d) directly [Mo 4, ch. 27, thm. 2].
        (d) $2 \cdot 3 = 1 \cdot 2 \cdot 3$ $14 \cdot 15 = 5 \cdot 6 \cdot 7$

CHAPTER X

(10.2) [We 5]
(10.3) This is due to Châtelet. See [Ca 7, thm. 11.1].
(10.4) (c) [Ca 7, cor. to lemma 10.3]
(10.6) This is due to Lang, see [La–Ta].
(10.8) [La–Ta]
(10.9) (a) [Mo 4, ch. 16, thm. 6]
(10.10) [Ca 2]
(10.11) (c), (d) Due to Lang–Tate and Shafarevich. See [Ca 7, lemma 12.2].
        (e) [Ca 3.5]
        (f) [Ca 2]
(10.19) (d) Due to Fueter [Fue].

Bibliography

[Ahl]   L. Ahlfors, *Complex Analysis*, 2nd. ed., McGraw-Hill, 1966.

[Ap]   T. Apostol, *Modular Functions and Dirichlet Series in Number Theory*, Springer-Verlag, 1976.

[Ar]   E. Artin, *Galois Theory*, Univ. of Notre Dame Press, 1942.

[Arth]   N. Arthaud, On the Birch and Swinnerton-Dyer conjecture for elliptic curves with complex multiplication, *Compositio Math.* **37** (1978), 209–232.

[A–M]   M. Atiyah and I. Macdonald, *Introduction to Commutative Algebra*, Addison-Wesley, 1969.

[A–W]   M. Atiyah and C. Wall, Cohomology of groups. In *Algebraic Number Theory*, J. W. S. Cassels and A. Fröhlich, eds., Academic Press, 1967, 94–115.

[Ba]   A. Baker, *Transcendental Number Theory*, Cambridge Univ. Press, 1975.

[Ba–C]   A. Baker and J. Coates, Integer points on curves of genus 1, *Proc. Camb. Philos. Soc.* **67** (1970), 595–602.

[Bi]   B. Birch, Cyclotomic fields and Kummer extensions. In *Algebraic Number Theory*, J. W. S. Cassels and A. Fröhlich, eds., Academic Press, 1967, 85–93.

[B–K]   B. Birch and W. Kuyk, *Modular Functions of One Variable IV*, Lecture Notes in Math. **476**, Springer-Verlag, 1975.

[B–Sw 1]   B. Birch and H. P. F. Swinnerton-Dyer, Notes on elliptic curves (I) and (II), *J. Reine Angew. Math.* **212** (1963), 7–25 and **218** (1965), 79–108.

[B–Sw 2]   ————, Elliptic curves and modular functions. *Modular Functions of One Variable IV*, Lecture Notes in Math. **476**, Springer-Verlag, 1975, 2–32.

[Bo–Sh]   Z. I. Borevich and I. R. Shafarevich, *Number Theory*, Academic Press, 1966.

[Br–C]   A. Bremner and J. W. S. Cassels, On the equation $Y^2 = X(X^2 + p)$, *Math. of Comp.* **42** (1984), 257–264.

[Br–K]   A. Brumer and K. Kramer, The rank of elliptic curves, *Duke Math. J.* **44** (1977), 715–743.

[Ca 1]   J. W. S. Cassels, A note on the division values of $\wp(u)$, *Proc. Camb. Philos. Soc.* **45** (1949), 167–172.

[Ca 2] ————, Arithmetic on curves of genus 1 (III). The Tate–Shafarevich and Selmer groups, *Proc. London Math. Soc.* **12** (1962), 259–296.

[Ca 3] ————, Arithmetic on curves of genus 1 (IV). Proof of the Hauptvermutung, *J. Reine Angew. Math.* **211** (1962), 95–112.

[Ca 3.5] ————, Arithmetic on curves of genus 1 (V). Two counter-examples, *J. London Math. Soc.* **38** (1963), 244–248.

[Ca 4] ————, Arithmetic on curves of genus 1 (VI). The Tate–Shafarevich group can be arbitrarily large, *J. Reine Angew. Math.* **214/215** (1964), 65–70.

[Ca 5] ————, Arithmetic on curves of genus 1 (VII). The dual exact sequence, *J. Reine Angew. Math.* **216** (1964), 150–158.

[Ca 6] ————, Arithmetic on curves of genus 1 (VIII). On the conjectures of Birch and Swinnerton-Dyer, *J. Reine Angew. Math.* **217** (1965), 180–189.

[Ca 7] ————, Diophantine equations with special reference to elliptic curves, *J. London Math. Soc.* **41** (1966), 193–291.

[Ca 8] ————, Global fields. In *Algebraic Number Theory*, J. W. S. Cassels and A. Fröhlich, eds., Academic Press, 1967, 42–84.

[Chi] T. Chinberg, An introduction to Arakelov intersection theory. In *Arithmetic Geometry*, G. Cornell and J. Silverman, eds., Springer 1986.

[Cle] C. H. Clemens, *A Scrapbook of Complex Curve Theory*, Plenum Press, New York, 1980.

[Co] J. Coates, Construction of rational functions on a curve, *Proc. Camb. Philos. Soc.* **68** (1970), 105–123.

[Co–W] J. Coates and A. Wiles, On the conjecture of Birch and Swinnerton-Dyer, *Invent. Math.* **39** (1977), 223–251.

[Cox] D. Cox, The arithmetic-geometric mean of Gauss, *L'Enseign. Math.* **30** (1984), 275–330.

[Dav] H. Davenport, On $f^3(t) - g^2(t)$, *Norske Vid. Selsk. Forrh.* **38** (1965), 86–87.

[Dan] L. V. Danilov, The Diophantine equation $x^3 - y^2 = k$ and Hall's conjecture, *Math. Notes Acad. Sci. USSR* **32** (1982), 617–618.

[Del] P. Deligne, La conjecture de Weil, *Publ. Math. IHES* **43** (1974), 273–307.

[De 1] M. Deuring, Die Typen der Multiplikatorenringe elliptischer Funktionenkörper, *Abh. Math. Sem. Hamburg* **14** (1941), 197–272.

[De 2] ————, Invarianten und Normalformen elliptischer Funktionenkörper, *Math. Zeit.* **47** (1941), 47–56.

[De 3] ————, Die Zetafunktion einer algebraischen Kurve vom Geschlechte Eins, I, II, III, IV, *Gott. Nach.*, 1953, 1955, 1956, 1957.

[Dw] B. Dwork, On the rationality of the zeta function of an algebraic variety. *Amer. J. of Math.* **82** (1960), 631–648.

[Ev] J.-H. Evertse, On equations in S-units and the Thue–Mahler equation, *Invent. Math.* **75** (1984), 561–584.

[Ev–S] J.-H. Evertse and J. Silverman, An upper bound for the number of solutions to the equation $y^n = f(x)$, to appear.

[Fa 1] G. Faltings, Endlichkeitssätze für abelsche Varietäten über Zahlkörpern, *Invent. Math.* **73** (1983), 349–366.

[Fa 2] ————, Calculus on arithmetic surfaces, *Annals of Math.* **119** (1984), 387–424.

[Frö 1] A. Fröhlich, Local fields. In *Algebraic Number Theory*, J. W. S. Cassels and A. Fröhlich, eds., Academic Press, 1967, 1–41.

[Frö 2] ————, *Formal Groups*, Lecture Notes in Math. 74, Springer-Verlag, 1968.

[Fue] R. Fueter, Über kubische diophantische Gleichungen, *Comm. Math. Helv.* **2** (1930), 69–89.

[Ful]   W. Fulton, *Algebraic Curves*, Benjamin, 1969.
[Gre]   R. Greenberg, On the Birch and Swinnerton-Dyer conjecture, *Invent. Math.* **72** (1983), 241–265.
[G–H]   P. Griffiths and J. Harris, *Principles of Algebraic Geometry*, Wiley, 1978.
[Gr–Z 1]   B. Gross and D. Zagier, On singular moduli, *J. Reine Angew. Math.* **355** (1985), 191–219.
[Gr–Z 2]   ————, Points de Heegner et dérivées de fonctions *L*, *C. R. Acad. Sc. Paris* **297** (1983), 85–87.
[Gr–Z 3]   ————, Heegner points and derivatives of *L*-functions, in preparation.
[Gru]   K. Gruenberg, Profinite groups. In *Algebraic Number Theory*, J. W. S. Cassels and A. Fröhlich, eds., Academic Press, 1967, 116–127.
[H–W]   G. Hardy and E. Wright, *An Introduction to the Theory of Numbers*, Oxford Univ. Press, 1960.
[Hal]   M. Hall, The Diophantine equation $x^3 - y^2 = k$. In *Computers in Number Theory*, A. Atkin and B. Birch, eds. Academic Press, 1971.
[Har]   R. Hartshorne, *Algebraic Geometry*, Springer-Verlag, 1977.
[Haz]   M. Hazewinkel, *Formal Groups and Applications*, Academic Press, 1978.
[Hur]   A. Hurwitz, Über ternäre diophantische Gleichungen dritten Grades, *Vierteljahrschrift d. Naturf. Ges. Zürich* **62** (1917), 207–229.
[Ig]   J.-I. Igusa, Class number of a definite quaternion with prime discriminant, *Proc. Nat. Acad. Sci. USA* **44** (1958), 312–314.
[Ka]   N. Katz, An overview of Deligne's proof, *Proc. Symp. Pure Math. AMS* **28** (1976).
[Ka–M]   N. Katz and B. Mazur, *Arithmetic Moduli of Elliptic Curves*, Princeton Univ. Press, 1985.
[Ke]   M. Kenku, On the number of ℚ-isomorphism class of elliptic curves in each ℚ-isogeny class, *J. Number Theory* **15** (1982), 199–202.
[Ko]   N. Koblitz, *Introduction to Elliptic Curves and Modular Forms*, Springer-Verlag, 1984.
[Ko–T]   S. V. Kotov and L. A. Trelina, *S*-ganze Punkte auf elliptischen Kurven, *J. Reine Angew. Math.* **306** (1979), 28–41.
[Kr]   K. Kramer, Arithmetic of elliptic curves upon quadratic extension, *Trans. AMS* **264** (1981), 121–135.
[Ku]   D. Kubert, Universal bounds on the torsion of elliptic curves, *Proc. London Math. Soc.* **33** (1976), 193–237.
[Lal]   M. Lal, M. F. Jones, and W. J. Blundon, Numerical solutions of the Diophantine equation $y^3 - x^2 = k$, *Math. Comp.* **20** (1966), 322–325.
[La 1]   S. Lang, Les formes bilinéaires de Néron et Tate, *Sém. Bourbaki* **274**, 1963/64.
[La 2]   ————, *Algebraic Number Theory*, Addison-Wesley, 1970.
[La 3]   ————, *Elliptic Functions*, Addison-Wesley, 1973.
[La 4]   ————, Division points of elliptic curves and abelian functions over number fields, *Amer. J. of Math.* **97** (1972), 124–132.
[La 5]   ————, *Elliptic Curves: Diophantine Analysis*, Springer-Verlag, 1978.
[La 6]   ————, *Introduction to Algebraic and Abelian Functions*, 2nd ed., Springer-Verlag, 1982.
[La 7]   ————, *Fundamentals of Diophantine Geometry*, Springer-Verlag, 1983.
[La 8]   ————, *Algebra* (2nd ed.), Addison-Wesley, 1984.
[La 9]   ————, Conjectured Diophantine estimates on elliptic curves, *Progress in Math.* **35**, Birkhäuser, 1983.
[La 10]   ————, *Complex Multiplication*, Springer-Verlag, 1983.
[La–Ta]   S. Lang and J. Tate, Principal homogeneous spaces over abelian varieties, *Am. J. of Math.* **80** (1958), 659–684.

[L–T]   S. Lang and H. Trotter, *Frobenius Distributions in GL_2-Extensions*, Lecture Notes in Math. **504**, Springer-Verlag, 1976.

[Las 1] M. Laska, An algorithm for finding a minimal Weierstrass equation for an elliptic curve, *Math. Comp.* **38** (1982), 257–260.

[Las 2] ———, Elliptic curves over number fields with prescribed reduction type, *Aspects of Math.* **4**, Friedr. Vieweg & Sohn, Braunschweig/Wiesbaden, 1983.

[Lau]   M. Laurent, Minoration de la hauteur de Néron–Tate, Sém. Th. Nombres, Paris 1981–82, *Progress in Math.* **38**, Birkhäuser, 137–152.

[Lin]   C.-E. Lind, Untersuchungen über die rationalen Punkte der ebenen Kubischen Kurven vom Geschlecht Eins, Uppsala, thesis.

[Liou]   J. Liouville, Sur des classes très-étendues de quantités dont la irrationelles algébriques, *C. R. Acad. Paris* **18** (1844), 883–885 and 910–911.

[Lut]   E. Lutz, Sur l'equation $y^2 = x^3 - Ax - B$ dans les corps p-adic, *J. Reine Angew. Math.* **177** (1937), 237–247.

[Mah]   K. Mahler, On the lattice points on curves of genus 1, *Proc. London Math. Soc.* **39** (1935), 431–466.

[Man 1] Ju. Manin, The Hasse–Witt matrix of an algebraic curve, *AMS Transl.* **45** (1965), 245–264.

[Man 2] ———, The p-torsion of elliptic curves is uniformly bounded, *Izv. Akad. Nauk SSSR* **33** (1969), AMS Transl., 433–438.

[Mas]   D. Masser, *Elliptic Functions and Transcendence*, Lecture Notes in Mathematics **437**, Springer-Verlag, 1975.

[M–W]   D. Masser and G. Wüstholz, Fields of large transcendence degree generated by the values of elliptic functions, *Invent. Math.* **72** (1983), 407–464.

[Mat]   H. Matsumura, *Commutative Algebra*, 2nd ed., Benjamin/Cummings, 1980.

[Maz 1] B. Mazur, Modular curves and the Eisenstein ideal, *IHES Publ. Math.* **47** (1977), 33–186.

[Maz 2] ———, Rational isogenies of prime degree, *Invent. Math.* **44** (1978), 129–162.

[Mes 1] J.-F. Mestre, Courbes elliptiques et formules explicites, Sém. Th. Nombres, Paris 1981–82, *Progress in Math.* **38**, Birkhäuser, 179–188.

[Mes 2] ———, Construction of an elliptic curve of rank $\geqslant 12$, *Comptes Rendus* **295** (1982), 643–644.

[Mig]   M. Mignotte, Quelques remarques sur l'approximation rationelle des nombres algébriques, *J. Reine Angew. Math.* **268/269** (1974), 341–347.

[Mo 1]   L. J. Mordell, On the rational solutions of the indeterminate equations of the third and fourth degrees, *Proc. Camb. Philos. Soc.* **21** (1922), 179–192.

[Mo 2]   ———, *A Chapter in the Theory of Numbers*, Cambridge Univ. Press, 1947.

[Mo 3]   ———, The Diophantine equation $x^4 + my^4 = z^2$, *Quart. J. Math.* (2) **18** (1967), 1–6.

[Mo 4]   ———, *Diophantine Equations*, Academic Press, 1969.

[Mum]   D. Mumford, *Abelian Varieties*, Oxford Univ. Press, 1974.

[M–F]   D. Mumford and J. Fogarty, *Geometric Invariant Theory*, 2nd ed., Springer-Verlag, 1982.

[Nag]   T. Nagell, Solution de quelque problémes dans la théorie arithmétique des cubiques planes du premier genre, *Wid. Akad. Skrifter Oslo I*, **1935**, Nr. 1.

[Né 1]   A. Néron, Problèmes arithmétiques et géométriques rattachés a la notion de rang d'une courbe algébrique dans un corps, *Bull. Soc. Math. France*

80 (1952), 101–166.

[Né 2] ———, Modèles minimaux des variétés abéliennes sur les corps locaux et globaux, *IHES Publ. Math.* **21** (1964), 361–482.

[Né 3] ———, Quasi-fonctions et hauteurs sur les variétés abéliennes, *Annals of Math.* **82** (1965), 249–331.

[Og 1] A. Ogg, Abelian curves of 2-power conductor, *Proc. Camb. Philos. Soc.* **62** (1966), 143–148.

[Og 2] ———, Abelian curves of small conductor, *J. Reine Angew. Math.* **226** (1967), 204–215.

[Og 3] ———, Elliptic curves and wild ramification, *Am. J. of Math.* **89** (1967), 1–21.

[Og 4] ———, *Modular Forms and Dirichlet Series*, Benjamin, 1969.

[Ols] L. Olson, Torsion points on elliptic curves with given *j*-invariant, *Manuscripta Math.* **16** (1975), 145–150.

[Pa] A. N. Parshin, Algebraic curves over function fields, *Math. USSR Isv.* **2** (1968), 1145–1170.

[Pi] R. G. E. Pinch, Elliptic curves with good reduction away from 2, *Proc. Camb. Philos. Soc.* **96** (1984), 25–38.

[Rei] H. Reichardt, Einige im Kleinen überall lösbare, im Grossen Unlösbare diophantische Gleichungen, *J. Reine Angew. Math.* **184** (1942), 12–18.

[Rob] A. Robert, *Elliptic Curves*, Lecture Notes in Math. **326**, Springer-Verlag, 1973.

[Roh] D. Rohrlich, On *L*-functions of elliptic curves and anti-cyclotomic towers, *Invent. Math.* **75** (1984), 383–408.

[Roq] P. Roquette, *Analytic Theory of Elliptic Functions Over Local Fields*, Vandenhoeck & Ruprecht, 1970.

[Ru] K. Rubin, Elliptic curves with complex multiplication and the conjecture of Birch and Swinnerton-Dyer, *Invent. Math.* **64** (1981), 455–470.

[Sat] P. Satgé, Une généralisation du calcul de Selmer, Sém. Th. Nombres, Paris 1981–82, *Progress in Math.* **38**, Birkhäuser, 245–266.

[Scha] S. Schanuel, Heights in number fields, *Bull. Soc. Math. France* **107** (1979), 443–449.

[Schm 1] W. Schmidt, Thue's equation over function fields, *J. Austral. Math. Soc.* **25** (1978), 385–422.

[Schm 2] ———, *Diophantine Approximation*, Lecture Notes in Math. **785**, Springer-Verlag, 1980.

[Sel 1] E. Selmer, The diophantine equation $ax^3 + by^3 + cz^3 = 0$, *Acta Math.* **85** (1951), 203–362 and **92** (1954), 191–197.

[Sel 2] ———, A conjecture concerning rational points on cubic surfaces, *Math. Scand.* **2** (1954), 49–54.

[Se 1] J.-P. Serre, Géométrie algébrique et géométrie analytique, *Ann. Inst. Fourier* **6** (1956), 1–42.

[Se 2] ———, *Groupes Algébriques et Corps de Classes*, Hermann, 1959.

[Se 3] ———, *Groupes de Lie l-Adiques Attachés aux Courbes Elliptiques*, Colloque de Clermont-Ferrand, IHES, 1964.

[Se 4] ———, Complex Multiplication. In *Algebraic Number Theory*, J. W. S. Cassels and A. Fröhlich, eds., Academic Press, 1967, 292–296.

[Se 5] ———, *Abelian l-Adic Representations and Elliptic Curves*, Benjamin, 1968.

[Se 6] ———, Propriétés galoisiennes des points d'ordre fini des courbes elliptiques, *Invent. Math.* **15** (1972), 259–331.

[Se 7] ———, *A Course in Arithmetic*, Springer-Verlag, 1973.

[Se 8] ———, *Cohomologie Galoisienne*, Lecture Notes in Math. **5**, Springer-Verlag, 1973.

[Se 9]   ————, *Local Fields*, Springer-Verlag, 1979.

[Se 10]   ————, Diophantine Geometry, unpublished notes from a course given at Harvard University, 1979.

[Se 11]   ————, Quelques application du théorème de densité de Chebotarev, *Publ. Math. IHES* **54** (1981), 123–202.

[Se–T]   J.-P. Serre and J. Tate, Good reduction of abelian varieties, *Annals of Math.* **68** (1968), 492–517.

[Set]   B. Setzer, Elliptic curves over complex quadratic fields, *Pacific J. of Math.* **74** (1978), 235–250.

[Sha 1]   I. R. Shafarevich, Algebraic number fields, *Proc. Int. Cong. Stockholm* (1962), A. M. S. Translations **31**, 25–39.

[Sha 2]   ————, *Basic Algebraic Geometry*, Springer-Verlag, 1977.

[Sha–T]   I. R. Shafarevich and J. Tate, The rank of elliptic curves, *AMS Transl.* **8** (1967), 917–920.

[Shi 1]   G. Shimura, *Introduction to the Arithmetic Theory of Automorphic Functions*, Princeton Univ. Press, 1971.

[Shi 2]   ————, On the zeta-function of an abelian variey with complex multiplication, *Ann. Math.* **94** (1971), 504–533.

[Shi 3]   ————, On elliptic curves with complex multiplication as factors of the Jacobians of modular function fields, *Nagoya Math. J.* **43** (1971), 199–208.

[Shi–T]   G. Shimura and Y. Taniyama, Complex multiplication of abelian varieties and its application to number theory, *Publ. Math. Soc. Japan* **6**, 1961.

[Sie]   C. L. Siegel, Über einige Anwendungen diophantischer Approximationen (1929), *Collected Works*, Springer-Verlag, 1966, 209–266.

[Sil 1]   J. Silverman, Lower bound for the canonical height on elliptic curves, *Duke Math. J.* **48** (1981), 633–648.

[Sil 2]   ————, Integer points and the rank of Thue elliptic curves, *Invent. Math.* **66** (1982), 395–404.

[Sil 3]   ————, Heights and the specialization map for families of abelian varieties, *J. Reine Angew. Math.* **342** (1983), 197–211.

[Sil 4]   ————, Integer points on curves of genus 1, *J. London Math. Soc.* **28** (1983), 1–7.

[Sil 5]   ————, Divisibility of the specialization map for families of elliptic curves, *Amer. J. Math.* **107** (1985), 555–565.

[Sil 6]   ————, Weierstrass equations and the minimal discriminant of an elliptic curve, *Mathematika* **31** (1984), 245–251.

[Sil 7]   ————, A quantitative version of Siegel's theorem, to appear.

[Sta]   H. Stark, Effective estimates of solutions of some Diophantine equations, *Acta Arith.* **24** (1973), 251–259.

[Ste]   N. M. Stephens, The Diophantine equation $x^3 + y^3 = Dz^3$ and the conjectures of Birch and Swinnerton-Dyer, *J. Reine Angew. Math.* **231** (1968), 121–162.

[Ta 1]   J. Tate, WC groups over p-adic fields, *Sém. Bourbaki* (1957/58), exp. 156.

[Ta 2]   ————, Duality theorems in Galois cohomology over number fields, *Proc. Intern. Cong. Math. Stockholm* (1962), 288–295.

[Ta 3]   ————, Global class field theory. In *Algebraic Number Theory*, J. W. S. Cassels and A. Fröhlich, eds., Academic Press, 1967, 162–203.

[Ta 3a]   ————, Residues of differentials on curves, *Ann. Sci. de l'E. N. S.* (4) **1** (1968), 149–159.

[Ta 4]   ————, Letter to J.-P. Serre, 1968.

[Ta 5]   ————, The arithmetic of elliptic curves, *Invent. Math.* **23** (1974), 179–206.

[Ta 6]   ————, Algorithm for determining the type of a singular fiber in an elliptic pencil. *Modular Functions of One Variable IV*, Lecture Notes in Math. **476**, Springer-Verlag, 1975, 33–52.

[Ta 7]   ————, Endomorphisms of abelian varieties over finite fields, *Invent. Math.* **2** (1966), 134–144.

[Ta 8]   ————, Variation of the canonical height of a point depending on a parameter, *Amer. J. Math.* **105** (1983), 287–294.

[VdW]   B. L. Van der Waerden, *Algebra*, 7th ed., Fred. Ungar Publ. Co, 1970.

[Ve]   J. Velu, Isogénies entre courbes elliptiques, *C. R. Acad. Sc. Paris* (1971), 238–241.

[Voj]   P. Vojta, A higher dimensional Mordell conjecture. In *Arithmetic Geometry*, G. Cornell and J. Silverman, eds., Springer-Verlag, 1986.

[Wa]   R. J. Walker, *Algebraic Curves*, Dover, 1962.

[We 1]   A. Weil, L'arithmétique sur les courbes algébriques, *Acta Math.* **52** (1928), 281–315.

[We 2]   ————, Sur un théorème de Mordell, *Bull. Sci. Math.* **54** (1930), 182–191.

[We 3]   ————, Number of solutions of equations in finite fields, *Bull. AMS* **55** (1949), 497–508.

[We 4]   ————, Jacobi sums as Grössencharaktere, *Trans. Amer. Math. Soc.* **75** (1952), 487–495.

[We 5]   ————, On algebraic groups and homogeneous spaces, *Am. J. of Math.* **77** (1955), 493–512.

[We 6]   ————, *Dirichlet Series and Automorphic Forms*, Lecture Notes in Math. **189**, Springer-Verlag, 1971.

[Wh–Wa]   E. T. Whittaker and G. N. Watson, *A Course in Modern Analysis*, 4th ed., Camb. Univ. Press, 1927.

[Wü 1]   G. Wüstholz, *Recent Progress in Transcendence Theory*, Lecture Notes in Math. **1068**, Springer-Verlag, 280–296.

[Wü 2]   ————, Multiplicity estimates on group varieties, to appear.

[Zi]   H. Zimmer, On the difference of the Weil height and the Néron–Tate height, *Math. Zeit.* **147** (1976), 35–51.

List of Notation

$I(V)$	homogeneous ideal of V, 11
V/K	(projective) V is defined over K, 11
$V(K)$	K-rational points of (projective) V, 11
f^*	homogenization of f, 13
$\bar{V}$	projective closure of V, 13
$K(V), \bar{K}(V)$	function field of (projective) V, 14
$\phi(P)$	value of a rational map, 16
C/K	(curve) C is defined over K, 21
ord_P	valuation on the local ring at P, 22
$f(P) = \infty$	f has a pole at P, 22
ϕ^*	induced map on function fields, 24
$\deg \phi$	degree of the map ϕ, 25
$\deg_s \phi$	separable degree of ϕ, 25
$\deg_i \phi$	inseparable degree of ϕ, 25
ϕ_*	norm map on function fields, 25
$e_\phi(P)$	ramification index of ϕ at P, 28
$f^{(q)}$	f with coefficients raised to q^{th} power, 29
$C^{(q)}$	image of qth power Frobenius map, 29
$\text{Div}(C)$	divisor group of C, 31
$\deg D$	degree of the divisor D, 31
$\text{Div}^0(C)$	degree 0 part of the divisor group, 31
$\text{Div}_K(C)$	group of divisors defined over K, 31
$\sim$	linear equivalence of divisors, 32
$\text{div}(f)$	divisor of a function f, 32
$\text{Pic}(C)$	divisor class group (Picard group) of C, 32
$\text{Pic}_K(C)$	group of divisor classes defined over K, 32
$\text{Pic}^0(C)$	degree 0 part of the divisor class group, 33
ϕ^*, ϕ_*	maps induced by ϕ on divisor groups, 33
ϕ^*, ϕ_*	maps induced by ϕ on divisor class groups, 34
Ω_C	space of differential forms on C, 34
dx	differential of x, 34
ϕ^*	map induced by ϕ on differential forms, 35
df/dt	derivative of f with respect to t, 35
$\text{div}(\omega)$	divisor of a differential ω, 36
$D \geqslant 0$	positive (effective) divisor, 37
$\mathscr{L}(D)$	space of functions associated to D, 38
$\ell(D)$	dimension of the space of functions associated to D, 38
a_1, a_2, a_3, a_4, a_6	coefficients of a Weierstrass equation, 46
O	basepoint (origin) on an elliptic curve, 46
b_2, b_4, b_6, b_8	b-coefficients associated to a Weierstrass equation, 46
c_4, c_6	c-coefficients associated to a Weierstrass equation, 46
$\triangle$	discriminant of a Weierstrass equation, 46
j	j-invariant of an elliptic curve, 46
ω	invariant differential for a Weierstrass equation, 46
E_λ	Legendre equation, 54
$\oplus$	addition on an elliptic curve, 55
$\ominus$	inverse of a point on an elliptic curve, 57
$+, -$	addition and inverse on an elliptic curve, 57
$[m]$	multiplication-by-m map, 57, 71, 116

E_{ns}	non-singular part of E, 60
σ	summation map from Div^0 (or Pic^0) to E, 66
κ	map from E to Pic^0, 66
$\mathrm{Hom}(E_1, E_2)$	group of isogenies from E_1 to E_2, 71
$\mathrm{End}(E)$	ring of endomorphisms of E, 71
$\mathrm{Aut}(E)$	group of automorphisms of E, 71
$\mathrm{Hom}_K(E_1, E_2)$	group of isogenies from E_1 to E_2 defined over K, 71
$\mathrm{End}_K(E)$	ring of endomorphisms of E defined over K, 71
$\mathrm{Aut}_K(E)$	group of automorphisms of E defined over K, 71
$E[m]$	m-torsion subgroup of E, 73
E_{tors}	torsion subgroup of E, 73
$E_{\mathrm{tors}}(K)$	group of torsion points defined over K, 73
ϕ_q	q^{th} power Frobenius (endo)morphism, 74
$E^{(q)}$	image of the qth power Frobenius morphism, 74
τ_Q	translation-by-Q map, 75
E/Φ	quotient of E by a finite subgroup, 78
$\hat{\phi}$	dual isogeny, 86
$T_\ell(E)$	Tate module of E, 90
ϱ_ℓ	ℓ-adic representation, 91
μ_{ℓ^n}	group of $(\ell^n)^{\mathrm{th}}$ roots of unity, 91
$T_\ell(\mu)$	Tate module of K, 91
e_m	Weil e_m-pairing, 96
M_2	ring of 2×2 matrices, 94, 102
$\mathrm{inv}_p \mathscr{K}$	invariant of a quaternion algebra, 102
$\hat{\mathbb{G}}_a$	formal additive group, 116
$\hat{\mathbb{G}}_m$	formal multiplicative group, 116
$\hat{E}$	formal group associated to an elliptic curve, 116
$\mathscr{F}(\mathcal{M})$	group associated to a formal group, 117
$\oplus_{\mathscr{F}}, \ominus_{\mathscr{F}}$	group operations associated to a formal group, 117
$ht(f)$	height of a homomorphism of formal groups, 126
$ht(\mathscr{F})$	height of a formal group, 126
q	a prime power, 130
χ	real-valued character on a finite field, 132
$Z(V/K; T)$	zeta function of a variety over a finite field, 133
$\zeta_{E/K}(s)$	zeta function of an elliptic curve, 136
$H_p(t)$	polynomial for testing supersingularity, 140
Λ	lattice for an elliptic curve over $\mathbb{C}$, 149, 150
$\mathbb{C}/\Lambda$	complex torus, 149, 150
E_Λ	elliptic curve corresponding to Λ, 150
$\mathbb{C}(\Lambda)$	field of elliptic functions, 150
$\bar{D}$	closure of a fundamental parallelogram, 150
$\mathrm{ord}_w(f)$	order of f at w, 151
$\mathrm{res}_w(f)$	residue of f at w, 151
$\displaystyle\sum_{w \in \mathbb{C}/\Lambda}$	sum over a fundamental parallelogram, 151
∂D	boundary of a fundamental parallelogram, 151
$\mathrm{Div}(\mathbb{C}/\Lambda)$	divisor group of a complex torus, 152
$\deg D$	degree of the divisor D, 152
$\mathrm{Div}^0(\mathbb{C}/\Lambda)$	group of divisors of degree 0, 152

$\mathrm{div}(f)$	divisor of the function f, 153
sum	summation map, 153
$\wp(z, \Lambda), \wp(z)$	Weierstrass $\wp$-function, 153
$G_{2k}(\Lambda), G_{2k}$	Eisenstein series, 153
$\sigma(z, \Lambda), \sigma(z)$	Weierstrass σ-function, 156
$g_2(\Lambda), g_2$	normalized value of G_4, 157
$g_3(\Lambda), g_3$	normalized value of G_6, 157
$\Delta(\Lambda)$	discriminant associated to a lattice, 158
ϕ_α	scalar multiplication by α, 159
$H_1(E, \mathbb{Z})$	homology of an elliptic curve, 161
$A(\Lambda)$	area of a fundamental parallelogram, 166
$\zeta(z)$	Weierstrass ζ-function, 166
$\eta(\omega)$	quasi-period function, 166
$G(z)$	Green's function, 167
$j(\Lambda)$	j-invariant associated to a lattice, 167
k	modulus of an elliptic integral, 168
$K(k)$	complete elliptic integral of the first kind, 168
$T(k)$	complete elliptic integral, 168
$M(a, b)$	arithmetic-geometric mean, 169
$I(a, b)$	elliptic integral, 170
$\mathcal{M}$	maximal ideal of a local ring, 171
π	a uniformizer for a local ring, 171
k	residue field of a local ring, 171
$\sim$	reduction modulo π, 173
$\tilde{E}$	reduction of E modulo π, 173
$E_0(K)$	points of E with non-singular reduction, 173
$E_1(K)$	kernel of reduction modulo π, 173
$E(K)[m]$	points of order m defined over K, 176
K^{nr}	maximal unramified extension of K, 178
I_v	inertia group at v, 178
M_K	set of absolute values on K, 189
M_K^∞	set of archimedean absolute values on K, 189
M_K^0	set of non-archimedean absolute values on K, 189
v	logarithm of an absolute value, 189
ord_v	normalized valuation associated to v, 189
K_v	completion of K at v, 190
R_v	ring of integers of K_v, 190
$\mathcal{M}_v$	maximal ideal of R_v, 190
k_v	residue field of R_v, 190
κ	Kummer pairing, 191
$\tilde{E}_v$	reduction of E modulo v, 193
R_S	ring of S-integers, 194
$H(t)$	height of a rational number, 202
$h_x(P)$	height on $E(\mathbb{Q})$, 202
$M_\mathbb{Q}, M_K$	standard absolute values on $\mathbb{Q}$ and K, 206
n_v	local degree at v, 206
H_K	height relative to K, 207
H	absolute height function, 208
$H(x), H_K(x)$	height of an algebraic number, 211

$O(1)$	bounded function, 215	
h	logarithmic height function, 215	
h_f	height function associated to a rational function, 215	
Δ_v	discriminant of a minimal equation at v, 224	
$\mathscr{D}_{E/K}$	minimal discriminant of a elliptic curve, 224	
$\mathfrak{a}_\Delta$	ideal of a Weierstrass equation, 224	
$\bar{\mathfrak{a}}_{E/K}$	Weierstrass ideal class of an elliptic curve, 224	
$\hat{h}, \hat{h}_E$	canonical (Néron–Tate) height on an elliptic curve, 228	
$\langle\ ,\ \rangle$	Néron–Tate pairing, 229, 232	
$R_{E/K}$	elliptic regulator, 233	
$L_{E/K}$	L-series of an elliptic curve, 234	
$v(N)$	number of prime divisors of N, 236	
$v_K(N, C)$	number of points of bounded height, 236	
$\zeta(s)$	Riemann zeta function, 236	
$P * Q$	Segre embedding, 236	
$P^{(d)}$	d-uple embedding, 237	
$\tau(d)$	approximation exponent, 243	
$d_v(P, Q)$	v-adic distance from P to Q, 245	
$\tau(\kappa)$	exponent for linear forms in logarithm, 257	
$m(\alpha)$	size of α, 258	
ϱ_S	S-regulator homomorphism, 258	
$K(S, m)$	group of locally-away-from-S m^{th}-powers, 278	
$\text{Isom}(C)$	group of isomorphisms of C, 284	
$\text{Isom}_K(C)$	group of K-isomorphisms of C, 284	
$\text{Twist}(C/K)$	set of twists of C, 284	
$\bar{K}(C)_\xi$	twist of a function field by a cocycle, 286	
μ	action of an elliptic curve on a homogeneous space, 288	
ν	subtraction map on a homogeneous space, 288	
$WC(E/K)$	Weil–Châtelet group of an elliptic curve, 290	
$E[\phi]$	kernel of an isogeny, 296	
G_v	decomposition group at v, 296	
$S^{(\phi)}(E/K)$	ϕ-Selmer group of an elliptic curve, 297	
$\text{III}(E/K)$	Shafarevich–Tate group of an elliptic curve, 297	
$H^1(G_{\bar{K}/K}, M; S)$	group of unramified cohomology classes, 299	
$S^{(m,n)}(E/K)$	relative Selmer group, 305	
$T_{(m,r)}(E/K)$	rational point group, 305	
Γ	Cassels' pairing on the Shafarevich–Tate group, 306	
$\text{Twist}((E, O)/K)$	twists of an elliptic curve, 307	
$v(D)$	number of prime divisors of D, 310	
$\dim_2$	dimension of an $\mathbb{F}_2$ vector space, 310	
$(a\vert b)$	Legendre symbol, 317	
M^G	group of G-invariants, 330	
$H^0(G, M)$	zeroth cohomology group, 330	
$C^1(G, M)$	group of one-cochains, 331	
$Z^1(G, M)$	group of one-cocycles, 331	
$B^1(G, M)$	group of one-coboundaries, 331	
Res	restriction map on cohomology, 332	
Inf	inflation map on cohomology, 332	
$Z^1_{\text{cont}}(G, M)$	group of continuous one-cocycles, 334	

$\mathbb{Q}^{ab}$	maximal abelian extension of $\mathbb{Q}$, 338
$\mathscr{Cl}(\mathscr{R})$	ideal class group of $\mathscr{R}$, 339
$\mathscr{H}$	Hilbert class field of $\mathscr{K}$, 339
Frob(p)	Frobenius for p, 341
$\mathscr{K}^{ab}$	maximal abelian extension of $\mathscr{K}$, 341
ϕ_E	Weber function, 341
$\mathbb{H}$	upper half-plane, 343
Λ_τ	lattice generated by 1 and τ, 343
$G_{2k}(\tau)$	Eisenstein series, 343
$\Delta(\tau)$	discriminant function, 343
$j(\tau)$	j-invariant function, 343
$SL_2(\mathbb{Z})$	modular group, 343
$\mathscr{F}$	fundamental domain for $\mathbb{H}/SL_2(\mathbb{Z})$, 343
q	$\exp(2\pi i\tau)$, 344
$\zeta(s)$	Riemann zeta function, 345
$\sigma_\alpha(n)$	divisor function, 345
$\tau(n)$	Ramanujan function, 345
$\eta(\tau)$	Dedekind eta function, 346
$\varepsilon(a,b,c,d)$	transformation factor for Dedekind eta function, 346
$q^{\mathbb{Z}}$	cyclic group generated by q, 346
M_k	space of modular forms of weight $2k$, 347
$M_{k,0}$	space of cusp forms of weight $2k$, 347
M	graded ring of modular forms, 347
$T(n)$	Hecke operator, 347
$\lambda(n)$	eigenvalue for Hecke operator, 348
$\mathbb{H}^*$	extended upper half plane, 349
$\Gamma_0(N), \Gamma_1(N), \Gamma(N)$	congruence subgroups, 350
$X_0(N), X_1(N), X(N)$	modular curves, 353
$X(\Gamma)$	modular curve, 354
$Y(\Gamma)$	modular curve, 354
$Y_0(N), Y_1(N), Y(N)$	modular curves, 354
E_q	Tate curve, 356
$\mathscr{C}$	curve over Spec(R), 357
$\mathscr{E}$	Néron model of an elliptic curve, 358
$\tilde{\mathscr{E}}, \tilde{\mathscr{C}}$	special fiber of $\mathscr{E}$ and $\mathscr{C}$, 358
$\tilde{\mathscr{E}}^0$	identity component of special fiber of $\mathscr{E}$, 358
f_v	exponent of the conductor of an elliptic curve, 358, 361
$L_v(T)$	local factor of an L-series, 360
$L_{E/K}(s)$	L-series of an elliptic curve, 361
δ_v	measure of wild ramification, 361
$N_{E/K}$	conductor of an elliptic curve, 361
$\xi_E(s)$	normalized L-series of an elliptic curve, 361
w	sign of the functional equation (equals ± 1), 362
Ω	(real) period of an elliptic curve defined over $\mathbb{Q}$, 362
c_p	order of the group of components modulo p, 362
λ_v	local height function, 365
$B_2(T)$	second Bernoulli polynomial, 366
E_t	specialization of an elliptic curve, 367
P_t	specialization of a point, 367
σ_t	specialization homomorphism, 367

Index

Graduate Texts in Mathematics

continued from page ii